World Soils Book Series

Series editor

Prof. Alfred E. Hartemink
Department of Soil Science, FD Hole Soils Laboratory
University of Wisconsin–Madison
Madison
USA

Aims and Scope

The *World Soils Book Series* brings together soil information and soil knowledge of a particular country in a concise and reader-friendly way. The books include sections on soil research history, geomorphology, major soil types, soil maps, soil properties, soil classification, soil fertility, land use and vegetation, soil management, and soils and humans.

International Union of Soil Sciences

More information about this series at http://www.springer.com/series/8915

Selim Kapur · Erhan Akça
Hikmet Günal
Editors

The Soils of Turkey

Springer

Editors
Selim Kapur
Department of Soil Science and Plant
  Nutrition
University of Çukurova
Adana
Turkey

Hikmet Günal
Department of Soil Science and Plant
  Nutrition
Gaziosmanpaşa University
Tokat
Turkey

Erhan Akça
School of Technical Sciences
University of Adıyaman
Adıyaman
Turkey

ISSN 2211-1255 ISSN 2211-1263 (electronic)
World Soils Book Series
ISBN 978-3-319-64390-8   ISBN 978-3-319-64392-2 (eBook)
https://doi.org/10.1007/978-3-319-64392-2

Library of Congress Control Number: 2017947833

© Springer International Publishing AG 2018
This work is subject to copyright. All rights are reserved by the Publisher, whether the whole or part of the material is concerned, specifically the rights of translation, reprinting, reuse of illustrations, recitation, broadcasting, reproduction on microfilms or in any other physical way, and transmission or information storage and retrieval, electronic adaptation, computer software, or by similar or dissimilar methodology now known or hereafter developed.
The use of general descriptive names, registered names, trademarks, service marks, etc. in this publication does not imply, even in the absence of a specific statement, that such names are exempt from the relevant protective laws and regulations and therefore free for general use.
The publisher, the authors and the editors are safe to assume that the advice and information in this book are believed to be true and accurate at the date of publication. Neither the publisher nor the authors or the editors give a warranty, express or implied, with respect to the material contained herein or for any errors or omissions that may have been made. The publisher remains neutral with regard to jurisdictional claims in published maps and institutional affiliations.

Printed on acid-free paper

This Springer imprint is published by Springer Nature
The registered company is Springer International Publishing AG
The registered company address is: Gewerbestrasse 11, 6330 Cham, Switzerland

To my Great Teachers and Friends in Pedology and Geology, E.A. FitzPatrick (Aberdeen, Scotland), H.H. Eswaran (Arlington, Virginia, USA), U. Dinç (Adana, Turkey), O. Erol (İstanbul, Turkey), N. Munsuz (Ankara, Turkey), G. Kelling OBE (Keele, United Kingdom) and J. Ryan (Tipperary, Ireland), and to my dear Family for their everlasting encouragement and patience during the period of editing/writing and our grandchildren Demir and Derin for their devoted love.

Ewart A. FitzPatrick (1926–) (Aberdeen, Scotland)

Hari Eswaran (1941–2014) (Arlington, Virginia, USA)

Ural Dinç (1937–2015) (Adana, Turkey)

Oğuz Erol (1926–2014) (İstanbul, Turkey)

Nuri Munsuz (1933–2006) (Ankara, Turkey)

Gilbert Kelling O.B.E (1933–) (Keele, England)

John Ryan (1944–) (Tipperary, Ireland)

Demir and Derin Kapur (Adana, Turkey)

Selim Kapur

Preface

There is broad agreement that, as far as possible, soils should be protected against inappropriate use by other expanding sectors of a national economy, primarily in order to sustain the natural resources and the social welfare of individual countries and of the global community. This is especially important for those countries that have already attained a broadly acceptable standard of living but are ambitious to attain even higher levels. In this context Turkey offers a salutary example, since the rapid growth in the Turkish economy and industry over the past few decades has been achieved, at least in part, at the expense of a national loss of prime soil and related resources through urban expansion and the enhanced space and resource requirements of large-scale industry.

Moreover, both gigantic and smaller scale projects based on irrigation have been implemented widely over recent decades within many regions of Turkey. Many of these have been justified as a means of alleviating rural poverty and enhancing regional economies and income generation. Regrettably, however, the planning and implementation of these irrigation projects have been based primarily on the manifold requirements and availability of a reliable water supply but, in general, less attention has been devoted to soil properties data. Such an approach effectively ignores the bulk of the substantial body of data and maps concerning soil properties that have been generated previously to assist in crop and sustainable water management. Thus, the ultimate aim of this book is to provide an explanatory review of the soils of this country and to summarize the knowledge accumulated over many decades concerning their formation and contemporary management.

Furthermore, some of the topics described in this updated version provide important clues and criteria that relate to a future Sustainable Land Management approach partly established on the historical ecosystem-based soil use, more precisely on the 'Soil Ecosystems' and/or the 'Anthroscapes'. The soil-nomenclature team that attempted to achieve this goal was chaired by Prof. Dr. Mehmet Ali Çullu of the Harran University and comprised the Editors and a selection of the article authors, chosen on the basis of their substantial research record on pertinent topics, both in the field and in the GIS environment. This group comprised Profs. Erhan Akça (Adıyaman University), Salih Aydemir (Harran University), Ertuğrul Aksoy (Uludağ University), Gönül Aydın (Adnan Menderes University), Orhan Dengiz (Ondokuz Mayıs University), Sabit Erşahin (Çankırı Karatekin University), Mahmut Dingil (Çukurova University), Özden Görücü (Kahramanmaraş Sütçü İmam University), Yusuf Kurucu (Ege University), Hasan Özcan (Onsekiz Mart University), and Mustafa Sarı (Akdeniz University) in alphabetical order.

The editorial board is also highly indebted to the following reviewers: Profs. Joselito Arocena (University of Northern British Columbia, Canada), Angel Faz Cano (Polytechnic University of Cartagena, Spain) Orhan Dengiz (Ondokuz Mayıs University), Emin Bülent Erenoğlu (Çukurova University), Muhsin Eren (Mersin University), Sabit Erşahin (Çankırı Karatekin University), Selahattin Kadir (Osmangazi University), Salvatore Madrau (Sassari University, Italy), Hasan Özcan (Onsekiz Mart University), Franco Previtali (University of Milan-Bicocca, Italy), Mustafa Sarı (Akdeniz University), Suat Şenol (Çukurova University), Pandi Zdruli (MAI-Bari, Italy), and Claudio Zucca (ICARDA, Amman, Jordan). We gladly acknowledge Prof. Ahmet R. Mermut for his help in peer reviewing a number of the submitted articles and also for his valuable suggestions on the reconstruction of the chapters related to the soils. Ms. Corina van der Giessen (Springer editorial) is highly acknowledged for her continuous encouragement and valuable efforts devoted during the preparation of the book. Thanks are due to Suresh Rettagunta (Springer Chennai), Yaprak Tanrıverdi and Osman Dolaş (Adıyaman University) for their efforts in designing the book layout. Finally, we are highly indebted for the valuable suggestions raised by Prof. Alfred Hartemink (Editor of the "Soils of the World" series) which made this book a worthwhile achievement.

Adana, Turkey	Selim Kapur
Adıyaman, Turkey	Erhan Akça
Tokat, Turkey	Hikmet Günal

Contents

Historical Perspective of Soil Research in Turkey 1
Hikmet Günal, Koray Haktanır, and Selim Kapur

Vegetation 15
İbrahim Atalay

Climate 25
Ufuk Utku Turunçoğlu, Murat Türkeş, Deniz Bozkurt, Barış Önol,
Ömer Lütfi Şen, and Hasan Nüzhet Dalfes

Climate Change and Soils 45
Selim Kapur, Mehmet Aydın, Erhan Akça, and Paul Reich

Geology 57
Mehmet Cemal Göncüoğlu

Pedogeomorphology 75
İbrahim Atalay, Cemal Saydam, Selahattin Kadir, and Muhsin Eren

Soil Geography 105
Mehmet Ali Çullu, Hikmet Günal, Erhan Akça, and Selim Kapur

Cambisols and Leptosols 111
Erhan Akça, Sevda Polat, Somayyeh Razzaghi, Nadia Vignozzi,
Zülküf Kaya, and Selim Kapur

Fluvisols 129
Hasan Özcan

Calcisols and Leptosols 139
Erhan Akça, Salih Aydemir, Selahattin Kadir, Muhsin Eren, Claudio Zucca,
Hikmet Günal, Franco Previtali, Pandi Zdruli, Ahmet Çilek, Mesut Budak,
Ahmet Karakeçe, Selim Kapur, and Ewart Adsil FitzPatrick

Vertisols 169
Hasan Özcan, Salih Aydemir, Mehmet Ali Çullu, Hikmet Günal,
Muhsin Eren, Selahattin Kadir, Hüseyin Ekinci, Timuçin Everest,
Ali Sungur, and Ewart Adsil FitzPatrick

Alisols-Acrisols 207
Hasan Özcan, Orhan Dengiz, and Sabit Erşahin

Podzols 217
Hüseyin Ekinci, Hasan Özcan, Orhan Yüksel, and Sabit Erşahin

Kastanozems 223
Ertuğrul Aksoy, Gökhan Özsoy, Ekin Ulaş Karaata, and Duygu Boyraz

Luvisols 231
Mustafa Sarı, Yusuf Kurucu, Erhan Akça, Muhsin Eren, Selahattin Kadir,
Hikmet Günal, Claudio Zucca, İbrahim Atalay, Zülküf Kaya, Franco Previtali,
Pandi Zdruli, Selim Kapur, and Ewart Adsil FitzPatrick

Regosols 251
Yusuf Kurucu, Mustafa Tolga Esetlili, Erhan Akça, and Mehmet Ali Çullu

Rendzic Leptosols 259
Yusuf Kurucu, and Mustafa Tolga Esetlili

Solonchaks and Solonchak-Like Soils 267
Hasan Özcan, Mehmet Ali Çullu, Hikmet Günal, Hüseyin Ekinci, Mesut Budak,
Ali Sungur, and Timuçin Everest

Lixisols 285
Orhan Dengiz, Hasan Özcan, and Sabit Erşahin

Arenosols 291
Hasan Özcan, Hüseyin Ekinci, Erhan Akça, Osman Polat, Muhsin Eren,
Selahattin Kadir, Ali Sungur, Timuçin Everest, Franco Previtali, and Selim Kapur

Gleysols 313
Hasan Özcan, Hüseyin Ekinci, Ali Sungur, and Timuçin Everest

Histosols 321
Orhan Dengiz, Hasan Özcan, and Zülküf Kaya

Solonetz Soils–Solonchaks (Solonetz-Like Soils) 331
Mehmet Ali Çullu, and Hikmet Günal

Andosols/Andosol-Like Soils 347
Mahmut Dingil, Muhsin Eren, Selahattin Kadir, and Alhan Sarıyev

Management 359
Hakkı Emrah Erdoğan

Index 367

Contributors

Erhan Akça School of Technical Sciences, University of Adıyaman, Adıyaman, Turkey

Ertuğrul Aksoy Department of Soil Science and Plant Nutrition, Uludağ University, Bursa, Turkey

İbrahim Atalay Department of Geography and Forestry, Karabük University, Karabük, Turkey

Salih Aydemir Department of Soil Science and Plant Nutrition, University of Harran, Şanlıurfa, Turkey

Mehmet Aydın Kangwon National University, Chuncheon, South Korea

Duygu Boyraz Department of Soil Science and Plant Nutrition, Namık Kemal University, Tekirdağ, Turkey

Deniz Bozkurt Center for Climate and Resilience Research, University of Chile, Santiago, Chile; Eurasia Institute of Earth Sciences, İstanbul Technical University, İstanbul, Turkey

Mesut Budak Department of Soil Science and Plant Nutrition, Siirt University, Siirt, Turkey

Ahmet Çilek Department of Landscape Architecture, University of Çukurova, Adana, Turkey

Mehmet Ali Çullu Department of Soil Science and Plant Nutrition, University of Harran, Şanlıurfa, Turkey

Hasan Nüzhet Dalfes Eurasia Institute of Earth Sciences, İstanbul Technical University, İstanbul, Turkey

Orhan Dengiz Department of Soil Science and Plant Nutrition, Ondokuz Mayıs University, Samsun, Turkey

Mahmut Dingil School of Technical Sciences, University of Çukurova, Adana, Turkey

Hüseyin Ekinci Department of Soil Science and Plant Nutrition, Çanakkale Onsekiz Mart University, Çanakkale, Turkey

Hakkı Emrah Erdoğan General Directorate of Agrarian Reform, Turkish Republic Ministry of Food, Agriculture and Livestock, Ankara, Turkey

Muhsin Eren Department of Geological Engineering, Mersin University, Mersin, Turkey

Sabit Erşahin Department of Forestry Engineering, Çankırı Karatekin University, Çankırı, Turkey

Mustafa Tolga Esetlili Department of Soil Science and Plant Nutrition, Ege University, İzmir, Turkey

Timuçin Everest Department of Soil Science and Plant Nutrition, Çanakkale Onsekiz Mart University, Çanakkale, Turkey

Ewart Adsil FitzPatrick Department of Plant and Soil Science, University of Aberdeen, Aberdeen, UK

Mehmet Cemal Göncüoğlu Department of Geological Engineering, Middle East Technical University, Ankara, Turkey

Hikmet Günal Department of Soil Science and Plant Nutrition, Gaziosmanpaşa University, Tokat, Turkey

Koray Haktanır Department of Soil Science and Plant Nutrition, Ankara University, Ankara, Turkey

Selahattin Kadir Department of Geological Engineering, Eskişehir Osmangazi University, Eskişehir, Turkey

Selim Kapur Department of Soil Science and Plant Nutrition, University of Çukurova, Adana, Turkey

Ekin Ulaş Karaata Department of Soil Science and Plant Nutrition, Uludağ University, Bursa, Turkey

Ahmet Karakeçe Ministry of Food, Agriculture and Livestock, Dir. of the Koçaş Agricultural Establishment, Aksaray, Turkey

Zülküf Kaya Department of Soil Science and Plant Nutrition, University of Çukurova, Adana, Turkey

Yusuf Kurucu Department of Soil Science and Plant Nutrition, Ege University, İzmir, Turkey

Barış Önol Aeronautics and Astronautics Faculty, Meteorological Engineering, İstanbul Technical University, İstanbul, Turkey

Hasan Özcan Department of Soil Science and Plant Nutrition, Çanakkale Onsekiz Mart University, Çanakkale, Turkey

Gökhan Özsoy Department of Soil Science and Plant Nutrition, Uludağ University, Bursa, Turkey

Osman Polat Ministry of Forestry and Water Affairs, Eastern Mediterranean Forestry Research Institute, Tarsus, Turkey

Sevda Polat Ministry of Forestry and Water Affairs, Eastern Mediterranean Forestry Research Institute, Tarsus, Turkey

Franco Previtali Department of Geosciences, University of Milan (Bicocca), Milan, Italy

Somayyeh Razzaghi Department of Soil Science and Plant Nutrition, University of Çukurova, Adana, Turkey

Paul Reich USDA Natural Resources Conservation Service, Washington, DC, USA

Mustafa Sarı Department of Soil Science and Plant Nutrition, Akdeniz University, Antalya, Turkey

Alhan Sarıyev Department of Soil Science and Plant Nutrition, University of Çukurova, Adana, Turkey

Cemal Saydam Department of Environmental Engineering, Hacettepe University, Ankara, Turkey

Ömer Lütfi Şen Eurasia Institute of Earth Sciences, İstanbul Technical University, İstanbul, Turkey; Mercator-IPC Fellowship Program, İstanbul Policy Center, Sabanci University, İstanbul, Turkey

Ali Sungur Department of Soil Science and Plant Nutrition, Çanakkale Onsekiz Mart University, Çanakkale, Turkey

Murat Türkeş Center for Climate Change and Policy Studies, Boğaziçi University, İstanbul, Turkey

Ufuk Utku Turunçoğlu Informatics Institute, İstanbul Technical University, İstanbul, Turkey

Nadia Vignozzi CRA Agricultural Research Council, Rome, Latium, Italy

Orhan Yüksel Department of Soil Science and Plant Nutrition, Onsekiz Mart University, Çanakkale, Turkey

Pandi Zdruli Mediterranean Agronomic Institute of Bari, CIHEAM, Bari, Italy

Claudio Zucca International Center for Agricultural Research in the Dry Areas, ICARDA, Amman, Jordan

Historical Perspective of Soil Research in Turkey

Hikmet Günal, Koray Haktanır, and Selim Kapur

1 Introduction

Soils have always attracted and continued to attract the attention of people, due to their essential functions in food and fiber production. The use of terraces to control erosion, attempts to increase the fertility of soils and create higher yielding soils are the indications for such attempts performed by the ancient Greeks and Romans. The efforts during the Middle Ages, the Renaissance, and Enlightenment periods were responsible for the significant contributions to the knowledge of soils. However, soil science became a branch of science on its own by the works of Vasili V. Dokuchaev in the nineteenth century (Brevik 2005).

The history of soil science and related studies in Turkey have two important pillars, the Governmental Agencies composed of research with many efforts related to soil science and the Universities which have offered research and education.

2 Soil Research by Relevant Topics

2.1 Soil Survey Studies and Spatial Analyses

Soil survey studies aimed to identify our renewable resources and to evaluate soils for land management. It has been almost a century since the first soil report of Turkey was published by İhsan Abidin (1882–1943) in 1928 entitled 'Agriculture in Anatolia and the Status of Plant Growth' (Şahin 2012). Soils were classified based on the three geographical regions of the country, namely the Black Sea, the Mediterranean, and the central-east. This first report was important to document the level of knowledge on soils in Turkey at that time. Shortly after the report of Abidin, Giesecke published two important papers in 1929 and 1930. The title of the first report was 'The Plant Nutrition Status of the Turkish Soils'. Giesecke traveled in the Mediterranean, Aegean, southern Thrace regions and central Turkey and prepared the first soil map for the half of western Turkey. He classified the soils into ten groups based on the observations on soils, geology, and climate (Şahin 2012).

Modern soil science in Turkey was established by Kerim Ömer Çağlar from Ankara University (AU) in the 1950s. Çağlar was the pioneering author of the 'Soil Science' publications in Turkey printed by the University of Ankara. A schematic map of Turkey's Soil Classification was prepared by Çağlar based on the 1938 American soil classification system (Baldwin et al. 1938) and soils were considered under 11 different classes via soil color as the main property. These soils were the dry and Chestnut Dark Yellowish Soils, the Mediterranean–Aegean and south east zone Red Soils, the north eastern and eastern Black Sea Region Black Soils (Çağlar et al. 1956). The efforts of Çağlar et al. (1956) to map the soils of Iğdır (east Turkey) were the first field-based effort to employ the contemporary soil survey methodology of the Soil Survey Manual (Mermut et al. 1981).

The first nationally organized, large-scale soil survey effort was undertaken by Harvey Oakes and the personnel of the Ministry of Agriculture of Turkey. The surveys started in 1952 and ended in 1954 yielding the publication of the General Soil Map (1/800,000) of Turkey (Oakes 1954). The practices developed in the United States had strong influence throughout the history of the soil surveys in Turkey. Soil survey was primarily considered after the foundation of the General Directorate of Soil and Water (TOPRAKSU) with

H. Günal
Department of Soil Science and Plant Nutrition, Gaziosmanpaşa University, Tokat, Turkey
e-mail: hikmet.gunal@gmail.com

K. Haktanır (✉)
Department of Soil Science and Plant Nutrition, Ankara University, Ankara, Turkey
e-mail: koray.haktanir@gmail.com

S. Kapur
Department of Soil Science and Plant Nutrition, University of Çukurova, Adana, Turkey
e-mail: kapurs@cu.edu.tr

© Springer International Publishing AG 2018
S. Kapur et al. (eds.), *The Soils of Turkey*, World Soils Book Series,
https://doi.org/10.1007/978-3-319-64392-2_1

significant funding available in the 1960s. The major purpose of the surveys conducted by the TOPRAKSU was to identify and map the main soil types to plan the agricultural production and practices according to the nature of the soils. Soil survey reports provided information on land suitability for specific uses. The classification system of the US Bureau of Soils was used in all soil survey reports (Dinç et al. 2001). However, funding and soil survey activities sharply declined after the abolishment of the TOPRAKSU, which were later initiated and enhanced by the large-scale training and research project in the semi-aridGreat Konya Basin (central Turkey) that was organized under the direction of Prof. P. Buringh from the Agricultural University in Wageningen (AUW) with the cooperation of the Turkish Government. The purpose of the AUW was to facilitate practical training and offer experience to their soil science students from several foreign countries. Four reports were published as the outcomes of this project.

The first report was related to the results of the reconnaissance soil surveys carried out by more than 30 participants, competently correlated and edited by De Meester (map 1:200,000) (De Meester 1970). The doctoral theses submitted to the AUW within this initiative dealt respectively with the morphology of some highly calcareous soils (De Meester 1970). Soil salinity and alkalinity (Driessen and Meester 1969) and soil fertility (Janssen 1970) in the Great Konya Basin (De Meester 1970) were the third and the fourth publications of the project. The other detailed large-scale soil survey studies were conducted by Groneman (1968) in Karapınar (central Turkey) and Boxem and Wielemaker (1972) in the Küçük Menderes Valley (western Turkey) (Dinç et al. 2001).

After the 1990s, the universities, the General Directorate of Agrarian Reform and other state agencies started using new technologies such as the global positioning system, geographic information system, remote sensing, on-site geophysical instrumentation and associated data loggers, and the statistical and geostatistical techniques in soil surveys (Tekinel 1991; Çullu 2003; Yüksel et al. 2008; Dengiz et al. 2009; Budak 2012). These new technologies have increased the ability to collect and analyze the data compiled from all over the country.

The Soil Science Department of the Çukurova University (ÇU) initiated its soil survey activities in the 1980s with the support of the provincial departments of agriculture and the financial assistance of the Ministry of Agriculture under the leadership of Ural Dinç (deceased staff member of the ÇU in 2016). His scholar Suat Şenol is at present successfully in charge of a country-wide soil survey training course supported by the GD of the Agrarian Reform of the Turkish Ministry of Food, Agriculture and Livestock and the Chamber of Agricultural Engineers of Turkey. The ÇU primarily carried out the soil surveys of the Çukurova region (337,000 ha, 1:25,000 scale) and later of the northern Cyprus (326,000 ha, 1:25,000 scale). The detailed soil maps (1:25,000 scale) of the Turkish Republic of Northern Cyprus were prepared using the contemporary soft- and hardware technology by the soil survey and remote sensing research group of the ÇU from 1998 to 2000 (Dinç et al. 2005). Moreover, the ÇU has accomplished the detailed soil surveys of 24 State Farms of the Ministry of Agriculture (total area of 361,980 ha, 1:25,000 scale) together with 14 basins (853,188 ha, 1:25,000 scale) of the Southeastern Anatolian Irrigation Project area (Turkish acronym: Güneydoğu Anadolu Projesi, the GAP Project). The detailed soil surveys of the GAP irrigation projects in southeast Turkey started in 1985 in the Harran Plain and ended in 1998.

Some other individual soil surveys have been carried out by Ekinci (1990), Şenol (1993), Aydemir (2001), Çullu (2002), Özcan et al. (2003), Kurucu et al. (2004) Günal et al. (2004), Başayiğit and Dinç (2010), Dengiz (2010) and Gündoğan et al. (2011).

The need for reliable soil data in terms of sustainable land management is increasing in Turkey as well as in other parts of the world with the increase of the populations, land degradation, and threat of global warming. In the last three decades, new technologies emerged to facilitate data sampling and acquisition with the use of the geographic information system (GIS) which is defined as a tool to combine various spatial information (Burrough and McDonnel 1998). Integration of geostatistics into the GIS improved the understanding of uncertainty and helped to reduce the error GIS-based spatial analyses (Burrough 2001). Geostatistical techniques have recently been widely used in determining the spatial characteristics of soil properties, processes, and planning the best management practices in Turkey (Erşahin 2001; Erşahin and Brohi 2006; Ardahanlıoğlu et al. 2003; Öztaş2003; Günal et al. 2012; Sürücü et al. 2013; Özgöz et al. 2013).

2.2 Soil Fertility and Plant Nutrition

Modern soil fertility and plant nutrition in world research were initiated by the studies of Justus von Liebig who stated that several nutrients were required in different amounts for better crop yield. In this context, the ability of the soils of Turkey to provide nutrients to crops has always been of great concern to the researchers from both Universities and Governmental Agencies of the country. The earliest report concerning soil fertility in Turkey was published by the studies conducted in the 'Cotton Research Institute' of Nazilli (west Turkey) on cotton cultivation and fertilization of cotton in 1934 (Mermut et al. 1981).

Starting from 1949 onwards, numerous experiments on the use of several plant nutrients on various crops have been conducted. The use of stable isotopes in plant nutrition, soil fertility, and environmental studies in Turkey had been stimulated by the efforts of Nurinnisa Özbek, who had formerly been the faculty member of the AU. Initial investigations mainly involved the biological nitrogen fixation gained through the N15 isotope technique and uptake of fertilizer and soil phosphorus by test plants using P32 as tracer (Özbek 1984). Özbek (1984) estimated biological N_2 fixation under Mediterranean field conditions in the Çukurova Region using N15 labeled fertilizer growing call and soybean and corn as the test plants. The rate of fertilizer and soil phosphorus uptake by test plants were investigated by Özbek and Akdeniz in several studies (1965, 1966). Özbek et al. (1974) also tested the isotope technology in the identification of clay minerals via Rb^{86} as a tracer for potassium and clay minerals. Nitrogen is the most frequently used nutrient in crop production in Turkey as in the rest of the world. The role and fate of nitrogen in crop production have also attracted the attentions of soil scientists in Turkey. Significant nitrogen losses occur through the process of volatilization in soils of Turkey due to the high pH of soils. Bayraklı (1990) and Bayraklı and Gezgin (1996) from the Selcuk University team conducted leading studies to better understand the fate of nitrogen fertilizers in agricultural ecosystems. Moreover, Özbek's (1971) study on the effect of nitrogen on the development of pyro-catechin humin acids in the soil was one of the pioneering studies in understanding the nitrogen-related chemical quality of the Turkish soils. Further, in the context of soil chemical and biological fertility, Başkaya's (1975) (AU) meticulous study on the sub-microscopic morphology of the humic substances undertaken following alkali extraction of the Turkish tea soils lit the light on the will to search the fate of organic matter in the soils of the other soil ecosystems of Turkey. The significance of the effect of the chemical properties in soils and irrigation/drinking waters were highlighted at several occasions by Oruç (2008) and Oruç and Alpman (1976) from the Ataturk University team. The studies of Oruç (1971) bear high credit due to the widespread representation of the soils studied in the country.

As in many other countries of the world, phosphorus among the other plant nutrients was also highlighted by the soil scientists of Turkey in their studies (Güzel and İbrikci 1994; Erdal et al. 2002; Özturk et al. 2005; Güneş et al. 2006). Burhan Kacar (AU) published several papers from 1962 onwards on the principles underlying particularly phosphorus and sulfur analysis, phosphorus fractions, the use of fertilizers for various crops and sources and composition of fertilizers (Kacar 1964a; Kacar and Akgül 1966). Kacar determined the status of plant available soil phosphorus for the various regions of Turkey (Kacar 1964b). One of his earliest studies was the determination of the phosphorus status of the Çukurova soils (south Turkey) and the appropriate methods of analysis (Kacar 1964b). He tested four biological and 16 chemical methods to determine the phosphorus status of 15 soil series in the Çukurova region. Many of the related studies in Turkey refer to his books containing invaluable information on soil and plant analyses (Kacar 1994, 2009; Kacar and İnal 2008) as reference guides for the characterization of soil and plant phosphorus. Significant contributions to enlighten the chemical status and quality of the Turkish soils, especially related to N and P, were also accomplished by Ünal (1966, 1967) as AU publications and handbooks for the Turkish soil scientist.

The post-pioneer studies on plant nutrition identified the role of zinc in plant growth and quality in the world by the studies conducted by Çakmak and Marshner (1988). In this context, the role of zinc in plant physiology (Çakmak and Marschner 1988), the significance of zinc in crops for human health (Çakmak et al. 1999), the effects of zinc application methods on grain yield (Yılmaz et al. 1997), the enrichment of cereal grains with zinc (Çakmak 2008) and many more related topics have been studied and published concerning the Turkish soils. Current research on plant nutrition focuses on the genetic and agronomic bio-fortification of cereal crops with Zn, Fe, and I as world-wide stated by Çakmak et al. (2010). Moreover, the HarvestPlus Zinc Fertilizer Project currently coordinated by İsmail Çakmak (Sabancı University) himself is investigating the potential of various Zn-containing fertilizers for increasing the Zn concentration of cereal grains and improving the yield in different target countries such as India, China, Pakistan, Thailand, Laos, Turkey, Zambia, Mozambique, and Brazil (Çakmak 2012).

2.3 Soil/Land and Water Management and Conservation

A preliminary study on the soil moisture conditions of the Turkish soils was published by Kerim Ömer Çağlar in 1937. Çağlar investigated the soil–water relationships under the conditions of central Turkey. The hydrology, climate, soil conditions, microorganisms and soil–water relations, the quality of water used for irrigation and characteristics of the waters in Turkey had also been discussed in this publication. The second study on soil moisture was conducted by Evliya in 1940. The significance of water in crop production, crop yield increases through the application of fertilizers with and without water, measurement of soil water and the significance of fallow in land management was discussed in this paper (Mermut et al. 1981). Other relevant studies on the physical and technological properties of the Turkish soils were conducted by Ataman (Avcı and Ataman 1994), Özkan (1999) and Ünver (Çelik and Ünver 1999) from the AU providing significant outcomes in the productive use of soils. Detailed studies in soil physics undertaken by Aydın (1994,

2008) (Aydın et al. 2004, 2005) in the following periods included the mathematical modeling to understand the processes of water movement and evaporation. Aydın (2015, 2016) also conducted modeling studies on the sensitivity of evaporation/evapotranspiration to climate variables in Turkey and Korea.

Soils in the semiarid regions of Turkey are vulnerable to degradation, either because they have poor resistance to erosion, or because of their chemical and physical properties. Therefore, the majority of the publications on soil water conservation studies in Turkey have been concentrated on soil erosion. Hayati Çelebi (Atatürk University) published many papers defining the severity of erosion problems in Turkey and also published papers explaining the conservation practices against erosion (Çelebi 1970, 1971). From the beginning of the 1950s, İlhan Akalan published papers on soil and water erosion (Mermut et al. 1981) and studied the measures to prevent the hazard of water and wind erosion on Turkish soils. The paper entitled "Land classification on the basis of soil conservation in the Marmara Region, Turkey" with special reference to the Yalova district was one of the earliest contributions on soil and water conservation studies in Turkey (Yamanlar 1956). The terminology and measures to be used in erosion mapping in the country were identified in this study (Mermut et al. 1981).

The Karapınar district, located in one of the scarce arid (mild) regions of Turkey, with the lowest precipitation, tackled with severe wind erosion problems since the 1960s. The conservation practices to mitigate the problem and reverse the degradation was initiated in 1962 by the TOPRAKSU and continued for 10 years. The area allocated for erosion control was recently converted to an International Erosion Control Centre of Education. The efforts of this center covered a 16,000 ha area and practices applied brought the success against land degradation (Çevik and Berkman 1990) by enhancing soil development (Akça 2001). Similar efforts for sand dune rehabilitation were undertaken for the sand dune areas in Adana and Mersin-Tarsus (south Turkey) enhancing carbon sequestration and soil formation at the stone pine canopies/root-zones of Arenosols (Akça et al. 2010; Polat and Kapur 2010). These coastal and inland dune areas were recommended for allocation as extensive carbon sequestration lands within the National Action Plan of Desertification of Turkey (NAP-D 2006). Consequently, extensive research work is underway for determining the appropriate crops and management methods of the dunes to mitigate climate change and enhance soil formation and quality for crop production within the NAP-D of the country (NAP-D 2006). At this point, Cemil Cangir's unique endeavors developed together with the Namık Kemal University (NKU) team (C Cangir, D Boyraz, H Sarı) working on land/soil allocation have been of utmost impetus in creating awareness on the significance of soil resources among public, governmental agencies, and decision makers. Cangir and Boyraz (1999, 2000) explained the extent of the inappropriate use of agricultural soils, land degradation, and desertification issues (Cangir et al. 2000) in many occasions related to SLM discussions. Moreover, the NKU team expressing the need for an updated legislation prepared especially for the appropriate use of the Thracian soils and Ergene River polluted by the growing industry of Turkey was also highly regarded by the relevant governmental bodies and universities (Cangir and Boyraz 1999; Cangir et al. 2010).

The contemporary efforts of the AU team (İ. Bayramin, G. Erpul, A. Namlı, S. Okay) deserve high merit in developing nation-wide soil awareness, capacity building, and desertification indicators in collaboration with the Ministries of Forestry and Water Affairs and Food, Agriculture and Livestock. The AU team is also active in developing world-wide assessments of sustainable soil and land managements, land degradation and restoration, erosion control and mitigation and climate change and its impacts via the activities of UNCCD, LADA DS-SLM (Land Degradation Assessment in Drylands, Decision Support for Mainstreaming and Scaling Out Sustainable Land Management), IPBES (Intergovernmental Science-Policy Platform on Biodiversity and Ecosystem Services), GSP-ITPS (Global Soil Partnership-The Intergovernmental Technical Panel on Soils) and the updated version of the Soil Atlas of Europe.

Further contributions to land management studies were conducted by the members of the former Working Group on Land Degradation and Desertification (formal body of the IUSS, secretarial office in the ÇU-Adana) following the 1st International meeting organized by the ÇU on Land Degradation and Desertification (WGLDD) in 1996. The members of the former working group LDD (H. Eswaran chair, S Kapur Sec., AR Mermut, P Zdruli and E Akça, later M Pagliai Chair, AF Cano Dep. Chair, S Kapur Sec. P Zdruli Dep. Sec.) pursued their contemporary attempts to develop the appropriate land management approaches via extended scientific and governmental collaborations/contributions from Turkey (S Berberoğlu, C Dönmez and M Dingil from the ÇU, C Cangir and D Boyraz from NKU, E Özevren and E Yazıcı from the Ministry of Forestry and Water Affairs and G Küsek, Y Yılmaz and E H Erdoğan from the Ministry of Food, Agriculture and Livestock), Italy (L Montanarella and M Cherlet from IES-JRC Ispra, Milan, F Previtali from Milan Bicocca University, C Zucca from ICARDA and S Madrau from Sassari University), United Kingdom (EA FitzPatrick from Aberdeen University, G Kelling OBE from Keele University, P Bullock from Cranfield University), Spain (A F Cano from Cartagena Polytechnic University) and Japan (T Watanabe, T Nagano and T Kume from Kyoto, Kobe and Ehime Universities) over the 'Anthroscape' approach. The international efforts aimed to develop the Anthroscape approach were primarily undertaken by this extended research group attempting to assess the mitigation measures for land degradation within the human-reshaped landscapes. These measures were based on the combinations of the appropriate traditional technologies renovated by contemporary scientific know-how leading to the establishment of "New Agricultural Towns (NATs)" (Zdruli et al. 2010; Kapur et al. 2013). These concern the soil ecosystems allocated to special crop areas such as the olive groves, stone pine forests, and pasture soil/crop ecosystems or NATs (Kapur et al. 2013). Part of the members of the former WGLDD was also responsible for the preparation and publication of the National Action Plan of Desertification of Turkey of the UNCCD based on the Basin-wide Anthroscape solutions (NAP-D 2006) together with the SSST (Soil Science Society of Turkey).

2.4 Soil Mineralogy and Micromorphology

Studies on the clay minerals of the Turkish forest soils throughout the country were initiated by Mitchell and Irmak (1957) soon after followed by Gülçür (1958) who sought to understand the chemical and mineralogical characteristics of the clay fractions of some virgin soils developed under humid climatic conditions in the Black Sea Region. Gülçür (1958) classified soils up to 600 m above sea level (asl) as Krasnozems and soils above this elevation as Brown Forest soils.

In the 1960s, the number of studies on soil mineralogy have increased via İ Akalan (AU), F Gülçür (İstanbul University IU), N Munsuz (AU), F Saatçi (Ege University EU), B Öztan (Soil and Fertilizer Res. Ins., Ankara), and M Ş Yeşilsoy (Soil and Fertilizer Res. Ins, Ankara, ÇU) as the major contributors (Mermut et al. 1981). Saatçi mostly conducted studies around the İzmir province of Turkey aiming to determine the clay mineral contents of some great soil groups and soils formed over alluvial materials deposited in the Büyük Menderes Basin (Mermut et al. 1981). Yeşilsoy had two papers in the 1960s and both were related to the soils of the Thrace Region of Turkey (Mermut et al. 1981) with X-ray diffractometry, and differential thermal analysis techniques used to determine the clay mineral contents of Grumusols, Non-calcic Brown and Rendzina soils (Yeşilsoy 1966). Akalan also published papers in the same period with Saatçi and Yeşilsoy, however the majority of his studies were published in the 1970s. The mineralogy of the soils, particularly the Grumusols (Vertisols), in the Thrace region was studied by Akalan and Başer (1972) and Akalan (1976).

Studies in soil mineralogy have increased with the increase in the number of the faculties of Agriculture in the 1970s. The pillars of soil mineralogy in Turkey since the 1970s were İ Akalan (AU), Ü Altınbaş (EU), C Cangir (NKU), S Kapur (ÇU) (of Mediterranean studies), A R Mermut (Ankara) (of world-wide studies), N Munsuz (Ankara), G Şimşek (Erzurum) and especially M Sayın (studies on obtaining internal clay standards of soils for quantitative clay analysis) (Sayın 1982). Altınbaş (1976) investigated the mineralogy of sand and clay fractions of Terra Rossa, Non-calcic brown and Rendzina soils in Çeşme, İzmir. He also identified the minerals in sand and clay fractions, percentages, some of optical and physical properties using polarized microscopy. Mermut and Pape (1971, 1973) published two papers on the micromorphology of clay cutans (coatings) discussing the illuviation of ferriargillans and associated papules, and also reported the formation of ferriargillans from the in situ weathering of mica for the first time. The degradation of the argillic horizons by means of swelling and shrinkage was also discussed in these papers. Munsuz (1967, 1969) studied the clay mineral contents of Terra Rossas (Tarsus) and Hydromorphic Saline Alluvial (Alifakılı), Solonchak (Malya), Brown (Balgat), Reddish Brown (Suruç), Podzolic Gray Brown (Rize), Grumusol (Akçakoca) and Rendzina (Lüleburgaz) soils by infrared spectroscopy, X-ray diffraction, differential thermal analysis, and electron microscopy techniques. The author's study on the determination of clay minerals by the Imbibometry method was also a milestone in the clay mineral studies of the country (Munsuz et al. 1970).

Emphasis on micromorphology and mineralogy has gained momentum with the papers accomplished on contemporary soils, ceramics, and raw material sources by the ÇU team after the 1990s and especially after the International meetings held in Adana on the Red Mediterranean Soils (the 3rd Int. meeting, 1993), Land Degradation and Desertification (the 1st Int. meeting, 1996) and Soil Micromorphology (the 12th Int. meeting, 2004). At this point, Kapur et al. (1997) explained the processes of microstructure development in the soils (Vertisols) of Turkey and Israel. Further, Akça et al. (2009a, b) attempted to enlighten the historical pottery production technologies of the Neolithic Çatalhöyük (central Turkey) and of the late Hittite site in Karatepe (south Turkey) respectively. The questions on the source materials of the Neolithic pottery of Çatalhöyük (one of the oldest settlement sites in the world dating from almost 7500 BC), the clues on the production technologies of the İznik ceramics (Seljuk and Ottoman), and the Karatepe late Hittite site basalts and basaltic ceramics have also been revealed using micromorphological and mineralogical methods by Akça et al. (2009a), Çambel et al. (1994) and Kapur et al. (1995, 1998). The studies conducted by the interdisciplinary pedology-sedimentology-archeology-geomorphology team of Çukurova and Mersin (Turkey), Aberdeen (Scotland), Keele (England), Sassari and Milan (Italy) Universities and ICARDA (Amman, Jordan) have also yielded significant outcomes in explaining the relations between past climate changes of the Quaternary and the Luvisols/Calcisols-calcretes/paleosols of Turkey, Lebanon and Italy (Sardinia) by the use of soil analyses, clay mineral contents, soil dating, and especially micromorphology (Kapur et al. 1987, 1990, 1998; Ryan et al. 2009, Bal et al. 2003; Andreucci et al. 2012; Zucca et al. 2012, 2013, 2014a, b; Küçükuysal et al. 2013; Kaplan et al. 2013, 2014; Küçükuysal and Kapur 2014; Eren et al. 2014,2015; Yeşilot-Kaplan et al. 2014). Further, the basic principles and concepts of soil micromorphology developed for the other relevant subjects were also applied to describing processes occurring in contemporary and ancient concrete/mortars belonging to the Ottoman and Andalus periods by the pedology and material sciences interdisciplinary team established by the Çukurova, Keele-England-, Kahramanmaraş and Pamukkale Universities (Kelling et al. 2000; Binici et al. 2007, 2008, 2009, 2010, 2012, 2014; Binici and Kapur 2016).

2.5 Soil Microbiology

The main global interests of soil microbiology have been the metabolic activities of microorganisms and their roles in the energy flow of nutrient cycling (Paul 2014). In this context, especially scientists working in the related disciplines of the Faculties of Agriculture, Departments of Soil Science, initiated the studies related to soil microbiology in Turkey in the 1970s.

One of the earliest publications on soil microbiology in Turkey concerned the study of nodosity bacteria of groundnut roots (Şahinkaya 1969) with well-developed nodules. Forty-six bacterial cultures of different colony morphologies were isolated from the soil samples of the nine provinces and 18 districts of Turkey. In almost the same period, Aksoy (1971) investigated the inoculation effect of seeds with Azotobacter chroococcumon versus the yields of wheat, potato, and corn under the irrigated conditions of the Erzurum province. Aksoy (1973) also conducted studies on the effects of inoculation with microorganisms and fumigation at different moisture levels on the soils from the east Black Sea Coast Region, east and south east Turkey. Gürbüzer (1973) and Göktan (1974) investigated the efficiency of Rhizobium meliloti and Rhizobium phaseli, respectively isolated from Turkish soils as the follow-up studies of the AU. Moreover, studies were conducted to understand the causes of nitrogen limitations and to investigate the mechanisms of biological nitrogen fixation. Studies revealed that high rates of nitrogen in arid ecosystems are lost by denitrification (Gök and Ottow 1988; Gök et al. 2006).

Modern soil microbiology is an interdisciplinary study, involving agricultural production, aquatic and biogeochemical sciences, bioremediation, environmental quality, biodiversity, and global climate change (Anonymous 2016a). In this respect, the microbiologists of Turkey have also recently started conducting interdisciplinary studies as stated by Okur et al. (2010) investigating the effects of organic insecticides on soil microbial biomass and enzymatic activities. The effects of heavy metals and other pollutants on the activities of soil enzymes were among the popular research topics of Turkish soil microbiologists. At this extent, the pioneering enzyme activity studies in Turkey involved the 'Enzyme activities and their relation to soil characteristics in the soils of Ankara' (Rashid 1971) and the 'Effects of fallow-wheat-legume rotation on important soil enzyme activities under Ankara conditions' (Haktanır 1973). Further, studies on soil enzymes of different aspects have been studied by Ünal (1967), Özbek and Rasheed (1972, 1973), Karaca et al. (2002), Kızılkaya and Bayraklı (2005), Turgay et al. (2010) and Arcak et al. (2011) in Turkey. Studies conducted by the Haktanır scholars (Sözüdoğru et al. 1996; Arcak et al. 1999; Karaca and Haktanır 2000; Turgay et al. 2007; Karaca et al. 2009, 2010; Kızılkaya et al. 2011) concerning the effects of environmental pollutants on soil enzyme activities have been significant in documenting the degradation of the soil biological quality in Turkey.

The effects of mycorrhiza application on the growth and nutrient uptake of several crops and soil physical properties have also been documented in the country by Özcan and Taban (2000), Çelik et al. (2004), Ortaş et al. (2002, 2011) and Ortaş (2003, 2012).

3 Governmental Bodies

3.1 State Universities

The initial accomplishment of the Republic of Turkey (1923) in the grounds of Agriculture was to establish the "Higher Institute of Agriculture" as the core of the AU to serve the farmers of the country in 1933. In the aftermath, the progress of soil science in Turkey was directed by the Soil Science Departments of the universities, graduate programs, and State research institutes. As the first soil science department was established in 1933 in Ankara, many others followed, and new departments were founded in the different parts of the country. The Faculty of Agriculture of the Ege University was the second faculty with a Soil Science Department established in 1955 fallowed by the Atatürk University of Erzurum in 1958 and the AU, Faculty of Agriculture of Adana (the University of Çukurova, Faculty of Agriculture after 1973) in 1967.

The earliest member of the Soil Science Department of the Faculty of Agriculture, Kerim Ömer Çağlar (1903–1972), made many significant contributions to the advancement of soil science in Turkey. Çağlar completed his education in agriculture in the Institute National

Agronomique of Paris and his advanced training in the Berlin School of Agriculture. Çağlar published numerous papers on several issues in soil science during his career. The report on soil analysis of the rice fields of the Silifke Atatürk farm in 1933 was one of Çağlar's first reports, a year later followed by the investigation of the steppe soils of central Turkey (Mermut et al. 1981).

3.2 Agricultural Research Centers

The soil fertility and plant nutrition research of the Government Agencies in Turkey other than the Universities emerged with the establishment of the Agricultural Experimental Stations. The experimental stations aimed the development of research dealing with soil fertility, soil, and water conservation.

The Eskişehir Dry Farming Experiment Station of the Ministry of Agriculture was established in 1932 to conduct field trials on soil fertility. The first Irrigated Agriculture Experimental Station (IAES) of Turkey was established in Tarsus in 1947. The IAES was particularly aimed to improve the irrigation knowledge of farmers in Ceyhan, Seyhan, and Berdan Plains and also produce the necessary data for the irrigation infrastructure to be built in the region. In this context, the Cotton Research Institutes in Nazilli (west of Turkey) and Adana (south of Turkey) were established in 1934 and 1937, respectively, and have been conducting intense soil fertility experiments since then. The positive feedback of the IAES in Tarsus led to the establishment of the Irrigated Agriculture Experiment Stations at Menemen in 1949, Kadınhanı in 1950, Çumra in 1953 and Eskişehir in 1955.

3.3 State Hydraulic Works (DSİ -Devlet Su İşleri- Turkish Acronym)

The DSİ is a state agency of the Ministry of Environment and Forestry of Turkey and deals with energy and agricultural and environmental services. The establishment of the DSİ dates back to the establishment of the 'General Directorate of Public Works' by the Ottoman Government in 1914. The name of this establishment was changed to 'Waters Directorate' soon after the foundation of the Turkish Republic in 1923. Studies on feasibility, planning, gaging, and water level recording had been reorganized under the name of the 'Water Works General Directorate' in 1939. The institution was finally named as the 'General Directorate of State Hydraulic Works' in 1954 (Anonymous 2016b). The DSİ prepared and published numerous projects on land reclamation, irrigation, land classification, reconnaissance, and land management. The irrigation projects of DSİ cover 10% of the total agricultural area and 57% of the irrigated agricultural land of Turkey.

3.4 Soil–Water General Directorate (TOPRAKSU–Turkish Acronym)

The milestone of Turkish soil science was the establishment of the TOPRAKSU in 1960. Reconnaissance and detailed soil surveys, classifications and mapping of soils in Turkey were carried out by the TOPRAKSU. The TOPRAKSU prepared many detailed reports on agricultural irrigation, soil conservation, and land reclamation. Soil survey reports were prepared from 1966 to 1971 using 1:25,000 scaled topographic maps as base maps which significantly improved the soil surveyors' ability to accurately record spatial soil information.

Soils were classified at Great Group levels according to the classification system of Baldwin et al. (1938) to be used for land use planning. The reports were later converted for provincial (1:100,000) and basin mapping purposes (1:200,000) (Dinç et al. 2001). The TOPRAKSU was later abolished and the General Directorate of Rural Services (GDRS) was founded in 1984. Subsequently, the GDRS continued the soil survey, classification, and mapping activities of the TOPRAKSU. The GDRS activities and responsibilities in the area of soils were to sustain soil and water management, monitor soil fertility of agricultural soils, make environmental impact assessments, and to create the national soil database.

3.5 Central Research Institute for Soil, Fertilizers and Water Resources

The Institute was established on March 1, 1954 in Ankara as a National Specialization Institute under the General Directorate of Agricultural Affairs for undertaking the tasks on "Soil Surveys, Classification and Mapping of Turkey, Soil, Fertilizer, Water and Plant Analyses, Determining the suitability of Soils in terms of Irrigation and Soil Salinity, Alkalinity and Reclamation, Protection of Soil Fertility and Utilization of the Soils at Optimum Level" in accordance with the relevant law (number 3203) concerning the organization and duties of the Ministry of Agriculture.

3.6 General Directorate of Agricultural Research and Policies (GDARP)

The GDARP conducts an applied Agricultural research program via its 58 research institutes throughout the country.

Studies of GDARP mainly concentrate on climate change, biodiversity, soil fertility, soil degradation, land use planning, rural developments, natural resource contaminations, and integrated watershed management. The GDARP compiled a huge soil data base through several projects conducted nation wide.

3.7 General Directorate of Agricultural Reform (GDAR)

The GDAR is the responsible institution of the country on taking the necessary measures for protecting soil and water resources. The GDAR established the National Soil Information System (NSIS, 1:5000 scale) which has compiled the detailed information on soil chemical and physical properties, current land use, parent material, and land capability classes. The institute funds large-scale projects to consolidate fragmented lands which are not suitable for sustainable economic production. Since the quality of land consolidation largely depends on the quality of the soil data base, detailed soil surveys are the prerequisite to complete the projects. Thus, soil profiles are described and sampled along with the samples collected from check points between the profiles. The consolidation of the fragmented lands includes "changes in ownership rights to land and other real estate property, exchange of parcels among owners, changes in parcel borders, parcel size and shape, joining and dividing of parcels, changes in land use, construction of roads, bridges and water structures" (Erdoğan et al. 2013).

4 Non-governmental Organizations (NGOs)

The two most prominent NGOs of Turkey devoting active efforts and services for the protection of soils are TEMA (The Turkish Foundation for Combating Soil Erosion, for Reforestation and the Protection of Natural Habitats) and the SSST (The Soil Science Society of Turkey).

4.1 The Turkish Foundation for Combating Soil Erosion, for Reforestation and the Protection of Natural Habitats (TEMA)

The TEMA Foundation (holder of the Land for Life Award of 2012 of the UNCCD) is the largest (450,000 volunteers) and leading environmental NGO in Turkey that was founded in 1992 by Hayrettin Karaca. Hayrettin Karaca, the founder of the Karaca Arboretum and holder of the UN Environment Award, and Nihat Gökyiğit (co-founder of TEMA), through their life-long efforts in TEMA created national- and world-wide awareness on soils via education and numerous experimental soil/crop research projects. TEMA has been fulfilling its mission in creating an effective and conscious public opinion on environmental problems, deforestation, desertification, biodiversity loss, and climate change with special reference to soils. TEMA implements model projects of various scales and intervenes on government policies for proper environmental and soil protection. TEMA also conducts legal activities by legislation drafting and through its lawsuits. It influences and accelerates the processes at local and national level, via its strong international links. TEMA succeeds in linking the global perspective in soil protection to the national by closely following the international agenda (Anonymous 2016c).

4.2 The Soil Science Society of Turkey (SSST)

The SSST has many members from universities, the public sector, private institutions and 21 focal points located in different cities of the country. It is an NGO established in 1964 seeking to develop, disseminate, and introduce the theoretical and applied soil science in Turkey. The SSST has welcomed members from all professions and scientific communities who deal with soil. The EUROSOIL Meeting (16–22 October 2016) and the First Int. Carbon Summit held in Istanbul (3–5 April 2013) by the SSST were immemorable scientific events for the world of soils. As the NGO representative for the UNCCD within the Ministry of Environment and Forestry, the SSST has been the leading contributing body for the preparation of NAP-D of Turkey (2006) and its renovated version for the 10-year strategic plan of the UNCCD. The major tasks of the SSST are

a. to guide and lead efforts for the protection, use and evaluation of the Turkish soils/lands by taking into account the scientific experiences and developments on soil and the misapplications of the past,
b. to establish committees consisting of specialists in the field of soil science to support training and research and to organize soil survey activities. Support authorship for books and essays written in order to increase soil awareness,
c. to support relevant institutes on related subjects of soil science,
d. to cooperate with other public, private, and non-governmental organizations in order to increase the awareness on soils and
e. to organize national and international soil science congresses, symposia, and workshops.

In conclusion, the expectations of the SSST, from all these attempts and activities, is to raise the consciousness that the soil as the source of life, is vulnerable, and is in need of protection (Anonymous 2016d).

References

Akalan I (1976) Some physical and chemical characteristics and clay mineralogy of typical Grumusol profiles in Thrace, Turkey. Ankara Univ Ann Fac Agric 26(2):243–260 (in Turkish)

Akalan I, Başer S (1972) Clay mineralogy of grumusols (Vertisol) of Thrace, Turkey. TUBİTAK TOAG No: 12. TOAG project. No: 96. 74 P (in Turkish)

Akça E (2001) Determination of the soil development in Karapınar erosion control station following rehabilitation. Ph.D. Thesis. University of Çukurova, Institute of Natural and Applied Sciences, Adana, Turkey (in Turkish)

Akça E, Kapur S, Özdöl S, Hodder I, Poblome J, Arocena J, Bedestenci Ç (2009a) Clues of production for the Neolithic Çatalhöyük (central Anatolia) pottery. Sci Res Essay 4(6):612–625

Akça E, Arocena J, Kelling G, Nagano T, Degryse P, Poblome J, Çambel H, Büyük G, Tümay T, Kapur S (2009b) Firing temperatures and raw material sources of ancient hittite ceramics of Asia Minor. Trans Indian Ceram Soc 68:35–40

Akça E, Kapur S, Tanaka Y, Kaya Z, Bedestenci HÇ, Yaktı S (2010) Afforestation effect on soil quality of sand dunes. Polish J Environ Stud 19(6):1109–1116

Aksoy N (1971) The effect of inoculation of seeds with *Azotobacter chroococcum* on the yields of wheat, potato and corn under irrigated conditions in Erzurum. Ataturk University, J Fac Agric 2(3):15–23. Erzurum (in Turkish)

Aksoy N (1973) The effects of inoculation with microorganisms and fumigation at different moisture levels on the soils from East Black Sea Coast Region, East and South East Turkey. Ataturk University, Publications of Faculty of Agriculture. No: 56. 122 p. Erzurum (in Turkish)

Altınbaş Ü (1976) The mineralogy of sand and clay fractions of Terra Rossa, Non-calcic brown and Rendzina soils in Çeşme, İzmir. Thesis for Associate Professor Position. 20 p. (in Turkish)

Andreucci S, Bateman MD, Zucca C, Kapur S, Akşit İ, Dunajko A, Pascucci V (2012) Evidence of Saharan dust in upper Pleistocene reworked palaeosols of NW Sardinia, Italy: palaeo-environmental implications. Sedimentology 59:917–938

Anonymous (2016a) An historical perspective of soil microbiology. Chapter 1: http://www4.ncsu.edu/~lagillen/SSC%20532/Chapter%201/Chap1historicaloverview.html. Accessed in 11 March, 2016

Anonymous (2016b) General directorate of state hydraulic works. The History. http://en.dsi.gov.tr/about-dsi-/history Accessed in 21 Feb 2016

Anonymous (2016c) Soil Science Society of Turkey. http://www.toprak.org.tr/en/ Accessed in 21 April 2016

Anonymous (2016d) The Turkish foundation for combating soil erosion, for reforestation and the protection of natural habitats. http://www.tema.org.tr/web14966-22/index.aspxAccessed in 21 April 2016

Arcak S, Karaca A, Kaplan M, Turgay OC, Haktanır K (1999) Heavy metal distribution in green house soils of Antalya and their relationship between biological properties of soils. Selçuk Univ J Fac Agric 20(13):138–155 (in Turkish)

Arcak S, Kütük AC, Haktanır K, Çaycı G (2011) Effects of tea wastes on soil enzyme activities and nitrification. Pamukkale Univ J Eng Sci 3(1):261–266 (in Turkish)

Ardahanlıoğlu O, Öztaş T, Evren S, Yılmaz H, Yıldırım ZN (2003) Spatial variability of exchangeable sodium, electrical conductivity, soil pH and boron content in salt-and sodium-affected areas of the Iğdır plain (Turkey). J Arid Environ 54(3):495–503

Avcı Y, Ataman Y (1994) The effect of soil management systems in spring lentil-wheat rotation on soil physical properties and crop yields. J Field Crops Res Inst 3(3–4):1–18 (in Turkish)

Aydemir S (2001) Properties of palygorskite-influenced vertisols and vertic-like soils in the Harran Plain of Southeastern Turkey. PhD Thesis. Texas A & M University, 518 p

Aydın M (1994) Hydraulic properties and water balance of a clay soil cropped with cotton. Irrig Sci 15(1):17–23

Aydın M (2008) A model for evaporation and drainage investigations at ground of ordinary rainfed-areas. Ecol Model 217(1):148–156

Aydın M, Yano T, Kılıç Ş (2004) Dependence of zeta potential and soil hydraulic conductivity on adsorbed cation and aqueous phase properties. Soil Sci Soc Am J 68(2):450–459

Aydın M, Yang SL, Kurt N, Yano T (2005) Test of a simple model for estimating evaporation from bare soils in different environments. Ecol Modell 82(1):91–105. ISO 690

Aydın M, JungY-S Yang JE, Kim S-J, Kim K-D (2015) Sensitivity of soil evaporation and reference evapotranspiration to climatic variables in South Korea. Turk J Agric For 39:652–662

Aydın M, Watanabe T, Kapur S (2016) Sensitivity of reference evapotranspiration and soil evaporation to climate change in the Eastern Mediterranean Region. Springer (in Press), ICCAP Book

Bal Y, Kelling G, Kapur S, Akça E, Çetin H, Erol O (2003) An improved method for determination of holocene coastline changes around two ancient settlements in Southern Anatolia: a geoarchaeological approach to historical land degradation studies. Land Degrad Dev 14(4):363–376

Baldwin M, Kellogg CE, Thorp J (1938) Soil Classification. Yearbook of Agriculture, Soils and Men, pp 979–1160

Başayiğit L, Dinç U (2010) Prediction of soil loss in lake watershed using GIS: a case study of Egirdir Lake, Turkey. J Nat Environ Sci 1(1):1–11

Başkaya H (1975) Untersuchungen über die organischen Stoffe in Türkischen Teeböden sowie deutschen Basalt- und Lockerbraunerden, Dissertation, Universitaet Göttingen, 125 p

Bayraklı F (1990) Ammonia volatilization losses from different fertilizers and effect of several urease inhibitors, $CaCl_2$ and phosphogypsum on losses from urea. Fertilizer Res 23(3):147–150

Bayraklı F, Gezgin S (1996) Controlling ammonia volatilization from urea surface applied to sugar beet on a calcareous soil. Commun Soil Sci Plant Anal 27(9–10):2443–2451

Binici H, Kapur S (2016) The physical, chemical and microscopic properties of masonry mortars from Alhambra Palace (Spain) in reference to their earthquake resistance. Front Architectural Res 5:101–110

Binici H, Aksoğan O, Bodur MN, Akça E, Kapur S (2007) Thermal isolation and mechanical properties of fibre reinforced mud bricks as wall materials. Constr Build Mater 21:901–906

Binici H, Çağatay İH, Shah T, Kapur S (2008) Mineralogy of plain Portland and blended cement pastes. Build Environ 43:1318–1325

Binici H, Arocena J, Kapur S, Aksoğan O, Kaplan H (2009) Microstructure of red brick dust and ground basaltic pumice blended cement mortars exposed to magnesium sulfate solutions. Can Civ Eng J 36:1784–1793

Binici H, Arocena J, Kapur S, Aksoğan O, Kaplan H (2010) Investigation of the physico-chemical and microscopic properties of Ottoman mortars from Erzurum (Turkey). Constr Build Mater 24:1995–2002

Binici H, Kapur S, Arocena J, Kaplan H (2012) The sulfate resistance of cements containing red brick dust and ground basaltic pumice with sub-microscopic evidence of intra-pore gypsum and ettringite as strengtheners. Cement Concr Compos 34:279–287

Binici H, Kapur S, Rızaoğlu T, Kara M (2014) Resistance to thaumasite form of sulfate attack of blended cement mortars. Br J Appl Sci Technol 4(31):4356–4379

Boxem HW, Wielemaker WG (1972) Soils of the Küçük Menderes Valley, Turkey. Agricultural Research Reports. No. 785, Wageningen: Pudock

Brevik EC (2005) A brief history of soil science. Global sustainable development, Theme, 1. Land Use, Land Cover and Soil Sciences-VI: 1–10

Budak M (2012) Genesis and classification of Saline Alkaline Soils and mapping with both classical and geostatistical techniques. (Ph.D Dissertation) Gaziosmanpasa University, Tokat, Ins. of Science, No: 322692 (in Turkish)

Burrough PA (2001) GIS and geostatistics: Essential partners for spatial analysis. Environ Ecol Stat 8(4):361–377

Burrough PA, McDonnell RA (1998) Principles of geographical information systems. Oxford University Press, Oxford

Çağlar KÖ, Hızalan E, Akalan İ, Saatçı F, Ergene A (1956). The soils of Iğdır Plain. Ankara University Faculty of Agriculture. Chair of soil science. 326 p (in Turkish)

Çakmak İ (2008) Enrichment of cereal grains with zinc: agronomic or genetic biofortification? Plant Soil 302(1–2):1–17

Çakmak I (2012) HarvestPlus zinc fertilizer project: HarvestZinc. Better Crops 96(2):17–19

Çakmak İ, Marschner H (1988) Increase in membrane permeability and exudation in roots of zinc deficient plants. J Plant Physiol 132(3):356–361

Çakmak İ, Kalaycı M, Ekiz H, Braun HJ, Kılınç Y, Yılmaz A (1999) Zinc deficiency as a practical problem in plant and human nutrition in Turkey: a NATO-science for stability project. Field Crops Research 60(1):175–188

Çakmak İ, Pfeiffer WH, McClafferty B (2010) Review: biofortification of durum wheat with zinc and iron. Cereal Chem 87(1):10–20

Çambel H, Kapur S, Karaman C, Akça E, Kelling G, Şenol M. Yeğingil Z, Yaman S (1994) Source Determination of the Late Hittite Basalts at Karatepe-Arslantaş and Domuztepe. In: Özer AM, Demirci Ş and GD Summers (eds) The Proceedings of the 29th international symposium on Archaeometry, Middle East Technical University, Ankara. 9–14 May 1994. TUBİTAK, pp 575–584

Cangir C, Boyraz D (1999) Regulations on agricultural land/soils and their technical evaluation. 3. Meeting on Thrace in the verge of the 21st Century Industrialization. 11–13 Nov (in Turkish)

Cangir C, Boyraz D (2000) Inappropriate land use in Turkey. 5. Tech Conf Chamber Turkish Agr Eng 17–19 (in Turkish)

Cangir C, Kapur S, Boyraz D, Akca E, Eswaran H (2000) An assessment of land resource consumption in relation to land degradation in Turkey. J Soil Water Conserv 55(3):253–259

Cangir C, Boyraz D, Sarı H (2010) Formation and combatting desertification. Symposium on Combatting Desertification, Çorum, Turkey 17–18 June, 36-43 (in Turkish)

Çelebi H (1970) The soil losses induced by wind erosion and some physical and chemical properties of soils of Sivas Ulaş State Farm. Güven publisher, Ankara. 46 p (in Turkish)

Çelebi H (1971) Study on the resistance to soil erosion in the Kars Meadow and Sarıkamış Forest soils. Agric Eng 53:18–22 (in Turkish)

Çelik YM, Ünver İ (1999) Investigation of optimum tillage depth for sunflower in crop rotation under Central Anatolian Conditions. Turkish J Agric Forest 23(5):1087–1094 (in Turkish)

Çelik İ, Ortaş İ, Kılıç S (2004) Effects of compost, mycorrhiza, manure and fertilizer on some physical properties of a Chromoxerert soil. Soil Tillage Res 78(1):59–67

Çevik B, Berkman A (1990) Sand dune stabilization practices implemented in the Great Konya Basin. In: sand transport and desertification in arid lands: proceedings of the international workshop 375 P. World Scientific

Çullu MA (2003) Estimation of the effect of soil salinity on crop yield using remote sensing and geographic information system. Turkish J Agric Forest 27(1):23–28

Çullu MA, Almaca A, Şahin Y, Aydemir S (2002) Application of GIS monitoring soil salinization in the Harran Plain. In: proceedings of international conference of sustainable land use and management. Çanakkale, Turkey. 326–332

De Meester TD (1970) Soils of the Great Konya Basin, Turkey. Agric Res Rep No 740, Pudoc, Wageningen

Dengiz O (2010) Morphology, physico-chemical properties and classification of soils on terraces of the Tigris River in the south-east Anatolia region of Turkey. Tarım Bilimleri Dergisi 16(3):205–212

Dengiz O, Yakupoğlu T, Başkan O (2009) Soil erosion assessment using geographical information system (GIS) and remote sensing (RS) study from Ankara-Güvenç Basin. Turkey J Environ Biol 30(3):339–344

Dinç U, Kapur S, Akça E, Özden M, Şenol S, Dingil M, Öztekin E, Kızılarslanoğlu HA, Keskin S (2001) History and status of soil survey programs in Turkey and suggestions on land management. Soil resources of Southern and Eastern Mediterranean Countries. Options Mediterraneennes. Series B: Studies and Research No. 34, pp 263–276

Dinç U, Şenol S, Cangir C, Dinç AO, Akça E, Dingil M, Kapur S (2005) Soil survey and soil database of Turkey. Soil resources of Europe, pp 371–375

Driessen PM, De Meester (1969) Soils of the Çumra area, Turkey. Agricultural reports 720, Wageningen. 105 p

Ekinci H (1990) Possibility to arrange general soil maps of Turkey according to soil taxonomy, an example of Tekirdağ Region. PhD Thesis, Çukurova University, Institute of Science, Soil Science Department, Adana, Turkey (in Turkish)

Erdal I, Yılmaz A, Taban S, Eker S, Torun B, Çakmak I (2002) Phytic acid and phosphorus concentrations in seeds of wheat cultivars grown with and without zinc fertilization. J Plant Nutr 25(1):113–127

Erdoğan EH, Şahin M, Yüksel S (2013) National soil information system in Turkey. EU General Assembly 2013, held 7–12 April 2013 in Vienna, Austria, id. EGU 9962 p

Eren M, Kadir S, Zucca C, Akşit İ, Kaya Z, Kapur S (2014) Pedogenic manganese oxide coatings (calcium buserite) on fracture surfaces in Tortonian (Upper Miocene) red mudstones, southern Turkey. Catena 116:149–156

Eren M, Kadir S, Kapur S, Huggett J, Zucca C (2015) Color origin of Tortonian red mudstones within the Mersin area, southern Turkey. Sed Geol 318:10–19

Erşahin S (2001) Assessment of spatial variability in nitrate leaching to reduce nitrogen fertilizers impact on water quality. Agric Water Manag 48(3):179–189

Erşahin S, Brohi AR (2006) Spatial variation of soil water content in topsoil and subsoil of a Typic Ustifluvent. Agric Water Manag 83(1):79–86

Gök M, Ottow J (1988) Effect of cellulose and straw incorporation in soil on total denitrification and nitrogen immobilization at initially aerobic and permanent anaerobic conditions. Biol Fertil Soils 5:317–322

Gök M, Doğan K, Coşkan A (2006) Effects of drivers organic substrate application on denitrification and soil respiration under different plant vegetation in Çukurova region. In: international symposium on water and land management for sustainable irrigated agriculture, Adana. Turkey, 48. April 2006, 2530 p

Göktan D (1974) Studies on the efficiency of Rhizobium phaseli strains isolated from some cultivated Turkish soils. PhD Thesis. Ankara University. Faculty of Agriculture. Soil Science Department. 67 p (in Turkish)

Groneman AF (1968) The soils of the wind erosion control camp area, Karapinar, Turkey. Ph.D. Thesis, Agric University Wageningen, Netherlands, 160 p

Gülçür F (1958) Research on chemical and mineralogical properties of the clay fractions of some virgin soils developed under humic climatic conditions in the Black Sea Region (Rize), Turkey. Ist Univ Jour Forest Fac 8(2):35–104 (in Turkish)

Günal H, Durak A, Akbaş F, Kılıç S (2004) Genesis and classification of vertisols as affected by rice cultivation. International Soil Congress. Natural resource management for sustainable development. 7–10 June, Erzurum, Turkey

Günal H, Acır N, Budak M (2012) Heavy metal variability of a native saline pasture in arid regions of Central Anatolia. Carpathian J Earth Environ Sci 7(2):183–193

Gündoğan R, Özyurt H, Akay AE (2011) The effects of land use on properties of soils developed over ophiolites in Turkey. Int J Forest Soil Erosion (IJFSE), 1(1): 36–42

Güneş A, İnal A, Alpaslan M, Çakmak I (2006) Genotypic variation in phosphorus efficiency between wheat cultivars grown under greenhouse and field conditions. Soil Sci Plant Nutr 52(4):470–478

Gürbüzer E (1973) Study on the characteristics of Rhizobium meliloti strains and their efficiency. PhD Thesis. Ankara University. Faculty of Agriculture. Soil Science Department. 62 p (in Turkish)

Güzel N, İbrikçi H (1994) Distribution and fractionation of soil phosphorus in particle-size separates in soils of Western Turkey. Commun Soil Sci Plant Anal 25(17–18):2945–2958

Haktanır K (1973) Effects of fallow-wheat-legume rotation on important soil enzyme activities under Ankara conditions. Annals of the Ankara University, Faculty of Agriculture Publications, 613 p (in Turkish)

Janssen BH (1970) Soil fertility in the Great Konya Basin, Turkey. PhD Thesis, Wageningen Agricultural University, Pudoc, 113 p

Kacar B (1964a) Phosphorus status of Çukurova soils and studies on the methods to be used to determine the phosphorus contents of these soils. Habilitation Thesis in Plant Nutrition. 146 p (in Turkish)

Kacar B (1964b) Estimation of plant available phosphorus by the combination of different H_2SO_4 and NH_4F concentrations in Çukurova soils. Ann Ankara Univ 10:103–131 (in Turkish)

Kacar B (1994) Chemical analysis of plant and soil. Foundation of Ankara University Faculty of Agriculture Education, Research and Development, Ankara, Turkey (in Turkish)

Kacar B (2009) Soil Analysis. Nobel Publications and Distribution, Ankara, Turkey (in Turkish)

Kacar B, Akgül ME (1966) Evaluation of various methods for the estimation plant available phosphorus in the soils of Shiraz (Iran). Ann Ankara Univ 6:3–14 (in Turkish)

Kacar B, Inal A (2008) Plant Analysis. Nobel Publications and Distribution. 1241:891. Ankara, Turkey (İn Turkish)

Kaplan MY, Eren M, Kadir S, Kapur S (2013) Mineralogical, geochemical and isotopic characteristics of Quaternary calcretes in the Adana region, southern Turkey: implications on their origin. Catena 101:164–177

Kaplan MY, Eren M, Kadir S, Kapur S, Huggett J (2014) A microscopic approach to the pedogenic formation of palygorskite associated with Quaternary calcretes of the Adana area, southern Turkey. Turkish J Earth Sci 23:559–574

Kapur S, Çavuşgil VS, FitzPatrick EA (1987) Soil-calcrete (caliche) relationship on a Quaternary surface of the Çukurova Region, Adana (Turkey). In: Fedoroff N, Bresson LM, Courty MA (eds) Micromorphologie des Sols/Soil Micromorphology. AFES, Plaisir, France, pp 597–603

Kapur S, Çavuşgil VS, Şenol M, Gürel N, FitzPatrick EA (1990) Geomorphology and pedogenic evolution of Quaternary calcretes in the northern Adana Basin of southern Turkey. Zeitschrift für Geomorphologie 34:49–59

Kapur S, Sakarya N, Karaman C, Fitzpatrick E, Pagliai M (1995) Micromorphology of basaltic ceramics. Br Ceram Trans 94:33–37

Kapur S, Karaman C, Akça E, Aydın M, Dinç U, FitzPatrick EA, Mermut AR (1997) Similarities and differences of the spheroidal microstructure in Vertisols from Turkey and Israel. CATENA 28(3):297–311

Kapur S, Saydam C, Akça E, Çavuşgil VS, Karaman C, Atalay I, Özsoy T (1998a) Carbonate pools in soils of the Mediterranean: a case study from Anatolia. In: Lal R, Kimble JM, Stewart BA (eds) Global climate change and pedogenic carbonates. CRC Press, Boca Raton, Florida, pp 187–212

Kapur S, Sakarya N, FitzPatrick E, Pagliai M, Kelling G, Akça E, Karaman C, Sakarya B (1998) Mineralogy and micromorphology of iznik ceramics. Anatolian Stud 48:81–189

Kapur S, Gök T, Erk N, Fisunoğlu MH, Berberoğlu S, Akça E, Evrendilek F, Dingil M, Yılmaz E, Polat S, Polat O, Özevren E (2013) Anthroscape, the "human ecosystem and multifunctional integrated satellite town/new Agricultural town systems" approach in land/carbon planning. TÜBİTAK Project Report no: 110 Y 120. Ankara, 209 p

Karaca A, Haktanır K (2000) Effects of sewage sludge on available lead and dehydrogenase enzyme activity. Ankara University, Faculty of Agriculture. Agric Sci J 6(3):13–19 (in Turkish)

Karaca A, Naseby DC, Lynch JM (2002) Effect of cadmium contamination with sewage sludge and phosphate fertiliser amendments on soil enzyme activities. Microbial Fertility Soils 35(6):428–434

Karaca A, Çetin SC, Turgay OC, Kızılkaya R (2009) Effects of heavy metals on soil enzyme activities. In: Varma A (Editor-in-Chief), & Sherameti I (eds) Springer-Soil Biology Series: Soil Heavy Metals. 19(11): 237–262

Karaca A, Çetin SC, Turgay OC, Kızılkaya R (2010) Soil enzymes as indication of soil quality. Soil Enzymology. Springer, Berlin Heidelberg, pp 119–148

Kelling G, Kapur S, Sakarya N, Akça E, Karaman C, Sakarya B, Robinson P (2000) Basaltic Tephra: potential new resource for ceramic industry. Br Ceramic Trans 99(3):129–136

Kızılkaya R, Bayraklı F (2005) Effects of N-enriched sewage sludge on soil enzyme activities. Appl Soil Ecol 30(3):192–202

Kızılkaya R, Karaca A, Turgay OC, Çetin S (2011) Earthworm interactions with soil enzymes. Biology of earthworms. Springer, Berlin, pp 141–158

Küçükuysal C, Kapur S (2014) Mineralogical, geochemical and micromorphological evaluation of the Plio-Quaternary paleosols and calcretes from Karahamzalı, Ankara (central Turkey). Geol Carpath 65:241–253

Küçükuysal C, Türkmenoğlu A, Kapur S (2013) Multiproxy evidence of Mid-Pleistocene dry climates observed in calcretes in central Turkey. Tur J Earth Sci 22(3):469–483

Kurucu Y, Bolca M, Altınbaş U, Esetlili MT (2004) A study on the determination of the land use, elevation and slope of the land to the west of Söke by forming a digital elevation model and satellite image. J Appl Sci 4(4):542–546

Mermut AR, Pape T (1971) Micromorphology of two soils from Turkey, with special reference to in-situ formation of clay cutans. Geoderma 5(4):271–281

Mermut AR, Pape T (1973) Micromorphology of "In Situ" formed clay Cutans in soils. Leitz Sci Technical Inf West Germany 2:147–150

Mermut AR, Tanju Ö, Hatipoğlu F, Başkaya HS, Aktaş M (1981) Annotated bibliography on Turkish soil science 1928–1976. Documentation of Turkish Research and Technical Centre. 1007 p (in Turkish)

Mitchell WA, Irmak A (1957) Studies on clay mineralogy of forest soils in Turkey. Ist Univ J Forest Fac VII Seri B 2:4–15 (in Turkish)

Munsuz N (1967) Determination of clay minerals in some Turkish soils by infrared spectroscopy method. Thesis for Assoc. Prof. position. Ankara University Publications of Fac of Agric No: 305. 46 p (in Turkish)

Munsuz N, Nouri K (1969) The clay mineralogy of certain soils that have been formed from a variety of parent materials. Univ of Ankara. Annals of the Faculty of Agriculture, pp 1–8 (in Turkish)

Munsuz N, Rasheed MA, Başkaya H (1970) Determination of clay minerals by the Imbibometry method. Ankara University Publications of Fac. of Agriculture, No: 461. 25 p (in Turkish)

NAP-D (2006) Turkey's national action program of combating desertification of Turkey. In: Düzgün M, Kapur S, Cangir C, Boyraz D, Akça E, Özevren E and Gülşen N (eds) Ministry of forestry and water affairs, 89 p

Oakes H (1954) Soils of Turkey. Min. of Agriculture, Soil Con. and Farm Irrigation Division Pub. No.1, Ankara, Turkey

Okur N, Tuna AL, Okur B, Altunlu H, Kayıkçıoğlu HH, Civelek HS (2010) Non-target effect of organic insecticides: effect of two plant extracts on soil microbial biomass and enzymatic activities in soil. Environ Monit Assess 165(1–4):389397

Ortaş İ (2003) Effect of Selected Mycorrhizal Inoculation on Phosphorus Sustainability in Sterile and Non-sterile Soils in the Harran Plain in South Anatolia. J Plant Nutr 26(1):1–17

Ortaş I (2012) The effect of mycorrhizal fungal inoculation on plant yield, nutrient uptake and inoculation effectiveness under long-term field conditions. Field Crops Res 125:35–48

Ortaş İ, Ortakçı D, Kaya Z, Çınar A, Önelge N (2002) Mycorrhizal dependency of sour orange in relation to phosphorus and zinc nutrition. J Plant Nutr 25(6):1263–1279

Ortaş İ, Sarı N, Akpınar G, Yetişir H (2011) Screening mycorrhiza species for plant growth, P and Zn uptake in pepper seedling grown under greenhouse conditions. Sci Hortic 128(2):92–98

Oruç N (1971) Investigation of negative charge load depending on effective and pH values in acid soils of Rize region. J Fac Agric 2(3):56–67 (in Turkish)

Oruç N (2008) Occurrence and problems of high fluoride waters in Turkey: an overview. Environ Geochem Health 30(4):315–323

Oruç N, Alpman N (1976) Hydrogeology of spring waters with high fluoride content around Tendürek volcano. Bull Geol Soc Turkey 19:1–8

Özbek H (1971) The effect of nitrogen on the development of the pyro-catechin humin acid and its bonding in soil. Ann Ankara Univ Fac Agric pub (516):75 p (in Turkish)

Özbek H, Rasheed MA (1972) The activity of enzymes in soils of Black Sea Coast. University of Çukurova. Ann Ankara Univ Fac Agric 1:89–102 (in Turkish)

Özbek H, Rasheed MA (1973) Enzymatic activities and the role of pH in some of the Turkish soils. University of Ankara. Ann Ankara Univ Fac Agric 8:237–244 (in Turkish)

Özbek N, Munsuz N. Ahmed F (1974) Suitability of using Rb86 as a tracer for potassium for soils having different types of clay minerals. Annals of Ankara University, Faculty of Agriculture 13:131–140

Özbek N (1984) Biological denitrogen fixation and its economic importance for agriculture, Ankara: Ankara Nuclear Agriculture Research Center, Project No: 311- DISR-96

Özbek N, Akdeniz A (1965) A study on the uptake of fertilizers and soil phosphorus by test plant as influenced by the rate of applied phosphorus, using P-32 as a tracer. Ann Ankara Univ Fac Agric 15(3):40–64 (in Turkish)

Özbek N, Akdeniz A (1966) Study on uptake of fertilizer and soil phosphorus by test plants at different stages of growth, using P-32 as tracer. Ann Ankara Univ Fac Agric 16(3):32–49 (in Turkish)

Özcan H, Taban S (2000) Effects of VA-mycorrhiza on growth and phosphorus, zinc, iron, copper and manganese concentrations of maize grown in acid and alkaline soils. Turkish J Agric Forest 24(5):629–636

Özcan H, Çetin M, Diker K (2003) Monitoring and assessment of land use status by GIS. Environ Monit Assess 87(1):33–45

Özgöz E, Günal H, Acir N, Gökmen F, Birol M, Budak M (2013) Soil quality and spatial variability assessment of land use effects in a typic Haplustoll. Land Degrad Dev 24(3):277–286

Özkan İ (1999) Soil physics. University of Ankara. Publication of agricultural faculty: 946, Lecture notes. Ankara. 270 p (in Turkish)

Öztaş T, Koç A, Çomaklı B (2003) Changes in vegetation and soil properties along a slope on overgrazed and eroded rangelands. J Arid Environ 55(1):93–100

Öztürk L, Eker S, Torun B, Çakmak İ (2005) Variation in phosphorus efficiency among 73 bread and durum wheat genotypes grown in a phosphorus-deficient calcareous soil. Plant Soil 269(1–2):69–80

Paul EA (2014) Soil microbiology, ecology and biochemistry. Academic press, Cambridge

Polat O, Kapur S (2010) Soil quality parameters in Arenosols under Stone Pine (*Pinus Pinea L.*) Plantations in the Turan Emeksiz sand dune Area. Çukurova University, Institute of Science. J Sci Eng 3(2):167–174. Adana (in Turkish)

Rashid MA (1971) Enzymatic activities and their relation to soil characteristics in Ankara soils. PhD Thesis. Ankara University. Soil Science Department. 130 p (in Turkish)

Ryan J, Kapur S, Akça E (2009) Application of soil analyses as markers to characterize a middle-eastern Chalcolithic-Late Bronze age mound. TÜBA-AR (Publication of the Academy of Sciences-Archaeology of Turkey) 12:66–76

Şahin G (2012) The Study of Soil in Turkey from Past to Date. "Soil in Our Culture". In: Gürsoy E, Hilal N, Altun O (eds) Acta Turcica. 4(1):102–118 (in Turkish)

Şahinkaya H (1969) An investigation on the nodosity bacteria of earth nut. Soil Science Society of Turkey, vol 1. Papers of the 1st, 2nd and 3rd Scientific Meetings, pp 107–113 (in Turkish)

Sayın M (1982) Quantitative soil mineral analyses by calculating mass absorption coefficients (Habilitation thesis), University of Çukurova, Faculty of Agriculture, Department of Soil Science, Adana. 13. 150 p

Şenol S (1993) Application of a quantitative land use planning system in the south eastern project. The future of land: mobilizing and integrating knowledge for land use options, vol 1. Anniversary conference of the Wageningen Agricultural University, London, p 1

Sözüdoğru S, Karaca A, Haktanır K (1996) Effects of Chicken manure on nitrogen mineralization and urease enzyme activity, Ankara University, Publication of Agricultural Faculty No, 1445, Scientific Research and Investigations Ankara 798 p (in Turkish)

Sürücü A, Günal H, Acır N (2013) Importance of spatial distribution in reclamation of boron toxic soils from Central Anatolia of Turkey. Fresenius Environ Bull 22(11):3111–3122

Tekinel O, Yeğingil İ, Dinç U, Kapur S, Şenol S (1991) Remote sensing applications for the southeastern Anatolian project. Options Mediterraneennes. Series A: Seminaires Mediterraneens (CIHEAM), pp 95–102

Turgay OC, Erdoğan EE, Karaca A, Bilen S (2007) Evaluation of soil enzyme activities and biochemical index for monitoring the impacts of humic substances induced biostimulation on hydrocarbon degradation in a crude oil contaminated Soil. In: the third international conference Viterbo/Italya, 15–19 July 2007

Turgay OC, Erdoğan EE, Karaca A (2010) Effect of humic deposit (leonardite) on degradation of semi-volatile and heavy hydrocarbons and soil quality in crude-oil- contaminated soil. Environ Monit Assess 170(1–4):45–58

Ünal H (1966) Phosphorus status and some other characteristics of soils used for sugar beet production in Devrenkani. Ankara Univ Fac Agric Pub No 267 168:53 p (in Turkish)

Ünal H (1967) Enzymatic activities and their relationships between these activities and important soil properties in Rize tea plantation soils. Ankara Univ Fac Agric Pub No 306 191:79 p (in Turkish)

Yamanlar O (1956) Land classification on the basis of soil conservation in Marmara Region, Turkey with special reference to Yalova district. İstanbul Univ Pub. No: 697. Fac Forest No 42:98 p (in Turkish)

Yeşilot-Kaplan M, Eren M, Kadir S, Kapur S, Huggett J (2014) Palygorskite formation within Quaternary calcretes (Adana, southern Turkey). In: 8th international symposium on Eastern Mediterranean Geology, Muğla, Turkey, 15 p

Yeşilsoy MŞ (1966) The clay mineralogy of Grumusols, Non-calcic Brown and Rendzina soils, Thrace, Turkey. Ph.D Thesis. Ankara. 90 p (in Turkish)

Yılmaz A, Ekiz H, Torun B, Gültekin I, Karanlık S, Bağcı SA, Çakmak I (1997) Effect of different zinc application methods on grain yield and zinc concentration in wheat cultivars grown on zinc-deficient calcareous soils. J Plant Nutr 20(4–5):461–471

Yüksel A, Gündoğan R, Akay AE (2008) Using the remote sensing and GIS technology for erosion risk mapping of Kartalkaya dam watershed in Kahramanmaraş. Turkey Sens 8(8):4851–4865

Zdruli P, Pagliai M, Kapur S, Cano AF (2010) What we know about the saga of land degradation and how to deal with it. In: Zdruli P, Pagliai M, Kapur S and Cano AF (eds) Proceedings of the land degradation and desertification: assessment, mitigation and remediation, doi:10.1007/978-90-481-8657-0_1, Springer, Heidelberg

Zucca C, Andreucci S, Akşit İ, Koca YK, Shaddad MS, Madrau S, Pascucci V, Previtali F, Kapur S (2012) Genesis and paleoenvironmental implications of upper Pleistocene paleosols on the NW Sardinian coast. In: Poch RM, Casamitjana M, Francis ML (eds) Proceedings of the 14th international working meeting on soil micromorphology, IUSS Commission 1.1 Soil Morphology and Micromorphology, Lleida, 8–14 July, Spain, pp 235–238

Zucca C, Vignozzi N, Madrau S, Dingil M, Previtali F, Kapur S (2013) Shape and intra-porosity of top soil aggregates under maquis and pasture in the Mediterranean region. J Plant Nutr Soil Sci 4:229–239

Zucca C, Sechi D, Andreucci S, Shaddad SM, Deroma M, Madrau S, Pascucci V, Kapur S (2014a) Pedogenic and palaeoclimatic evidence from an Eemian calcrete in NW Sardinia. Eur J Soil Sci 65:420–435

Zucca C, Andreucci S, Akşit I, Kapur S (2014b) Buried palaeosols of NW Sardinia (Italy) as archives of the late quaternary climaticfluctuations. Catena 122:72–90

Vegetation

İbrahim Atalay

1 Introduction

Turkey possesses a rich variety of vegetation due to its geomorphological features and varieties of ecological conditions and is located at the intersection of phytogeographical regions of the Mediterranean, Euro-Siberian, and Irano-Turonian. The distinctive vegetation of the country reflects differences in climate, geology, topography, soils and floristic diversity. Ninety percent of the forests in Turkey are 'natural' in origin and contain over 450 species of trees and shrubs. Turkey has 21.2 million ha of forests, covering 27% of the land surface of the country. Most of these forests occur in the mountains. About 42% of the forests in Turkey are composed of coniferous species, 53% of broad-leaved species, and 5% are mixed of coniferous and broad-leaved forests. Furthermore, the mountainous landscapes of Turkey, with their remarkable bioclimatic, geomorphologic, and pedologic diversity, support a great many different high mountain vegetation types (Atalay 1994, 2014; Mayer et al. 1986; Öztürk et al. 1991; Quezel 1986; Yaltırık 1973; Zohary 1973).

Turkey takes place on the Alpine-Himalayan orogenic belt and has a mountainous topography. There are two main orogenic belts: the northern Turkey Mountains in the north and the Taurus Mountains in the south. The isolated volcanic cones more than 3000 m altitude are found in the central and eastern parts of Turkey. Horst and graben systems occur in the west and east Turkey. Orogenic belts have been deeply dissected by the rivers flowing into the seas (Atalay 1987c). These topographic forms are responsible for the formation of the various habitats. Three main climatic types prevail in Turkey, namely the northern part of Turkey under the effect of humid and cold–humid climates, the Mediterranean climate prevailing in the western and southern parts of the country, and the Continental climate prevailing in the central and eastern parts of Turkey.

2 The Forests of Turkey

The vegetation of Turkey is classified under eight eco-regions, namely the Black Sea, the Marmara Transitional, the Aegean, the Mediterranean, the Mediterranean Transitional, central Anatolian, the eastern Anatolian, and the south eastern Anatolian regions (Atalay 2014) (the geographical position of Turkey is designated in two parts, namely Asia Minor or Anatolia and Thrace—the European side of Turkey—in the Turkish earth sciences and geographical context).

Figure 1 illustrates the different properties of the vegetation compositions of the eco-regions of Turkey.

2.1 Forests of the Black Sea Region

The Black Sea ecoregion extends from the Istranca Mountains in the west to Artvin near the Georgian border in the east. The Black Sea region is characterized by a mild-humid and cold–humid climate that only prevails in this Region. The mean annual temperature varies from 14 to 10 °C along the coastal belt up to 1000 m elevation. The average temperature is about 10–6 °C from 1000 to 2000 m. The eastern part of the region is highly humid with an average of 1500–2500 mm annual precipitation. The western part of the region is less humid and has an annual precipitation from 1000 to 1500 mm. The rainy period covers all seasons of the year but the amount of the rainfall varies between seasons. The mean annual precipitation is over 1000 mm, and humid and perhumid climatic conditions are common in the region in general (Atalay 1984, 1987a, b, 1992, 1994, 2010, 2014).

Semiarid subhumid conditions, on the other hand, are dominant in the valleys such as the Çoruh, Gökırmak,

İ. Atalay (✉)
Department of Geography, Karabük University, Karabük, Turkey
e-mail: ibrahim.atalay@deu.edu.tr; atalay@mehmetakif.edu.tr

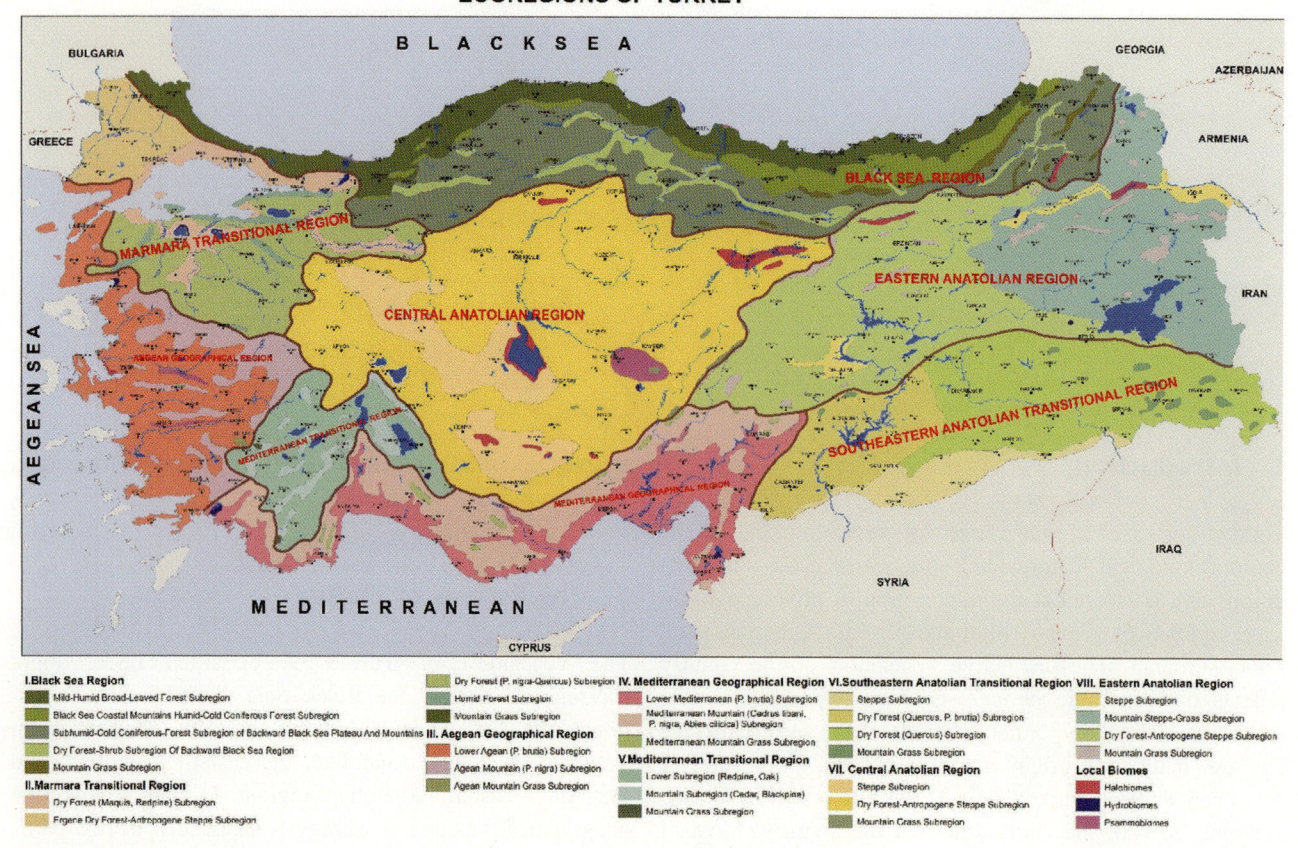

Fig. 1 The eco-regions and distribution of vegetation of Turkey (Atalay 2014)

Fig. 2 Pure oriental beech (*Fagus orientalis*) forest growing under humid mild conditions in the lower belt of the Black Sea Region

Fig. 3 Northern Anatolian Fir (*Abies nordmanniana*) forest in the foggy and cold–humid habitat of the upper level of the Black Sea Region

Kelkit, and Devrez, which are located in inland areas. The most arid part of the region is the Yusufeli district and its vicinity in the Çoruh valley. During summer, the northern slopes of the mountains are covered with fog. Foggy habitats lead to the decrease of transpiration. Thus, most of the plants are of hydrophytic and hydrophyllic character in these areas (Atalay 2014; Yaltırık 1973).

Humid Broad-Leaved Deciduous (Beech, Castanea, Alnus, Tilia and Quercus) Forests

The Black Sea region is very rich in terms of plant species and communities. Forests are composed of different species and floristic composition. Understory flora is associated with Rhododendron sp. in general. The main forest communities along the coastal belt of the Black Sea are defined under the

Fig. 4 Oriental spruce (*Picea orientalis*) and *Rhododendron ponticum* only grow under the heavy foggy and humid–cold conditions in the eastern part of the Black Sea Region

Fig. 5 A good stand of scotch pine (*Pinus sylvestris*) forest growing on the sunny, subhumid, and cold conditions in NE Anatolia

Fig. 6 Mixture of shrubby Mediterranean vegetation composed of maquis (*yellowish shrubs*) and garrique (in the *front* of photo) is widespread where *red pine* forests have been completely destroyed

Fig. 7 Garrique vegetation spreading on the abandoned agricultural fields in the Aegean Region

Fig. 8 A virgin *red pine* (*Pinus brutia*) forest with maquis lowerstory grows under direct solar radiation and is very resistant to summer droughts

Fig. 9 A good stand of Anatolian *black pine* (*Pinus nigra*) growing under subhumid and cold conditions in the vicinity of Beyşehir locality in the upper level of the Mediterranean Region

eastern Black Sea, central, western Black Sea, and The Coastal Belt of the Marmara subregions (Figs. 1 to 5).

2.1.1 The Eastern Black Sea Subregion

Humid Broad-Leaved Deciduous (Beech, Castanea, Alnus, Tilia and Quercus) Forests

The high mountains exercise a predominant influence on the climate, and thus on the vegetation. The broadleaf deciduous forests are the main vegetation type of the area. The main forest type of this region is *Fagus orientalis* from the seashore to an elevation of 1700 m (Fig. 2). However, the tree composition of these forests changes in relation to altitude. The *Quercus petraea* subsp. *iberica, Castanea sativa, Acer platanoides, Acer cappadocicum, Zelkova carpinifolia, Tilia rubra*, and *Pterocarya fraxinifolia* are other broad-leaved species in the oriental beech forests especially between 500 and 1200 meters elevation. The Oriental spruce (*Picea orientalis*) is the dominant tree species at elevations from 1200 to 1500 m. At this subregion, alongside the pure *F. orientalis* forests, other mixed forests composed of beech and oriental spruce, beech + oriental spruce + alder, beech + chestnut, beech, chestnut + red lime, beech + fir, and fir + beech + oriental spruce are also present. These populations have a rich understory of ferns and vines (Atalay 1984, 2014; Yaltırık 1973).

2.1.2 The Central Black Sea Region

The natural forests of the subregion are degraded considerably due to the densely populated rural settlements. Productive beech forests are common on the upper part of the Canik Mountains in the vicinity of Akkuş. Oriental beech forests occupying the southern part of the Bafra district are composed of *C. sativa, Alnus barbata, Alnus glutinosa, Prunus* sp., and *Carpinus* sp. The shrub layer of such forests is mainly composed of *Rhododendron flavum* (Atalay 1992).

2.1.3 The Western Black Sea Subregion

Coastal mountain ranges, deeply dissected river valleys such as the Gökırmak, Devrez, and the inland mountains of Bolu-Abant-Köroğlu determine the distribution of the vegetation cover of this subregion.

Coastal mountains have productive *F. orientalis* forests and their floristic compositions in the coastal belt are associated with *Pinus brutia, Laurus nobilis, C. sativa, Tilia sp*, and *C. betulus*. The forests of *P. sylvestris, Abies bornmulleriana* and mixed *F. orientalis* forests also occur at lower intensity.

2.1.4 The Coastal Belt of the Marmara Subregion

Broad-leaved deciduous forests are common in the Yıldız Mountains in Thrace, Çatalca, and Kocaeli Peninsulas, at the two sides of the Bosphorus. The leading forest populations in these regions are the *F. orientalis, C. betulus, C. orientalis*, and *Quercus* sp. The northern part of the Kocaeli Peninsula is the main occurrence areas of *Quercus, Fagus* and *Castanea* forests (Dönmez 1968, 1979).

The Humid-Subhumid Coniferous (Oriental Spruce, Fir, Black Pine and Scots Pine) Forests

Coniferous forests represent a response to a higher part of the mountains on which cold and humid climatic conditions prevail. The tree species and forest communities change depending on the fog formation level and sunny habitats of the mountains. Pure and mixed *Picea orientalis* (oriental spruce), *P. sylvestris*, and *Abies nordmanniana* forests are common in the east (Fig. 3), meanwhile in the middle and western part of the Black Sea subregions, *P. sylvestris, Abies bornmulleriana* and *Pinus nigra* occur on the slopes facing south of the Northern Anatolian Mountains (Atalay et al. 2010; Mayer et al. 1986).

Picea orientalis forests occur only in higher elevations between 1500 and 2000 meters in the places where upslope fog formation is common (Fig. 4). The sunny habitats of the northern Anatolian mountains are the main occurrence areas of *P. sylvestris*. In fact, pure *P. sylvestris* forests are common in the upper plateau surfaces of the Northern Anatolian Mountains and in the central part of the mountainous areas (Atalay et al. 2012) (Figs. 1, 5 and 9).

Abies bornmulleriana forests are widespread on the slopes facing north in the middle part of the Black Sea region. On the contrary, *Pinus nigra* forests, growing somewhat in the continental part, occur on the lowland part of the mountainous areas. The higher parts of the mountains with intense fog and the northern slopes of the Abant and Bolu mountains are the principal occurrence areas of the pure and mixed *Abies bornmulleriana* forests.

Dry (Oak, Black Pine, Red Pine) Forests

The effects of drought increase towards the inland parts of the Black Sea Region. So that, dry forests begin on the bottom lands of the tectonic corridors and lower levels of the wide river valleys and their south-facing slopes in the inland section of the Black Sea Region (Figs. 1 and 9).

The Shrub (Pseudomaquis and Maquis) Formation

This region contains both Mediterranean shrubs and the mild-humid Black Sea trees called the pseudomaquis (Dönmez 1968; Zohary 1973). Pseudomaquis are widespread as a narrow belt along the Black Sea coast. This shrub formation has been developed following the destruction of the broad-leaved deciduous forests. In other words, degraded forests are mostly replaced or covered by pseudomaquis. Maquis are also common in the bottom land of the valleys and the tectonic depressions lying in the inland part of the Black Sea Region.

2.2 The Marmara Transitional Region

This region surrounding the Marmara Sea is the transitional region between the Black Sea mild-cold–humid climate in the north and the Mediterranean climate in the south and west. Thus, it contains both Mediterranean and Black Sea region vegetation and can be divided into three subregions.

2.2.1 The Broad-Leaved Deciduous Forests

These forests that are mainly composed of *F. orientalis, Tilia tomentosa*, and *C. sativa* are found on the northern slopes of the mountains in this region. *Fagus orientalis* forests are present in the higher parts of the Samanlı mountains in the east to the lower (1000 to 1100 m) in the south. In this area, beech forests are partly composed of *C. betulus, Tilia tomentosa, P. nigra* and *P. sylvestris*.

The northern slopes, like the Ulu and Domaniç mountains are affected by humid air masses coming from the north. These mass fronts are being intercepted by the mountains, where the northern slopes receive more rainfall than the southern slopes. The upper parts of the mountains are covered by the coniferous forests. *Pinus nigra* forests are widespread on the sunny habitat of the mountains in the south.

Different vegetation covers also yield different soil types in the Euro-Siberian region. For example, acidic soils and/or Podzols are common under the Fagus, Abies, *P. sylvestris*, and *Picea orientalis* forests. Alkaline soils appear under the *Pinus nigra* and xerophytic shrub communities.

2.2.2 The *P. Brutia* and Quercus Forests

In the Marmara subregion, the forests are associated with *P. brutia* and Quercus, mainly found on the lowland and the south-facing slopes of the mountains. In the areas where *P. brutia* has been removed, *Quercus coccifera, Calicotome villosa, Paliurus spina-christii*, and *Erica arborea* occur as the dominant shrub vegetation. In these vegetation areas, alkaline soils are common.

2.3 The Mediterranean and Aegean Phytogeographical Region

This region covers the surrounding areas of the Marmara Sea in the NW part of Turkey. The Black Sea humid mild and Mediterranean climatic conditions prevail in the Marmara Transitional Region. Thus, this region contains both Mediterranean and Black Sea vegetation.

The mean annual temperature ranges from 18 to 14 °C from the south to the north. The mean January temperature changes between 10 and 5 °C from the Mediterranean coast to the north of the Aegean coast. The mean July temperature is over 20 °C. The mean yearly precipitation changes between 400 and 2500 mm most of which falls during the winter period. The amount of precipitation increases and temperature decreases towards the higher part of the mountain due to the altitude. The amount of the precipitation of the south-facing slopes of the Taurus Mountains and north-facing slopes of the Aegean Region Mountains is much more than the other slopes (Efe 1998, 2005; Kantarcı 1982; Atalay 2010). These climatic and topographic properties are responsible for the distribution of vegetation formations. Thus, oro-Mediterranean forests composed of oaks, black pine are present over 800 m in the Aegean Region, and mountain forests associated with cedar, fir, and black pine of the Taurus Mountains commence over 800 m (Fig. 8, 9 and 10). The karstic lands of the Mediterranean region are the main occurrence areas of Luvisols and/or Red Mediterranean soils.

2.3.1 Shrub *(Maquis* and *Garrigue)* Formation

Shrub vegetation, which is the secondary succession of the thermo-Mediterranean, occurs along the coastal belt of the region and it continues a few hundred kilometers towards the inland part of the area through the grabens of Gediz, Büyük, and Küçük Menderes.

2.3.2 Maquis Vegetation

In most cases, maquis have originated from the degradation of red pine (*P. brutia* Ten) forests via human activities. Maquis vegetation is common in the Aegean and Mediterranean geographical regions up to 600–800 m (Fig. 6). Maquis have a deep root system and can occur on a shallow soil cover and stony–rocky areas. Most of them are evergreen and fast-growing species. *Olea europea, Ceratonia siliqua, Q. coccifera, Pistacia palestina, Pistacia lentiscus,* and *Arbutus andrachne* are the principal species of the maquis communities. *Quercus coccifera, Quercus ilex, P. lentiscus, Cistus sp*, and *C. villosa* are the species resistant to forest fires, and they rapidly regenerate by root shoots after fire (Yaltırık 1973; Atalay 1994; Atalay et al 1998; Atalay 2014).

2.3.3 Garrigue (Phrygana) Vegetation

One of the most characteristic habitats of the Mediterranean region is Garrigue, a lowland vegetation community of poor soils, composed largely of spiny or aromatic dwarf shrubs. This short shrub vegetation which is termed as low matorral occupies all the Aegean and Mediterranean coastal areas of Turkey without any demand for selective properties related to parent materials. In fact, one can find the garrigue in the same places where both *P. brutia* and maquis grow. But, Garrigue vegetation is dominant in the areas where the natural vegetation/cultivated field were mostly degraded, burned, or abandoned (Fig. 7) (Öztürk 1995; Öztürk et al 2002; Efe 2005; Atalay 2014).

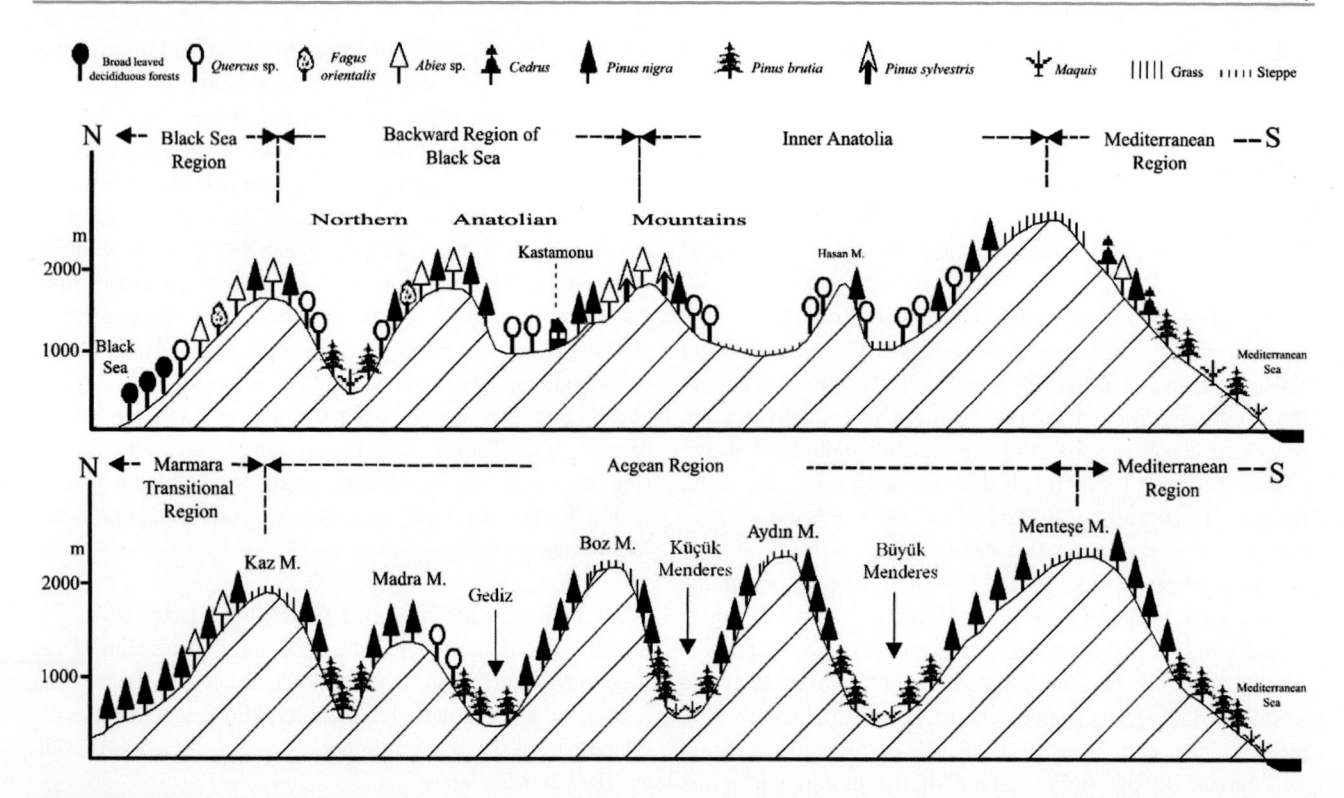

Fig. 10 Vegetation cross sections of Turkey

2.3.4 Forest Formations of the Mediterranean Region

The forest formations of the Mediterranean region can be divided into two main groups, namely the Eu-(Lower) Mediterranean and oro-Mediterranean according to climatic and altitudinal conditions.

Pinus brutia Forests

These are the climax forests of the lower belt of the Aegean and Mediterranean regions and are resistant against drought (Fig. 8). Moreover, the germination, regeneration, and fire resistance of their seeds are much higher than the other forest trees. Burned forest areas are very shortly occupied by red pine communities. Red pines are present at sea level and reach elevations up to 300–400 m in the Marmara, 700–800 m in the Aegean, and 1500 m in the Mediterranean Regions (Atalay et al. 1988).

Biomass productivity and the physiognomic features of *P. brutia* are determined by the physical and chemical properties of the underlying parent materials/rocks, the amount of precipitation, and the ground water level. The biomass productivity is low on the quartzite and peridotite-serpentine parent materials, such as the shrubby red pine stands on serpentine-peridotite parent materials in the Datça Peninsula and in the vicinity of Köyceğiz. In addition to this, poor stands of shrubs exist on the quartzite and siliceous parent materials around the Foça district, north of İzmir. The best productive and widespread regions of red pine forests are in the Mediterranean region. They rise above 1000 m on the south-facing slopes of the Taurus Mountains and produce the mixed forests of the oro-Mediterranean forest belt (Atalay et al 1998; Atalay et al 2008).

Areas covered by Mediterranean vegetation are also suitable for the formation of Luvisols/Red Mediterranean Soils on the slightly undulating and well-drained flat lands.

2.3.5 Mountain Forests of the Mediterranean Region

Forest Formations of the oro-Mediterranean Region

In the moist parts of the oro-belt of the Taurus Mountains, *Cedrus libani*, *P. nigra*, and *Abies cilicica* forests are dominant (Figs. 9, 11 and 12). Most of the forests are coniferous, comprising the Anatolian black pine *(P. nigra)*, the Lebanese or Taurus cedar *(C. libani)*, the Taurus fir *(Abies cilicica)*, and juniper *(Juniperus foetidissima* and *Juniperus excelsa)*, which form the tree line. *Cedrus libani* occurs in areas affected by the Mediterranean climate, while *P. nigra* prefers inland, continental sites. *Pinus nigra* forests are found between 1200 and 2000 m in the Taurus Mountains. They grow very well on the soft parent materials such as flysch and colluvial deposits and often associated to the *C. libani* and *A. cilicica* in the Taurus Mountains. Pure and good *P. nigra* stands occur around Lake Beyşehir, Lake Eğirdir (Fig. 9), and the Söğüt plateau extending from Antalya to Gazipaşa and to the eastern part of the Taurus Mountains (Karsantı province) (Mayer et al. 1986; Atalay 1987a, 1987b, 1988).

Fig. 11 A pure and productive cedar (*Cedrus libani*) forest on the north-facing slopes of Western Taurus Mountains. Here, the age of the cedar trees is more than 600 years

Fig. 12 A pure and productive Taurus fir (*Abies cilicica*) forest growing on the foggy and karstic areas in the vicinity of Çamlıyayla locality, Central Taurus Mountains

Abies cilicica forests occur between 1150 and 2000 m elevation on the slopes facing north where pure stands are only found on the north-facing foggy slopes (Fig. 12). The optimum growth areas are generally found between 1200 and 1800 m altitudes. *Abies cilicica* occurs rarely in pure stands but mostly mixed with *P. nigra* and *C. libani*. *Abies cilicica* forests are also found between 1300 and 1500 m in the Nur (Amanos) mountains, east of the Iskenderun Gulf.

Cedrus libani forests (Fig. 11), being one of the climax trees of the oro-Mediterranean belt begin at an elevation of 800 m and reach up to 2000 m on the southern slopes of the Taurus Mountains, and also continue to 2200 m in the inner section of the Mountains (Kantarcı 1982; Mayer et al. 1986; Atalay 1987b, 1988).

Cedar forests being in mixed and pure stands are present from 800 to 2000 m elevations in the vertical direction of the southern parts and from 1400 to 2100 m in the southern areas of the Taurus Mountains (Fig. 11). They are also present at 1800–2000 m in the Nur Mountains, and descend as low as 500–550 m in the eastern part of the Nur Mountains, west of the immense Kahramanmaraş-Antakya graben extending south to the Red Sea and the Rift Valley. The distribution of the cedar in Turkey clearly reveals that cedar stands do not occur at the extreme maritime and continental conditions. Thus, the optimum growing areas of cedar are found in the transitional region, the zonoecotone, extending from the Mediterranean to central Turkey.

Cedar grows on all parent materials, namely on marly deposits, schists, quartzite, and limestones belonging to the Tertiary, Mesozoic and Paleozoic eras. Pure *C. libani* stands only occur on the slopes facing north of the Taurus Mountains and they grow mostly on the karstic lands providing seed germination within the limestone cracks (Atalay 1987b; Efe 1998).

Juniper *(J. excelsa, J. foetidissima)* forests are common on the Taurus Mountains in places where coniferous forests, composed of cedar and black pine, were entirely cleared. This points out that the juniper communities can be considered as the regressive and/or secondary successions of this area. Indeed, the seeds originating from the excrements of the birds easily germinate in the destroyed coniferous forests. In the higher parts of the Taurus forest belt, *Juniperus communis* subsp. *nana* communities are prevalent Oak *(Quercus libani, Q. infectoria, Q. cerris)* forests that are mainly found between thermo and oro-Mediterranean belts and extend at an altitude of 800–1200 m, where they are common in the western and eastern parts of the Taurus.

As to the Aegean geographical region, oro-Mediterranean forests appear above the *P. brutia* forests on mountainous areas of the Aegean region such as Boz, Aydın, and Menteşe Mountains. Humid forests are found as small stands and only occupy the deep valleys of the northern slopes of Boz and Aydın Mountains. Several *C. sativa* communities are the prominent clusters in the area and are found on the northern slopes between 850 and 1000 m in the Yamanlar Mountain and 800–1300 m in the Boz Mountains.

Pinus pinea (stone pine) forests occur on the sandy soils derived from the weathering of granitic rocks in the Kozak Plateau, in the north and central parts of the Aegean region (Atalay 1994).

Dry forests are found on the south-facing slopes of the mountains. Their productivity is generally low and understory vegetation is poor. Dry forests composed of *P. nigra* and oak species such as *Q. infectoria, Quercus ithaburensis* subsp. *macrolepis, Quercus cerris, Quercus frainetto, Quercus pubescens* occur on the southern slopes of the Aegean mountains.

2.4 The Forests of the Mediterranean Transitional Region

This is the Lakes Region and lies between the Mediterranean and the central Turkish continental climatic areas. There are karstic-tectonic depressions and lakes and closed basins that are separated by the mountain chains in this region. As a whole, this region climatically reflects the semiarid continental influence. The mean annual precipitation decreases as low as 400 mm due to rain shadow. The mean annual temperature on the lowland is about 10–13 °C, the mean winter temperature is above the freezing point. The relative humidity during the summer season is very low (40–50%) (Atalay 2010, 2014).

2.4.1 Pinus Brutia, Cedrus Libani and Quercus Forests

These forests are widespread on the lowland between 800 and 1000 m elevations and the productivity of the *P. brutia* forests here is relatively lower than the *P. brutia* forests in the Mediterranean region. The shrub layer of the *P. brutia* forest is mainly composed of *Q. coccifera* and *Paliurus spina-christii*. The *Quercus* forests become dominant towards the north in the tectonic depression (Atalay et al 1998).

Cedrus libani and *P. nigra* forests of the Lakes Region are the dominant natural vegetation. The *P. nigra* forest starts after the *Quercus* sp. and replaced by *C. libani* forests on the upper part of the mountains. The productive *C. libani* forests are found on the north-facing slopes of the mountains in the southern part of the region because the northern slopes of the mountains are relatively more humid with cool breezes flowing from the north (Atalay 1987b; Atalay et al. 2010).

2.5 Central and Eastern Anatolian Regions

This region encompasses all parts of the central and eastern Anatolian geographical regions. Semiarid continental climatic conditions characterized by snowy and cold winters and hot and dry summers prevail in the region. These are the natural steppe areas of central Turkey and of the depressions of eastern Turkey due to the insufficient amount of annual precipitation (less than 400 mm). The mean annual precipitation of higher areas encircling the lowland region is more than 500–600 mm. Since the area is characterized by a dry continental climate, these forests are resistant to cold temperatures and water shortages and are therefore considered dry or xeric forests (Atalay 1994) with dominant oak species, including *Q. pubescens* and *Q. infectoria*, with steppe vegetation increasing to the east.

2.5.1 Dry Forests

The highlands surrounding central Turkey are the main areas of *P. nigra*, *Quercus*, and *Juniperus* (Fig. 13). The remaining areas are occupied with the anthropogenic steppe. *Quercus* stands are found as clusters in the transitional region between the steppe and the forest, whereas the *P. nigra* clusters occur on the mountains over 1200 m.

Pinus nigra subsp. *pallasiana* and *Q. pubescens* are the climax forest types of central Turkey. The degraded areas of *P. nigra* forests are occupied by *Q. pubescens* and *J. excelsa*, *J. oxycedrus* and *Q. pubescens*. Dry (*Pinus nigra-Quercus-Juniperus*) forests occur in the mountainous areas encircling central Turkey. Oak stands mainly composed of *Q. pubescens* are common, more or less, in the transitional areas extending between the anthropogenic steppe and the *P. nigra* forests. *Pinus nigra* forests are found at an altitude over 1200 m in the western part of central Turkey. *Pinus sylvestris* forest stands occur on the high plateau surfaces in the vicinity of Akdağmadeni, northeast of central Turkey and also in a good stand on the Sündiken Mountains, in the northwest of central Turkey (Louis 1939; Çetik 1986; Atalay et al. 2012).

The climax forests composed of *P. nigra*, *Quercus*, and *Juniperus* trees are mostly degraded in the higher parts and north-facing slopes of the mountains and is replaced by steppe vegetation. The remnant stands of the oak and black pines are present around Ankara (the Beynam Black Pine Forest) and in the Yozgat National Park.

The steppe vegetation takes place on the lowlands of the Konya and Tuz Lake basins and in Eskişehir. *Artemisia, Bromus, Achiella, Trifolium,* and *Astragalus* species are common in these areas (Çetik 1986; Öztürk et al 2002; Atalay 2014) where Cambisols are widespread (Fig. 14).

Fig. 13 Dry forest composed of oak (mainly *Q. pubescens*) and *black pine* (*Pinus nigra*) as the climax forest of the upland part of Central Anatolia

Vegetation

Fig. 14 Steppe vegetation common in the lowland part of Central Anatolia and in the depressions of East and lowlands of SE Anatolia

Fig. 15 Tall grass vegetation in the east of Lake Çıldır in the eastern part of East Anatolia

2.5.2 Quercus Forests of Eastern Turkey

The moist areas of the middle and western mountains of eastern Turkey are among the productive *Quercus* forest areas of the country. The prominent oak species are *Q. infectoria, Q. ithaburensis* subsp. *macrolepis (synonym Q. aegilops), Quercus brandii, Q. libani, Quercus robur* subsp. *Pedunculiflora, Q. brandii,* and *Q. petraea*. The Mercan (Munzur) and Bingöl Mountains are densely covered by oak forests. The majority of these oak forests have been degraded and/or completely destroyed in many places for fuel wood and fodder. Indeed, dried oak leaves are the main fodder for goats especially during the winter period. More than half of the oak forests in the vicinity of Bingöl were completely degraded in the last two decades (Atalay 1994) due to overgrazing.

2.5.3 Tall Steppe and Grass Vegetation

This vegetation is highly spread in the eastern part of Anatolia due to the short summer and long winter periods as well as the higher humidity than that of the central and southeast Anatolian lowlands (Fig. 15). One can see animal flocks in the tall grass vegetation areas where Chernozems most likely occur due to the relatively higher organic material accumulation under the colder climatic conditions.

2.6 The Southeastern Anatolian Region

This is the driest and the warmest region of Turkey partly resembling the eastern and central parts of the country in terms of plant communities and the Mesopotamian floristic region. The mean annual temperature and precipitation are about 17–18 °C and more than 500 mm, respectively. But the amount of annual evapotranspiration is about 1800–2000 mm in the southern lowland of southeast Turkey in turn with a relative humidity as low as 1%. The forests of southeast Turkey can be divided into two groups (Atalay 2014).

2.6.1 *Pinus Brutia* and *Quercus* Forests

These forests are both mixed and pure and found in the western part of southeast Turkey due to the Mediterranean climatic influence. As a general rule, the pine stands on the lowlands of this region are generally composed of individual trees with poor stem quality.

2.6.2 Pure *Quercus* Forests

These forests are found in the eastern part of the Karaca volcanic mountain. The hilly and deeply dissected mountainous areas of the southeast Taurus Mountains are the natural occurrence areas of oak forests. *Quercus brandii* and *Q. infectoria* subsp. *boissieri* are the dominant oak species in the area. But, productive oak forests are found in very limited localities due to the severe degradation (Çetik 1986).

Alpine vegetation

This vegetation is present above the natural timberline in the northeast, north and southeast Anatolian mountains

Fig. 16 Alpine grass vegetation in the eastern part of the North Anatolian Mountains

(Fig. 16). The short vegetation period supports the growth of the grasses notably belonging to the Alpine environment. The Alpine vegetation is the main grazeland of the animals during the summer period.

References

Atalay İ (1984) Regioning of seed transfer of *Picea orientalis* L. in Turkey. Ministry of Forestry, Directory of Seed Research Pub. No.2, Ankara, Turkey

Atalay İ (1987a) Vegetation formations of Turkey: Travaux de l' Inst. de Geographie de Reims 65(66): 17–30

Atalay İ (1987b) General Ecological Properties of the Natural Occurrence Areas of Cedar (Cedrus Libani A. Rich) and Regioning of Seed Transfer of Cedar in Turkey. Ankara: General Directory of Forestry Pub. No. 663. Ankara. 167 P

Atalay İ (1987c) Introduction to the geomorphology of Turkey. Aegean University Press, İzmir (in Turkish)

Atalay İ (1988) Vegetation levels of the Taurus Mountains of Mediterranean region in Turkey. Aegean Geogr J 4:88–122

Atalay İ (1992) The Ecology of Eastern beech (*F. orientalis* Lipsky) forests and their regioning in terms of seed transfer. Ministry of Forestry, Directory of Seed Research Pub. No. 5, Ankara (in Turkish)

Atalay İ (1994) Vegetation Geography of Turkey. Ege University Press, İzmir, Turkey, ISBN 9759552787.352 p

Atalay İ, Sezer İ, Çukur H (1998) Ecological properties of the natural occurrence areas and regioning of seed Transfer of red pine (*Pinus brutia* Ten.) in Turkey. Ministry of Forestry, Directory of Seed Research Pub. No.6, Ankara (in Turkish)

Atalay İ, Efe R, Soykan A (2008) Evolution and ecological properties of Taurus Mountains. in Environment and Culture in the Mediterranean Region. Part I, Chapter One, (R Efe, G Cravins, M Öztürk and İ Atalay Eds.). Cambridge Scholars Publishing, Newcastle, UK. 3–37

Atalay İ (2010) Applied climatology. Meta Press, İzmir, 250 p. (in Turkish)

Atalay İ, Efe R (2010) Ecology of Anatolian Black Pine *(Pinus nigra subsp. pallasiana (Lamb.) Holmboe)* and its dividing into regions in terms of seed transfer. Research Center for Forest Trees and Seed Breeding. Pub. No. 37. Ankara

Atalay İ, Efe R (2010) Ecology of scots pine *(Pinus sylvestris var. sylvestris)* forests and their dividing into regions in terms of seed transfer. Research Center for Forest Trees and Seed Breeding. Pub. No: 45. Ankara (in Turkish)

Atalay İ (2014) Eco-regions of Turkey. Meta Press, İzmir. Turkey, 496 p. (in Turkish)

Çetik R (1986) Vegetation of Turkey I: vegetation and ecology of Central Anatolia: Selcuk University Faculty of Arts and Sciences. Pub. No. 7. Konya (in Turkish)

Dönmez Y (1968) Plant Geography of Trace region. Ist University Institute of Geography Publications: 51, İstanbul (in Turkish)

Dönmez Y (1979) Plant Geography of Kocaeli Semi-island. Ist University Institute of Geography Pub. No. 112. İstanbul (in Turkish)

Efe R (1998) Ermenek river basin: natural environmental properties. Fatih University Publication, No. 1, 210 p

Efe R (2005) Land degradation in Taurus Mountains (Southern Turkey). European Geosciences Union, 2nd general assembly, EGU-Geophysical research abstracts. Volume 7, no. 00922

Kantarcı D (1982) The relationships between the distribution of natural tree and bushes and regional growing characteristics in the Mediterranean region. Ist University Faculty of Forestry Pub. No 330 (in Turkish)

Louis H (1939) Das naturliche Pflanzenkleid Anatoliens. Geographische Abhandlungen no. 12.Stuttgart, Germany

Mayer H, Aksoy H (1986) Walder der Türkei. Gustav Fischer Verlag, Sttutgart, Germany

Öztürk M, Gemici Y, Gork G, Seçmen O (1991) A general account of high mountain flora and vegetation of Mediterranean part of Turkey. Ege Univ J Fac Sci 13(1–2):51–59

Öztürk M (1995) Recovery and rehabilitation of Mediterranean type ecosystem: A case study from Turkish Maquis. Evaluating and monitoring the health of large-scale ecosystems. Springer, Berlin Heidelberg, pp 319–332

Öztürk M, Çelik A, Yarcı C, Aksoy A, Feoli E (2002) An overview of plant diversity, land use and degradation in the Mediterranean region of Turkey. Environ Manage Health 13(5):442–449

Quezel P (1986) The forest vegetation of Turkey. Proc Roy Soc Edinburgh 89B:113–122

Yaltırık F (1973) The floristic composition of major forests in Turkey. İstanbul University, Publication of Forestry Faculty. 1921/209 (in Turkish)

Zohary M (1973) Geobotanical foundations of the Middle East, vol 2. Gustav Fischer Verlag, Stuttgart

Climate

Ufuk Utku Turunçoğlu, Murat Türkeş, Deniz Bozkurt, Barış Önol, Ömer Lütfi Şen, and Hasan Nüzhet Dalfes

1 Introduction

Climate plays a crucial role in the development of soils on the surface of the earth. In fact, there is a broad agreement between the distributions of soil and climate types at a global scale. The direct impact of climate on the soil formation is through weathering caused by temperature and precipitation. Both temperature and precipitation affect the rates of the physical, chemical, and biological processes in the soil. Warmer and moister climates favor the weathering, leading to a richer and deeper soil layer. Colder and drier climates slow down the soil formation process, resulting in a relatively poor shallow soil layer. Leaching is enhanced in rainy climates, because of increased percolation rates. Climate also affects the soil formation indirectly via its influence on plant cover. Humid and warmer climates favor the plant growth and microbial activity in the soil, and the balance between them determines the organic matter content of the soil. Climate change is expected to make alterations in temperature and precipitation. Temperatures are projected to increase worldwide, however precipitation projections show increases and decreases for different regions. Moreover, not only its annually averaged values, but also its characteristics such as seasonality, intensity, duration, and frequency are projected to change. Such changes may have profound impacts on soil development processes in some areas, as they are highly effective in weathering. Soil formation rates could also be impacted by the northward and upward (to higher elevation) shifts in the biomes that are expected as a result of temperature rise.

Turkey lies in a transition zone between arid and temperate climates, comprising primarily Mediterranean and continental climate types. It has a mountainous topography, prone to significant erosion rates. The IPCC's (Intergovernmental Panel on Climate Change) Fifth Assessment Report (IPCC 2013) indicates that Turkey is located in a region that is highly vulnerable to climate change impacts. A northward shift in the storm (of Atlantic origin) tracks is expected to result in less precipitation for the southern parts of Turkey while more precipitation for the northern parts. Such changes, together with temperature increases, will certainly have important implications for the soils of Turkey. It is therefore important to, first, understand the climate of Turkey and the possible changes it could undergo. This chapter, thus, introduces what is known about the present and future climates of Turkey. The rest of this chapter is organized as follows: Next section introduces the present climate of the Turkey and its Köppen–Geiger classification. Then, the future climate of the region is given by the evaluation of CMIP3 and CMIP5 projections for future climate of Turkey. It is followed by a section that introduces CMIP3-based high-resolution projections for Turkey. This section also includes a discussion on the impacts of climate change from soil moisture and drought perspectives.

U.U. Turunçoğlu (✉)
Informatics Institute, İstanbul Technical University, İstanbul, Turkey
e-mail: ufuk.turuncoglu@itu.edu.tr

M. Türkeş
Center for Climate Change and Policy Studies, Boğaziçi University, İstanbul, Turkey

D. Bozkurt
Center for Climate and Resilience Research, University of Chile, Santiago, Chile

D. Bozkurt · Ö.L. Şen · H.N. Dalfes
Eurasia Institute of Earth Sciences, İstanbul Technical University, İstanbul, Turkey

B. Önol
Aeronautics and Astronautics Faculty, Meteorological Engineering, İstanbul Technical University, İstanbul, Turkey

Ö.L. Şen
Mercator-IPC Fellowship Program, İstanbul Policy Center, Sabanci University, İstanbul, Turkey

© Springer International Publishing AG 2018
S. Kapur et al. (eds.), *The Soils of Turkey*, World Soils Book Series,
https://doi.org/10.1007/978-3-319-64392-2_3

2 Present Climate of Turkey

The Anatolian peninsula is mainly characterized by subtropical climate, similar to other regions in the Mediterranean, except for minor differences arising from the fact that the region is surrounded by water bodies, the Black Sea in the north, the Aegean Sea in the west and the Mediterranean Sea in the south. Due to its unique location and complex topography, the different subregions or basins have varying climatic characteristics. The southern and western coastal areas have Mediterranean climate, which is mainly associated with the seasonal migrations of the mostly dynamic-originated pressure systems and the alternating tropical and polar air masses between summer and winter, and the dry summer subtropical Mediterranean climates can be found along the west coasts between about 25° and 40° latitudes as a natural part of the mid-latitude mild climates (Türkeş 2010). The region is mostly dominated in summer by dry, stable and subsiding air from the eastern portions of dynamically originated subtropical highs (e.g., Azores high), which are also associated with the sinking branch of the Hadley cell circulation. In winter, the wind and pressure systems shift equator-ward as a result of the equator-ward migration of the Rossby waves and associated polar jet streams and upper air westerlies with the polar front, and the Mediterranean climate regions are influenced by the westerlies with their trailing middle-latitude frontal cyclones (Türkeş 2010).

Besides having different climate types distributed across the country, Turkey is considered as the largest Mediterranean climate region found in the Mediterranean Basin, which can be thus called as the 'true' Mediterranean macroclimate. In the Atlantic-ward of the southern Europe and the Mediterranean Sea Basin, there are no north–south-oriented mountain chains and the Mediterranean climates penetrate a great distance from the west basin of the Mediterranean Sea with some parts of the Iberia Peninsula to the East Basin of the Mediterranean Sea and the Mediterranean coastal and inland regions of the Middle East region. Although the term Mediterranean is used to describe the climate synonymous with the region surrounding the Mediterranean Sea, it is not exclusive to this region. The Mediterranean climates are also found, for example, in the northern Iran, California, Chile, Australia, New Zealand and South Africa, in addition to the 'true' Mediterranean regions of Portugal, Spain, Coastal North-west Africa (Morocco, Tunisia, Algeria), France, Italy, Greece, Turkey, Lebanon and Israel (Türkeş 2010; Türkeş et al. 2011).

In addition, seasonality is the most dominant and distinctive character and factor of the Mediterranean climates. The Mediterranean climate tends to alternate wet and dry seasons, because it is located in the transitional zone between the dry west coast tropical desert mainly related to the subtropical high pressure systems and the descending segment of the tropical Hadley circulation cell, and the wet west coast climate mainly related with the polar front and associated mid-latitude cyclones.

As we shortly discussed above on the mid-latitude climates, there are many atmospheric features that can be considered as dominating the Mediterranean climates. For instance, the Rossby waves, which are formed by the upper air troughs and lows, and the upper air ridges and highs, controlled the penetration of polar air masses (continental polar—cP, maritime polar—mP, and very rarely continental Arctic—cA) towards equator in certain months of the year, and tropical air masses (continental tropical—cT, and maritime tropical—mT) towards poles in other certain months of the year. The seasonal movements arise from mainly the Sun's seasonal migration (i.e., Sun's apparent movement) and thus amount and intensity of the Sun's radiation result an energy exchange between the poles and the equatorial belt (Türkeş 2010). On the other hand, due to the seasonal movements of the inter-tropical convergence zone (ITCZ) associated closely with the movements of the Sun, the Rossby waves would be closer to the equatorial belt in winter than that in summer. Thus, polar originated weather systems can more strongly fluctuate further equator-ward in winter than in summer. This movement is also responsible for enhancing the seasonal energy contrasts globally particularly over the continents. In the mid-latitudes, while the planetary-scale Rossby waves and the upper air jet streams governing also westerlies and the frontal cyclones are the most powerful features of these climates, these atmospheric controls are also themselves controlled by topography, continentally, land–sea distribution and interactions, air masses and their thermodynamic and mechanical modifications arising from the physical geographical features of the Earth surfaces (Türkeş 2010). Consequently, there is no simple explanation of the mid-latitude climates including the Mediterranean climate as a whole.

An analysis of cyclone paths over the Anatolian Peninsula performed by Karaca et al. (2000) using ECMWF data shows a set of dominant cyclone trajectories: (1) the path extending from northern Turkey to the southwest of Russia (and Balkans), affecting Marmara and the Black Sea region, (2) the path starting in the Genoa Gulf and passing over the Aegean Sea, affecting all of Turkey and (3) the path which originates in western or central Mediterranean (and in some cases north of the Sahara), affecting southern Turkey. Almost all precipitation comes from the frontal cyclones, except for the late spring and early summer convective instability showers and thunderstorms in inland Mediterranean climates of the Anatolia Peninsula and the Middle East region. On the other hand, west-to-east oriented Alp mountains of the South Europe and the North Anatolia and Taurus mountains of Turkey create stronger influence on the westerlies and associated mid-latitude frontal cyclones by means of the orographic lifting of the moist air masses resulting adiabatic cooling and condensation, and occurrence of the precipitation over the mountains, particularly by the time of that faced to southerly and westerly moist air masses (Türkeş 1996, 1998, 2010).

Based on all these explanations above, the climate of the region have following five distinctive characteristics (Türkeş 2010): (1) about half of the modest annual precipitation amount falls in winter, whereas summers are mostly virtually rainless, (2) winter temperatures are unusually mild for the middle-latitudes except some eastern and inland regions, summer air temperatures vary from hot to warm, (3) cloudless skies and intensive sunshine (shortwave solar radiation) are typical particularly in summer months, (4) the seas (Mediterranean and Black Seas) have major influences on the climate of the region and land distribution and the interactions between sea and lands, in addition to the ocean–atmosphere interaction (Bozkurt and Sen 2011, Turunçoğlu 2015), during the year particularly in the 'true' or 'actual' Mediterranean macroclimate region, and (5) The major characteristic of the Mediterranean climate is of high temporal variability varying from seasonal and inter-annual to centennial scales due to following factors (Türkeş 2010):

i. It extends in a transition region between temperate and cold mid-latitudes and tropics (i.e., subtropical zone)
ii. It has been facing significant circulation (associated pressure and wind systems characterizing mid-latitude and tropical/monsoonal weather and climate, respectively) changes between winter and summer
iii. It is closely associated with several atmospheric oscillation and/or teleconnection patterns such as North Atlantic Oscillation (NAO), Arctic Oscillation (AO), Mediterranean Oscillation (MO), El Niño-Southern Oscillation (ENSO), and North Sea Caspian Pattern (NCP), etc. (Kutiel and Türkeş 2005; Tatlı 2007; Türkeş and Erlat 2003, 2005, 2008, 2009; Erlat and Türkeş 2012).

2.1 The Climate Classification of the Region

The idea of forming climate classes with the combination of long-term temperature and precipitation characteristics, along with setting limits and boundaries fitted to prevailing vegetation and soil distributions, was first suggested in 1918 by Wladimir Köppen. This classification was subsequently revised and improved by Köppen himself and his students, and it becomes the most widely used tool of climatic classifications for the physical geographical and environmental (climatological, biogeographical, ecological, etc.,) purposes. We have accepted here the criteria that follow Köppen's last publication about his classification system in the Köppen–Geiger Handbook (Köppen 1936), with the exception of the boundary between the temperate (C) and cold (D) climates, because we have used the Köppen–Geiger climate data computed by Peel et al. (2007) who also followed Russell (1931) and used the temperature of the coldest month >0 °C, rather than >−3 °C as used by Köppen in defining the temperate—cold climate boundary (Wilcock 1968; Essenwanger 2001). The Köppen climate system carries out a shorthand code of letters designating major climate groups, subgroups within the major groups, and also subdivisions to distinguish particular seasonal characteristics of temperature and precipitation. For instance, Table 1 indicates two-letter group climates of the Köppen–Geiger classification.

To determine the Köppen–Geiger climate classifications for Turkey, the long-term average monthly precipitation total and monthly mean temperature data were calculated by using time series of the station-based climate data of Turkey (Türkeş 1996, 1999; Türkeş et al. 2002b). Detailed information for meta-data and homogeneity analyses applied to long-term precipitation and temperature series of Turkey can be found in Türkeş (1996, 1999) and Türkeş et al. (2009) and Türkeş et al. (2002a), respectively.

The climate of Turkey according to the Köppen–Geiger climate system is very diverse (Fig. 1), as in many other climate classifications. Only the first and second-hand letters in the Köppen–Geiger climate classification system, following the major climate types can be separated (Türkeş 2010):

(1) Subtropical steppe climate BS (mostly BSk—cold steppe) is found in the mid-part of the continental Central Anatolia region and the Van-Iğdır district over most eastern part of the continental Eastern Anatolia region.
(2) Temperate rainy or humid temperate west coast climate without dry season Cf (mostly Cfa and Cfb—humid mesothermal) is dominant in the Black Sea coastal region of Turkey with the exception of the western subregion.
(3) The Marmara, Aegean, Mediterranean and southeastern Anatolia regions, and the western and southern parts of the continental central Anatolia region belong to the dry summer subtropical Mediterranean climate or temperate rainy climate with dry summer (humid mesothermal) Cs (mostly Csa). Csa classification indicates that summers are hot with midsummer monthly averages between 24 and 29 °C and high maximums above 38 °C. Average cold-month temperatures are about 10 °C with occasional minimums below freezing temperatures.

Table 1 Two-letter group climates of the Köppen–Geiger classification. The red solid boxes show climate types that are found in Turkey

- *Af*: Tropical rainforest climate
- *Am*: Tropical monsoon climate (a variant of *Af* with a short dry season)
- *Aw*: Tropical savanna climate
- *BW*: Desert climate
- *BS*: Steppe climate
- *Cs*: Temperate rainy climate with dry summer (humid mesothermal) or Mediterranean climate (dry summer subtropical)
- *Cw*: Temperate rainy climate with dry winter (humid mesothermal)
- *Cf*: Temperate rainy climate without dry season (humid mesothermal) or humid temperature west coast
- *Ds*: Cold snowy forest climate with dry summer (humid microthermal).
- *Dw*: Cold snowy forest climate with dry winter (humid microthermal)
- *Df*: Cold snowy forest climate humid in all seasons (humid microthermal)
- *ET* Polar tundra climate
- *EF* Polar forest climate

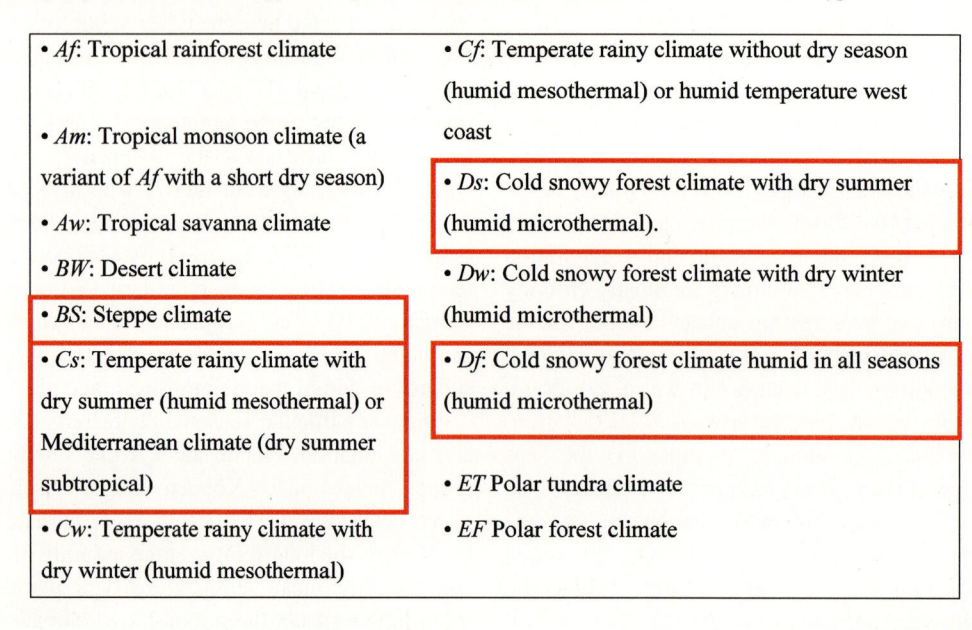

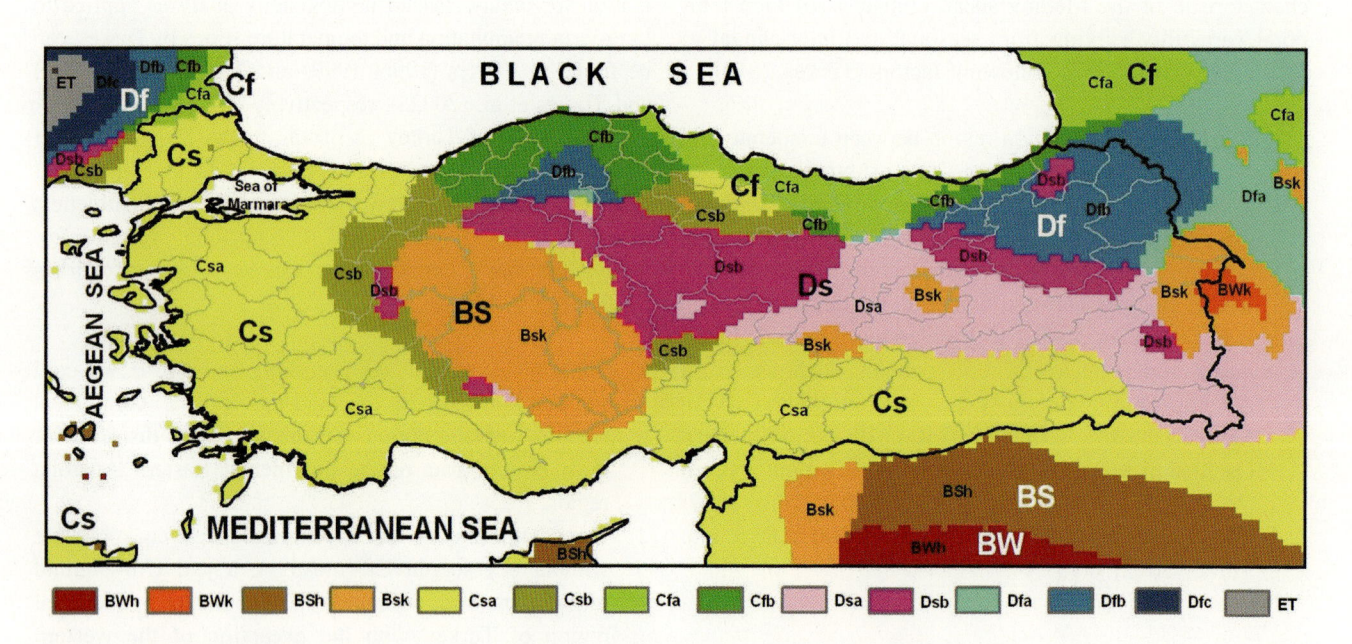

Fig. 1 Geographical distribution of climate types in Turkey based on the first-, second-, and third-hand letters classification of the Köppen–Geiger climate system (Türkeş 2010)

Yearly soil moisture deficiency is not characteristic in Group C climates in general, whereas seasonal soil moisture deficiency particularly in summer months is evident in the Mediterranean Csa and Csb climates due to changes in the hemispheric and regional circulation, air mass and pressure systems producing dry conditions in the summer months. In this case, Csb winters are slightly milder than Csa winters, the former having cold-month averages of about 13 °C. Csb regions also have higher humidity, frequent advection and evaporation fogs, and occasional low stratus, altostratus, and nimbostratus overcast.

(4) A cold snowy forest climate with dry summer (humid microthermal) Ds (mostly Dsa and Dsb) takes place over a relatively large zone lying in the mid-northern parts of the continental central and eastern Anatolia regions of Turkey, whereas a cold snowy forest climate, humid in all seasons (humid microthermal) Df (mostly Dfb) exists

over relatively small areas seen in the northern parts of the continental Central Anatolia region and the northeastern Anatolia subregion (mostly Erzurum-Kars subregion) of Turkey.

The maritime influences on the weather and climate conditions of Turkey characterized mainly by the maritime polar (mP) and maritime tropical (mT) air masses, and in winter Mediterranean air masses carried by the westerly air flows (W, SW, and NW) tend to decrease toward the continental inland regions including the central, eastern and southeastern Anatolia regions. Major changes of the inland Mediterranean regions concern decreases of precipitation amounts in winter and number of rainy days, an increase of precipitation amounts in spring months, particularly with the continentally such as in the mid-west and south parts of central Anatolia region. An increase in temperature range is observed as the continental effects are also enhanced. The Northern Anatolia Mountains, on the other hand, separate the Mediterranean, steppe, and cold snowy forest climates from the west coast temperate rainy (Black Sea in Turkey) climate, while the Taurus' and the Southeastern Taurus Mountains separate the Mediterranean climate (Csa) from the more continental Mediterranean climate (Csb), steppe and cold snowy forest climates to the north and the east of the Anatolian Peninsula (Fig. 1).

As a requirement of the United Nations Convention to Combat Desertification (UNCCD), arid, semiarid and dry subhumid climates were defined as "areas, other than polar and sub-polar regions, in which the ratio of annual precipitation to potential evapotranspiration falls within the range from 0.05 to 0.65" (UNCCD 1995). In the present study, the UNCCD Aridity Index (AI) is used as one of the base methods for determining dry-land types in Turkey and assessing their vulnerability to the desertification processes. Following the UNEP (1993), AI is written as follows:

$$A = (P/PE) \quad (1)$$

where P and PE are annual precipitation (mm) and potential evapotranspiration (mm) totals, respectively. The criteria in Table 2 were used to characterize the dry-lands of Turkey (Türkeş 1999). In theory, and according to the UNCCD Aridity Index (AF), AI values below 1.0 generally show an annual moisture (soil water) deficit in average climatic conditions, whereas AI values above 1.0 generally show an annual moisture surplus (Fig. 2). In Turkey, dry subhumid climatic conditions extend throughout most of the continental central Anatolia and southeastern Anatolia regions, some part of the eastern Mediterranean, and eastern and western parts of the continental eastern Anatolia region, while the semiarid climatic conditions are found only in the Konya Plain and the Iğdır district of the Eastern Anatolia region (Fig. 2).

The areas having values $0.65 < AI < 0.80$, where an annual soil moisture deficit exists, are also concentrated around the semiarid and dry subhumid areas of Turkey. The arid lands in Turkey having AI values between 0.20 and 0.80 are likely supposed to have been influenced by the desertification processes (Table 2), by considering the existing hydro-climatological conditions, human-induced land degradation, observed and projected climate change and variability, etc. (Öztürk et al. 2012, 2015; Topçu et al. 2010; Şen et al. 2012; Türkeş 1999, 2010; Tatlı and Türkeş 2011, 2014; Türkeş and Akgündüz 2011; Altınsoy et al. 2011; Türkeş et al. 2011).

3 Future Climate of Turkey: Evolution of CMIP3/CMIP5 Models

Human influence on Earth's climate is now accepted as a well-tested scientific hypothesis, therefore, for a sustainable environment both mitigation and adaptation are imperative at all levels of social organization. Starting point of any mitigation or adaptation effort is the generation of information concerning what the future will bring and how natural and human- built systems will behave under climate change. Such information, in turn, forms the basis for studies of the impacts of current climate variability and future climate change on all aspects of a country's resources, from water to health, from Agriculture to urban environment. Climate simulations, which are the basis for IPCC's Fourth and Fifth Assessment Reports (IPCC 2007, 2013), provide such information assuming several greenhouse gas emission scenarios for the future. Before giving details of the analysis of projected climate change over Turkey, it is useful to give the basic definitions behind the emission scenario design and the difference between different emission scenario approaches.

The emission scenarios basically describe the future concentrations of the main anthropogenic emissions of all relevant greenhouse gases (GHGs) and a small set of other

Table 2 Dry-land (arid climate) types in Turkey according to the Aridity Index (AI) and their vulnerability to desertification (Türkeş 1999)

Aridity criteria	Dry-land type	Assessment
$0.20 \leq AI < 0.50$	Semiarid areas	Vulnerable to desertification
$0.50 \leq AI < 0.65$	Dry subhumid areas	Vulnerable to desertification
$0.65 \leq AI < 0.80$	Subhumid areas	Vulnerable to desertification

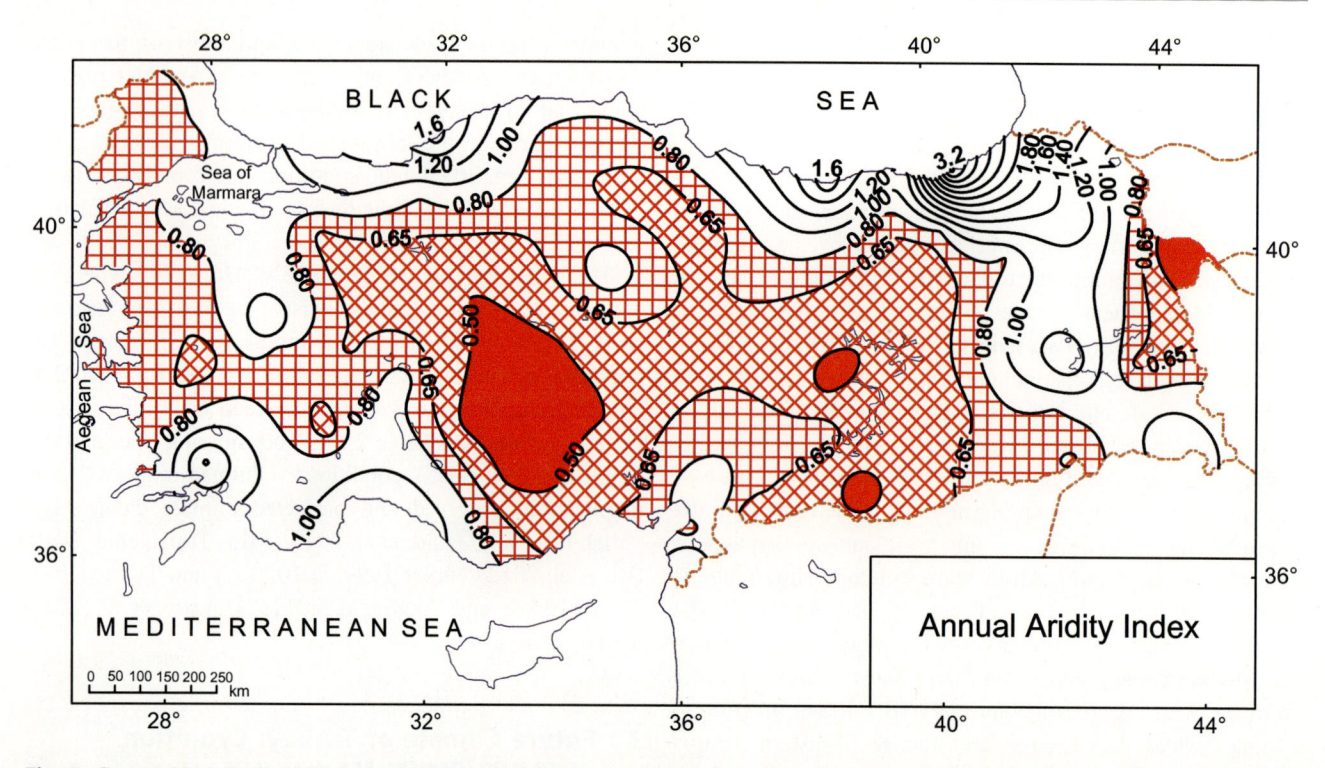

Fig. 2 Geographical distribution of the UNCCD aridity indices over Turkey, in which arid, semi-arid, dry subhumid and subhumid areas of the country are highlighted by hatching with the red color (Türkeş 1999, 2010)

gases such as SO_2, CO, and NO_x. From this point of view, emission scenarios can be identified as the main component of any assessment of climate change because they are used to drive global multi-component climate models for future climate change projections, such as those planned under the World Climate Research Programme's Coupled Model Intercomparison Projects (CMIP3 for AR4, Meehl et al. 2007; CMIP5 for AR5 Taylor et al. 2012). In parallel with the developments in climate science and based on the needs of the climate science community, the designed emission scenarios also evolved from non-mitigation Special Report on Emission Scenarios (SRES) (Nakicenovic and Swart 2000) to the Representative Concentration Pathways (RCPs) (Meinshausen et al. 2011), which are based on scenarios that consider possible approaches to climate change mitigation. In AR4, the 40 different SRES emission scenarios were defined based on the range of global energy related CO_2 emissions from the literature (Nakicenovic and Swart 2000), and they are categorized into four distinct scenario families such as A1, A2, B1 and B2. The scenario family A1 is also subdivided into four scenario groups (i.e., A1C, A1G or A1FI, A1B, and A1T) based on the different future directions of technological change. According to the classification of the SRES scenarios, B1 is defined as the low-forcing scenario (CO_2 concentration of about 550 ppm by 2100), which assumes rapid economic development, low population growth, and introduction of new energy sources and/or technologies. A1B is designed as medium forcing scenario (CO_2 concentration of about 700 ppm by 2100) and A2 is defined as high forcing scenario (CO_2 concentration of about 820 ppm by 2100), which assumes a very heterogonous world with high population growth. The projected changes in the concentrations of the two main GHGs (CO_2 and CH_4) used in SRES scenarios can be seen in Fig. 3a, b, respectively.

The AR5 differs from the previous report (AR4) in terms of the definition of emission scenarios. The new report builds upon a new concept, called representative concentration pathways (RCPs), which includes approaches to both climate change mitigation and adaptation. Basically, the RCPs assume policy actions that will be taken to achieve certain emission targets (Taylor et al. 2012). For CMIP5 project, the four-emission scenarios or RCPs (RCP2.6, RCP4.5, RCP6.0, and RCP8.5) are defined by considering future population growth, technological development and community responses. According to the definition, the radiative forcing for 2100, which is relative to the preindustrial condition, is estimated for each RCPs, and emission scenarios are created based on these assumptions. For example, radiative forcing reaches about 8.5 Wm^{-2} in RCP8.5 high emission scenario with a rising pathway. Unlike RCP8.5, the RCP4.5, and RCP6.0 can be considered as intermediate level emissions scenarios whose radiative forcing's reach 4.5 and 6.0 Wm^{-2}, respectively.

Fig. 3 Concentrations of **a** CO_2 (ppm) and **b** CH_4 (ppb) under SRES and RCP scenarios

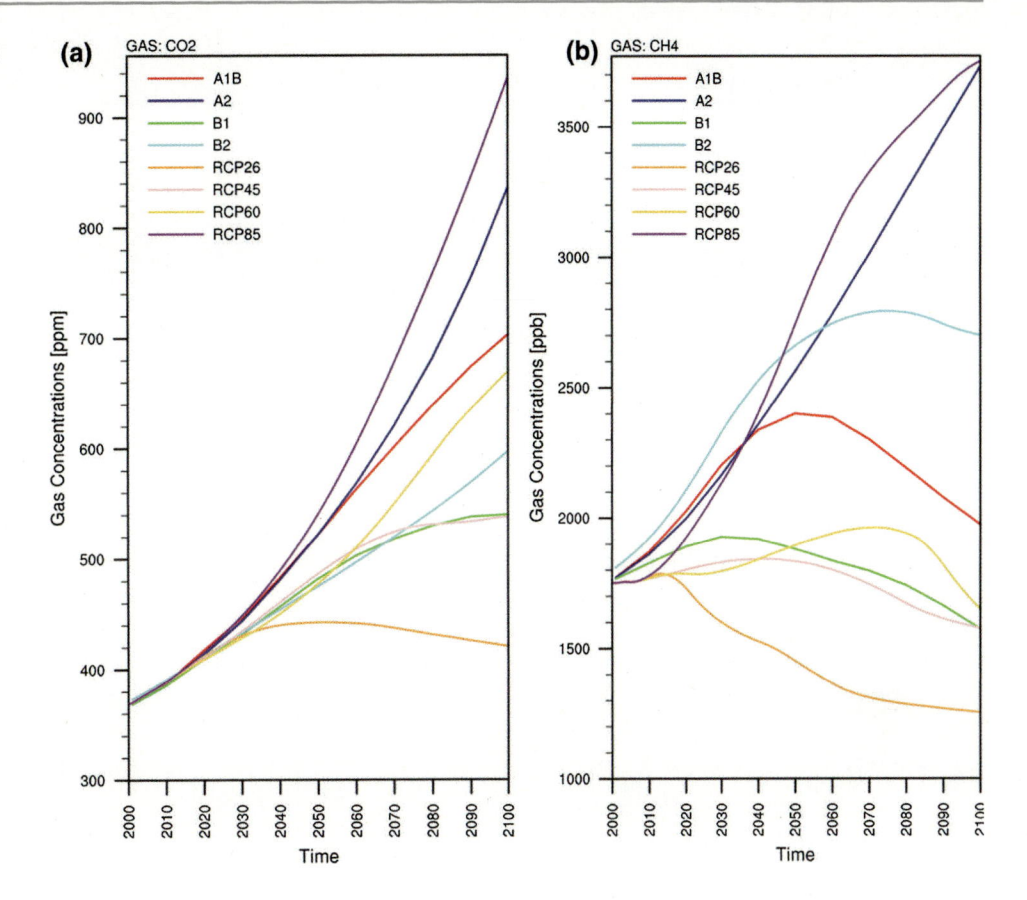

The low-level scenario, RCP2.6, has its maximum in the middle of the twenty-first century and starts to decrease, thereafter, to reach 2.6 Wm^{-2} in 2100. The comparison of CO_2 and CH_4 gas concentrations of the different RCPs and SRES scenarios can be seen in Fig. 3. It is clearly shown that RCP8.5 and SRES A2 scenarios have very similar temporal patterns for CO_2 and CH_4 gas concentrations. Likewise, RCP4.5 and SRES B1 emission scenarios are similar to each other.

The SRES and RCP scenarios are used to force multi-component (i.e., atmosphere, ocean, land, ice) global circulation models (hereafter referred as GCM) to have an assessment of future climate change and its direct and indirect effects on the water resources, agriculture, soil, human health, etc. This section provides information on the climate change projections for Turkey using available data from the downscaled CMIP3 simulations as well as the original outputs of CMIP3 and CMIP5.

As mentioned previously, GCMs are the primary tools to obtain climate change projections based on emission scenarios. The main shortcoming of the information one can derive from the global circulation modeling studies is related to their spatial resolution, which is usually on the order of a few hundreds of kilometers. However, they can still provide a crude estimation of the regional climate change and its impacts over a studied region. Given the fact that the number of available global simulations is comparatively high, it becomes possible to use an ensemble approach that allows uncertainty analysis. Therefore, the use of the multi-model and/or multi-scenario average approach might give more robust and reliable results. This section basically aims at identifying the future climate change over Turkey by estimating the multi-model ensemble average of surface temperature and precipitation fields from the outputs of the available CMIP3 and CMIP5 simulations.

Figure 4a shows the ensemble average of surface air temperatures over Turkey for three different emission scenarios (B1, A1B, and A2). The ensemble average is calculated from the outputs of the 15 different CMIP3 models after interpolating them into a common 0.5° × 0.5° grid. The reader should also note that the colored regions show the inter-model uncertainty for each emission scenario. As it can be seen from the figure, the surface air temperature has an increasing linear trend for all SRES emission scenarios, and the change in average surface air temperatures by 2100 ranges from 1.94 to 4.31 °C. As expected, the B1 scenario shows the lowest change in surface air temperature, and the maximum change occurs for the A2 scenario. The medium emission scenario (A1B) indicates a 3.33 °C increase by the end of the twenty-first century. Unlike surface air temperatures, associated precipitation projections show decreasing trends (Fig. 4b). The SRES A2 emission scenario shows the

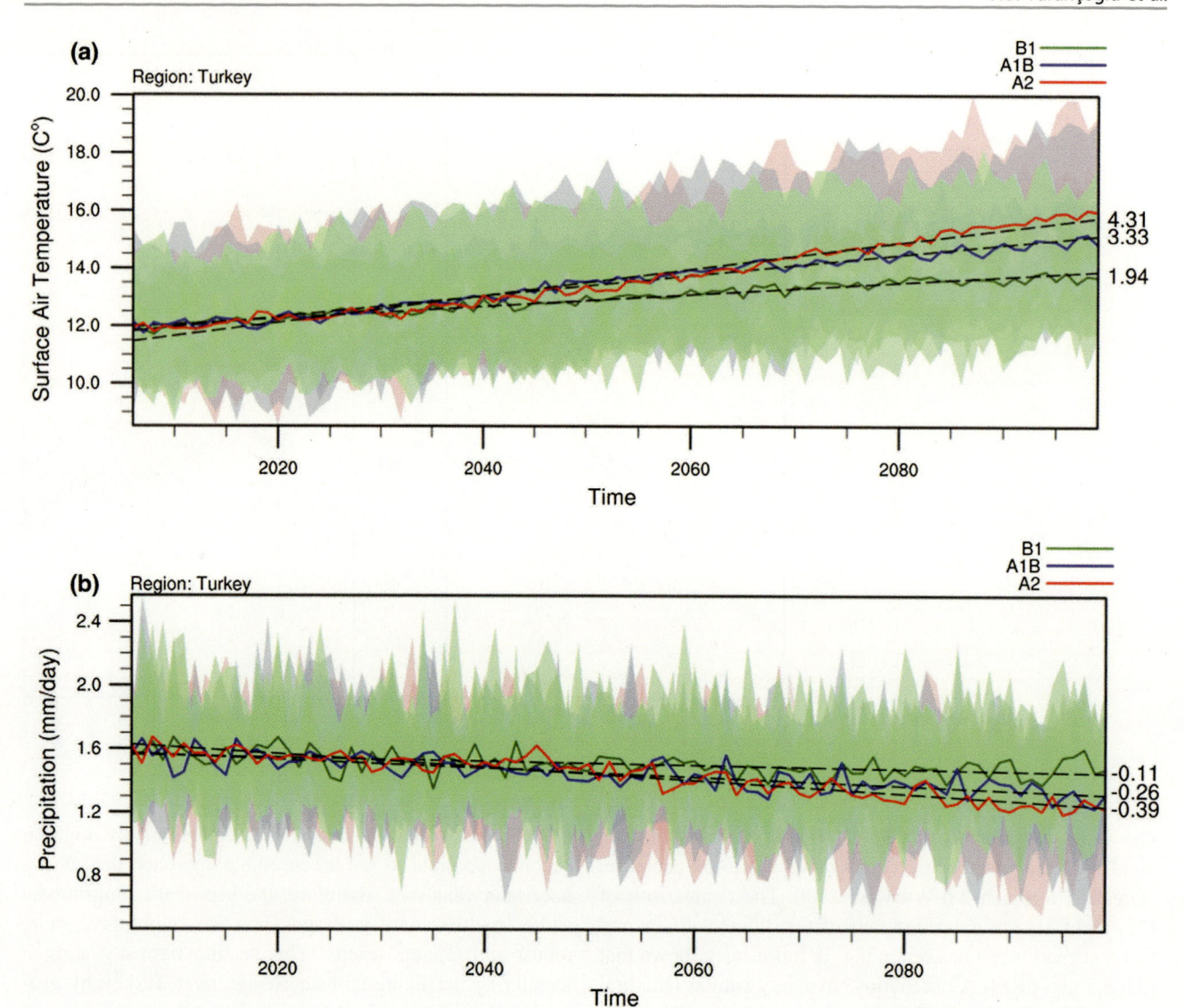

Fig. 4 Time series of annual **a** surface air temperature (°C), and **b** precipitation (mm/day) of CMIP3 model averages over Turkey for the period 2006–2099. The *colors* indicate different emission scenario simulations (B1: *green*, A1B: *blue*, and A2: *red*). The *colored regions* show the inter-model uncertainty for SRES scenario simulations. The *solid lines* represent the multi-model ensemble average and *dashed lines* indicate the trend. The numbers give the absolute changes by the end of the twenty-first century

highest significant reduction with 0.39 mm/day in 2100, and A1B and B1 follow it with 0.26 and 0.11 mm/day respectively (Fig. 4b). The combination of precipitation loss with increasing surface air temperature might adversely affect the water availability and soil moisture, especially in the regions that are under the risk of aridity (southeast and central Turkey). Kömüşçü et al. (2010) estimated the change in soil moisture as 4–43% loss for 2 °C warming without a change in the amount of precipitation and 8–91% for a 4 °C warming.

The CMIP5 models can be assumed superior to the CMIP3 models considering their improved model physics (Taylor et al. 2012), finer spatial resolution (1°–2.8°—but still coarse for constructing assessments at a regional scale), integrated dynamic vegetation models and better representation of the components of the Earth system (included interactive atmosphere- chemistry models, representation of glaciers, etc.). The multi-model ensemble of surface air temperature (Fig. 5a) and precipitation (Fig. 5b) for CMIP5 models is calculated using the same methodology that is used in CMIP3 models previously. In this case, 32 models are used to calculate the multi-model ensemble for each RCPs.

The CMIP5 simulations show similar results to the CMIP3 simulations in terms of the direction of the change

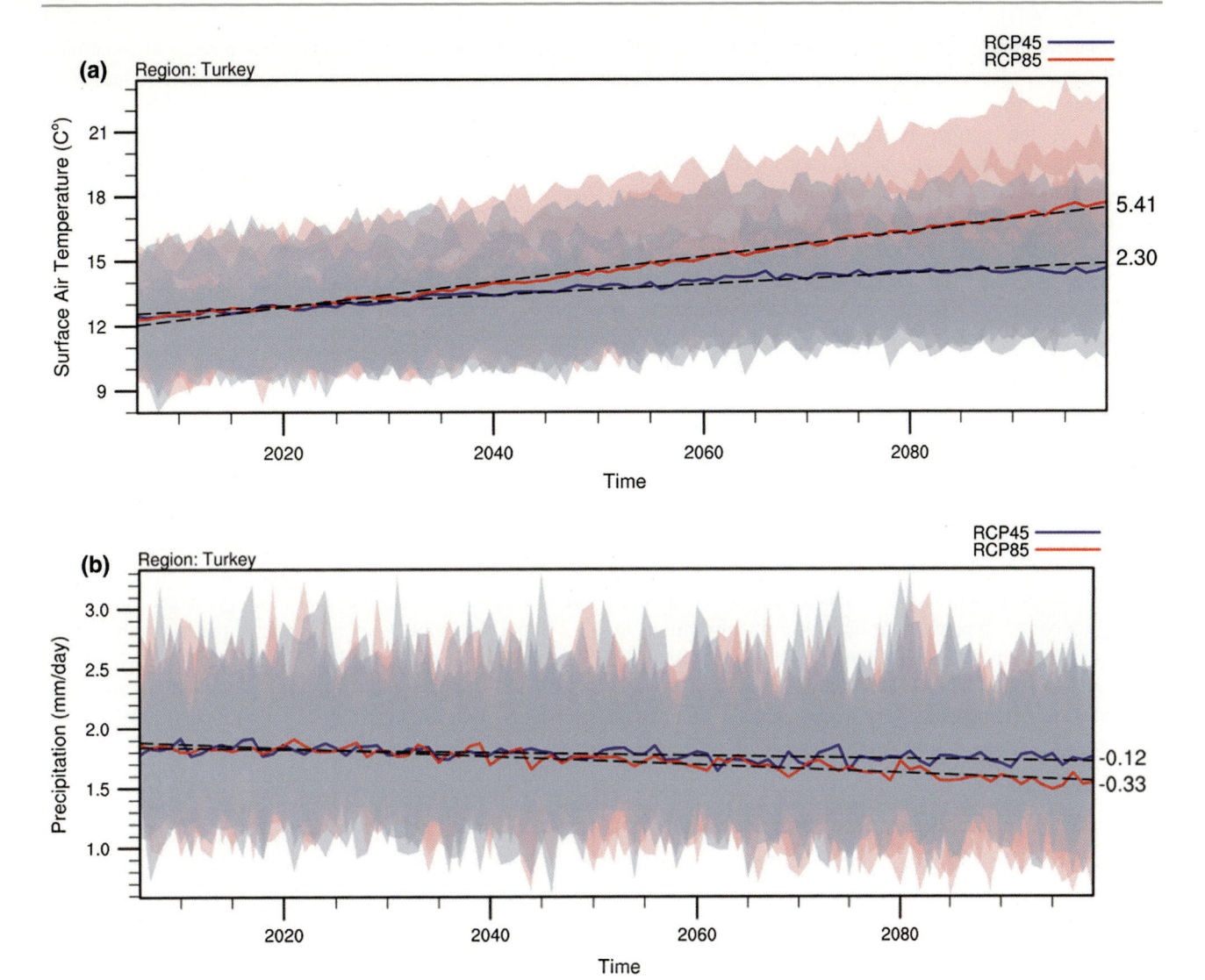

Fig. 5 Time series of annual **a** surface air temperature (°C), and **b** precipitation (mm/day) of CMIP5 model averages over Turkey for the period 2006–2099. The *colors* indicate different emission scenario simulations (RCP45: *blue*, and RCP85: *red*). The *colored regions* show the inter-model uncertainty for both RCP scenarios. The *solid lines* represent the multi-model ensemble average and *dashed lines* indicate the trend. The numbers give the absolute changes by the end of the twenty-first century

(i.e., increase in temperature and decrease in precipitation). The changes in temperature are, however, larger than those in CMIP3. Based on the analysis of the CMIP5 model results, changes in average surface air temperature for Turkey indicates 2.30 and 5.41 °C warming by the end of the twenty-first century for RCP4.5 and RCP8.5 respectively (Fig. 5a). Unlike the surface air temperature, precipitation shows similar signals to CMIP3 models. The maximum end of century change in precipitation is −0.33 mm/day for RCP8.5 scenarios. The RCP4.5 also depicts slight negative slope, and reaches a 0.12 mm/day deficiency by the end of the century (Fig. 5b).

3.1 Dynamical Downscaling Using CMIP3 Models: Future Climate Projections

As mentioned briefly in the previous section, the coarse-resolution GCMs may lead to substantial uncertainties in projected regional climate indicators such as surface air temperature, precipitation and soil moisture. To overcome such problems, regional climate models (RCMs) are used to downscale GCM model outputs to produce higher resolution (both spatial and temporal scale) representations of regional effects over the region including Turkey (Önol and Semazzi 2009; Önol 2012). The first step in a model-based climate change study is the assessment of the model's performance in reproducing present-day climate conditions (Giorgi and Mearns 1999). The performance analysis of the perfect boundary condition experiment (driving the regional model at the lateral boundaries with fields obtained from reanalysis of observations) and three selected GCMs, ECHAM5, CCSM3, and HadCM3 whose climate change projections used in this study are given in Bozkurt et al. (2012). In their study, RegCM3 (Pal et al. 2007) is used as the regional climate model component. In addition, the projected seasonal and daily changes in these three GCM-driven simulations are also extensively discussed for the wider model domain covering the eastern Mediterranean–Black Sea region by Önol et al. (2014). They reported that winter runoff over Turkey's mountainous areas increases in the second half of the twenty-first century, because the snowmelt process accelerates where the elevation is higher than 1500 m.

Bozkurt et al. (2012) suggests that the selected two GCMs (ECHAM5 and CCSM3) are highly skilled in simulating the winter precipitations and surface temperatures in Turkey. However, the CCSM3 model produces relatively drier and warmer summer conditions compared to the observations. In general, the two models could be used in the climate change and impact assessment studies as long as their strengths and weaknesses are taken into account.

The rest of this chapter includes mainly the results of the future climate change projections based on the dynamically downscaled ECHAM5 and CCSM3 model results (Önol et al. 2014). The analysis is extended to examine the extreme drought events and their future changes by using the outputs of the downscaled simulations.

3.2 Temperature and Precipitation

Figure 6 shows the annually averaged surface air temperature anomalies with respect to the 1961–1990 reference period for both CCSM3 (A1FI and A2 emission scenario) and ECHAM5 (A2 emission scenario) simulations. As can be seen from the figure, all simulations show similar warming trends in selected future periods (2010–2039, 2040–2069 and 2070–2099). The CCSM3-A2 simulation has, on the average, around 1 °C stronger warming signal than the ECHAM5-A2 simulation. This result is also consistent with the reference simulations of the same models. The CCSM3 model is generally drier and warmer than ECHAM5 (Bozkurt et al. 2012). Substantial increases in surface temperature start to appear in the second selected period (2040–2069) over Turkey and reach a level around 3.5–6.0 °C at the end of the twenty-first century. The model simulations also suggest that the increase in surface temperature over Turkey will not be uniform. The eastern interior, southern, and southeastern parts will experience greater rises in temperatures. The real added value of the dynamical downscaling can be also seen when the RCM results are compared with the GCM's raw surface air temperature data. In this case, RCM simulation driven by the same GCM, gives a 1.5–3.0 °C stronger warming signal than the raw data (Fig. 4).

As Table 3 indicates, there are a total of three downscaled simulations (ECHAM5 A2, CCSM3 A2, and A1FI) for Turkey for the period between 2071 and 2099. All three simulations exhibit similar behaviors in the changes of surface temperatures. For instance, the changes are relatively small in winter but they increase in transition seasons and reach a peak in the summer. They mostly indicate larger increases in eastern rather than in western Turkey. The differences between model simulations arise mostly in the magnitudes of their projections. The CCSM3's estimation of fall temperature increase for Turkey (about 5.4 °C) is larger than ECHAM5 simulation results (about 4.2 °C). A1FI simulation of CCSM3 yields 0.5–1.4 °C larger values than A2 simulation of the same model. It produces an average summer increase of 7.3 °C for the eastern Turkey.

Figure 7 shows the simulated precipitation anomalies relative to their 30-year climatology (1961–1990) of reference runs. We first note the large differences between ECHAM5 and CCSM3 simulations for the A2 emission scenario. ECHAM5 shows precipitation increase (15–20%) in the first defined future period (2010–2039) while CCSM3 mostly shows decreases. In the mid-century (2040–2069), both models show similar patterns in precipitation change: increase in the Black Sea region (10–20%) and decreases in interior and southern parts of Turkey (15–25%).

Table 4 provides the seasonal changes in precipitation from the three different simulations for the 2071–2099 periods. There are broad agreements between the model estimations of the precipitation changes for the same scenario (i.e., A2). However, the magnitude of the changes may not be fully consistent because areas outside of 'hot spots' may show different sensitivity to the increased emissions in different models, and this affects the average values. The projected changes in precipitation are usually stronger in the CCSM3's A1FI simulation than those in its A2 simulation, especially in fall and spring, which are wet seasons. All models broadly agree that Turkey will have less annual precipitation in the last 30-year period of the twenty-first century compared to the present times.

3.3 Soil Moisture

The soil moisture can be defined as the amount of water stored in the soil. The Biosphere-Atmosphere Transfer Scheme (BATS, Dickinson et al. 1993) represents the

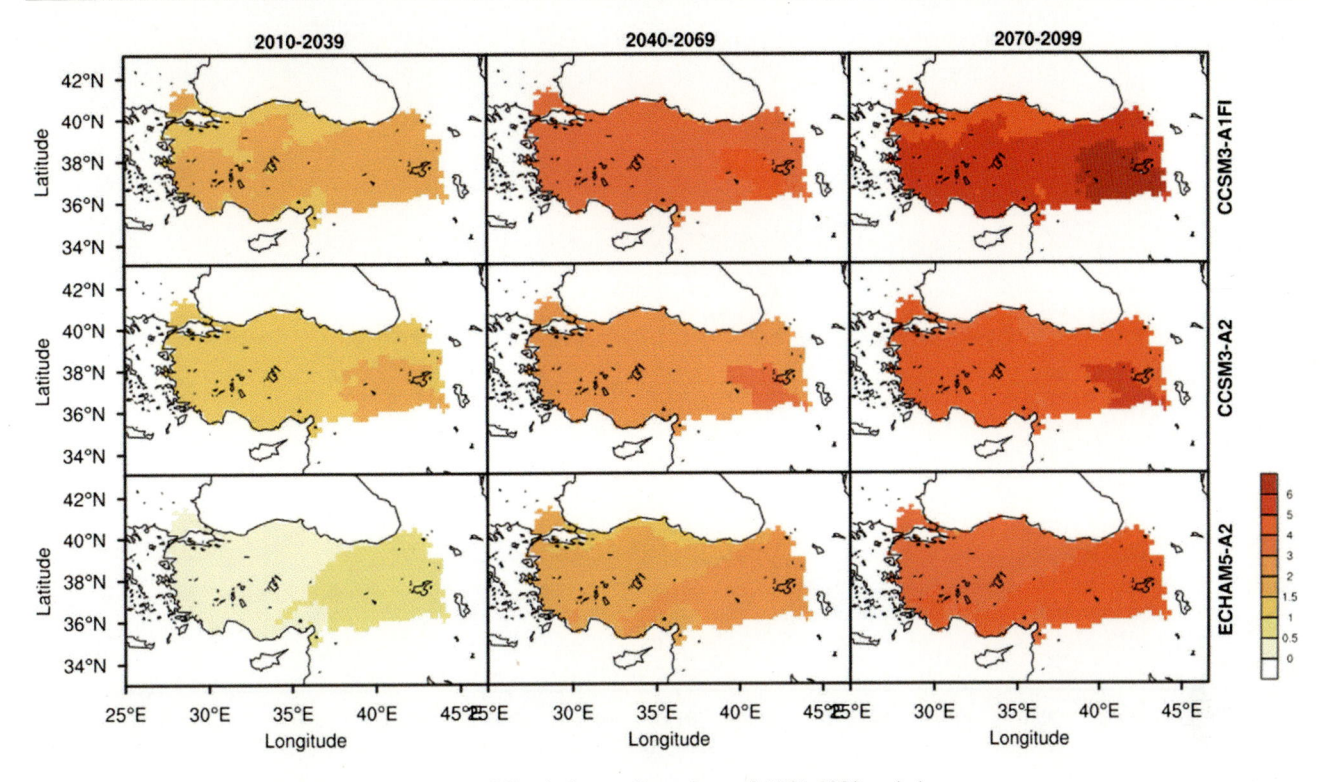

Fig. 6 Annual changes in surface air temperature (°C) relative to climatology of 1961–1990 period

Table 3 Projected seasonal surface temperature changes (°C) in the 2071–2099 period over 1961–1990 period based on different scenario simulations

Scenario	GCM	Winter		Spring		Summer		Fall	
		West	East	W	E	W	E	W	E
A2	ECHAM5	2.9	3.4	3.1	4.1	4.7	5.2	4.0	4.4
	CCSM3	2.5	2.9	3.6	3.5	6.4	6.8	4.9	5.9
A1FI	CCSM3	3.5	4.0	4.8	4.9	6.9	7.3	5.5	6.8

W indicates the western half of Turkey and *E* indicates the eastern half of Turkey

land-surface processes in RegCM3. In BATS, the soil is characterized by three layers to calculate soil moisture and temperature. In addition, it also includes a one-layer vegetation scheme and simple surface runoff component. The change in soil moisture has an important effect on agriculture, potential evaporation and surface runoff (IPCC 2001). Changes in soil moisture are predominantly related to two major factors: regional climate change and soil characteristics. This section presents the future change in soil moisture under different SRES emission scenarios (A1FI and A2).

The change in simulated soil moisture with respect to the 1961–1990 periods is illustrated in Fig. 8. As can be seen from the figure, the change signal with decreasing trend in soil moisture is much stronger along the Taurus mountain range compared to the surrounding region. The difference reaches 20% in the CCSM3 A1FI and 10% in the ECHAM5 and CCSM3 A2 emissions scenario, but the spatial extent of the change is larger in A1FI simulation. This result is also consistent with the change in surface runoff, which is not shown here (Önol et al. 2014). This is most likely an indication of early snow melting in response to the increased surface temperatures and decreased precipitation in the higher elevation regions. The projected change in soil moisture also reaches higher values (>30%) in the last 30-year period (2070–2099).

The soil moisture analysis is extended to analyze the changes in the soil moisture amounts for the 26 major basins in Turkey (Fig. 9). The figure indicates that the CCSM3 and ECHAM5 models (with A2 emission scenario) show different behavior in the first 30-year of the twenty-first century. The decreasing signal in CCSM A2 and A1FI emission scenario (0.5–1.0 mm) is reversed in ECHAM5 A2 scenario, but all the models show similar results after 2040s. The Çoruh and Aras basins also behave differently compared to the other basins. Soil moisture in these basins continues to increase (∼0.5 mm) until 2070s before it starts to decrease. As could be seen from the figure, the change in the Seyhan basin is stronger than the other basins. This also suggests

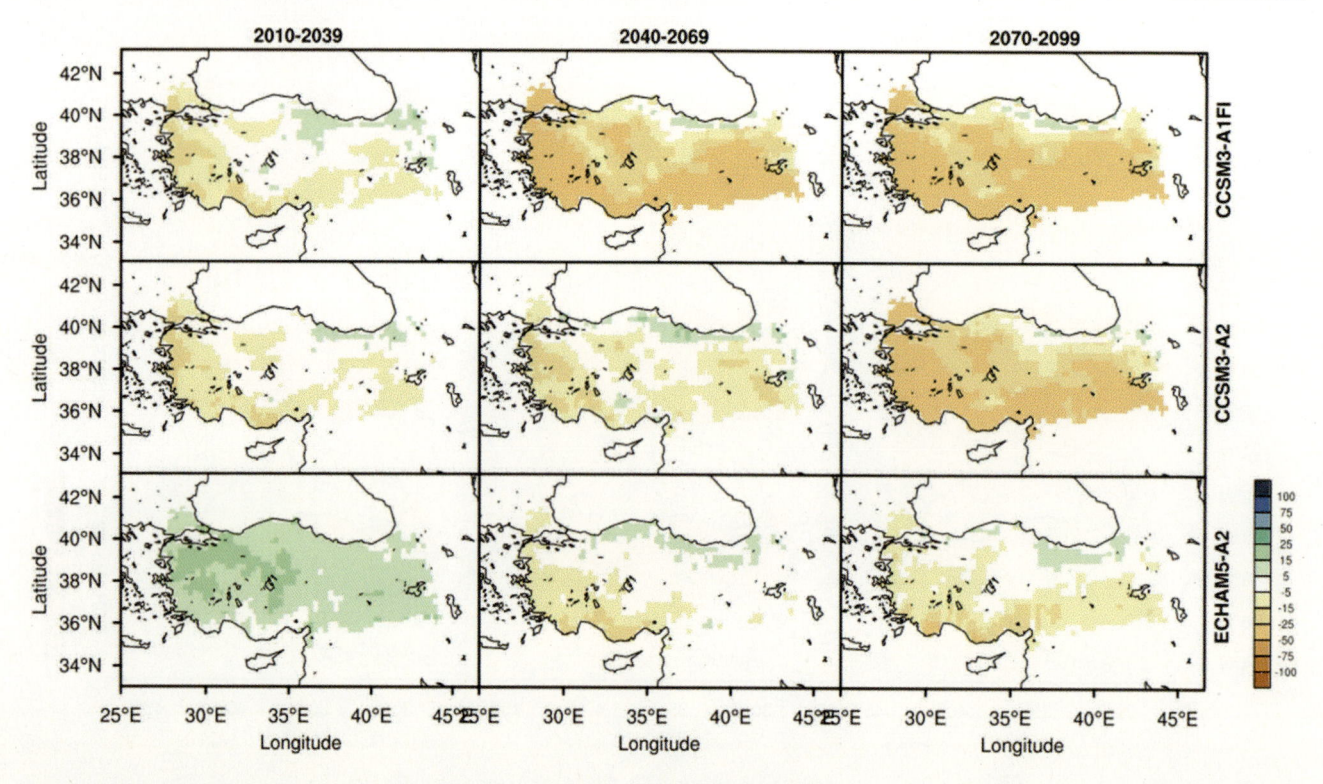

Fig. 7 Annual changes in precipitation (%) relative to climatology of 1961–1990 period

Table 4 Projected seasonal precipitation changes (%) in 2071–2099 period over the 1961–1990 period based on different scenario simulations

Scenario	GCM	Winter		Spring		Summer		Fall	
		North	South	N	S	N	S	N	S
A2	ECHAM5	+13	−17	+1.5	−23	−23	−30	−4	+4
	CCSM3	−6	−32	−21	−36	−33	−62	−6	−23
A1FI	CCSM3	−0.6	−35	−30	−47	−57	−70	−1.5	−10

N indicates the northern half of Turkey and *S* indicates the southern half of Turkey

that the Seyhan basin is more sensitive (>1.0–1.5 mm) to the change in the regional climate.

3.4 Drought Indices: Palmer Drought Severity Index (PDSI)

Climate indices provide information on the extreme events such as drought and floods that might affect the daily life in a negative way. A drought is an extended period of months or years when a region notes a deficiency in its water supply. Drought, which can be also characterized by a decrease in precipitation over a period of months to years, is one of the most important climate hazards due to its impact on agriculture, water resources and also human health (Dai 2011a). Recent studies investigating drought events in Turkey have mainly focused on their historical trends (Tatlı and Türkeş 2011) and their effects on crop yields. In this respect, studies on corn (Durdu 2012), wheat (Özdoğan 2011), and olives (Tunalıoğlu and Durdu 2012) predict a decrease in production. Thus, this section aims to provide future projections about drought events over Turkey.

The Palmer drought severity index (PDSI, Palmer 1965) is a widely used measure of drought events (Heim 2002). PDSI and its variants have been used to investigate long-term changes of aridity over land in both global (Dai et al. 2004; Dai 2011a, b) and also regional scale, such as the Mediterranean Basin (Sousa et al. 2011). In this study, we used a modified version of the PDSI, i.e., the self-calibrating PDSI (scPDSI). The scPDSI basically replaces the empirically derived climatic characteristic and duration factors with automatically calculated values based upon the historical climatic data of a location (Wells et al. 2004). The scPDSI is selected in this study because it provides more

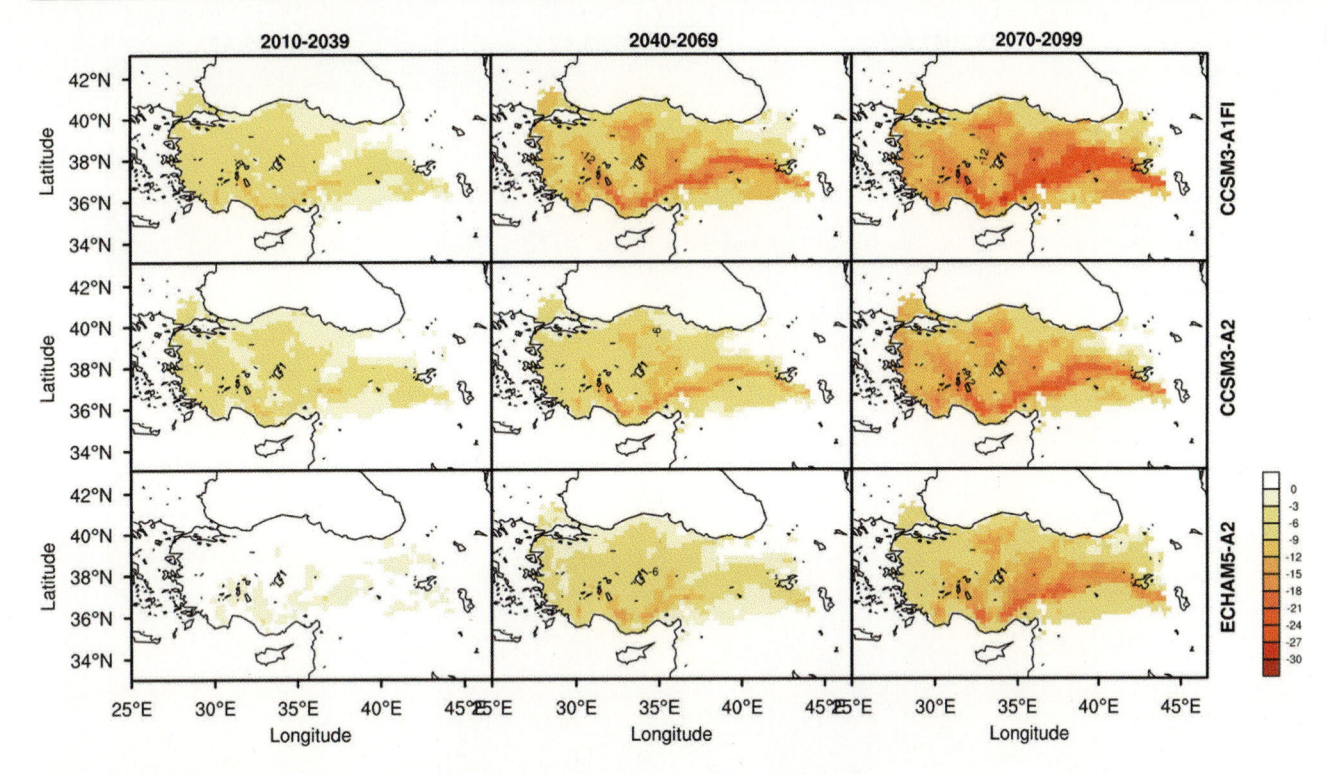

Fig. 8 Annual changes in soil moisture (%) relative to the climatology of 1961–1990 period

geographically comparable classification of climate divisions than the original PDSI (Wells et al. 2004).

The monthly scPDSI is calculated using both RCM simulated and observed monthly surface air temperature and precipitation. The potential evapotranspiration (PET) is estimated using Thornthwaite's method (Thornthwaite 1948) rather than Penman-Monteith parameterization, but Van der Schrier et al. (2011) showed that PDSI values based on these two different PET calculations give similar results in terms of regional averages and trends. In addition to the surface air temperature and precipitation, the calculation of scPDSI values also closely depend on the available water-holding capacity of the soil. A soil texture-based water-holding-capacity map from Webb et al. (1993) is used for this purpose. The original dataset of water-holding capacity is in 1° × 1° resolution but it is interpolated into high-resolution RCM grids (27 km) to perform scPDSI calculations. The scPDSI calculations were started 10 years earlier than the beginning of the RCM simulations (1950 for historical and 1990 for future simulations) using climatological monthly mean values for temperature and precipitation at each grid box. This eliminates any spin-up problems related to the scPDSI calculation. The classification used in scPDSI can be seen in Table 5. The values from 0.5 to −0.5 are considered as normal conditions and those smaller than −0.5 are associated with drought events with different severity.

The simulated scPDSI statistics are assessed based on the results of the RCM simulation driven by the NCEP/NCAR Reanalysis dataset (Kalnay et al. 1996) for the 1961–1990 period. The simulated PDSI values are compared to the observation-based scPDSI values, which are calculated using data provided by the Climate Research Unit (CRU) of the University of East Anglia, UK (Mitchell and Jones 2005). The time series of the monthly scPDSI values averaged over Turkey for the model simulation (NCEP.RF) and CRU-based scPDSI values can be seen in Fig. 10.

The correlation between simulated and CRU-based scPDSI values are around 0.6. As can be seen from the figure, the scPDSI values are generally less than −0.5, which indicate that slight dry conditions are generally dominant over Turkey. The reader should also note that the simulated scPDSI values show a stronger drought signal after the first quarter of the 80s, which is not seen in the CRU-based drought index calculations. Despite this, it could be said that RegCM3 is able to satisfactorily reproduce observed drought events and can be used to produce future drought projections for Turkey.

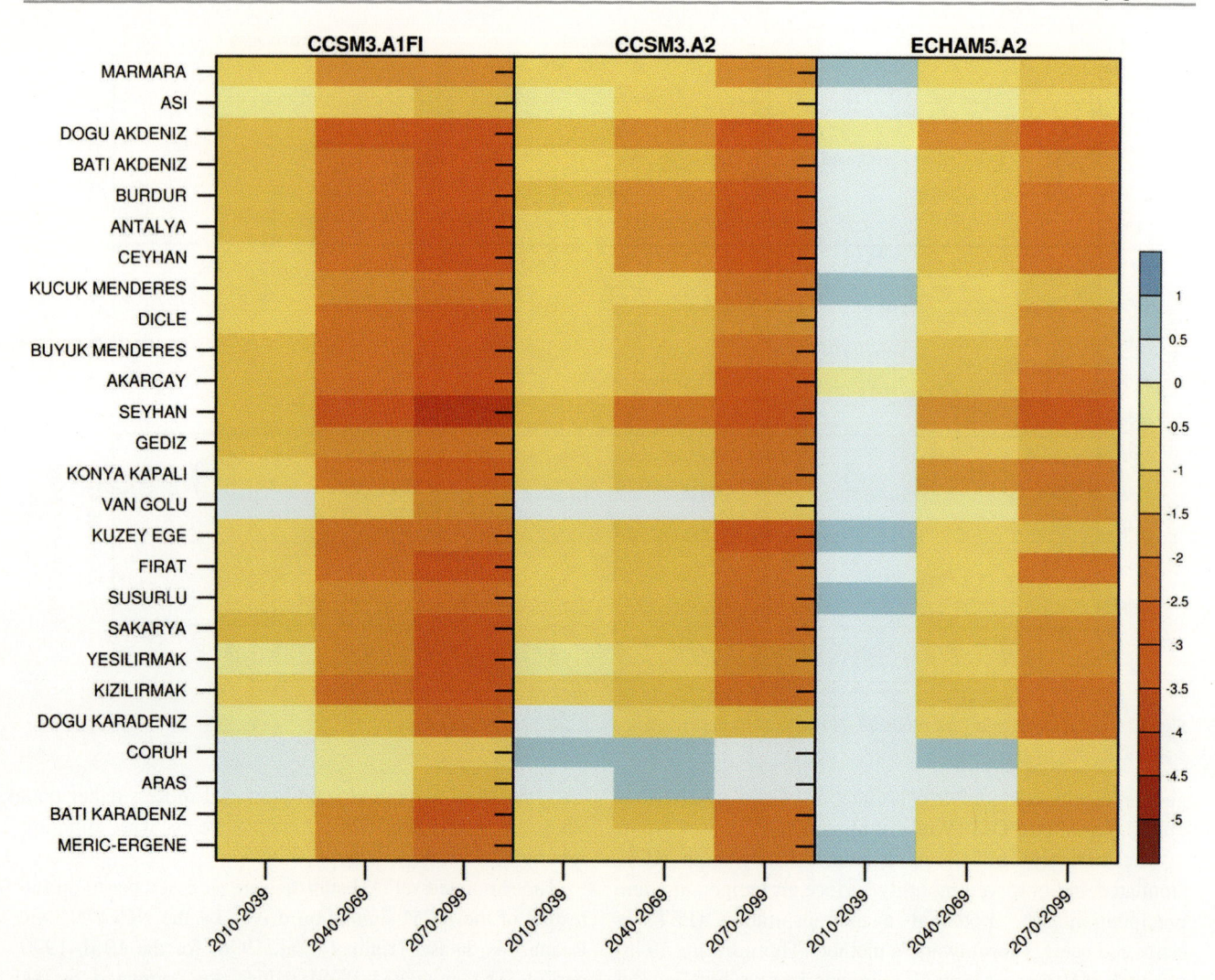

Fig. 9 Changes in basin averaged soil moisture (mm) relative to climatology of the 1961–1990 period

Table 5 Classification of dry and wet conditions as defined by Palmer (1965) for the PDSI

PDSI range	Class
PDSI ≥ 4.0	Extremely wet
4.0 > PDSI ≥ 3.0	Severally wet
3.0 > PDSI ≥ 2.0	Moderately wet
2.0 > PDSI ≥ 1.0	Slightly wet
1.0 > PDSI ≥ 0.5	Incipient wet spell
0.5 > PDSI ≥ −0.5	Near normal
−0.5 > PDSI ≥ −1.0	Incipient dry spell
−1.0 > PDSI ≥ −2.0	Slightly dry
−2.0 > PDSI ≥ −3.0	Moderately dry
−3.0 > PDSI > −4.0	Severally dry
PDSI ≤ −4.0	Extremely dry

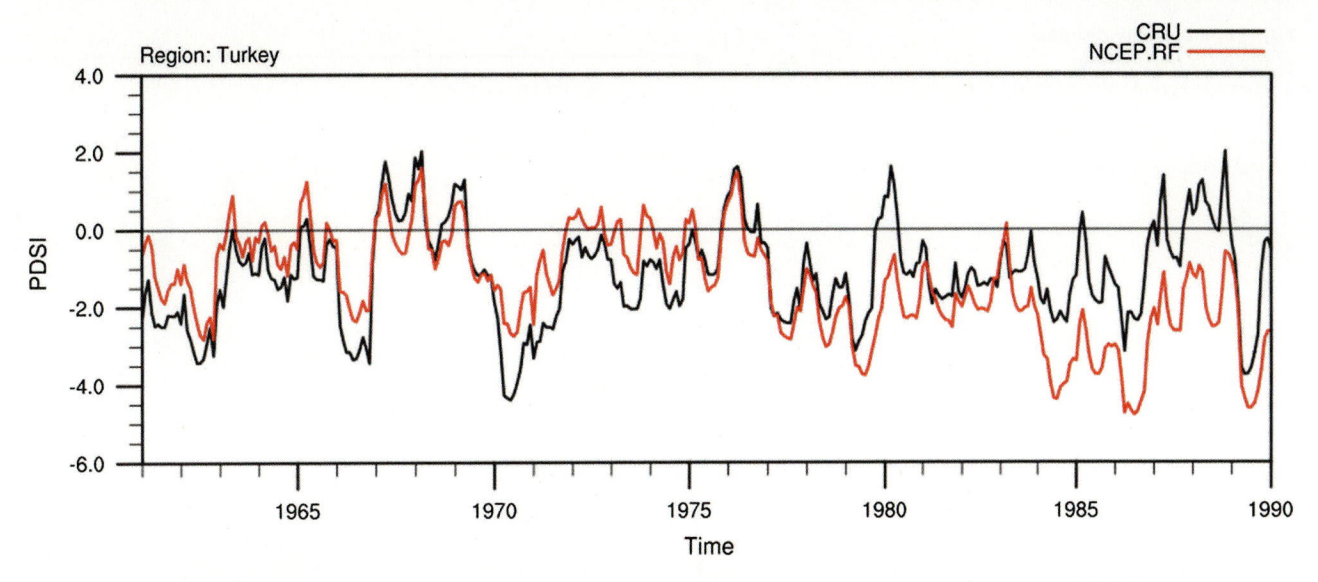

Fig. 10 The time series of the monthly PDSI estimated by CRU observational data and RCM simulation forced with NCEP/NCAR data during the 1961–1990 period

The probability distribution or frequencies of drought events over Turkey in terms of the metric of scPDSI can be seen in Fig. 11. The topmost plot (Fig. 11a) shows the probability of wet and dry years between 1961 and 1990 for model simulations and also observation (CRU)-based scPDSI. It is clearly seen that models are broadly able to reproduce the frequency of the drought events when they are compared with the NCEP/NCAR and CRU-based scPDSI. In general, the frequent occurrence of slight (25%) and moderate (20%) dry events are overriding in Turkey. It should also be noted that the frequency of extreme drought events is around 5% in the reference period.

Figure 11b–d displays the future projections of drought events for three different time periods (2010–2039, 2040–2069, and 2070–2099). The probability distribution of scPDSI shows a completely different behavior in the first selected period (2010–2039), and the shift to the relatively wet years can be clearly seen in Fig. 11b. The frequency of drought events is around 10–15% for slight and moderate cases and less than 2% for extreme cases in this period. As for the drought events, floods are also very important extreme events. The occurrence of extremely to moderate wet years start to increase and reach 8–15% for the ECHAM5 A2 emission scenario and 2–8% for CCSM3 A2. The A1FI scenario indicates lower probability of wet years than the ECHAM5 A2 scenario, which is also consistent with the analysis of the reference period. In fact, the CCSM3 model is generally drier than the ECHAM5 model (Bozkurt et al. 2012; Önol et al. 2014).

The probability distributions of scPDSI values in the 2040–2069 periods (Fig. 11c) are very similar to the reference period (Fig. 11a). In this case, two different emission scenarios of the CCSM3 model (A2 and A1FI) also behave very similarly, but extreme drought events are doubled in A1FI.

The most significant change in the probability distribution of scPDSI is seen in the last 30 years of the twenty-first century (Fig. 11d). The frequency of extreme drought events reaches up to 40% in ECHAM5 A2 and around 30% in CCSM3 A2 and A1FI simulation in this period. Based on the projections of all the RegCM3 simulations, the occurrence of the strong drought events will be dominant, and occurrence of the dry years will increase in the future.

The spatial extents of the drought events are also important. Figure 12 shows the probability of drought events (scPDSI < −2) for different time slices. Similar to the Fig. 11b–d, the probability of the dry years increases and reaches its maximum value in the 2070–2099 periods. In the same period, the southwestern, southeastern and central parts of Turkey will be affected mostly by extreme drought events. Again, the ECHAM5 A2 simulation shows stronger signal in probability of drought events than the CCSM3 A1FI in the last 30-year.

Fig. 11 Probability distribution of PDSI (in fraction) for **a** reference period (1961–1990), **b** 2010–2039, **c** 2040–2059 and **d** 2070–2099 for A1FI and A2 future SRES scenarios

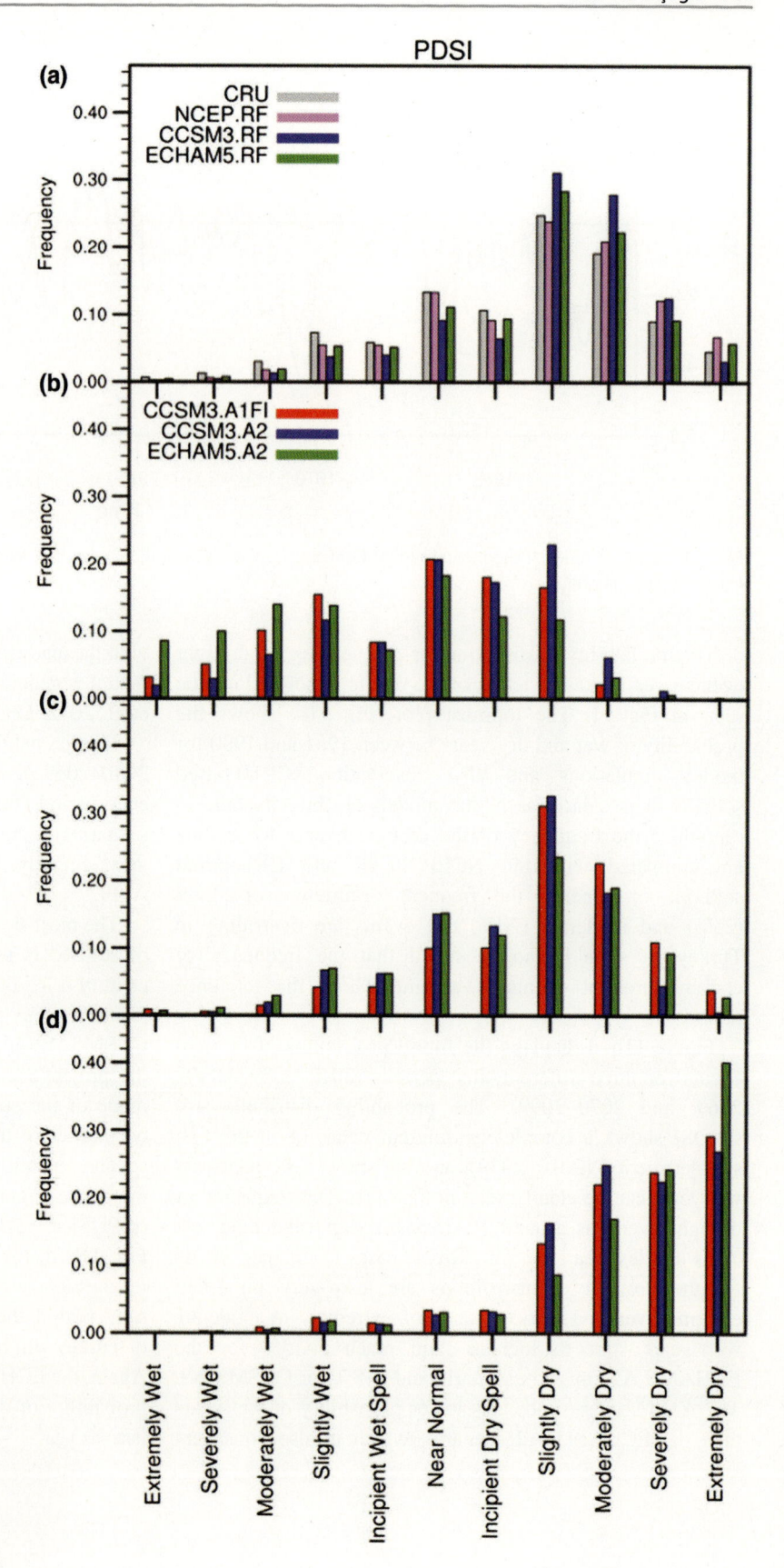

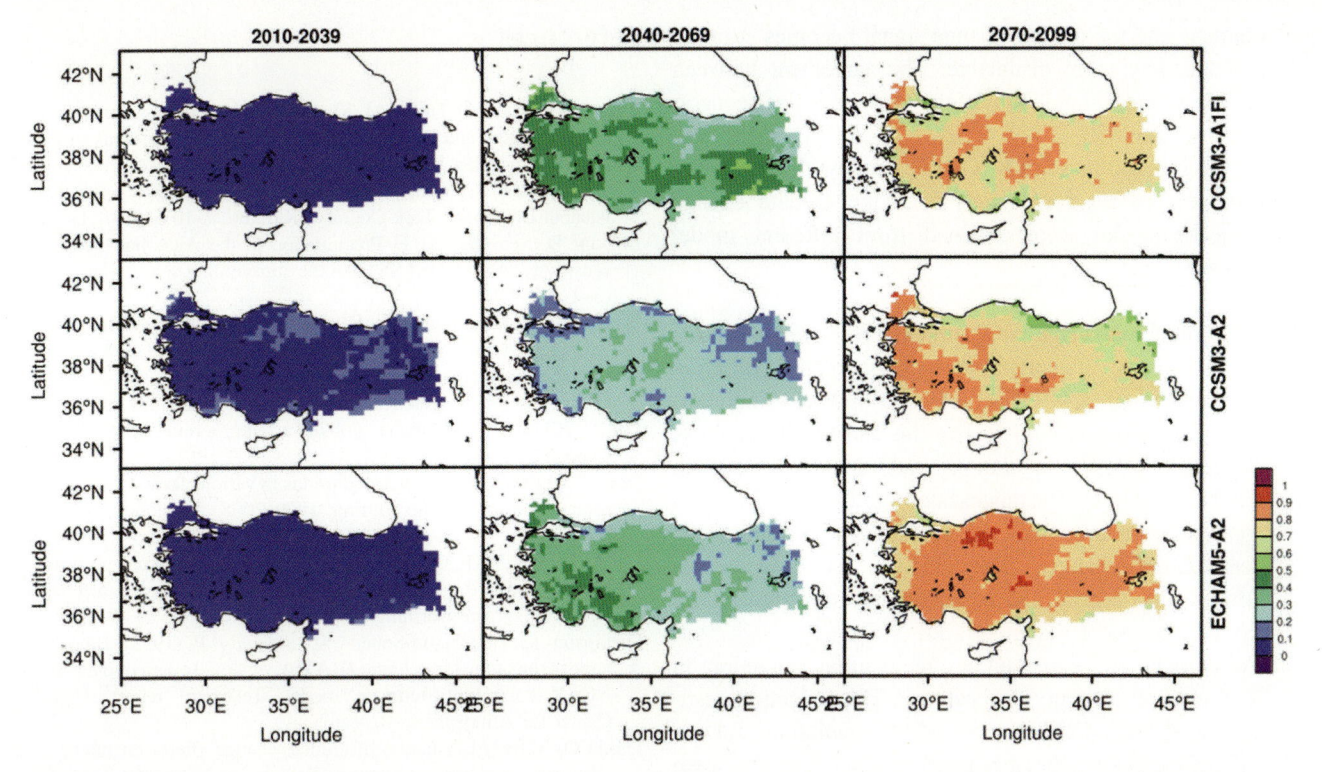

Fig. 12 The probability of the drought events (PDSI < −2) estimated by RCM simulations forced by CCSM3 A2, A1FI and ECHAM5 A2 data for 2010–2039, 2040–2069, and 2070–2099 periods

4 Conclusions

This chapter aims to provide detailed information concerning the main features of the climate of Turkey, which is mainly effected by the surrounding large water masses (i.e., Mediterranean and Black Sea) and Atlantic and Mediterranean (mainly the Genoa Bay and Adriatic Sea) originated cyclonic systems. In addition to the detailed information about the regional climate system and its main drivers, the Köppen-Geiger climate classification system and an Aridity Index analysis are also presented to extend the given analysis from the perspective of land surface and soil. The results indicated that the climate of Turkey according to the Köppen- Geiger climate system is very diverse. The sub-tropical steppe climate (BS) is found in the central Anatolia region as well as the farthest eastern part of eastern Anatolia. The coastal Black Sea region is mainly dominated by a temperate rainy or humid temperate west coast climate without a dry season (mostly Cfa and Cfb). The regions around the coast of the Mediterranean Sea (Marmara, Aegean, and southwestern Anatolia regions) belong to the dry summer subtropical Mediterranean climate or temperate rainy climate with dry summer (mostly Csa). Lastly, a cold snowy forest climate with dry summer (mostly Dsa and Dsb) takes place over a relatively large zone lying along the mid-northern portions of the continental regions of Turkey (central and eastern Anatolia), whereas a cold snowy forest climate humid in all seasons (humid microthermal, mostly Dfb) exists over relatively small areas seen in the northern portions of the continental central Anatolia region and the northeastern Anatolia subregion of Turkey.

As for future climate of Turkey, an analysis involving the century-long changes obtained directly from the CMIP3 and CMIP5 GCM simulations and a more detailed analysis based on the dynamically downscaled outputs of some of the CMIP3 simulations are included. The number of the scenario/model simulations in the latter is limited, which hinders a thorough analysis including the estimations of the uncertainty metrics. Nevertheless, given the fact that these simulations provide much higher resolution climate parameters, they are extremely beneficial for the illustration of the spatial distributions for Turkey that has a very heterogeneous topography and landscape, which cannot be resolved by the coarse-resolution General Circulation Models. It is also important to mention here the changes occurring in projections for the early periods of the 21st century. Because the projected emissions in these periods are still close to the present-day emissions, the changes (i.e., the signals) should be considered with caution. The signal-to-noise ratio improves toward the end of the century, and the climate change signal becomes stronger and spatially more distinctive. The agreement between different model simulations of the same emission scenario also increases.

Having said these, it is possible to draw the following relatively robust conclusions from the climate change projections that were obtained from different model simulations:

1. All simulations agree on a temperature increase in Turkey in the twenty-first century. The different scenario-based CMIP3 simulations indicate increases between 1.94 and 4.31 °C by the end of the century, while the CMIP5 simulations yield increases between 2.50 and 5.41 °C. The downscaled simulations of CMIP3 give information about the spatial distribution of the temperature changes. They indicate larger increases towards the central and eastern parts compared to the coastal parts of Turkey.
2. All simulations agree on a precipitation reduction in Turkey in the twenty-first century. The reductions estimated by both CMIP3 and CMIP5 simulations for the end of the century broadly agree in magnitude (between 0.11 and 0.39 mm/day), as well. For regional changes, the high-resolution simulations indicate reductions in the Mediterranean region of Turkey while they consistently exhibit increases for the eastern Black Sea region.
3. Although not shown here, all simulations agree on a reduction in the total runoff of Turkey by the end of the century. Moreover, the simulations indicate a reduction in the spring runoff while an increase in winter runoff in eastern Anatolia in response primarily to increased temperatures is foreseen. These results may have important implications for the irrigation of the agricultural fields in southeastern Anatolia as well as the energy production that depend on the snow-fed rivers of the region.
4. In parallel to the increased surface air temperature and decreased amount of precipitation, the frequency of the extreme drought events is projected to increase by the end of the twenty-first century. Consequently, the assessment of the impacts of climate change on water resources and agriculture will become crucial for a sustainable future.

Acknowledgements This research has partly been accomplished by funds from the United Nations Development Program through 'MDG-F 1680' project entitled "Enhancing the Capacity of Turkey to Adapt to Climate Change". We are grateful to the National High Performance Computing Center at İstanbul Technical University for providing the computational resources for performing the regional climate simulations.

References

Altınsoy H, Öztürk T, Türkeş M, Kurnaz ML (2011) Projections of future air temperature and precipitation changes in the Mediterranean Basin by using the global climate model. In: Proceedings of the National Geographical Congress with international participation (CD-R), İstanbul University, Türk Coğrafya Kurumu, 7–10 Sept 2011

Bozkurt D, Sen OL (2011) Precipitation in the Anatolian Peninsula: sensitivity to increased SSTs in the surrounding seas. Clim Dyn 36 (3–4):711–726

Bozkurt D, Turunçoglu UU, Şen OL, Önol B, Dalfes HN (2012) Downscaled simulations of the ECHAM5, CCSM3 and HadCM3 global models for the eastern Mediterranean-Black Sea region: evaluation of the reference period. Clim Dyn 39(1–2):207–225

Dai A (2011a) Drought under global warming: a review. WIREs Clim Change 2:45–65

Dai A (2011b) Characteristics and trends in various forms of the palmer drought severity index during 1900–2008. J Geophys Res 116 (D12115)

Dai A, Trenberth KE, Qian T (2004) A global dataset of palmer drought severity index for 1870–2002: relationship with soil moisture and effects of surface warming. J Hydrometeorology 5:1117–1130

Dickinson R, Henderson-Sellers A, Kennedy P (1993) Biosphere-atmosphere transfer scheme (BATS) version 1e as coupled to the NCAR community climate model. Technical report, National Center for Atmospheric Research

Durdu OF (2012) Evaluation of climate change effects on future corn (Zea mays L.) yield in western Turkey. Int J Climatol 33(2):444–456

Erlat E, Türkeş M (2012) Analysis of observed variability and trends in numbers of frost days in Turkey for the period 1950–2010. Int J Climatol 32(12):1889–1898

Essenwanger OM (2001) Classification of climates. World survey of climatology volume 1C (General climatology), Epilogue by Landsberg HE. Elsevier, Amsterdam, The Netherlands

Giorgi F, Mearns LO (1999) Introduction to special section: regional climate modeling revisited. J Geophys Res 104:6335–6352

Heim RRJr (2002) A review of twentieth-century drought indices used in the United States. Bull Amer Meteor Soc 83:1149–1165

IPCC (2001) In: McCarthy JJ, Canziani OF, Leary NA, Dokken DJ, White KS (eds) Climate change 2001: impacts, adaptation and vulnerability, contribution of working group II to the third assessment report of the Intergovernmental Panel on Climate Change. Cambridge University Press, Cambridge, UK, and New York, USA, 1032 p

IPCC (2007) In: Solomon S, Qin D, Manning M, Chen Z, Marquis M, Averyt KB, Tignor M, Miller HL (eds) Climate change 2007: the physical science basis. Contribution of working group I to the fourth assessment report of the Intergovernmental Panel on Climate Change, 2007, Cambridge University Press, Cambridge, UK and New York, NY, USA, 996 p

IPCC (2013) In: Stocker TF, Qin D, Plattner GK, Tignor M, Allen SK, Boschung J, Nauels A, Xia Y, Bex V, Midgley PM (eds) Climate change 2013: the physical science basis. contribution of working group I to the fifth assessment report of the Intergovernmental Panel on Climate Change, Cambridge University Press, Cambridge, UK and New York, NY, USA, 1535 p. doi:10.1017/CBO9781107415324

Kalnay E, Coauthors (1996) The NCEP/NCAR 40-year reanalysis project. Bull Amer Meteor Soc 77:437–471

Karaca M, Deniz A, Tayanç M (2000) Cyclone track variability over Turkey in association with regional climate. Int J Climatol 20:1225–1236

Kömüşçü AU, Erkan A, Öz S (2010) Possible impacts of climate change on soil moisture availability in the southeast Anatolia Development Project Region (GAP): an analysis from an agricultural drought perspective. Clim Change 40(3–4):519–545

Köppen W (1936) Das geographisca System der Klimate. In: Köppen W, Geiger G (eds) Handbuch der Klimatologie, 1. C. Gebr, Borntraeger, pp 1–44

Kutiel H, Türkeş M (2005) New evidence about the role of the North Sea-Caspian Pattern (NCP) on the temperature and precipitation regimes in continental central Turkey. Geogr Ann Ser A Phys Geogr 87:501–513

Meehl GA, Covey C, Taylor KE, Delworth T, Stouffer RJ, Latif M, McAvaney B, Mitchell JF (2007) The WCRP CMIP3 multimodel dataset: a new era in climate change research. Bull Am Meteor Soc 88:1383–1394

Meinshausen M, Smith S, Calvin K, Daniel J, Kainuma M, Lamarque J-F, Matsumoto K, Montzka S, Raper S, Riahi K, Thomson A, Velders G, Vuuren DP (2011) The RCP greenhouse gas concentrations and their extensions from 1765 to 2300. Clim Change 109(1):213–241

Mitchell TD, Jones PD (2005) An improved method of constructing a database of monthly climate observations and associated high-resolution grids. Int J Climatol 25:693–712

Nakicenovic N, Swart R (2000) IPCC special report on emissions scenarios. Cambridge University Press, Cambridge

Önol B (2012) Understanding the coastal effects on climate by using high resolution regional climate simulation. Clim Res 52:159–174

Önol B, Semazzi FHM (2009) Regionalization of climate change simulations over the Eastern Mediterranean. J Clim 22:1944–1961

Önol B, Bozkurt D, Turunçoğlu UU, Sen OL, Dalfes HN (2014) Evaluation of the twenty-first century RCM simulations driven by multiple GCMs over the Eastern Mediterranean-Black Sea region. Clim Dyn 42(7):1949–1965

Özdoğan M (2011) Modeling the impacts of climate change on wheat yields in Northwestern Turkey, Agriculture. Ecosyst Environ 141:1–12

Öztürk T, Altınsoy H, Türkeş M, Kurnaz ML (2012) Simulation of temperature and precipitation climatology for central Asia CORDEX domain by using RegCM 4.0. Clim Res 52:63–76

Öztürk T, Ceber ZP, Türkeş M, Kurnaz ML (2015) Projections of climate change in the Mediterranean Basin by using downscaled global climate model outputs. Int J Climatol 35:4276–4292

Pal JS, Giorgi F, Bi X, Elguindi N, Solmon F, Gao X, Rauscher SA, Francisco R, Zakey A, Winter J, Ashfaq M, Syed FS, Bell JL, Diffenbaugh NS, Karmacharya J, Konare A, Martinez D, da Rocha RP, Sloan LC, Steiner AL (2007) Regional climate modeling for the developing world: the ICTP RegCM3 and RegCNET. Bull Am Meteorol Soc 88(9):1395–1409

Palmer WC (1965) Meteorological drought. Research paper 45, U.S. Department of Commerce, 58 p

Peel MC, Finlayson BL, McMahon TA (2007) Updated world map of the Köppen-Geiger climate classification. Hydrol Earth Syst Sci 11:1633–1644

Russell RJ (1931) Dry climates of the United States: I climatic map. Univ Calif Publ Geography 5:1–41

Şen B, Topçu S, Türkeş M, Şen B, Warner JF (2012) Projecting climate change, drought conditions and crop productivity in Turkey. Clim Res 52:175–191

Sousa PM, Trigo RM, Aizpurua P, Nieto R, Gimeno L, Garcia-Herrera R (2011) Trends and extremes of drought indices throughout the 20th century in the Mediterranean. Nat Hazards Earth Syst Sci 11:33–51

Tatlı H (2007) Synchronization between the North Sea-Caspian pattern (NCP) and surface air temperatures in NCEP. Int J Climatol 27:1171–1187

Tatlı H, Türkeş M (2011) Empirical orthogonal function analysis of the Palmer drought indices. Agric Meteorol 151(7):981–991

Tatlı H, Türkeş M (2014) Climatological evaluation of Haines forest fire weather index over the Mediterranean basin. Meteorol Appl 21(3):545–552

Taylor KE, Stouffer RJ, Meehl GA (2012) An overview of CMIP5 and the experiment design. Bull Am Meteor Soc 93(4):485–498

Thornthwaite CW (1948) An approach toward a rational classification of climate. Geogr Rev 38:55–94

Topçu S, Şen B, Türkeş M, Şen B (2010) Observed and projected changes in drought conditions of Turkey. In: Options Méditerranéennes, series A, mediterranean seminars 2010, No. 95—Economies of drought and drought preparedness in a climate change context, CIHEAM, Paris, pp 123–127

Tunalıoğlu R, Durdu OF (2012) Assessment of future olive crop yield by a comparative evaluation of drought indices: a case study in western Turkey. Theor Appl Climatol 108:397–410

Türkeş M (1996) Spatial and temporal analysis of annual rainfall variations in Turkey. Int J Climatol 16:1057–1076

Türkeş M (1998) Influence of geopotential heights, cyclone frequency and Southern Oscillation on rainfall variations in Turkey. Int J Climatol 18:649–680

Türkeş M (1999) Vulnerability of Turkey to desertification with respect to precipitation and aridity conditions. Turk J Eng Environ Sci 23:363–380

Türkeş M (2010) Climatology and meteorology, First edn. Kriter Publisher, Publication No. 63, Physical geography series no. 1, ISBN: 978-605-5863-39-6, İstanbul, 650 p (in Turkish)

Türkeş M, Akgündüz AS (2011) Assessment of the desertification vulnerability of the Cappadocian district (Central Anatolia, Turkey) based on aridity and climate-process system. Int J Hum Sci 8:1234–1268

Türkeş M, Erlat E (2003) Precipitation changes and variability in Turkey linked to the North Atlantic Oscillation during the period 1930–2000. Int J Climatol 23:1771–1796

Türkeş M, Erlat E (2005) Climatological responses of winter precipitation in Turkey to variability of the North Atlantic Oscillation during the period 1930–2001. Theoret Appl Climatol 81:45–69

Türkeş M, Erlat E (2008) Influence of the Arctic Oscillation on variability of winter mean temperatures in Turkey. Theoret Appl Climatol 92:75–85

Türkeş M, Erlat E (2009) Winter mean temperature variability in Turkey associated with the North Atlantic Oscillation. Meteorol Atmos Phys 105:211–225

Türkeş M, Sümer UM, Demir I (2002a) Re-evaluation of trends and changes in mean, maximum and minimum temperatures of Turkey for the period 1929–1999. Int J Climatol 22:947–977

Türkeş M, Sümer UM, Demir İ (2002b) Re-evaluation of trends and changes in mean, maximum and minimum temperatures of Turkey for the period 1929–1999. Int J Climatol 22:947–977

Türkeş M, Koç T, Sarış F (2009) Spatiotemporal variability of precipitation total series over Turkey. Int J Climatol 29:1056–1074

Türkeş M, Kurnaz ML, Öztürk T, Altınsoy H (2011) Climate changes versus security and peace' in the Mediterranean macroclimate region: are they correlated? In: Proceedings of international human security conference on human security: new challenges, new perspectives, İstanbul, pp 625–639, 27–28 Oct 2011

Turunçoğlu UU (2015) Identifying the sensitivity of precipitation of Anatolian peninsula to Mediterranean and Black Sea surface temperature. Clim Dyn 44(7):1993–2015

UNCCD (1995) The united nations convention to combat desertification in those countries experiencing serious drought and/or desertification, particularly in Africa. Text with Annexes, United Nations Environment Programme (UNEP), Geneva

UNEP (1993) World atlas of desertification. United Nations Environment Programme (UNEP), London

van der Schrier G, Jones PD, Briffa KR (2011) The sensitivity of the PDSI to the Thornthwaite and Penman-Monteith parameterizations for potential evapotranspiration. J Geophys Res, 116:D03106. doi:10.1029/2010JD015001

Webb RS, Rosenzweig CE, Levine ER (1993) Specifying land surface characteristics in general circulation models: soil profile data set and derived water-holding capacities. Global Biogeochem Cycles 7:97–108

Wells N, Goddard S, Hayes MJ (2004) A self-calibrating palmer drought severity index. J Clim 17:2335–2351

Wilcock AA (1968) Köppen after fifty years. Ann Assoc Am Geogr 58:12–28

Climate Change and Soils

Selim Kapur, Mehmet Aydın, Erhan Akça, and Paul Reich

1 Introduction

Soil scientists have had a great interest in the last decade concerning the change in climate and its effects on the current and possible future changes in the soil. Globally averaged precipitation is projected to increase, but at the regional scale both increases and decreases have been reported. During the past 100 years, the average temperature in the world has risen about 0.74 °C.

Climate change is attributed directly or indirectly to human activity that alters the composition of the global atmosphere and ecosystem. The reliable data of greenhouse gas concentrations in the atmosphere show a positive trend, and for some gases, the concentrations have surpassed levels during the last few millennia. During the last 200,000 years CO_2 has increased from a low level of about 180 ppmv to a present-day level of about 360 ppmv with an annual increase of about 1.5 ppmv (IPCC 1996).

The GISS (Goddard Institute for Space Studies-NASA) model was used to simulate precipitation and temperature conditions for the synoptic stations under a twofold CO_2 increase scenario. The resultant data serve as input to the soil moisture and temperature model. Information on current and predicted soil moisture and temperature regimes were plotted on a map using the Geographic Information System (GIS).

The most significant problem that directly or indirectly affects the quality of life in Turkey will be the quantity and supply of water for agriculture and human consumption. Precipitation is not likely to increase in Turkey where the effects of climate change on soil water balance are of major concern as the increased temperature stimulates the evaporative demand of the atmosphere. As much of the country has limestone bedrock or is hilly and mountainous, water supply situation becomes precarious. With time, agriculture and human requirements for water will also increase placing high demands on the reduced water supply. Several sectors of agriculture will also experience changes as a result of climate change. Those cropping systems that depend on cool winters will either show lower productivity or be phased out. This specifically applies to wheat. The southern coastal plains will have to change from wheat to other warmer crops. Agriculture can and will adapt to these climatic changes as they have done when economics of production forced them to change. A national strategic plan to cope with the impending changes should include studies on long-term water needs, networks of data monitoring sites, assessments for environmentally friendly Agriculture, systematic research for alternative land use, and monitoring of land degradation with improvement of early warning indicators.

Climate change is a long-term issue with short-term risks for the agricultural production which is strongly dependent on soil resources. The updated report of IPCC (2013) indicates that increased temperature and droughts, and decreased precipitation, as well as declined soil moisture would dramatically influence agricultural production. The research highlights the importance of understanding the dynamics of soil processes when addressing climate change impacts on agriculture. Rapid soil responses to climate change (e.g., soil water, organic carbon, and erodibility) have been widely investigated. However, it is important that longer term processes (e.g., pedogenesis) are not ignored by the research community as they may have potentially important implications for long-term agricultural land use. Perhaps the greatest impact of climate change on soils will arise from climate-induced changes in land use and management

S. Kapur (✉)
Department of Soil Science and Plant Nutrition, University of Çukurova, Adana, Turkey
e-mail: skapur@cu.edu.tr

M. Aydın
Kangwon National University, Chuncheon, South Korea

E. Akça
School of Technical Sciences, University of Adıyaman, Adıyaman, Turkey

P. Reich
USDA Natural Resources Conservation Service, Washington, DC, USA

(Rounsevell et al. 1999). Research on the impacts of climate on soil and its genesis continues to date, but it has been overtaken by research that focuses on the effects of soil on the climate. Such research began in the early 1980s when it was realized that the soil was a source and sink for greenhouse gases and that these gases affect the climate (Hartemink 2014).

In 1998, the Global Climate Change Convention was agreed to by most nations of the world at the United Nations sponsored meeting in Kyoto, Japan. The Convention emphasizes the urgency of preserving and protecting global environment and is a challenge to human ingenuity to find solutions to these problems. The ability to predict the near future is still speculative and consequently, the spectrum of opinion ranges from doomsday scenarios to formidable benefits. Current events portray a warming trend. These include the cracking and melting of polar ice flows, the receding of glaciers in northern latitudes, the erratic behavior of precipitation with more frequent storms, tornadoes, and hurricanes and all manifesting outside the normal periods, and a general rise of a few tenths of degrees in surface temperatures. These changes have been described by some as the normal aberration in earth's climate when evaluated on a geological scale.

Other than CO_2, the other gases that are also infrared absorbers and thus contributors to warming include methane, nitrous oxide, chlorofluorocarbons, and ozone. Some of these have an important agricultural origin, such as methane from waterlogged soils and enteric fermentation in ruminants, and soil emissions of N_2O from nitrogen fertilizers and biomass burning.

The effects of warming will not be uniform at all points on the earth's surface, though the general trend is that it will be significantly warmer at higher latitudes. Changes in surface temperatures will influence global wind patterns and so different kinds of shifts may be anticipated. Site-specific studies are required to evaluate this. Increased temperature will result in enhanced evapotranspiration and depending on the new pattern of precipitation, the net change could be anything from severe drought to excess water. As a result, the site-specific impact will vary. As ecosystem conditions have changed, floral and faunal compositions will also change. Plants require CO_2 for photosynthesis, some CO_2-induced fertilization enhances vegetative growth with different plants responding differently (Allen et al. 1996). This "negative feedback" is one of the beneficial effects, however, there is a general feeling that "crop-zones" will be shifted northwards in northern latitudes.

The impacts on animal, particularly human, life are less well documented. The effect of heat stresses can have several negative effects including perhaps greater incidence of forms of cancer. Due to changes in ambient conditions, insect populations will seek more favorable habitats, which imply a different assemblage of pests and diseases for agriculture. The net effect again is that agriculture in some localities will be devastated while others may be reaping bumper harvests. The great uncertainty in all these is according to Rosenzweig and Hillel (1998), "how much warming will occur, when and at what rate, and according to what geographical and seasonal patterns?"

The other unknown parameter in all the scenarios is that there are very few studies with respect to site-specific changes. Such studies are fraught with more difficulties and are more speculative. Several leading studies with respect to impact on soils (Scharpenseel et al. 1990; Lal et al. 1995; Bazzaz and Soinbroek 1996; Rounsevell et al. 1999) provide a good overview of expected changes. The present study attempts to evaluate the impact of doubling of atmospheric CO_2 on the soil resource conditions. Unlike previous studies, we look at current soil moisture and temperature conditions and changes in these subsequent to global doubling of CO_2. Soil processes are then inferred from the changes depicted by the model. This chapter is based on the material published by Eswaran et al. (1998) and updated with more recent information.

The study uses Turkey as a location for the current assessment. Soil moisture and temperature conditions of the country are estimated using a water balance model with average monthly climatic data from 45 synoptic meteorological stations. A soil map of the country (Oakes 1957; updated by Dinç et al. 1997 and by the map refined over the soil map of the TOPRAKSU-TRGM 2013 for this book) and the many previous studies on soil genesis and classification of Turkey's soil resources provide an adequate knowledge base for this evaluation. Current ideas of pedogenesis are used to link soil climate features to changes in soil conditions.

A General Circulation Model (GCM) developed by the NASA-Goddard Institute for Space Studies (GISS) which simulates climate using algorithms that relate climate to its variables was utilized. Such models are complex and validation studies point to their reliability (Kalkstein 1991). A good example is their ability to predict El Nino Southern Oscillation (Meehl 1990). Some of the predictions of GCMs include:

- The high latitudes in the northern hemisphere will experience the highest increase in temperature, with a concomitant warmer winter,
- Mid-latitudes will have extended drought periods, soil moisture stress in winter will become pronounced and for longer periods,
- Land masses will warm up faster than oceans,
- Due to melting of polar ice caps, sea level will rise and the IPCC (1996) best guess is that the rate will be about 5 cm per decade.

The GISS model was used to simulate precipitation and temperature conditions for the synoptic stations under a twofold CO_2 increase scenario. The resultant data serve as input to the soil moisture and temperature model. Information on current and predicted soil moisture and temperature regimes were plotted on a map using G1S. The soil moisture model also provided information on length of the period when the soil moisture control section is dry, partly dry, or moist and this information is used to assess impacts on soil processes. Here, in addition to GISS datasets, the projections of a high-resolution (about 20 km grid) GCM developed at the Meteorological Research Institute (MRI) of Japan were included for complementarities.

2 Soil Resources of Turkey

The soil survey of Turkey was undertaken from 1965 to 1970 for the purpose of preparing a general soil management plan for land use and particularly, to combat soil erosion. Wind and water erosion were considered to be the major land degradation hazards. The plan included management unit descriptions of slope, soil depth, erosion level, dominant land use, and other properties such as stoniness, water table, drainage, texture and soil moisture contents, and protective measures. Thus, the base information for this task—the soils—was classified according to Baldwin et al. (1938) by the General Directorate of Rural Services (GDRS 1987). The zonality concepts of Baldwin et al. (1938) classification yielded five different soil zones in the study conducted by GDRS (1987) based on the climatic and vegetative cover (Fig. l). Figure 1 comprises the recent equivalents (Soil Survey Staff 1996) of the soil classes according to Baldwin et al. (1938) classification.

Dinç et al. (1997) revised the earlier soil map and provided the first approximation comprising soil order associations according to the Soil Survey Staff (1975). The major additions consist of more recent information from south east Turkey soil surveys (Dinç et al. 1991; GDRS 1990, 1991, 1992, 1994, 1996, 1997) and the State Farm Land Use reports throughout Turkey (Turkish State Farms-TSF 1980, 1992, 1995a, 1995b). The Aridisols in this revised map were mainly Inceptisols and/or Mollisols according to Soil Survey Staff (1996). Further modifications were undertaken on the Dinç et al. (1997) soil map (see chapter of Soil Geography) for use in the chapters of this book and this chapter (Fig. 2). The widely distributed Aridisols in southeastern and central Turkey (Western and Southern parts) are to be classified as Calcisols or Cambisols in the recently revised map of this book (Fig. 2). The soil moisture regime codes calculated by Newhall (1972) appearing in Fig. 3 and Table 1 were utilized in order to determine the probable pedo-climatic changes predicted to occur under a twofold CO_2 increase scenario in the future as seen in Table 2.

3 Changes in Climatic Conditions

Globally, the average precipitation is projected to increase but at the regional scale, both increases and decreases have been reported. During the past 100 years, the average temperature in the world has risen 0.74 °C (IPCC 2007). In the arid and semiarid regions, climate is dramatically changed from low rainfall to extreme events. The frequency and intensity of droughts in parts of Asia has been observed to increase in the recent decades.

The projected decrease in precipitation is concentrated in the coastal areas of southern Turkey. This decrease in precipitation is mainly projected in the winter and spring. On the other hand, the annual precipitation is projected to increase in the future over the eastern part of the Black Sea and northeastern regions of the country (Fig. 4). This increase in precipitation is projected mainly in fall. Year-to-year variations in precipitation are larger over most areas where an increase in mean precipitation is projected. At the end of the twenty-first century, the potential evaporation generally increases in Turkey almost throughout the years due to warmer surface temperatures. Therefore, even in the areas where precipitation increases, an increase in evaporation may over compensate for the increase in precipitation leading to the decreased surface runoff. For this reason, trends in the streamflow are not always the same as that of the precipitation (Kitoh et al. 2008). Actual evaporation has a different seasonal cycle to that in precipitation, with its peak in May to June. The precipitation minus evaporation (P-E) is an important indicator in the study of long-term climate changes of the moisture. The advantage of using the P-E term is that it shows the total moisture sinks or sources as determined by the sign of P-E (Alpert and Jin 2017).

The projected changes in P-E, thus water availability is significantly negative throughout the year, resulting in less surface runoff and less soil moisture in the future. Runoff and soil moisture will decrease in the future in all months in most parts of Turkey, where decreasing precipitation is responsible for less surface runoff and less soil moisture. Over the north east corner of the country, increasing evaporation detains increasing precipitation, and thus changes in runoff and soil moisture become smaller. The future climate would be characterized with a longer dry period but with a more intense rainfall season in western Turkey (Kitoh 2017).

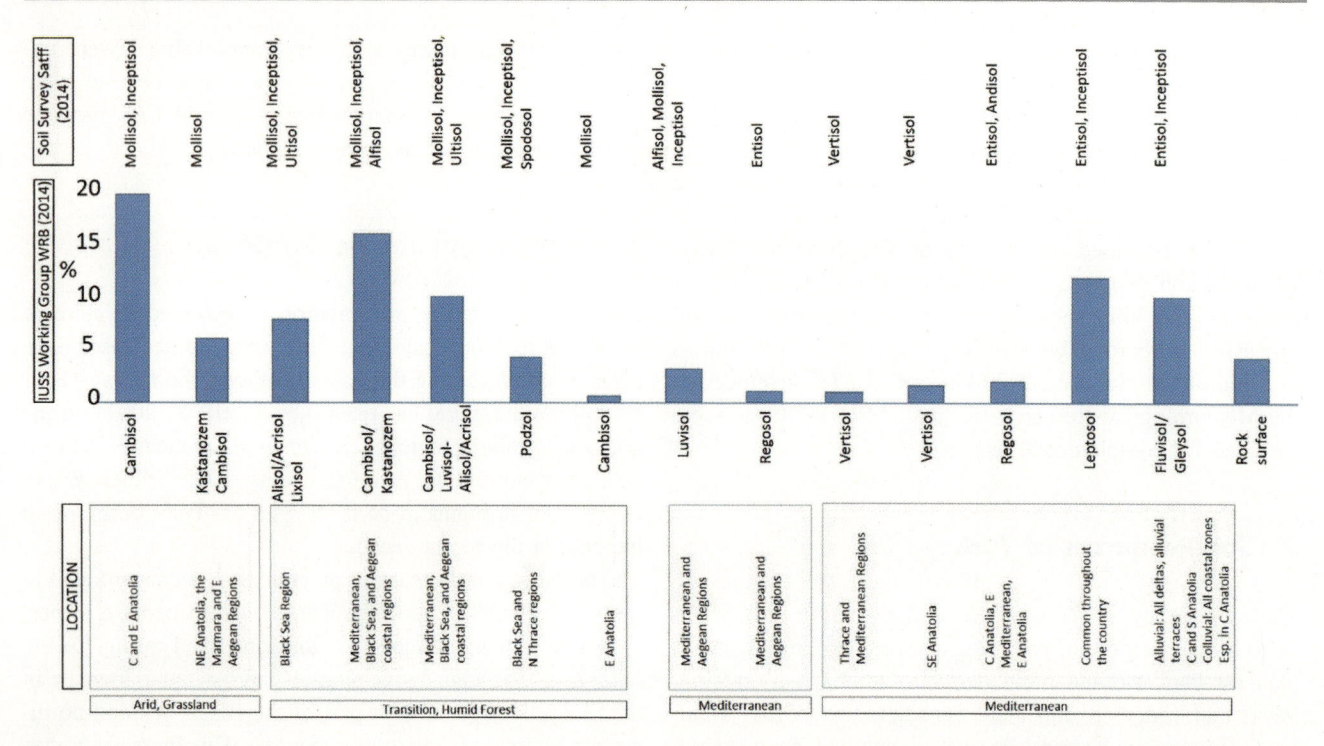

Fig. 1 Distribution of Great Soil Groups of Turkey with their equivalents in Soil Survey Staff (2014) and the IUSS Working Group WRB (2015)

4 Soil Climate Conditions

Table 2 presents several parameters that characterize atmospheric and soil climate conditions for current conditions and for a twofold CO_2 increase scenario. The annual precipitation and evapotranspiration data for the double CO_2 scenario are estimated with the GISS model. The Mean Annual Soil Temperature (MAST) is estimated for a depth of 50 cm. The Moisture Stress Severity Index (MSSI) which varies from 0 (no stress) to 1 (maximum stress) is estimated by the water balance model based on the number of days that the soil moisture is held at tensions greater than 15 kP. The length of the Growing Season (GS) is estimated based on the period for adequate moisture for normal plant growth. Terms for the soil moisture regimes are defined in Soil Taxonomy (Soil Survey Staff 1996).

Adana and İzmir represent stations on the Mediterranean coastal areas. The surface temperature is estimated to increase by 4.2 °C based on the average of projections of three different GCMs by 2100 (Yano et al. 2007). Similarly, there is a 4 °C increase in soil temperature subsequent to doubling of CO_2, which results in a 20–25% increase in evapotranspiration. These results are consistent with the published data for the physical processes controlling evaporation (Aydın et al. 2008). With increased moisture stress in summer, the length of the growing season is reduced.

Central Turkey, as represented by Konya, becomes slightly drier. The central part of the country, being generally at higher elevation than the coastal regions, experiences only small changes. The eastern part of the country appears to become more humid. Muradiye on the east changes from a weak Aridic to a Typic Xeric SMR. The length of the growing season increases from 81 to 106 days, which is due to the increased amount of rain that falls here. Much of the Black Sea Coast, on the other hand, becomes drier. The stations Ünye and Bartın (Table 2) illustrate this. A marked decrease in the length of the growing season is observed and this is due to increased dry periods during summer.

The change in the western part of the country as illustrated by the station Edirne is different from the rest of the country. The winters are milder but with more rains and the summers are shorter. The length of the growing season of Edirne increases from 172 days to about 259 days based on data projected by the GISS.

Table 2 illustrates the variations in changes that could be expected in different parts of the country. A systematic analysis of many climatic stations is necessary to demarcate areas for comprehensive studies and the development of tactical developmental actions.

5 Changes in Soil Processes by Climate Change and Human Intervention

Many of the soil properties are quite resistant in relation to short-term variations in climate with effects that are difficult to detect, due to the great impact of land use and land use changes, especially if we consider the great spatial

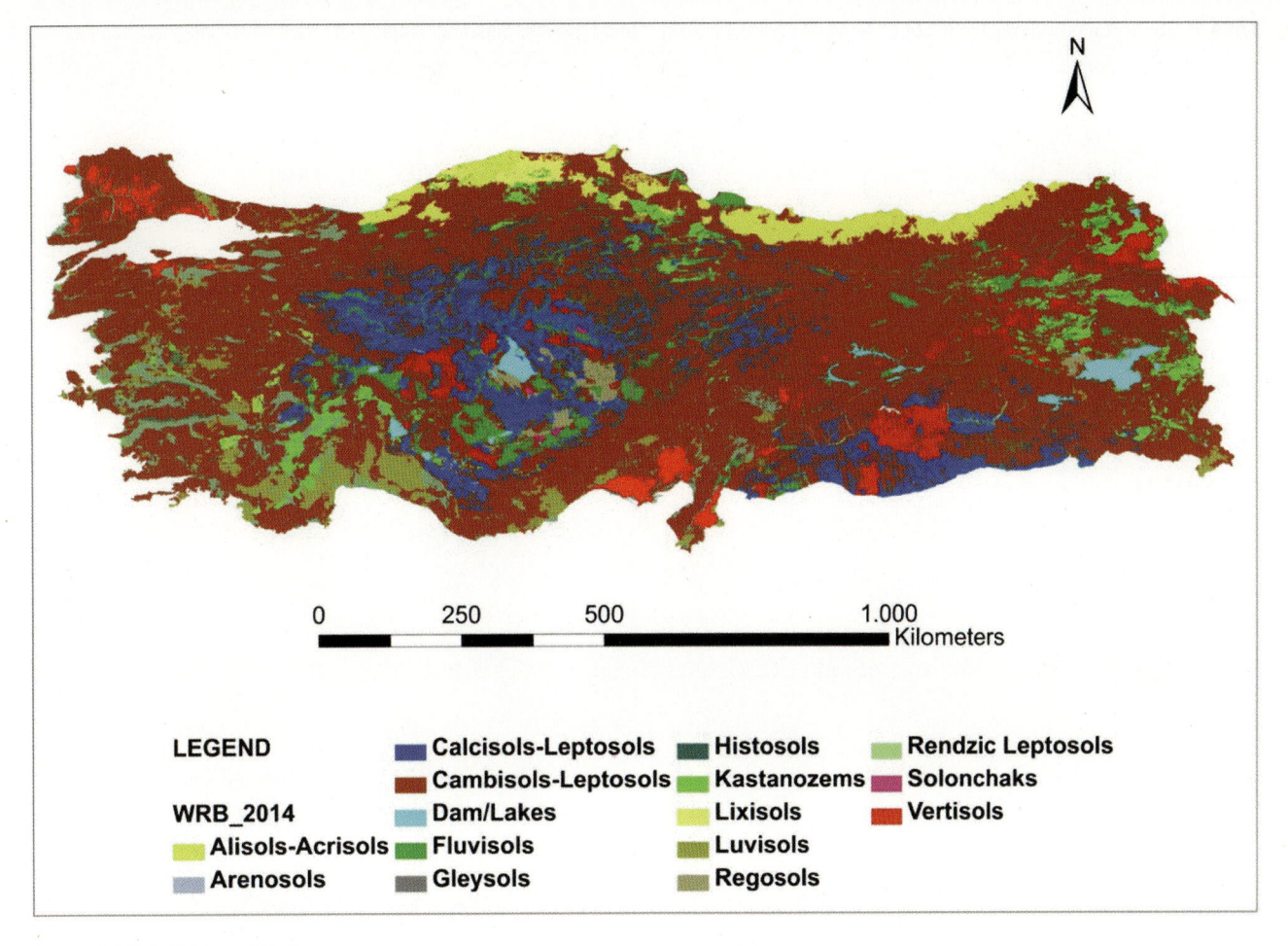

Fig. 2 WRB Soil Map of Turkey

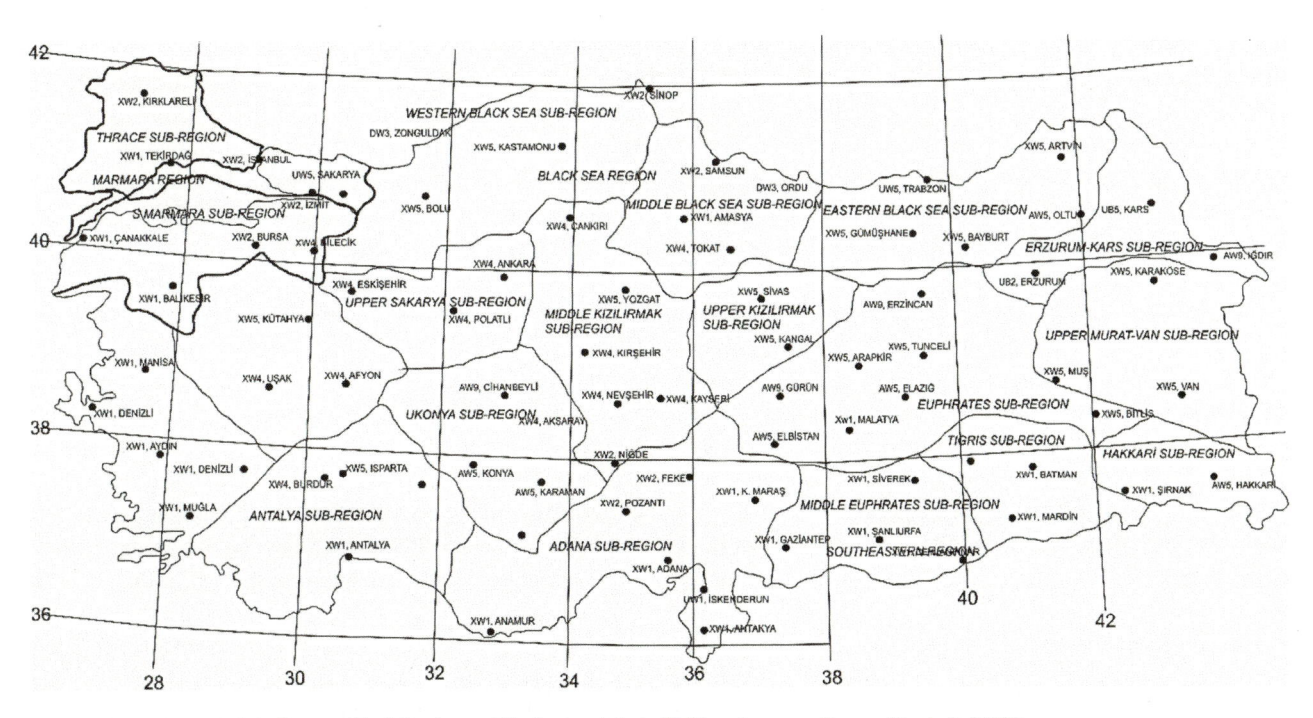

Fig. 3 The Major and Sub-Geographical Regions of Turkey and their SMR codes according to Newhall (1972)

Table 1 The soil moisture regime codes of Newhall (1972) for Arid, Mediterranean, Semiarid Tropical and Semiarid *Temperate regions*

	STR	Tropical				Temperate			
		Isomega-thermic	Isohyper-thermic	Iso-thermic	Iso-mesic	Mega-thermic	Hyper-thermic	Thermic	Mesic
Arid	*SMR*								
	Extreme Aridic	AM1	AT1	AT4	AT7	AG1	AW1	AW4	AW7
	Typic Aridic	AM2	AT2	AT5	AT8	AG2	AW2	AW5	AW8
	Weak Aridic	AM3	AT3	AT6	AT9	AG3	AW3	AW6	AW9
Mediterranean	Dry Xeric							XW1	XW4
	Typic Xeric							XW2	XW5
	Udic Xeric							XW3	XW6
Semiarid Tropical	Aridic Tropustic	SM1	ST1	ST4	ST7				
	Typic Tropustic	SM2	ST2	ST5	ST8				
	Udic Tropustic	SM3	ST3	ST6	ST9				
Semiarid Temperate	Xeric Tempustic					UG1	UW1	UW4	UW7
	Wet Tempustic					UG2	UW2	UW5	UW8
	Typic Tempustic					UG3	UW3	UW6	UW9

variability of soils (Vallejo et al. 2005). In the next few decades, no major changes in the assemblage of soils can be expected. However, there are a few processes which though may have a negligible effect on soil morphology and classification can have effects on crop production. A major difficulty in such an assessment is that changes induced by climate change cannot be differentiated from those resulting from human-induced land degradation processes. It is therefore difficult, with the knowledge available, to determine the sensitivity of the soils to presently perceived climate changes in an accurate and quantitative manner, but some examples can be given in which these relationships between climate and soil are evident (Vallejo et al. 2005). In fact, both sets of processes are complementary and feed one upon the other. This set of mutually enforcing processes accelerates desertification.

Studies on climate change indicate that erratic precipitation and intense storm may characterize the rainy period. This implies greater water erosion in the wet season with wind erosion in the dry summers for some parts of Turkey. As ground cover is already poor in this winter-rainfall pail of the world and coupled with overgrazing by small ruminants, rampant erosion is not abnormal. Removal of the already small organic matter content of many of the soils through erosion reduces not only the ability to withstand erosion but also reduces the water holding capacity of the soil. A general decline in soil quality may be expected.

In closed basins, salinization is accelerated. Even on gently undulating landscapes, salt build-up can be expected with increased aridity. Turkey is blessed with a large supply of fresh water and consequently the potential to irrigate large areas of land, such as in the GAP region. The propensity of the soils to salinize is very high and becomes higher with increasing aridity. The GAP area has historically been dominated by a system characterized by grazing via small ruminants and rain-fed grain crops of valley bottoms. When a source of water was available (tube wells or aquifers), local irrigated crops were developed and a larger variety of crops (cotton, chickpeas, lentil, and tree crops) were grown. The dramatic increases in performance on irrigation are evident. However, the stability of the system is uncertain. After harvest of grains, the straw is removed as fodder for animals, and small ruminants graze the remaining stubble. With the establishment of irrigation facilities and accompanying subsidies available, there is excessive use of agro-chemicals to maintain the high production. Water-logging and development of salinity have already been experienced in some irrigated areas in Turkey. Both these are yield depressing factors which are perhaps the greatest hazards of irrigation. Not all soils have the same propensity to develop salinity and thus an important task is to develop maps of salinity hazard. An equally important task is to monitor the soils for salinity or develop early warning indicators. In this context, salinity monitoring studies were carried out in the plain (the largest basin of the GAP area with irrigated agriculture) by the Harran University and the Ministry of Food, Agriculture and Livestock of Turkey. Salinity monitoring is coupled with periodic soil surveys and soil/salinity mapping of the

Table 2 Soil climate conditions at selected stations, current and under a twofold CO_2 scenario (Eswaran et al. 1998)

	Soil moisture regime	Precipitation (mm)	Evapotranspration	Mean annual soil temp (°C)	Moisture stress severity index	Length growing season (days)	Soil moisture regime	Precipitation (mm)	Evapotranspration	Mean annual soil temp (°C)	Moisture stress severity index	Length growing season (days)
	Current						*Under twofold CO_2 scenario*					
Adana	Dry Xeric	611	989	20.6	0.37	226	Xeric Tempustic	624	1256	24.6	0.45	191
İzmir	Dry Xeric	648	907	19.3	0.35	256	Xeric Tempustic	587	1167	23.3	0.42	234
Gaziantep	Dry Xeric	577	819	16.3	0.36	149	Dry Xeric	546	1036	20.7	0.4	217
Edirne	Typic Xcric	592	773	15.4	0.29	172	Dry Xeric	636	979	19.8	0.32	259
Konya	Weak Aridic	315	695	13.5	043	84	Weak Aridic	276	879	17.5	0.54	73
Erzurum	Weak Aridic	327	632	11.6	0.37	86	Weak Aridic	366	784	15.9	0.46	103
Muradiye	Weak Aridic	457	619	10.6	0.33	81	Typic Xeric	487	769	14.9	0.39	106
Ünye	Typic Udic	1092	753	16.2	0.02	360	Dry Tempudic	1137	925	20.5	0.14	287
Bartin	Typic Udic	1072	730	15.2	0.02	290	Weak Tempustic	1102	905	19.5	0.2	284

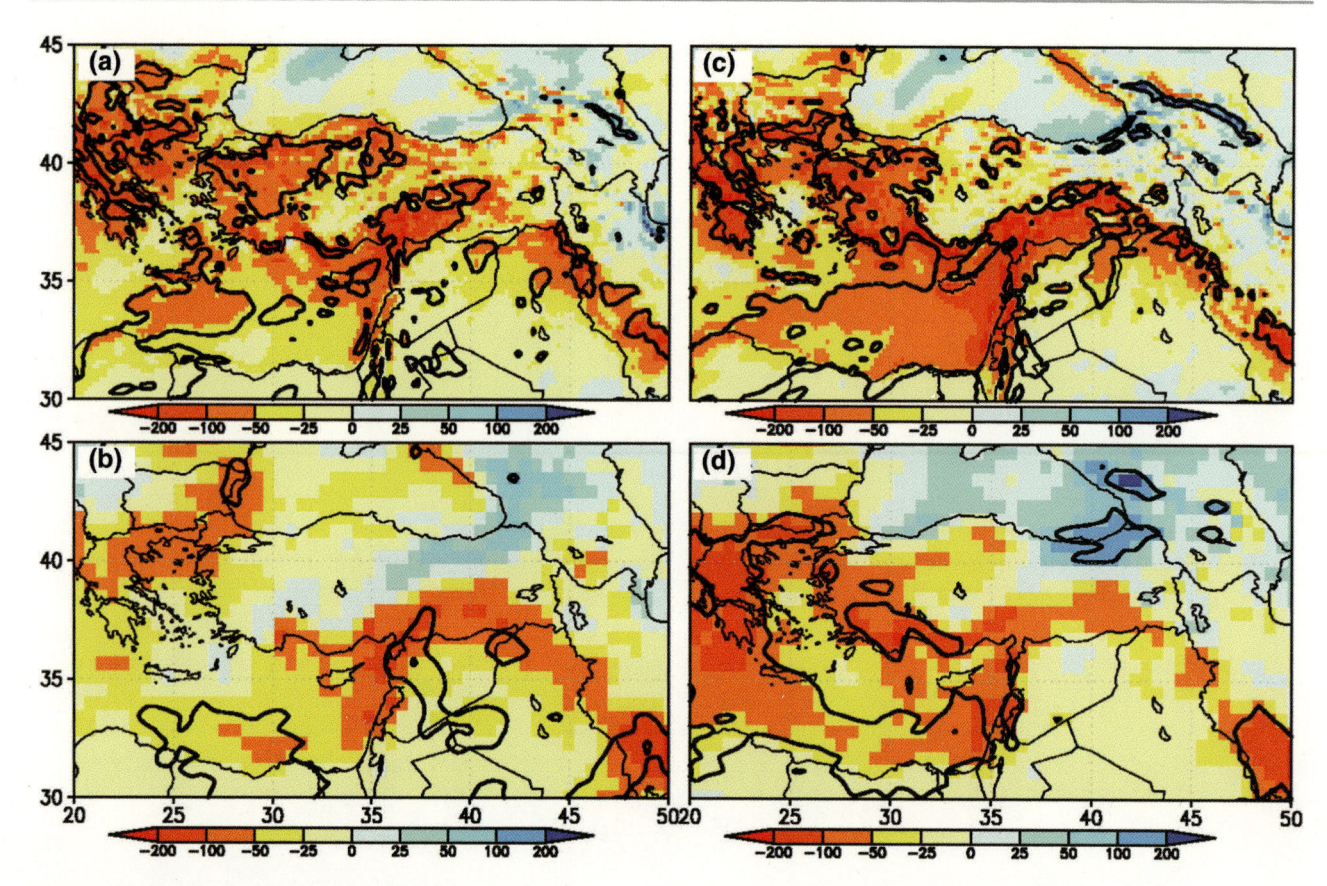

Fig. 4 The changes in annual mean precipitation (mm) projected by MRI with a grid of **a** 20 km and **b** 60 km until the year 2039, and **c** 20 km and **d** 60 km until the year 2099 (Kitoh 2017)

area in order to determine the changing conditions of the soils that would occur after irrigation and the subsequently established underground drainage infrastructure in especially Akçakale, the southern and the lowermost part of the Harran plain.

Soil change is being determined after amelioration in the salt contents and Bnt/Bt horizons shift to the Bt throughout the Akçakale area (see Solonetz-like soils in this book). Nonetheless, the area has also been prone to wind and water erosion before the present irrigated system was established. However, the contemporary ground cover of the irrigated crops would remain for longer periods and reduce both kinds of erosion. In contrast to this advantage, irrigation bears a major snag where excess water use would frequently carry high silt loads which would hamper seedling emergence by crusting.

The drastic increase in the development of the greenhouse cover (by glass and low- tunnel plastic greenhouse constructions) for high-income generating vegetables especially on leveled and/or unlevelled sand dune areas (Arenosols) and coastal Luvisol ecosystems is also a major cause of soil change. The shift, from the natural vegetation of the sand dunes and from the traditional olive tree crops of the Luvisol ecosystems to the contemporary use, is prone to decrease the sequestration of carbon and increase nitrogen. This would cause the loss of soil quality in the fragile sand dune and Luvisol ecologies and ultimately change the climate. Consequently, if all the above-mentioned inappropriate soil use proceeds, as it is, it seems to be inevitable due to human greed to earn more than care for the environment, a more rapid shift to a change in global climate should be expected for the near future.

6 Impacts of Climate Change

The study shows that the whole country undergoes changes in precipitation and a general increase in temperature resulting in an increase in the evapotranspiration. The relative magnitude of the changes depends on the location. There is a net desiccation over the whole country as reflected in the longer and drier summers. The southern coast will have longer and drier summers, the winter rainfall will not be adequate to counteract the enhanced evapotranspiration of the summer. The rainfall along much of the Black Sea coast will diminish resulting in some soil moisture deficit conditions in summer. The period of moisture deficit is generally not long enough to impact crop performance though in some years, supplemental irrigation will become necessary. The western part of the country will experience general slightly more favorable conditions in terms of soil moisture based on data projected by the GISS. From the current deserted conditions, situation will improve to growing at least one annual crop in most years. The changes in the central part of the country are not very significant. A slight increase in temperature and an accompanying increase in evapotranspiration will be experienced but this will not change the cropping systems.

The most significant problem that directly or indirectly affects the quality of life in Turkey will be the quantity and supply of water for agriculture and human consumption. As much of the country has limestone bedrock or is hilly and mountainous, water supply situation becomes precarious. With time, agriculture and human requirements for water will also increase placing high demands on the reduced water supply. Snow on the higher elevations is expected to be lower in quantity and much of the snowmelt in spring may be expected to be lost through evapotranspiration. This affects the general hydrology of the catchments. The coupled effects of lesser water supply and higher evapotranspiration on the basins of Turkey are expected to enhance aridity.

As the soil moisture supply during the year is expected to decrease in much of the country, large areas of marginal grazing lands that are currently under nomadic grazing will be degraded later. Uncontrolled grazing is one of the major causes of desertification in the semiarid parts of the world. Greater intensities of storms as a result of climate change in some parts of the country prompt higher rates of wind and water erosion, accelerating desertification. In some parts of the country, the onset of desertification is already evident. Desertification becomes intensified in almost the whole of Turkey as a result of climate change.

Several sectors of agriculture will also experience changes consequent to climate change. Those cropping systems that depend on cool winters will either show lower productivity or be phased out. This specifically applies to wheat. The southern coastal plains will have to change from wheat to other winter's crops. In other words, the plant water stress would cause the irrigation in the presence of water supply to maintain the wheat production for the future (Kapur et al. 2017). Agriculture can and will adapt to these climatic changes as they have done when economics of production forced them to change. In the absence of water supply, agricultural productivity will decrease. Larger and more efficient irrigation projects will be needed and this makes the GAP project relevant and a sensible long-term strategy. In addition to irrigation, new crops or new species must be introduced for the changing conditions. Irrigation technology must be enhanced to reduce or prevent land being degraded through salinization.

Although the subject of this paper is land resources, the impact of land use on the marine resources cannot be ignored. The Mediterranean Sea is different from others due to its closed nature that limits the natural exchange and circulation of waters. The deltas of the large rivers that flow into the Mediterranean Sea control to some extent its biodiversity. In recent years, the amount of fresh water that is discharged into the sea is declining. The nutrient content of the river waters is also declining while the amount of pollutants has significantly increased. The quality and quantity of water entering the sea has changed, affecting the biotic composition with consequent impact on climate and the land itself. Demographic pressures along the coast with the increased recreational and commercial activities bring other stresses to the system. Managing urban wastes and treating wastewater places additional burdens on the land and environment. These present-day problems may be expected to exacerbate with climate change.

The land–sea–atmosphere link, though not obvious, is most relevant in the Mediterranean region. Climate warming, in the Mediterranean context, requires a careful evaluation of the quantity and quality of water emanating from land resources, particularly in large-scale irrigation projects. Since the late 70s, through the initiatives of the United Nations Environment Program (UNEP), regional organizations and national efforts, considerable progress has been made on conservation of the resources of the Mediterranean Sea (Tekinel 1993). However, much less work has been done on land-based pollution and the impact of agriculture on the marine ecosystem and probably studies on impact of global warming on this ecosystem have not received the attention it deserves.

7 Conclusions

Most global climate change studies have focused on regional impacts. Many of the generalizations for regions are valid but the variations within a region and more specifically, within a country must be evaluated to develop strategic plans. The use of good land management practices, as currently understood, provides the best strategy for adaptation to the impact of climate change on soils. However, it appears likely that farmers will need to carefully reconsider their management options, and land use change is likely to result from different crop selections that are more appropriate to the changing conditions (Rounsevell et al. 1999). The aggregate impact may not look serious or even worthy of research and developmental efforts. However, site-specific evaluations show the disparity between locations calling for location specific actions (Kapur et al. 2017). The other feature of this process is that the changes are imperceptible and the manifestations are slow. The cumulative effect may be devastating. Finally, not all changes may be negative. As indicated, some parts of the country may experience changes which are considered an improvement, such as the eastern part of the country.

A national strategic plan to cope with the impending changes should include:

1. Studies on the long-term water needs of the country, region by region, to ensure adequate supplies to counteract climatic irregularities,
2. A comprehensive network of data monitoring sites, this will enable an assessment of the nation's natural resources, their conditions, and changes,
3. A better assessment of environmental health and enhancing the technology of land use so that agriculture is made environmentally friendly,
4. Systematic research on crops, cropping systems, and farming systems to provide options for alternative land uses,
5. Monitoring of land degradation and development of early warning indicators to signal when limits exceed threshold.

Acknowledgements The authors are highly indebted to the everlasting leadership of the deceased Dr Hari Eswaran on this and all occasions since the 1990s.

References

Allen IH, Baker JT, Boote KJ (1996) The CO_2 fertilization: higher carbohydrate production and retention as biomass and seed yield. In: Bazzaz F, Sombroek W (eds) Global climate change and agricultural production -Direct and indirect effects of changing hydrological, pedological and plant physiological processes- Rome: FAO, and Chichester. Wiley, New York, pp 65–100

Alpert P, Jin F (2017) The atmospheric moisture budget over the eastern mediterranean based on the super high-resolution global model-effects of global warming at the end of 21st century. In: Watanabe T, Kapur S, Aydın M, Kanber R, Akça E and Brauch HG (eds.) Climate change impacts on basin agro-ecosystem (tentative). Proposed Book for Springer Hexagon Series (under preparation)

Aydın M, Yano T, Evrendilek F, Uygur V (2008) Implications of climate change for evaporation from bare soils in a Mediterranean environment. Environ Monit Assess 140:123–130

Baldwin M, Kellogg CE, Thorpe J (1938) Soil classification. In: Soil and man, USDA agriculture yearbook

Bazzaz F, Sombroek W (1996) Global climate change and agricultural production: direct and indirect effects of changing hydrological, pedological and plant physiological processes—Rome: FAO, and Chichester. Wiley, New York

Dinç U, Şenol S, Sayın M, Kapur S, Çolak AK, Özbek H, Kara EE (1991) The physical, chemical, and biological properties and classification- mapping of soils of the Harran plain. In: Dinç U, Kapur S (eds) Soils of the harran plain. TÜBİTAK-TOAG-534. Ankara, pp 1–10

Dinç U, Şenol S, Kapur S, Cangir C, Atalay İ (1997) Soils of Turkey. University of Çukurova, Faculty of Agriculture Pub. No. A-12, 78 p (in Turkish)

Eswaran H, Kapur S, Reich P, Akça E, Şenol S, Dinç U (1998) Impact of global climate change on soil resource conditions: a study of Turkey. In: M. Şefik Yeşilsoy International Symposium on Arid Region Soils (YISARS), THAEM, Menemen-İzmir, Turkey, pp 1–13. 21–24 Sep 1998

GDAR-TRGM (2013) General directorate of agricultural reform, ministry of food, agriculture and livestock. Soil database for Turkey. In the GIS Media. Ankara

GDRS (1987) The general soil management planning of Turkey. Publication of GDRS. Ankara, 120 p (in Turkish)

GDRS (1990) The soils of the Harran plain, SE Turkey. Report of the general directorate of rural services, Survey and Project Department, Ankara, 337 p (in Turkish)

GDRS (1991) The soils of the Şanlıurfa—Suruç plain, SE Turkey. Report of the general directorate of rural services, Survey and Project Department, Ankara, 300 p (in Turkish)

GDRS (1992) The soils of the Şanlıurfa—Hilvan plain, SE Turkey. The general directorate of rural services, Survey and Project Department, Ankara, 250 p (in Turkish)

GDRS (1994) The soils of the Şanlıurfa—Hilvan 2 plain, SE Turkey. Report of the general directorate of rural services, Survey and Project Department, Ankara, 220 p (in Turkish)

GDRS (1996) The soils of the Kemlin-Adıyaman plain, SE Turkey. Report of the general directorate of rural services, Survey and Project Department, Ankara, 240 p (in Turkish)

GDRS (1997) The soils of the Kahta-Adıyaman plain, SE Turkey. Report of the general directorate of rural services, Survey and Project Department, Ankara, 270 p (in Turkish)

Hartemink A (2014) On the relation between soils and climate. 20th World Congress of Soil Science. Jeju, South Korea, 8–13 June 2014

IPCC (1996) Climate change 1995. The science of climate change. In: Houghton JT, Meira Filho IG, Callander BA, Harris N, Katenberg A and Maskell K (eds) Intergovernmental panel on climate change 1995. Cambridge University Press, Cambridge, 339 p

IPCC (2007) Summary for policymakers. In: Solomon S, Qin D, Manning M, Chen Z, Marquis M, Averyt KB, Tignor M and Miller HL (eds) Climate change 2007: the physical science basis. Contribution of Working Group I to the Fourth Assessment Report of the Intergovernmental Panel on Climate Change. Cambridge University Press, Cambridge, UK, and New York, USA

IPCC (2013) Summary for policymakers. In: Stocker TF, Qin D, Plattner G-K, Tignor M, Allen SK, Boschung J, Nauels A, Xia Y, Bex V and Midgley PM (eds) Climate change 2013: the physical science basis. Contribution of Working Group I to the Fifth Assessment Report of the Intergovernmental Panel on Climate Change. Cambridge University Press, Cambridge, UK, and New York, USA

IUSS Working Group WRB (2015) World reference base for soil resources 2014, update 2015 international soil classification system for naming soils and creating legends for soil maps. World Soil Resources Reports No. 106. FAO, Rome, 203 p

Kalkstein LS (1991) Global comparisons of GCM control runs and observed climate data. U.S. Environmental protection agency office of policy, planning, and evaluation. 2IP- 2002. Washington DC, 251 p

Kapur B, Aydın M, Yano T, Koç M (2017) Impacts of climate change on wheat production in the mediterranean region (Invited Chapter). In: Watanabe T, Kapur S, Aydin M, Kanber R, Akça E and Brauch HG (eds) Climate change impacts on basin agro-ecosystem (tentative). Proposed Book for Springer Hexagon Series (under preparation)

Kitoh A (2017) Climate change projection over Turkey with high-resolution GCM. In: Watanabe T, Kapur S, Aydin M, Kanber R, Akça E and Brauch HG (eds) Climate changeImpacts on basin agro-ecosystem (tentative). Proposed Book for Springer Hexagon Series (under preparation)

Kitoh A, Yatagai A, Alpert A (2008) First super-high-resolution model projection that the ancient Fertile Crescent will disappear in this century. Hydrol Res Lett 2:1–4. doi:10.3178 HRL.2.1

Lal R, Kimble J, Levine E, Stewart BA (1995) Soils and global change. Advances in Soil Science. CRC Lewis Publishers, Boca Raton, Fl

Meehl GA (1990) Seasonal cycle forcing El Nino Southern Oscillation in a global, coupled ocean-atmosphere GCM. J Clim 3:72–98

Newhall F (1972) Calculation of soil moisture regimes from climatic records. Unpublished. Soil Conservation Service, USDA. Rev. 4. Washington, D.C

Oakes H (1957) The soils of Turkey. Ministry of agriculture. Soil conservation and farm irrigation division, Ankara, Turkey. Pub. No. I, 180 p

Rosenzweig C, Hillel D (1998) Climate change and the global harvest. Oxford University Press, New York, 324 p

Rounsevell MDA, Evans SP, Bullock P (1999) Climate change and agricultural soils: impacts and adaptation. Clim Change 43(4):683–709

Scharpenseel HW, Schemaker M, Ayoub A (1990) Soils on a warmer earth. Effects of expected climate change on soil processes, with emphasis on the tropics and sub- tropics. Developments in soil science 20, Elsevier Amsterdam, 274 p

Soil Survey Staff (1975) Soil taxonomy: a basic system of soil classification for making and interpreting soil surveys. Soil Conservation Service, USDA Agr. Handbook. No. 436. Washington DC

Soil Survey Staff (1996) Keys to soil taxonomy. 7th edn. USDA NRSC, US Government Printing Office. Washington DC, 644 p

Soil Survey Staff (2014) Keys to soil taxonomy, 12th edn. USDA-Natural Resources Conservation Service, Washington, DC, 372 p

Tekinel O (1993) The mediterranean action plan: achievements and prospective. International meeting on "Mediterranean Sea and its Catchment Basins", Fiera del Levante-Bari, Italy, Unpublished Report. 26 p

TSF (1980) Acıpayam State Farm soil survey and mapping. Directorate of State Farms. Ankara, 75 p (in Turkish)

TSF (1992) Ceylanpinar State Farm soil survey and mapping. Directorate of State Farms. Ankara, 250 p (in Turkish)

TSF (1995a) Malya State Farm soil survey and mapping. Directorate of State Farms. Ankara, 125 p (in Turkish)

TSF (1995b) Kumkale state farm soil survey and mapping. Directorate of State Farms. Ankara, 105 p (in Turkish)

Vallejo VR, Díaz-Fierros F, de la Rosa D (2005) Impacts on soil resources. In: Moreno-Rodríguez JM (ed) A preliminary general assessment of the impacts in spain due to the effects of climate change. Ministerio de Medio Ambiente, Spain, pp 345–384

Yano T, Aydin M, Haraguchi T (2007) Impact of climate change on irrigation demand and crop growth in a Mediterranean environment of Turkey. Sensors 7(10):2297–2315

Geology

Mehmet Cemal Göncüoğlu

1 Introduction

Turkey is composed of numerous fragments of continental and oceanic lithospheres (Şengör and Yılmaz 1981; Yılmaz 1990; Göncüoğlu et al. 1997; Okay and Tüysüz 1999; Stampfli 2000; Moix et al. 2008) that were derived from Gondwanan and Laurasian mega-continents in the South and North, respectively. They were rifted off from the main body and amalgamated to each other or to one of the mega-continents during the geological past. These rifting-drifting and collision events have followed distinct cycles, known as the orogenic events. In the Turkish realm, at least four orogenic events, known as the Pan-African/Cadomian, Variscan, Cimmerian, and Alpine cycles have controlled the distribution of the tectonic settings at which distinct igneous, sedimentary, and metamorphic rock-assemblages were formed. Each cycle disturbed and redistributed the configurations of the earlier ones resulting in a mosaic of products of all these events. The last main orogenic event, the Alpine orogeny of Mesozoic-Tertiary age related to the closure of various Neotethyan branches and the post-orogenic tectonic events, known as the "Neotectonic Period" directly controls the present distribution of these terranes (Fig. 1).

In this brief review, the geological products of the Alpine orogeny and those of the former ones variably preserved within the alpine tectono-stratigraphic units or terranes (sensu Göncüoğlu et al. 1997; Göncüoğlu 2010) will be described and their Alpine evolution will be discussed. The aim of this review is to summarize briefly the processes and the products of the very complex Alpine events. It includes many over-simplifications for the sake of better-understanding of the very complex geology of the Turkish area for the non-earth scientists. This review is mainly based on a more detailed study of the author (Göncüoğlu 2010) which includes a more or less complete list of previous studies of copious earth scientists working on this subject. For the references on the details of the information given in this brief review, the reader is referred to the study cited above.

2 Classification of the Tectonic Units

The classification of the Turkish tectonic units or terranes, as they will be named from hereon, will be mainly based on the last and most prominent orogenic period, which controlled the formation of the geological units of oceanic and continental crust origin (e.g., Göncüoğlu 2010). From North to the South these are: the Istranca Terrane, a suspect terrane of Laurasian affinity, the İstanbul-Zonguldak Terrane of continental crust origin, the Intra-Pontide Ophiolite Belt, remnants of the northern branch of the Neotethys, the Sakarya Composite Terrane, including products of at least two orogenic events, the İzmir-Ankara-Erzincan Ophiolite Belt, representing the allochthonous oceanic assemblages and the subduction-accretion complex of the Neotethyan Ocean, the Tauride-Anatolide Composite Terrane another Alpine unit of continental crust origin, the SE Anatolian Ophiolite Belt including the remnants of the southern branch of Neotethys and finally the SE Anatolian tectonic unit at the northern edge of the Gondwanan Arabian-Libyan Platform (Fig. 2).

3 The Istranca Terrane

The Istranca Terrane or the Strandja (Strandhza) "Massif" in NW Turkey is a NW–SE trending, almost 300 km long and 40 km wide unit. It straddles across the Turkish–Bulgarian border in the southeastern Balkan Peninsula and includes structural units (nappes) that differ in lithostratigraphy,

M.C. Göncüoğlu (✉)
Department of Geological Engineering, Middle East Technical University, Ankara, Turkey
e-mail: mcgoncu@metu.edu.tr

© Springer International Publishing AG 2018
S. Kapur et al. (eds.), *The Soils of Turkey*, World Soils Book Series,
https://doi.org/10.1007/978-3-319-64392-2_5

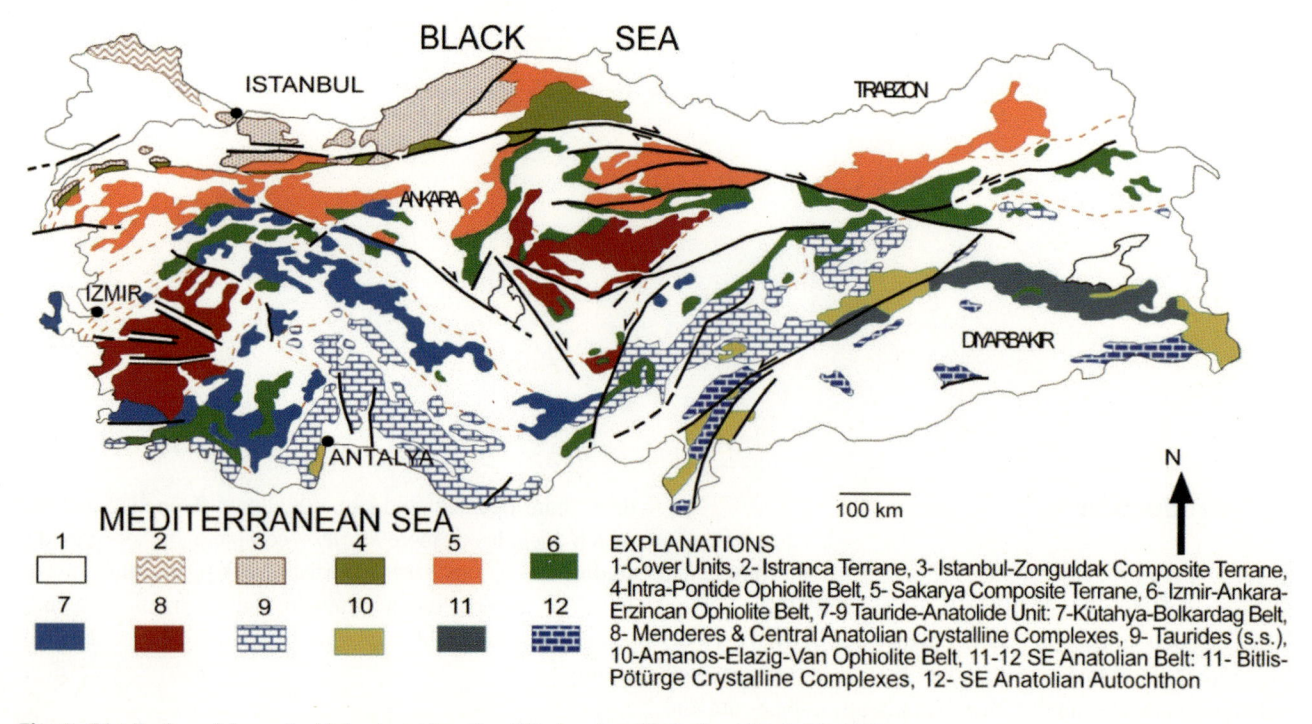

Fig. 1 Distribution of the main Alpine tectonic units of Turkey (modified after Göncüoğlu 2010)

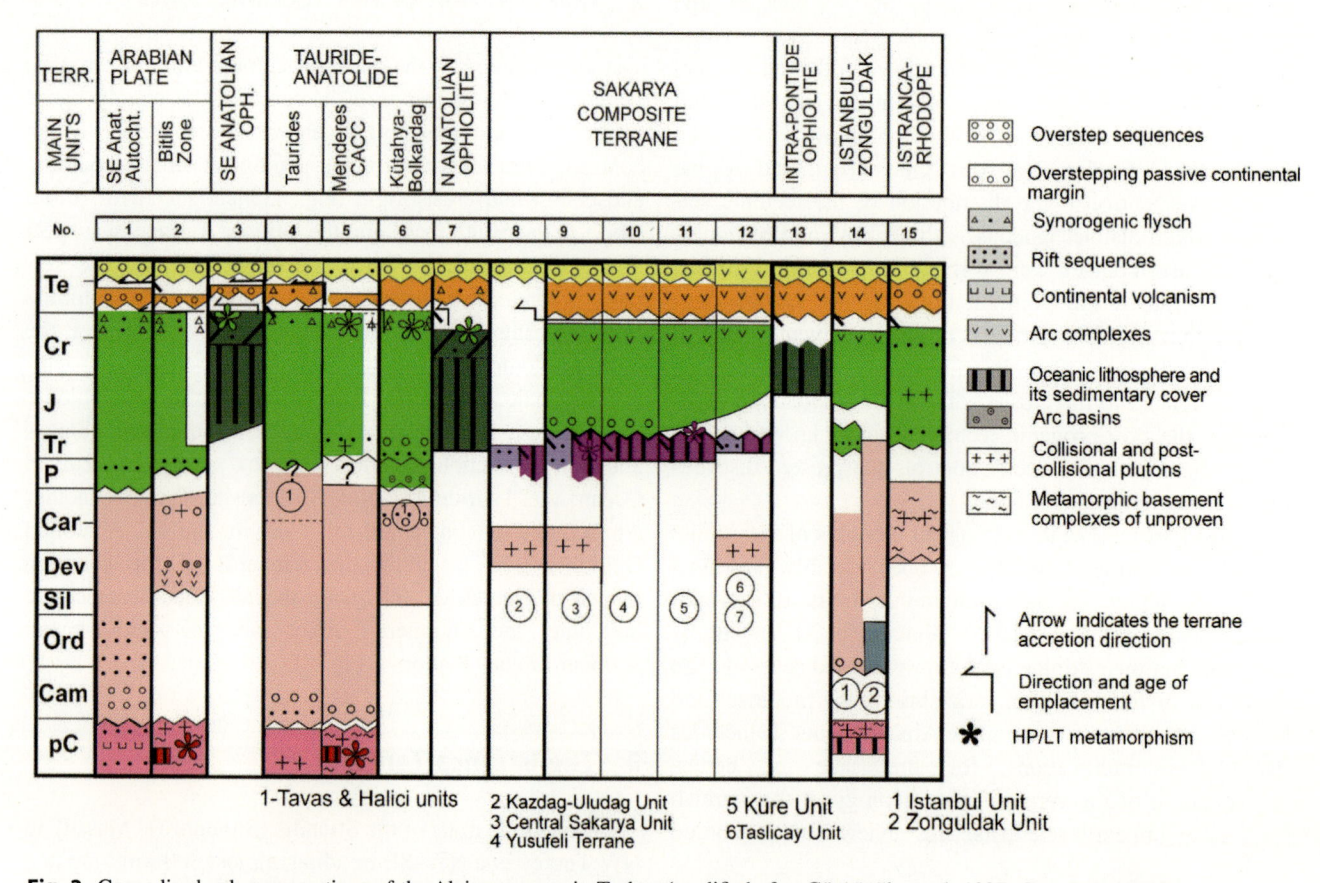

Fig. 2 Generalized columnar sections of the Alpine terranes in Turkey (modified after Göncüoğlu et al. 1997; Göncüoğlu 2010)

metamorphism, age and structural position (Bedi et al. 2013 and the references there in). They are collectively named as the "Istranca nappes". The oldest rocks within these nappes include Precambrian–Paleozoic metasediments, intruded by Late Carboniferous-Early Permian calcalkaline granitoids (Fig. 3a).

The disconformable Triassic metaclastic and metacarbonate rocks make up the main rock type. The nappes were juxtaposed at the end of the Triassic and transgressively overlain by Lower Jurassic coarse clastics. The Middle to Late Jurassic is represented by carbonates, black shales and an alternation of carbonates and siliciclastics.

The N and NW parts of the Istranca Terrane are covered by Cenomanian-Santonian volcano-sedimentary successions (Fig. 3b) intruded by Santonian-Campanian granitic to dioritic rocks.

The Tertiary cover of the Istranca Terrane is represented by the >5000 m thick sediments of the Thrace Basin, which also overlies a probable transform fold separating the Istranca and the İstanbul-Zonguldak terranes.

Fig. 3 *Upper image* is the field occurrence of the Late Carboniferous-Early Permian calcalkaline granitoids with the typical surface alteration resembling a frog, NW of Kırklareli, *lower image* shows the Late Cretaceous Volcano-sedimentary successions of the Istranca Terrane to the W of Igneada at the Turkish–Bulgarian border (Photographer Y. Bedi)

4 The İstanbul-Zonguldak Composite Terrane

The İstanbul-Zonguldak Composite Terrane consists of the İstanbul and Zonguldak units and their Mesozoic-Tertiary cover. They both include (Fig. 4) a basement composed of the remnants of Late Neoproterozoic oceanic lithosphere and a Cambrian intra-oceanic arc that were consolidated during the Cadomian events.

The basement is unconformably covered by a well-developed Ordovician sequence, mainly made of fluvial conglomerates followed by quartzites. The Silurian successions in the İstanbul Unit are mainly represented by siliciclastics and limestones, in the Zonguldak Unit, where graptolite-bearing black shales dominate. The Devonian to Carboniferous sediments in the İstanbul Unit are conformable over the Silurian and represent a deepening upwards sequence comprising Middle Devonian turbidites, Upper Devonian and Lower Carboniferous deep-marine sediments and Lower to Middle Carboniferous flyschoidal sediments (e.g., Özgül 2012).

The igneous product of the Variscan Orogeny in İstanbul is the Permian granitoids. In the Zonguldak Unit, the Early Devonian successions rest with an angular unconformity on the early Late Silurian shales. The Middle Devonian and Early Carboniferous succession of the Zonguldak Terrane is in contrast to the İstanbul Terrane mainly made up of shelf-type carbonates, followed by the well-known non-marine, coal bearing units of the Carboniferous (Fig. 4).

The Paleozoic successions of the İstanbul and Zonguldak units are affected by Variscan deformation and unconformably overlain by Late Permian-Early Triassic continental clastics (Fig. 5). Other than in the Zonguldak Unit, a well-developed alpine-type Triassic sequence is observed in the İstanbul Unit.

The post-Triassic cover of the İstanbul Unit is made up of Cretaceous volcanoclastic flysch-type sediments that unconformably cover earlier successions. In the Zonguldak Terrane, on the other hand, the Alpine cover sequences rest on an irregular (fault-controlled) basement, where Mid-Jurassic to Early Cretaceous basins (Fig. 5) were formed (Tüysüz 1999).

The Berriasian-Campanian volcano-sedimentary deposits here are interpreted as rift sediments, related to the opening of the Black Sea Basin to the north as a back-arc basin of the southerly located Intra-Pontide Ocean. The Early Tertiary deposits in the İstanbul-Zonguldak terrane are represented by Eocene and Oligocene shallow marine carbonates and clastics with volcanic intercalations that were deposited in E–W trending narrow basins. The Neogene sediments are mainly terrestrial and include limestones and lignite seams between clastic rocks.

Fig. 4 Paleozoic stratigraphy of the İstanbul and Zonguldak units (after Bozkaya et al. 2014)

Fig. 5 Field view of the Permo-triassic continental red clastics in Çakraz, NE of Bartın

5 The Intra-Pontide Ophiolite Belt

The Intra-Pontide Ophiolite Belt is a little-known suture between the Sakarya Composite Terrane in the South and Istranca/İstanbul-Zonguldak terranes in the North (Göncüoğlu et al. 2008). As the North Anatolian Transform Fault mainly follows this suture belt, these oceanic assemblages are disrupted and cryptic. It includes imbricated, more or less preserved ophiolitic assemblages, variably metamorphosed slices of accretion complexes and mélange complexes. The ophiolites are mainly represented by serpentinites.

The metamorphic slivers of the IPS are variably thick and are characterized by different metamorphic histories. Fine-grained amphibolites, coarse-grained banded amphibolites, garnet-bearing mica schists, marbles, quartzites, and paragneisses are the dominating rock types. Radiometric data suggest that they are metamorphosed during the Middle Jurassic and Early Cretaceous times, respectively. The mélange comprises blocks of serpentinites, gabbros, and pillow basalts, pelagic limestones, and radiolarian cherts. The radiolarian chert blocks have an age-range from Middle Triassic to mid Cretaceous. Neritic and pelagic limestone slide blocks are Late Jurassic to Early Cretaceous in age, whereas the sedimentary matrix consisting of shales, coarse-grained arenites, pebbly mudstones, and pebbly sandstones includes late Santonian nannofossils.

Overall the Intra-Pontide Ophiolite Belt characterizes the remnants of the Intra-Pontide oceanic branch of Neotethys. It may have developed from a relict branch of Paleotethys by back-arc spreading and closed prior to Middle Eocene, as the oldest overstep sequence starts with Lutetian shallow marine sediments and volcanic-volcanoclastic rocks.

6 The Sakarya Composite Terrane

The Sakarya Composite Terrane is a 100–200 km wide east–west trending belt covering almost the entire northern Anatolia (Fig. 1). It is considered as a composite terrane as it comprises several Variscan and Cimmerian continental as well as oceanic assemblages in its basement. The lower part of the overstep sequence is Early Jurassic in age, followed by a more or less continuous succession of Jurassic-Cretaceous platform sediments. From Late Cretaceous onward, slope-type sediments dominate, which in turn were covered by flysch-type deposits with ophiolitic blocks, derived from the northerly Intra-Pontide Ocean.

The pre-Jurassic basement of the Sakarya Composite Terrane comprises Variscan in layers, a Permian carbonate platform, destroyed by Triassic rifting and oceanic assemblages derived from the Paleotethys.

The Variscan Terranes are represented by metamorphic assemblages comprising ortho- and paragneisses, metamorphic mafic, and ultramafic rocks and marbles. The age of the protoliths are unknown, however the high-grade metamorphism in the amphibolite to granulite facies is Late Carboniferous in age. The Variscan terranes are characterized by Devonian and Late Carboniferous to Early Permian calk-alkaline magmatism. The oldest post-orogenic—in regard to Variscan—cover rocks are Lower to Middle Permian platform carbonates that grade into Late Permian olistostromes. They are unconformably covered by sediments of Triassic age. The Triassic sediments are structurally overlain and tectonically mixed with volcanic-volcano-sedimentary units. The available data is recently evaluated by Okay and Göncüoğlu (2004) and the following subunits were suggested to be tectonic members of this complex: i—a HP/LT metamorphic assemblage including metabasic rocks and sediments, ii—basalts of an oceanic plateau, its carbonate platform and slope sediments (Sayit and Göncüoglu 2009) and iii—flysch-type sediments that make the matrix of the mélange including various blocks, iv—a huge body of supra-subduction-type oceanic lithosphere (Ustaömer and Robertson 1999).

The mélange and the rift sediments of Triassic age are named as the Karakaya Complex (Tekeli 1981). Representatives of the Karakaya Complex and their Variscan basement outcrop as discontinuous metamorphic bodies from the Aegean coast in the West to the Armenian border in the East.

The stratigraphy of the Mesozoic cover units (Fig. 6) of the Sakarya Composite Terrane differs in the eastern and the western parts. In the western part, the oldest unconformable sediments are Liassic (in the west) or Late Jurassic-Early Cretaceous (in the central part) in age. In both areas the deposition is characterized by platform to slope-type carbonates during the Late Jurassic to Early Cretaceous interval. These sediments overstep both the İstanbul-Zonguldak and the Sakarya terranes. The Late Cretaceous sediments are mainly of flysch-type sediments (Göncüoğlu et al. 2000). They were formed on the passive margin of the alpine Sakarya platform and slope Ophiolite-bearing oceanic assemblages were then emplacement onto these flysch basins from north to south during the latest Cretaceous time, prior to the oblique collision of the Sakarya and İstanbul-Zonguldak plates. Contemporaneous with the ophiolite emplacement, arc-type plutons intruded the southern margin of the Sakarya Composite Terrane during the Paleocene-Eocene period.

In the eastern Pontides part of the Sakarya terrane, this igneous activity, known as the "Pontide Magmatic Arc", commenced earlier and has dominated the Northern Zone, whereas in the Southern Zone mainly sedimentary rock-assemblages (Okay and Şahintürk 1998) were formed from mid Jurassic to Late Cretaceous. The Turonian-Senonian series in the southern zone mainly comprise turbidites, unconformably overlain by Tertiary basic volcanic rocks. About 4/5 of the northern zone is covered by intrusive and extrusive distinct volcanic rocks. They form two distinct cycles. The lower and older one is Malm to Turonian in age and mainly tholeiitic in character. It starts with basalts, andesites, diabases, followed by dacite lavas-pyroclastics, dacite-rhyodacite lavas-pyroclastics, and volcano-sedimentary rocks. The upper one mainly includes calcalkaline rocks, followed by shoshonitic assemblages. It starts with basaltic lavas and dolerites, followed by mineralized dacite-rhyodacite lavas of Turonian to Paleocene. The younger igneous rocks in the south of this zone are interpreted as continental margin arcs.

7 The İzmir-Ankara-Erzincan Ophiolite Belt

The İzmir-Ankara-Erzincan Ophiolite Belt represents allochthonous assemblages of the İzmir-Ankara-Erzincan branch of the Neotethys Ocean, which were emplaced southward onto the Tauride-Anatolide Platform during its Late Cretaceous closure (e.g., Göncüoğlu et al. 2010). The units of the Sakarya Composite Terrane tectonically overlie the ophiolites in NW Anatolia. In Central and East Anatolia, the ophiolites are thrust along steep basement-thrusts onto Late Cretaceous or Tertiary basins (e.g., Çemen et al. 1999).

The İzmir-Ankara-Erzincan Ophiolite Belt consists of huge bodies of almost complete ophiolitic sequences and tectonic mélanges of the subduction-accretion complex (Göncüoğlu et al. 1997; Okay and Tüysüz 1999; Robertson 2002). Lithologically, the ophiolites consist of variably serpentinized peridotites, ultramafic and mafic cumulates, layered and isotropic gabbros. The pillow lavas mainly occur as huge thrust sheets or as blocks within the mélanges. They are mainly interlayered with radiolarian cherts and less frequently with micritic limestones or mudstones. Sub-ophiolitic metamorphic soles comprising mainly amphibolites are described from several outcrops all along the belt. The age of metamorphism in these rocks varies between Late Jurassic and Late Cretaceous, suggesting an intra-oceanic subduction as early as Late Jurassic within the İzmir-Ankara-Erzincan Ocean.

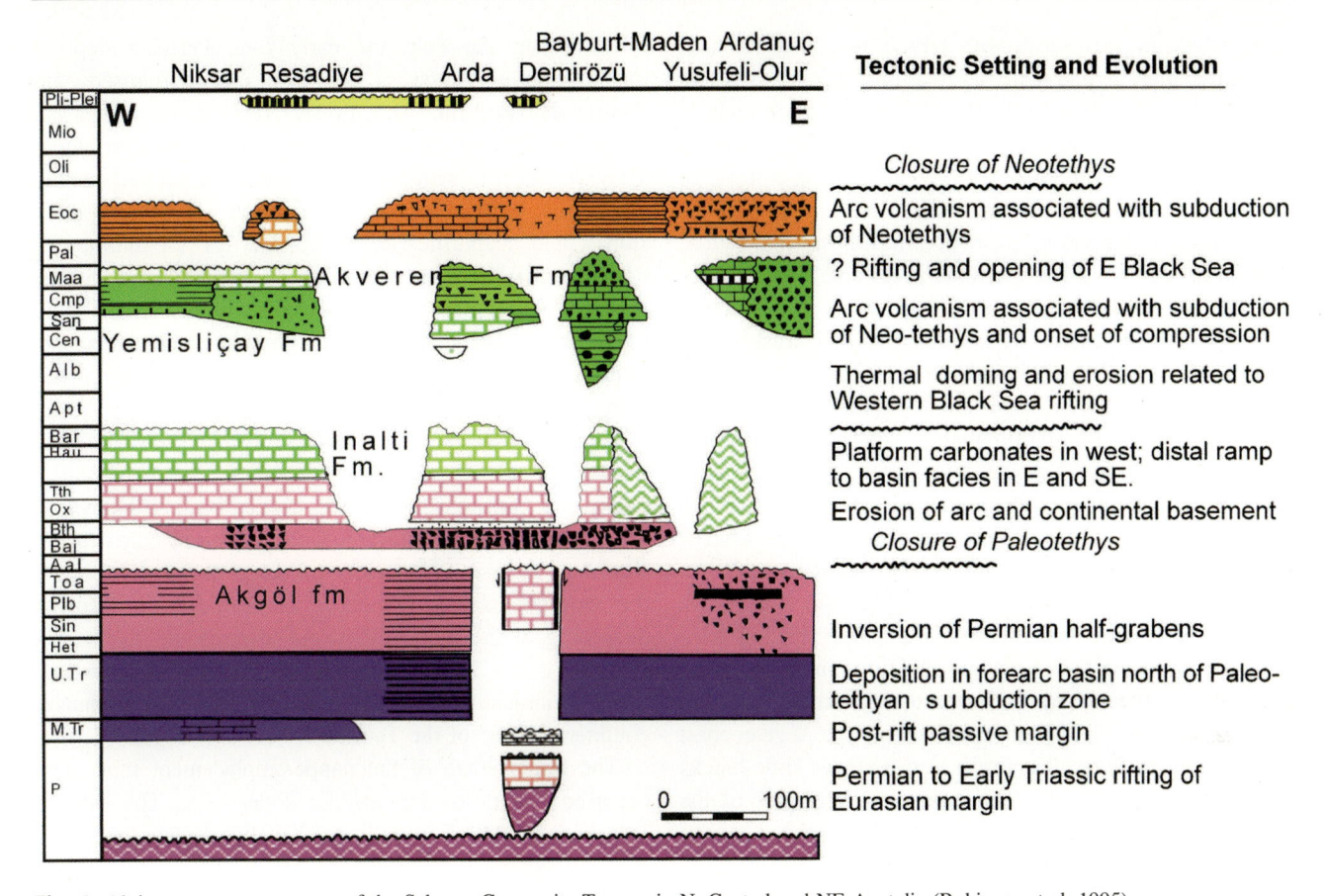

Fig. 6 Alpine overstep sequences of the Sakarya Composite Terrane in N, Central and NE Anatolia (Robinson et al. 1995)

The mélanges are both block–block and block-in-matrix type, where a deformed greywacke-dominated matrix is observed. The blocks in the mélange include, next to all members of the ophiolitic series, high-pressure–low-temperature metamorphic assemblages such as blueschists and eclogites. Variably recrystallized limestone and dolomitic limestone blocks of Mesozoic age, that were derived from the platform margin and slope sediments are also observed within the mélanges.

The ophiolites include island arc basalts, fore-arc and back-arc basalts (Yalınız et al. 2000). Their ages range from Middle Triassic to Late Cretaceous (Tekin et al. 2002). These ophiolitic units and the mélanges are transported for more than a hundred kilometers towards south onto the Tauride-Anatolide Platform and are observed as allochthonous units. The suture zone in situ is only observed in a Central Sakarya area.

8 The Tauride-Anatolide Unit

The Tauride-Anatolide Unit represents the continental platform between the Neotethyan İzmir-Ankara-Erzincan Ocean to the North and the southern Branch of Neotethys to the South. It comprises three main structural units. From north to south these are the Kütahya-Bolkardağ Belt representing the high-pressure metamorphic northern margin of the Tauride-Anatolide unit, the Menderes and Central Anatolian "core complexes", representing the high-temperature metamorphic central part of the platform and the Taurides (sensu *strictu*) characterizing the telescoped platform comprising mainly non-metamorphic nappes.

8.1 The Kütahya-Bolkardağ Belt

The Kütahya Bolkardağ Belt extends from the Aegean Sea to the Hınzır Mountains in eastern Central Anatolia. It includes numerous tectonic slices, formed during the closure of the İzmir-Ankara Oceanic branch of the Neotethys. The different tectonic slices mainly derived from the three different tectonic settings are namely the rocks of the oceanic lithosphere and subduction-accretion prism of the İzmir--Ankara Ocean (ophiolites and ophiolitic mélanges), the flysch-type deposits formed in foreland basins on the northern and passive edges of the Tauride-Anatolide platform in front of the southward advancing nappes (olistostromes with olistoliths, sedimentary mélanges), and the successions, in most cases with HP/LT metamorphism, of the slope margin and external platform of the northern Tauride-Anatolide margin. In W Anatolia, the rock-units of the Kütahya-Bolkardağ Belt surround the HT/LP Menderes Core Complex and are also observed as slices or klippen in the massif, or as nappes to the south.

The rocks of the İzmir-Ankara oceanic lithosphere occur as huge allochthonous bodies/tectonic slices and blocks within the mélanges and olistostromes of the belt. The radiolarian ages from the geochemically evaluated oceanic crust basalts suggest that the oldest "oceanic" lavas extruded during the middle Carnian. The mélanges are characterized by variable high-pressure metamorphism with a greenschist overprint. These oceanic lithologies are found within Middle Maestrichtian olistostromes with olistoliths formed in foreland basins in front of the nappes that include blocks of all kind of tectonic settings mentioned above. The flysch rocks are in depositional contact with the underlying platform and/or slope rocks of the Tauride-Anatolide passive margin (Fig. 7).

The Tauride-Anatolide external platform and basin slope deposits are affected by high-pressure metamorphism and occur as tectonic slices along the belt from the Aegean coast to the South of Sivas. They may also occur as slide blocks within the flysch basins. The internal stratigraphy of the slope and external platform sequences are variable.

All along the belt, the early Late Permian deposits unconformably cover a slightly metamorphosed and deformed basement. The Lower Triassic sequences are unconformable on a highly eroded Paleozoic basement. They commence with continental red clastics and include alkaline lavas within the variocolored sandstones, mudstones, and evaporitic carbonates. The earliest marine carbonate deposition is Anisian in age. In the allochtonous slide blocks (Fig. 7), presumably derived from more internal parts of the platform, the Middle Triassic-Lower Cretaceous sequences are represented with thick restricted to open shelf carbonates.

Typical slope sediments with olistostromes of Early Cretaceous age are sometimes associated with radiolarian cherts and mudstones indicating a deep basinal deposition. The transition from platform to slope and basin-type deposits from Late Jurassic to Early Cretaceous in the allochtonous of the external platform successions indicates a stepwise deepening of the platform margin towards the North. The presence of the HP/LT metamorphic platform margin sediments is indicative for a deep subduction of the attenuated continental crust of the Tauride-Anatolide margin.

The initial stage of the nappe-emplacement must have occurred prior to the Middle Paleocene. The Middle Paleocene-Middle Eocene in the Kütahya-Bolkardağ Belt is

Fig. 7 Mesozoic carbonate block within the Late Cretaceous mélange of the Kütahya-Bolkardağ Belt to the N of Manisa (after Göncüoğlu 2011)

characterized by shallow marine or continental molasse-type deposition in the remnant basins on the platform.

8.2 The Menderes Massif and the Central Anatolian Crystalline Complex (CACC)

These complexes are two huge crystalline culminations in western and central Turkey, respectively. Recent studies have shown that they both are lithostratigraphically made of successions almost identical to those of the non-metamorphic Taurides. The Menderes Massif is a metamorphic core complex comprising a Late Neoproterozoic "gneissic core" and a Paleozoic—Mesozoic "schist and marble envelope" (Fig. 8).

The core consists, of migmatites, para- and orthogneisses, granulites and amphibolites with relict eclogites (Candan and Dora 1998). Detrital zircon ages from the core metasediments is constrained between 570 and 550 Ma. The cover sediments start with metaconglomerates and consist mainly of kyanite + staurolite + garnet schists and garnet + mica schists with minor intercalations of meta-quartzites and garnet amphibolites.

Calc-schist and phyllite interlayers increase towards the top of the schist unit, which are followed by platform-type marbles, calc-schists, and dolomitic marbles of Mesozoic age. A conformable sequence represented by thin-bedded red marbles of Paleogene age forms the uppermost part of the Unit (Candan and Dora 1998).

Copious studies suggest that the Menderes Massif has been effected by extensional tectonic and represents a Miocene "core-complex" (e.g., Bozkurt and Oberhansli 2001). The E–W trending graben-structures in the massif indicate an ongoing extension in the eastern Aegean.

The CACC (sensu Göncüoğlu and Türeli 1994) is separated from the main trunk of the Taurides by the Tertiary Tuz Gölü, Ulukışla and Sivas basins. The lowermost unit of the CACC is composed of sillimanite-cordierite-bearing

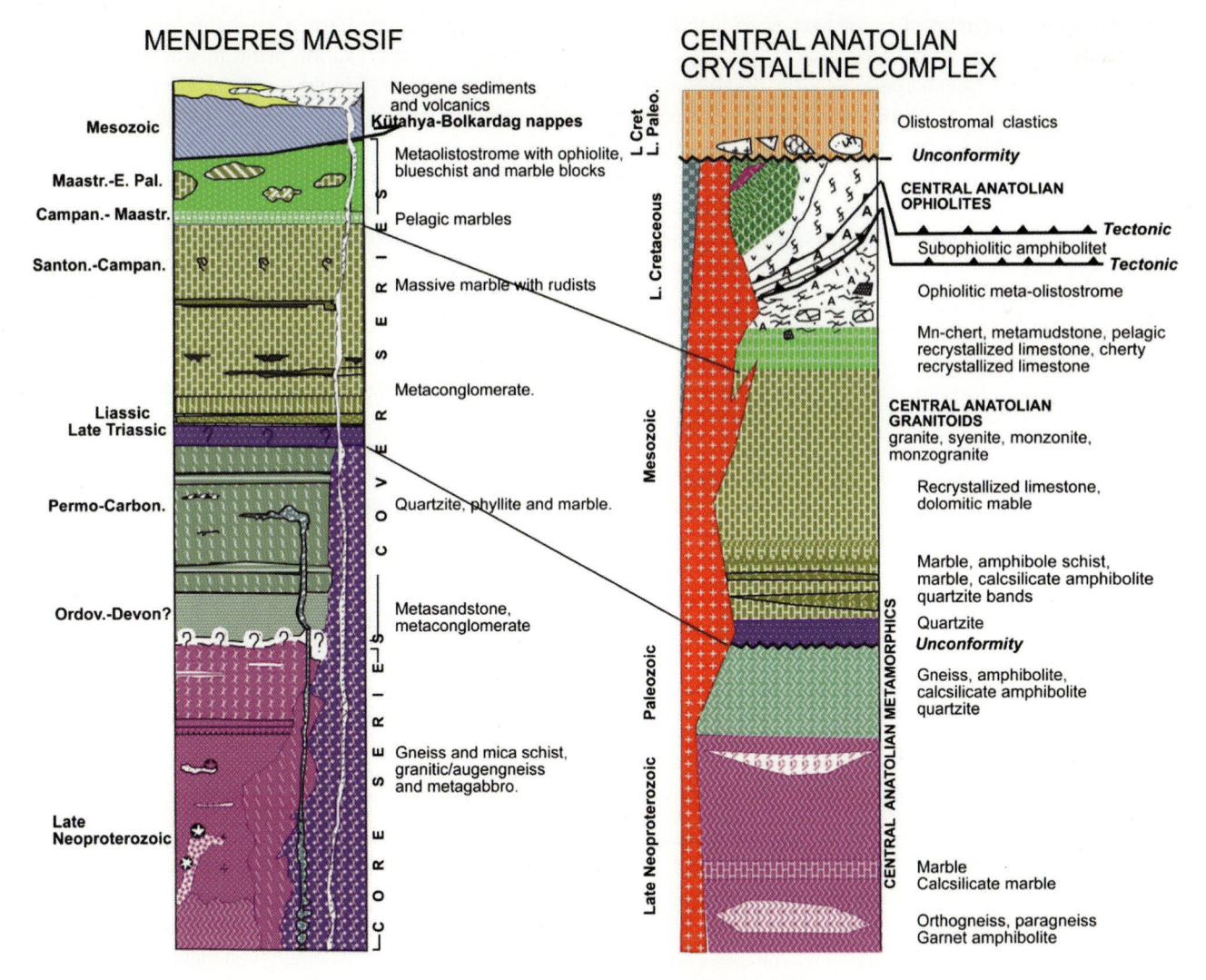

Fig. 8 Lithostratigraphy of the Menderes Massif and the Central Anatolian Crystalline Complex

gneisses, pyroxene gneisses, micaschists, amphibolites, bands and lenses of marbles/calc-silicate marbles and migmatites (Fig. 8). The xenocrystal zircon and sphene ages (Köksal et al. 2013) from S-type granitoids suggest a Precambrian basement of Gondwanan origin.

A thick quartzite band, followed by an alternation of marbles, sillimanite gneisses, amphibolites, calc-silicate amphibolites, and quartzites probably represents a Paleozoic succession. The upper unit of complex consists of a thick, Tauride-type Triassic-Early Cretaceous sequence with marbles passing upwards into cherty marbles and finally into cherts, a succession also known from the slope- to basin-type sediments of the Kütahya-Bolkardağ and Menderes units. These metasediments are followed by an ophiolite bearing meta-olistostrome and finally over thrust by disrupted supra-subduction zone-type ophiolites of early Late Cretaceous age (Yalınız et al. 2000). The Alpine metamorphism as dated by zircons is about 85 My (Whitney and Hamilton 2004). The metamorphic rocks as well as the ophiolites are intruded by I and S-type post-metamorphic collision-type granitoids with two groups of zircon ages of 85 and 75 my, respectively (Köksal et al. 2013).

The oldest post-metamorphic deposit disconformably overlying the CACC is a Late Maastrichtian-Early Paleocene cover succession. The youngest marine sediments on the crystalline complex are Middle-Upper Eocene in age. The Neogene-Quaternary cover is mainly fluvial and comprises in its upper part also the Cappadocian volcanics and volcanoclastics.

8.3 The Taurides

Taurides (sensu strictu) or the Tauride Belt is represented by a Neoproterozoic basement (Gürsu and Göncüoğlu 2006) and its non-metamorphic Paleozoic-Mesozoic cover, made of platformal sediments. The platform was surrounded during the Mesozoic by the İzmir-Ankara-Erzincan and Amanos-Elazığ-Van oceanic branches from North and South, respectively. The Late Cretaceous closure of these northern and southern branches of Neotethys gave way to a double-verging napped structure which consists of a number of tectono-stratigraphic units (Bozkır, Bolkar, Aladağ, Geyikdağ, Antalya, and Alanya units (Özgül 1976) with distinctive stratigraphic (Fig. 9) and structural features characterizing different depositional environments of the platform.

The initial thrusting of ophiolitic nappes and marginal sequences onto the Tauride platform has started during the Early Eocene. In Mid-Miocene the entire nappe-pile has been re-thrusted on the Late Tertiary cover.

Considering the overall stratigraphy, the pre-Paleozoic basement comprises slightly metamorphosed Late Neoproterozoic metaclastic rocks (slates, conglomerates and greywackes), stromatolithic limestones and lydites together with rhyolites and quartz-porphyries (Gürsu and Göncüoğlu 2006). Early Cambrian red clastics with trace fossils overlie the basement with a gentle unconformity. The platform-type deposition of Paleozoic varies in different tectono-stratigraphic units (Fig. 9), where the external platform sequences are characterized by local unconformities compared to the continuous deposition in the axial part (Geyikdağ unit) of the platform. An important regional unconformity during the Early Late Permian and the presence of the Carboniferous pyroclastics in the northern tectono-stratigraphic units in the Aladağ and Bolkardağ Units, suggests a Late Paleozoic event to the North of the Tauride Platform (Göncüoğlu et al. 2004). The Permian sequence represented by epicontinental carbonates is followed in the South and North of the platform by rift-related Early-Middle Triassic sediments and volcanics which indicate to the opening of Neotethyan basins and thus the beginning of the Alpine cycle.

The Middle Triassic-Early Cretaceous time interval in the central part of the Tauride platform was dominated by neritic carbonates, while in the northernmost margin, facing the İzmir-Ankara-Erzincan Ocean pelagic conditions continued.

Ophiolitic as well as marginal sequences were thrust from North onto the more external parts of the platform. The resulting crustal thickening generated a metamorphic zone (Menderes and Kütahya-Bolkardağ units) to the North of it. The arrival of these external nappes onto the Taurides is Early Eocene. In the South, however, the subduction of the Amanos-Elazığ-Van oceanic lithosphere beneath the Taurides gave way to the formation of a magmatic arc on the southern platform margin.

The oldest lithologies of the overstep sequences in the Central Taurides is Lutetian in age. The final re-thrusting of basement nappes in Western Taurides, however, is Middle Miocene.

9 The Amanos-Elazığ-van Suture Belt

The Southeast Anatolian Suture Belt or Amanos-Elazığ-Van Suture Belt is composed of imbricated structural units represented by dismembered bodies of the oceanic lithosphere together with pillow-lavas of arc-related assemblages (Fig. 10) and a huge subduction-accretion prism of the Southern Branch of Neotethys. It is separated from the northern Tauride-Anatolide Terrane and its marginal arc (Baskil Magmatic Arc, e.g., Parlak et al. 2004) by the pre-Maastrichtian S-verging thrusts, which were reactivated during Late Tertiary. The southern boundary towards the SE Anatolian Autochthon is an imbricated zone.

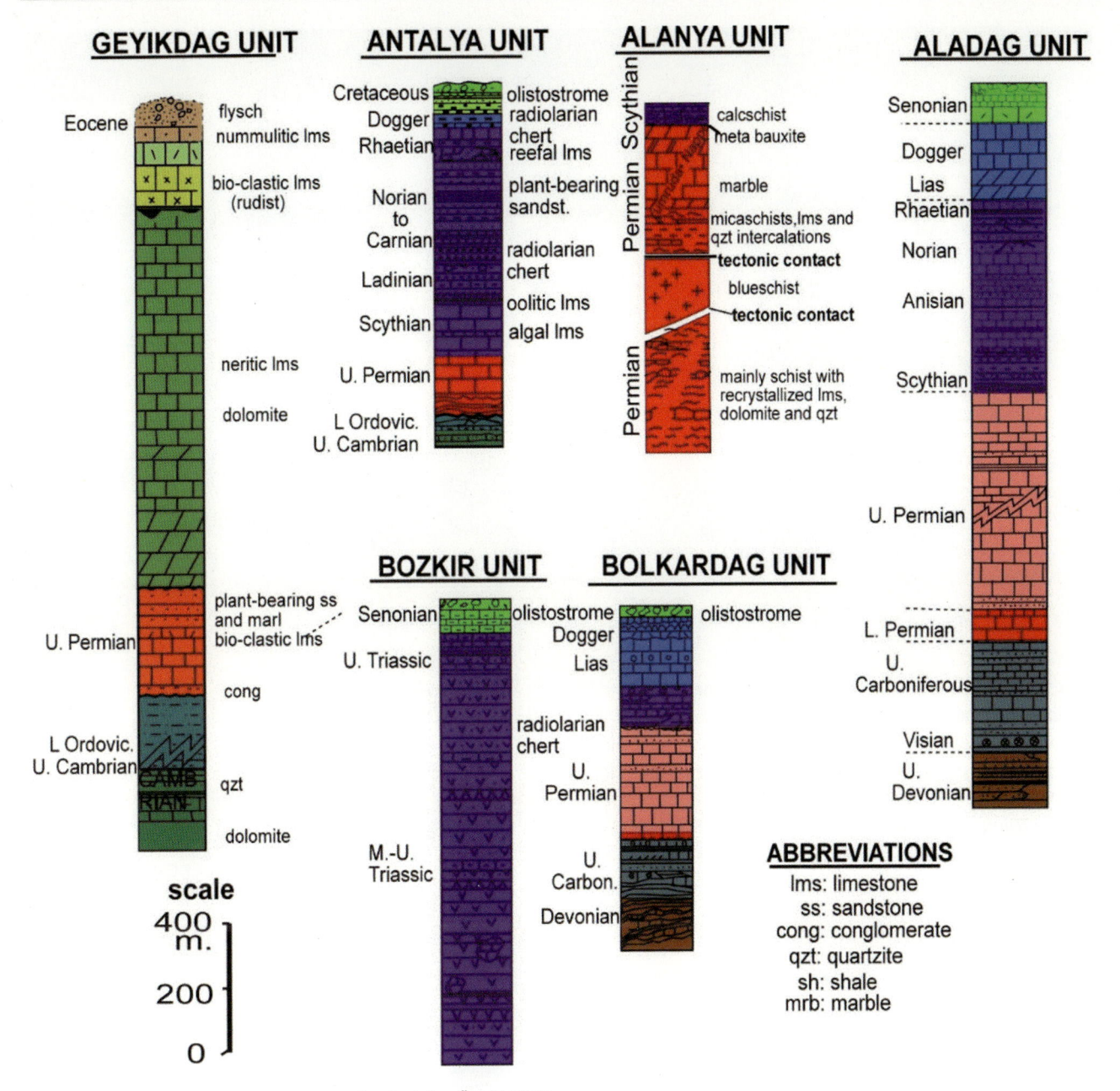

Fig. 9 Stratigraphy of the Tauride structural units (after Özgül 1976)

The ophiolites along the belt form up to 200 km long bodies with almost complete ophiolitic sequences. The Kızıldağ Ophiolite in Hatay, Guleman, Kömürhan, Ispendere ophiolites (in the central part), and Cilo Ophiolite in Hakkari are the best examples of the allochthonous ophiolitic bodies in this belt. In these bodies peridotites with podiform chromites dominate over lherzolites. Well-developed sequences of ultramafic- and mafic cumulates, gabbros, plagiogranites, variably preserved sheeted-dyke complexes and pillow lavas (Fig. 10) are common members.

The disrupted mélanges found from the Mediterranean coast to the Turkish–Iranian border and from there to SW Iran and Oman are members of this prism. Pieces of the mélange complex are also found as far-traveled allochthonous slide blocks on the Arabian platform and its metamorphic equivalents.

Considering their petrogenesis and formation ages the oldest oceanic volcanics within the belt are Late Triassic in age (Varol et al. 2011). A wide range of radiolarian ages from Middle Jurassic to Late Cretaceous was obtained from cherts of the oceanic basin sediments The Late Cretaceous supra-subduction-type oceanic basalts are mainly encountered in the Yüksekova Complex and its equivalents on the Arabian Platform. An oceanic volcanism younger than the

Fig. 10 Late Cretaceous pillow-lavas in the Amanos-Elazığ-Van Suture Belt to the NE of Elazığ

Late Maastrichtian is not found yet within the belt. Instead, the latest Cretaceous and Paleocene are marked by the formation of huge, E–W trending flysch basins on top of the accretional prism, dissected by arc-magmatics. The oldest post-tectonic event in regard to the alpine compressional tectonics is the formation of the Eocene extensional basins, known as the Maden Complex (Robertson et al. 2007) in SE Anatolia.

10 The SE Anatolian Autochthon

The northern edge of the Arabian plate comprises two main alpine tectonic units in southeast Anatolia: the Bitlis-Pütürge Metamorphic Belt and the Southeast Anatolian Autochthon (Fig. 11).

10.1 The Bitlis-Pütürge Metamorphic Belt

The Bitlis-Pütürge Metamorphic belt consists of a large number of post-Eocene S-verging slices of metamorphic rocks, ophiolitic melanges, and their sedimentary covers. The metamorphic rocks represent the northernmost edge of the Arabian Platform, which has been deformed and metamorphosed during the Alpine closure of the Southern Branch of Neotethys (Göncüoğlu and Turhan 1984). The slices of the metamorphic belt are thrust along a major structure

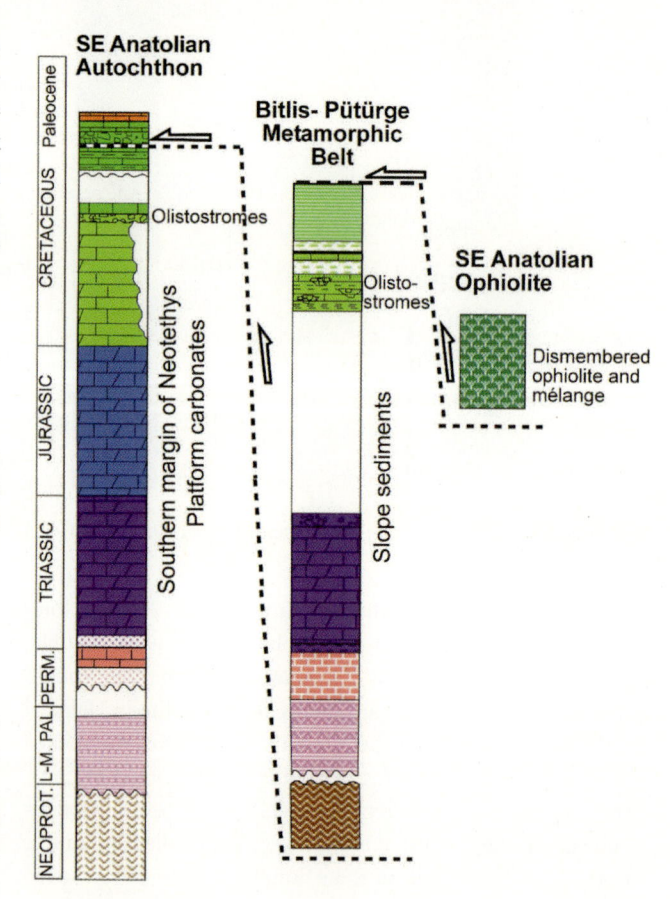

Fig. 11 Stratigraphy and structural relations of the SE Anatolian Autochthon

known as the Bitlis-Zagros Thrust. The Late Neoproterozoic basement rocks of the belt consist of various para- and orthogneisses, migmatites, amphibolites, and mica schists. Radiometric age data (e.g., Ustaömer et al. 2009) suggests a Late Neoproterozoic-Early Cambrian intrusion age for the orthogneisses.

The Paleozoic cover comprises a low-grade metamorphic clastic sequence overlain by Devonian carbonates, Carboniferous clastics and felsic metavolcanic/volcanoclastic rocks, respectively. The regional Permian unconformity and the presence of Carboniferous granitoids strongly suggest a Variscan event.

The Middle Triassic metavolcanic and volcanoclastics are conformably overlain by a condensed series, mainly consisting of metapelites interlayered with basic metavolcanic, metachert and metatuffs of Late Triassic-Early Cretaceous age (Fig. 11). This sequence is interpreted as the northern slope deposits of the Arabian passive margin. Ophiolites and ophiolitic olistostromes of Late Cretaceous age are observed as thrust sheets on the Bitlis-Pütürge metamorphics.

The metamorphic belt and the overthrusting ophiolites are covered by Middle Eocene shallow marine sediments and volcanic rocks. The final imbrication of the metamorphic rocks and ophiolites with their Eocene cover units, as well as their initial emplacement onto the Southeast Anatolian Autochthon is Middle Miocene in age.

10.2 The Southeast Anatolian Autochthon

The Autochthon comprises a Late Neoproterozoic volcanic-volcaniclastic basement (Gürsu et al. 2015), transgressively overlain by siliciclastics (Fig. 11) that grade into shelf-type carbonates and nodular limestones of Middle Cambrian age. They are followed by a thick package of Ordovician siliciclastic rocks. During Early Silurian a regional depositional break occurred in the region. The Late Silurian-Late Devonian deposition began unconformably with continental clastics and restricted marine sediments, followed by tidal-dominated clastics and terminated with regressive (fluvial) sediments in the central part of SE Anatolia. In the eastern part of SE Anatolia, however, the Ordovician clastics are overlain by Late Devonian-Early Carboniferous shallow marine sediments. A regional depositional break of Late Carboniferous-Early Permian age marks the far-field effects of a Variscan-time event (Fig. 11).

Late Permian shelf-type limestones are transitional to Triassic up to the Early Cretaceous shallow marine sediments, indicating to the stabilization of the platform conditions. During Late Cretaceous, a change to foreland deposition and arrival of northerly derived ophiolitic nappes are recorded. Eocene sediments and bimodal volcanics (e.g., Erler 1984) cover the Late Cretaceous allochthonous formations. The deposition of marine sediments in the southeast Anatolian Tertiary basin lasted until early Miocene. During the late early Miocene the second set of allochthons comprising the Bitlis-Pütürge Metamorphics were emplaced onto the Southeast Anatolian Autochthon due to ongoing N-ward movement of the Arabian Plate.

11 The Tertiary Basins of Turkey

The closure of the Neotethyan oceanic basins is associated with the formation of foreland fore-deep-type flysch basins, on the passive margins of Sakarya, Tauride-Anatolide and SE Anatolia during the latest Cretaceous-Early Tertiary period. This is followed by the formation of numerous huge transtensional and transpressional basins filled with molasse-type sediments (Görür and Tüysüz 2001) during the Paleogene.

Some of these basins are E–W trending, long and narrow troughs with marine sediments. They are controlled by oblique faults and include volcanic rocks. Typical examples of this type of Tertiary basins are the Eskipazar Basin in NW Anatolia, the Central Kızılırmak Basin in central Anatolia and the Muş Basin in SE Anatolia. Some others (e.g., the Thrace Basin in NW Anatolia, the Sivas–Çankırı–Tuz Gölü and Ulukışla basins in central Anatolia and the Maden and Hakkari basins in SE Anatolia) are of regional scale with several thousand meters-thick sediment accumulations.

In the Thrace Basin in NW Anatolia (Fig. 1) the basin-margins are dominated by carbonates during the Eocene, whereas turbidites were deposited in the basin-centers until Early Miocene (Fig. 12).

After a gap in Early-Middle Miocene the basin-fill is dominated by continental clastics and carbonates. The thickness of the sediment-fill in the depocenters of the Thrace Basin reaches up to 8 km.

The Sivas, Çankırı, Ulukışla, and Tuz Gölü basins in central Anatolia (e.g., Çemen et al. 1999; Göncüoğlu 2010) are related to post-collisional extension, where the basin development started above the suture belts with within plate-type alkaline volcanism (Fig. 13). Paleocene and Eocene are characterized by continental red clastics, evaporites, and lagoon and reef-type carbonates at the basin-margins. In the depocenters, turbiditic sandstones with olistoliths are common. Marine deposition has ended at the end of the Eocene or probably Early Oligocene.

A regional transpressional event resulted in the thrusting of the basin-margin successions and basement rocks onto the basin sediments. This period lasted until middle Miocene in the Sivas basin. Thereafter, Miocene-Pliocene continental (fluvial to lake sediments associated with lavas and volcanoclastic rocks) deposition was dominant in almost all of these basins.

The Tertiary basins in SE Anatolia are the remnants of the eastern Tethys gateway, which was the marine connection between the Mediterranean and Indian Ocean. Hence, the Paleogene deposits in this area are mainly shallow marine carbonates, that shallow upwards grading into evaporites and finally into continental clastics during the mid-Miocene (Tortonian).

The onset of the extensive volcanism in eastern Anatolia and the formation of the East Anatolian Fault Zone are coinciding in time with the uplifting in SE Anatolia, ascribed to the main events of the Neotectonic period in Turkey (Perinçek et al. 1991).

12 The Neotectonic Period

The Neotectonic Period covering the Late Miocene-Quaternary time-span is mainly triggered by the collision of the Anatolian plate and the Arabian promontory. The only active subduction zone during this period is the Hellenic Subduction Zone in southern Aegean, along which the remnant of the Southern Neotethyan plate was subducting towards north beneath the Aegean-Anatolian plates. Contemporaneously, earlier collided crustal units in East Anatolia were imbricated and domed to form the East Anatolian Accretionary Prism with a number of volcanic centers (e.g., Ağrı, Nemrut, Tendürek, and Süphan volcanoes of the East Anatolian Volcanic Province, e.g., Keskin 2003). Continuing convergence gave way to rupturing of the Anatolian continental crust along two main transform faults: the left-lateral East Anatolian Transform Fault (Koçyigit et al. 2001) trending in NE–SW direction as the northern continuation of the Dead Sea Transform (Fig. 14) and the right-lateral North Anatolian Transform Fault dissecting north Anatolia in the E–W direction from the Aegean to Azerbaijan (Saroglu et al. 1992).

The westward escape of Anatolia along these main transform faults is accompanied by the formation of second-grade strike-slip faults and fault-zones (e.g., Ecemiş and Tuz Gölü faults, e.g., Dirik and Göncüoğlu 1996) in Central Anatolia along which Late Neogene-Quaternary pull-apart basins were formed. Another implication of this event is the formation of the Central Anatolian Volcanic Province (Toprak and Göncüoğlu 1993) and volcanic centers (e.g., Erciyes, Melendiz, and Hasandağ stratovolcanoes) on the intersection of the main strands of these faults (Toprak and Göncüoğlu 1993).

The Neogene-Quaternary volcanism in central Anatolia is mainly characterized by calc-alkaline andesites-dacites, with subordinate tholeiitic-mildly alkaline basaltic volcanism of the monogenetic cones, whereas voluminous rhyolitic ignimbrite deposits are produced which cover the Cappadocian Volcanic Province (Fig. 14). Another main volcanic province related to the Neotectonic events is the Galatian Province in NW central Anatolia, where large volumes of trachyandesitic-dacitic lava flows and pyroclastics of Miocene age are formed in a transtensional tectonic setting associated with movement along the North Anatolian Fault zone.

During the Neotectonic Period West Anatolia was the side of extension where the Menderes Core Complex (Bozkurt 2001) and numerous intra-cratonic fault-controlled basins (e.g., Aegean Graben System) were generated. This extension was also accompanied by alkaline to tholeiitic magmatism (Yılmaz 1989).

The youngest products of the Neotectonic Period are the Quaternary pull-apart basins along the main transform faults and the grabens in western Anatolia. Several alluvial planes and deltas at the Mediterranean (e.g., the Çukurova alluvial plane) or the Black Sea coast (Kızılırmak delta) are sites of ongoing fluvial deposition.

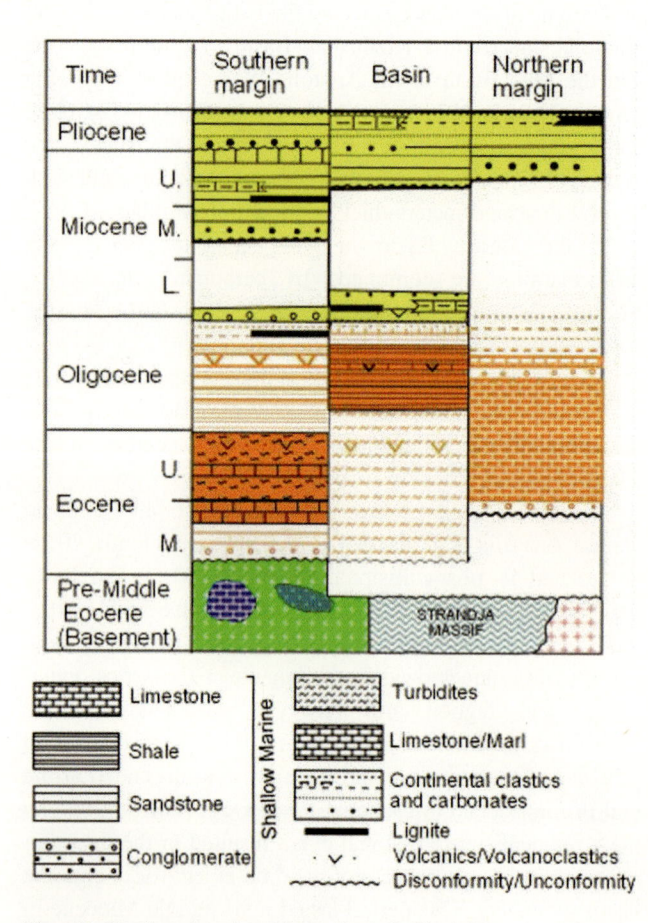

Fig. 12 Stratigraphy of the Thrace Basin in North West Anatolia (after Göncüoğlu 2010)

Geology

Fig. 13 Columnar jointing of the Eocene basalts at the southern Çankırı Basin

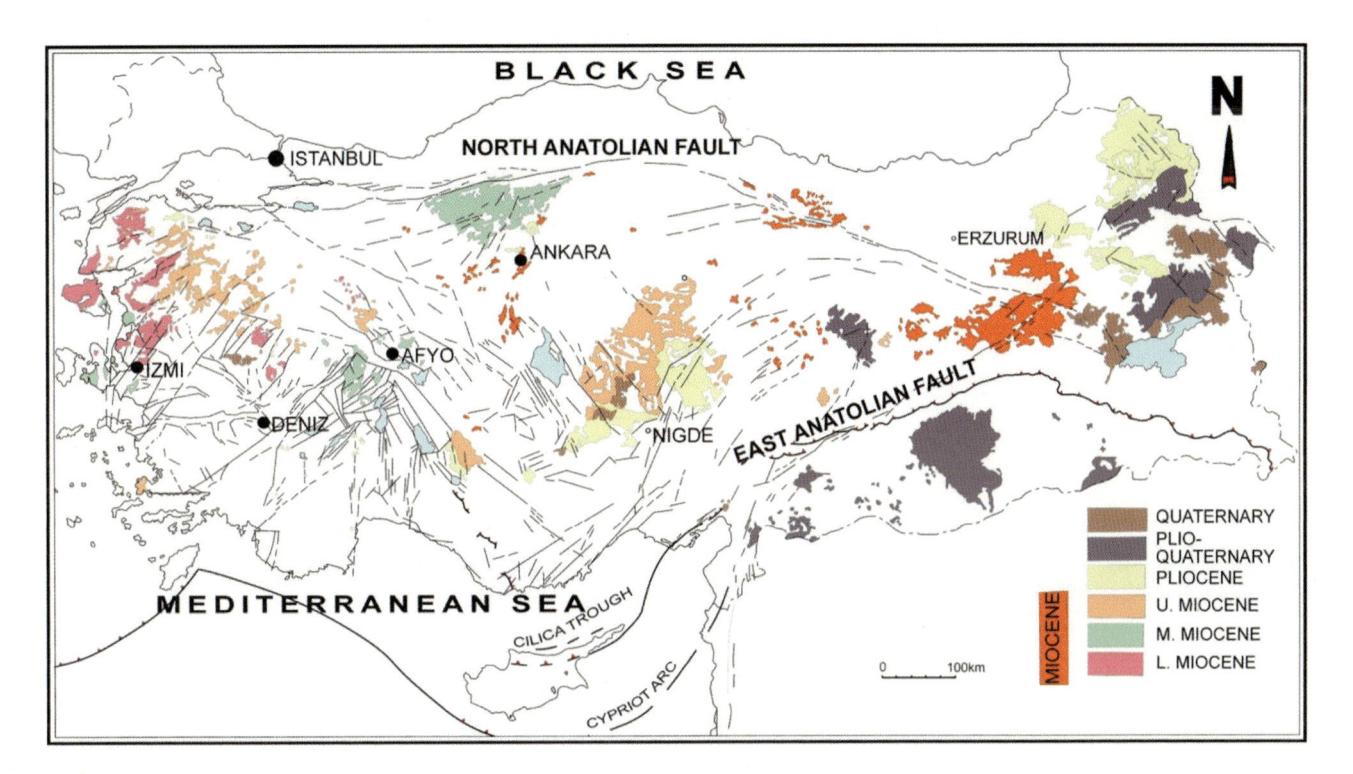

Fig. 14 Simplified tectonic map of Turkey showing the major Neotectonic elements and the sites of Neogene volcanism (after Akal 2001)

13 Conclusion

Turkey was located all along its geological history on a mobile belt between the Gondwanan and the Laurasian mega-continents. During the main orogenic cycles, pieces of their lithosphere were rifted off, drifted away from the mega-continents by the opening of the oceanic seaways at their rear, where large oceanic plates were created. The closure of these oceans resulted in a complex mosaic of terranes. Each configuration of a single orogenic cycle was reorganized by the following one. The cycles repeated at least four times (the Cadomian, Variscan, Cimmerian, and Alpine events) during the geological past in the Eastern Mediterranean area. The products of the last event, the Alpine orogeny, have been chosen as the basis for the tectonic classification of the major geological units of Turkey in this study. The geology of the Alpine period is relatively well studied. The products of earlier orogenic events, however, are not very well understood yet (for a detailed discussion see Göncüoğlu 2010) and needs additional effort and a multidisciplinary approach.

References

Akal C (2001) Occurrence, emplacement, and origin of high-potassium volcanics in the southern part of the Afyon region. Ph.D. thesis, Dokuz Eylül Üniversity, 239 p

Bedi Y, Vasilev E, Dabovski C, Ergen A, Okuyucu C, Doğan A, Tekin UK, Ivanova D, Boncheva I, Lakova I, Sachanski V, Kuscu I, Tuncay E, Demiray DG, Soycan H, Göncüoğlu MC (2013) New age data from the tectonostratigraphic units of the Istranca "Massif" in NW Turkey: a correlation with SE Bulgaria. Geol Carpath 64:255–277

Bozkaya O, Yalçın H, Göncüoğlu MC (2014) Diagenetic to very-low grade metamorphic characteristics and origin of the İstanbul-Zonguldak Terrane. In: Proceedings 4. Symposium on the geology of İstanbul, İstanbul, 26–28 Dec 2014, pp 173–183

Bozkurt E (2001) Neotectonics of Turkey-a syntheses. Geodin Acta 14:3–30

Bozkurt E, Oberhansli R (2001) Menderes Massif (Western Turkey): structural, metamorphic and magmatic evolution—a synthesis. Int J Earth Sci 89:679–708

Candan O, Dora OO (1998) Granulite, eclogite and blueschist relics in the Menderes massif: an approach to Pan-African—tertiary metamorphic evolution. Geol Soc Turk Bull 41:1–35

Çemen I, Göncüoğlu MC, Dirik K (1999) Structural evolution of the Tuz Gölü (Salt Lake) basin: evidence for late Cretaceous extension and Cenozoic inversion in central Anatolia, Turkey. J Geol 107:693–706

Dirik K, Göncüoğlu MC (1996) Neotectonic characteristics of the Central Anatolia. Int Geol Rev 38:807–817

Erler A (1984) Tectonic setting of the massive sulfide deposits of the Southeast Anatolian Thrust Belt. In: Tekeli O, Göncüoğlu MC (eds) International symposium on the geology of the taurus belt, MTA Publications, pp 237–244

Göncüoğlu MC (2010) Introduction to the geology of Turkey: geodynamic evolution of the pre-alpine and alpine terranes. General Directorate of Mineral Research Exploration, Monography series 5, pp 1–66

Göncüoğlu MC (2011) Geology of Kütahya-Bolkardağ Belt. Min Res Explor Bull 142:223–277

Göncüoğlu MC, Türeli TK (1994) Alpine collisional-type granitoids from the Central Anatolian crystalline complex. J Kocaeli Univ 1:39–46

Göncüoğlu MC, Turhan N (1984) Geology of the Bitlis metamorphic belt. In: Tekeli O, Göncüoğlu MC (eds) International symposium on the geology of the taurus belt, MTA Publications, pp 237–244

Göncüoğlu MC, Kozlu H, Dirik K (1997) Pre-Alpine and Alpine terranes in Turkey: explanatory notes to the terrane map of Turkey. Ann Geol Pays Hellen 37:515–536

Göncüoğlu MC, Turhan N, Şentürk K, Özcan A, Uysal Ş, Yalınız K (2000) A geotraverse across NW Turkey: tectonic units of the Central Sakarya region and their tectonic evolution. In: Bozkurt E, Winchester JA, Piper JA (eds) Tectonics and magmatism in Turkey and the surrounding Area. Geological Society, London, Special Publications 173, pp 139–161

Göncüoğlu MC, Göncüoğlu Y, Kozlu H, Kozur H (2004) Geological evolution of the Taurides during the Infra-Cambrian to Carboniferous period: a Gondwanan perspective based on new biostratigraphic findings. Geol Carpath 55:433–447

Göncüoğlu MC, Gürsu S, Tekin UK, Köksal S (2008) New data on the evolution of the Neotethyan oceanic branches in Turkey: late Jurassic ridge spreading in the Intra-Pontide branch. Ofioliti 33:153–164

Göncüoğlu MC, Sayıt K, Tekin UK (2010) Oceanization of the northern Neotethys: Geochemical evidence from ophiolitic mélange basalts within the İzmir-Ankara suture belt, NW Turkey. Lithos 116:175–187

Görür N, Tüysüz O (2001) Cretaceous to Miocene palaeogeographic evolution of Turkey: implications for hydrocarbon potential. J Petrol Geol 24:119–146

Gürsu S, Göncüoğlu MC (2006) Petrogenesis and tectonic setting of Cadomian felsic igneous rocks, Sandıklı area of the western Taurides, Turkey. Int J Earth Sci 95:741–757

Gürsu S, Möller A, Göncüoğlu MC, Köksal S, Demircan H, Toksoy KF, Kozlu H, Sunal G (2015) Neoproterozoic continental arc volcanism at the northern edge of the Arabian Plate, SE Turkey. Precambr Res 258:208–233

Keskin M (2003) Magma generation by slab steepening and breakoff beneath a subduction- accretion complex: an alternative model for collision-related volcanism in Eastern Anatolia, Turkey. Geophys Res Lett 30:1–4

Koçyiğit A, Yılmaz A, Adamia S, Kuloshvili S (2001) Neotectonics of East Anatolian Plateau (Turkey) and Lesser Caucasus: implication for transition from thrusting to strike-slip faulting. Geodin Acta 14:177–195

Köksal S, Toksoy-Köksal F, Göncüoğlu MC, Möller A, Gerdes A, Frei D (2013) Crustal source of the Late Cretaceous Satansari monzonite stock (central Anatolia-Turkey) and its significance for the Alpine geodynamic evolution. J Geodyn 65:82–93

Moix P, Beccaletto L, Kozur HW, Hochard C, Rosselet F, Stampfli GM (2008) A new classification of the Turkish terranes and sutures and its implication for the paleotectonic history of the region. Tectonophysics 451:7–39

Okay AI, Göncüoğlu MC (2004) The Karakaya Complex: A review of data and concepts. Turk J Earth Sci 13:75–95

Okay AI, Şahintürk O (1998) Geology of the Eastern Pontides. Am Assoc Petrol Geol Memoir 68:291–313

Okay AI, Tüysüz O (1999) Tethyan sutures of northern Turkey. Geol Soc Lond Spec Pub 156:475–515

Özgül N (1976) Basic geology of Taurids. Geol Soc Turk Bull 19:75–87 (in Turkish)

Özgül N (2012) Stratigraphy and some structural features of the İstanbul Paleozoic. Turk J Earth Sci 21:817–866

Parlak O, Höck V, Kozlu H, Delaloye M (2004) Oceanic crust generation in an island arc tectonic setting, SE Anatolian orogenic belt (Turkey). Min Mag 141:583–603

Perinçek D, Duran O, Bozdoğan N, Çoruh T (1991) Stratigraphy and paleogeographical evolution of the autochthonous sedimentary rocks in the SE Anatolia. In: Turgut S (ed) Ozan Sungurlu symposium proceedings, Ankara, pp 274–305

Robertson AHF (2002) Overview of the genesis and emplacement of Mesozoic ophiolites in the Eastern Mediterranean Tethyan region. Lithos 65:1–67

Robertson AHF, Parlak O, Rızaoglu T, Ünlügenç Ü, İnan N, Taşlı K, Ustaömer T (2007) Tectonic evolution of the South Tethyan ocean: evidence from the Eastern Taurus Mountains (Elazığ region, southeast Turkey). Geol Soc Lond Spec Pub 272:231–270

Robinson AG, Banks CJ, Rutherford MM, Hirst JPP (1995) Stratigraphic and structural development of the Eastern Pontides, Turkey. J Geol Soc Lond 152:861–872

Şaroğlu F, Emre Ö, Boray A (1992) Active fault map of Turkey. MTA General Directorate, Ankara (in Turkish)

Sayıt K, Göncüoğlu MC (2009) Geochemistry of mafic rocks of the Karakaya complex, Turkey: evidence for plume-involvement in the Palaeotethyan extensional regime during the Middle and Late Triassic. Int J Earth Sci 98:157–185

Şengör AMC, Yılmaz Y (1981) Tethyan evolution of Turkey: a plate tectonic approach. Tectonophysics 75:181–241

Stampfli GM (2000) Tethyan oceans. In: Bozkurt E, Winchester JA, Piper JD (eds) Tectonics and magmatism in Turkey and the surrounding area. Geological Society, London, Special Publications 173, pp 1–23

Tekeli O (1981) Subduction complex of pre-Jurassic age, N Anatolia, Turkey. Geology 9:68–72

Tekin UK, Göncüoğlu MC, Turhan N (2002) First evidence of Late Carnian radiolarian fauna from the İzmir-Ankara Suture Complex, Central Sakarya, Turkey: implications for the opening age of the İzmir-Ankara branch of Neotethys. Geobios 35:127–135

Toprak V, Göncüoğlu MC (1993) Tectonic control on the development of Neogene-Quaternary Central Anatolian Volcanic Province, Turkey. Geol J 28:357–369

Tüysüz O (1999) Geology of the cretaceous sedimentary basins of the Western Pontides. Geol J 34:75–93

Ustaömer T, Robertson AHF (1999) Geochemical evidence used to test alternative plate tectonic models for pre-Upper Jurassic (Palaeotethyan) units in the Central Pontides, N Turkey. Geol J 34:25–53

Ustaömer PA, Ustaömer T, Collins AS, Robertson AHF (2009) Cadomian (Ediacaran-Cambrian) arc magmatism in the Bitlis Massif, SE Turkey: magmatism along the developing northern margin of Gondwana. Tectonophysics 473:99–112

Varol E, Bedi Y, Tekin UK, Uzunçimen S (2011) Geochemical and petro-logical characteristics of Late Triassic basic volcanic rocks from the Kocali Complex, SE Turkey: implications for the Triassic evolution of southern Tethys. Ofioliti 36:99–133

Whitney DL, Hamilton MA (2004) Timing of high-grade metamorphism in central Turkey and the assembly of Anatolia. J Geol Soc 161:823–828

Yalınız MK, Göncüoğlu MC, Özkan-Altıner S, Parlak O (2000) Formation and emplacement ages of the SSZ-type Neotethyan Ophiolites in Central Anatolia, Turkey: Paleotectonic implications. Geol J 35:53–68

Yılmaz Y (1989) An approach to the origin of young volcanic rocks of Western Turkey. In: Şengör AMC (ed) Tectonic evolution of the Tethyan region. Kluwer Academic Publishers, Berlin, pp 159–189

Yılmaz Y (1990) Allochthonous terranes of Tethyan Middle East: Anatolia and the surrounding regions. Phil Trans R Soc Lond A 331:611–624

Pedogeomorphology

İbrahim Atalay, Cemal Saydam, Selahattin Kadir, and Muhsin Eren

1 Introduction

This chapter is the attempt to discuss the pedogeomorphology of Turkey in the contexts of geology and topography/geomorphic units related to the soils of the country. The soils occurring on some of the geomorphic units mentioned in this chapter as pedogeomorphic significant sites are dealt in more detail in the chapters named after these soils. For example, the wind-blown particle additions and soil re-calcification via the Saharan dust episodes are explained in the Luvisol chapter as part of its formation theory that underwent throughout the Tertiary and Quaternary, whereas the Saharan dust episodes in this chapter stand for the contemporary trends of particle additions to the surface soils.

1.1 Palaeozoic

Palaeozoic terrains are composed of both metamorphic and sedimentary rocks. The metamorphics are mica and quartzitic schists, gneiss, quartzite, and phyllites located in the Yıldız Mountains of the Thrace region, the Menderes massif in west Turkey, Biga Peninsula, northwest of Turkey, east Taurus Mountains (Bitlis massif), and the north part of central Turkey. Sandy soils are dominant on the mica schists and gneiss (Atalay 1987). The rocky landscapes are the sloping areas of Bitlis and Menderes masses where natural vegetation and especially forests have been completely destroyed and degraded.

Gneiss forming sandy soil supports the growth of stone pine (Pinus pinea) especially in the vicinity of the Koçarli town, in the Aegean Region. Sedimentary rocks, especially sandstones, siltstones, and the clayey schists mostly phyllites are common in both sides of the Bosphorus (İstanbul), in the Çatalca–Kocaeli Peninsula, the Ilgaz Mountains in the north of central Turkey, Anamur and Silifke in the middle part of the Taurus Mountains, Mardin–Derik and Çukurca localities in SE Turkey, and the Zonguldak coal province in the west part of the Black Sea region. Clayey-textured soils belonging of the Luvisols/Alfisols (acid brown forest soils) are dominant on the flat and undulating areas. Aqualfs develop on flat land where the ground water level is high.

1.2 Mesozoic

The orogenic ranges of the present landscape extending both in the north and south parts of Turkey were occupied by the Tethys Sea during the Mesozoic. The ocean crust composed of ultrabasic rocks of serpentine and peridotite were formed by the sea floor spreading. The Mesozoic terrains are mostly composed of limestones, sandstones, and flysch (alternating layers of sandstone, siltstone, and limestone) in the west parts of the north of Turkey mountains, whereas limestones are widespread in the west Taurus Mountains (Figs. 1 and 2). Sandy and silty-textured soils are widespread on the flysch terrains of the north of Turkey mountains. Volcano-sedimentary formations are common in the east part of north Turkey due to the submarine volcanic activities (Atalay 1987, 2014).

Limestone masses of more than a few thousand-meter thickness are the continual sedimentary structures of the Tethys Sea mostly occurring in the west part of the Taurus Mountains. Thin-layered limestones are common mostly in the north part of the west Taurus Mountains (Atalay 1987).

İ. Atalay (✉)
Department of Geography, Karabük University, Karabük, Turkey
e-mail: ibrahim.atalay@deu.edu.tr

C. Saydam
Department of Environmental Engineering, Hacettepe University, Ankara, Turkey

S. Kadir
Department of Geological Engineering, Eskişehir Osmangazi University, Eskişehir, Turkey

M. Eren
Department of Geological Engineering, Mersin University, Mersin, Turkey

© Springer International Publishing AG 2018
S. Kapur et al. (eds.), *The Soils of Turkey*, World Soils Book Series,
https://doi.org/10.1007/978-3-319-64392-2_6

Fig. 1 The Mesozoic limestones in the west part of the Taurus ranges (south Turkey)

Fig. 2 A flysch section consisting of thinly bedded sandstone, siltstone, marly, and argillaceous limestones in the west coastal part of the north of Turkey mountains

There is no soil cover on the bare steep slopes of the Taurus Mountains due to surface erosion, but there is widespread soil formation along the thin cracks and between the layers and/or bedding surfaces of the limestones due to high water infiltration (Atalay et al. 2008; Atalay 2011b, 2014) (Figs. 3, 4 and 5).

The texture of the soils that have developed in the limestone cracks and fissures is generally clayey due to the dissolution of the carbonates. Thus, the clay size particles in the cracks remain as the residual soils of the dissolved limestones (Atalay 2011b).

1.3 Senozoic

During the early Senozoic period, most of Turkey, especially the north and south parts of the country, were converted into a land mass via Alpine orogeny and during the Oligocene

Fig. 3 The Red Mediterranean soil (Luvisol/Xeralf, *arrow*) formed along the cracks/dissolution channels and bedding surfaces of the limestone (*Lm*) in the Karaburun Peninsula, west Turkey

Fig. 4 Red Mediterranean soil (Luvisol/Xeralf, *arrow*) formed along the bedding surface of the limestone in south of Elmalı Basin (west Turkey) (*Lm* limestone)

the shallow seas and lakes (Atalay 1987) occupied the lowlands of the intermontane basins. A significant example for the activities of this episode is the deposition of the calcareous and fine materials in the lowlands of Thrace by the Miocene sea. Most of the depression areas in Turkey were occupied by the lakes in which clay, sand, and silt-rich calcareous materials were deposited, and lignite was formed with the accumulation of organic-rich materials in the Neogene basins suggesting an anaerobic, shallow swampy lacustrine depositional environment and frequent fluctuations of the warm and wet climatic conditions (Kadir 2007).

On the other hand, in the closed basins extending from the east of Ankara to Iğdır in the east of the country, evaporitic sediments containing salt, gypsum, and alkaline compounds were formed under hot climatic conditions (Fig. 6).

Vertisols are widespread on the Neogene clayey deposits which are found in the Muş Basin in east Turkey and the

Fig. 5 Red Mediterranean soil (Luvisol/Xeralf, *arrow*) formed along the cracks/dissolution channels of Mesozoic limestone under forest vegetation (*Lm* limestone)

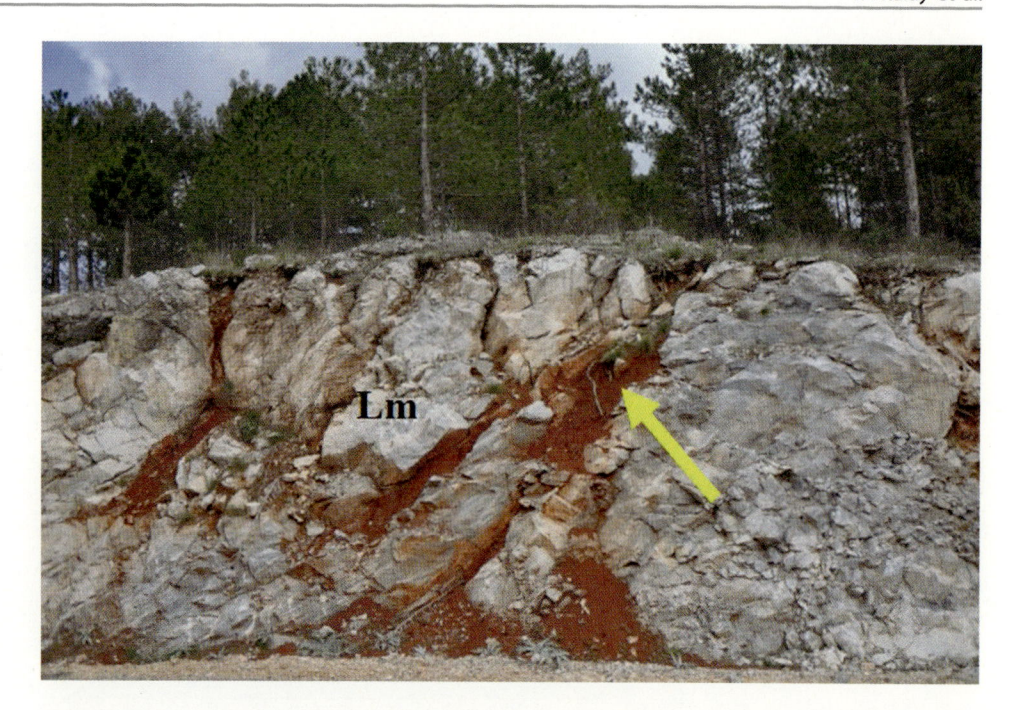

Fig. 6 Evaporitic formations containing salt-bearing material associated with red mudstone in the Oltu basin (north east Turkey)

Ergene basin in Thrace, whereas Cambisols/Mollisols mostly Rendolls are found in the lake deposits of the lowlands. Saline and highly alkaline Neogene deposits developing in the Oltu-Narman Basin and the Aras corridor in east Turkey are responsible for the Halomorphic soils (Gleysols). These areas are considered as desertified bare lands in terms of vegetation, except for sparse salt tolerant herbaceous plants (Atalay 2011b, 2014; Atalay and Mortan 2011; Cangir et al. 1990; Dinç et al. 1993) (Fig. 6).

Volcanic Activities

Turkey is rich in terms of magmatic activities. Igneous rocks are found mostly in the mountain areas, most prominently as the enormous granite batholiths outcropping in the Kaçkar and east Black Sea Mountains, the Ulu Mountain in the S–SE of Bursa city and Kozak plateau north of the Aegean Region. There are numerous smaller granite masses in central Turkey.

Deeply weathered granitic rocks form sandy soils sheltering productive stone pine *(Pinus pinea)* forests on the Kozak plateau, in the north Aegean region and south of the Samanlı Peninsula in the east of the Marmara Sea (Atalay 2014). Deeply weathered peridotite–serpentine masses, mainly found along ophiolitic suture zones, form fertile soils with varying properties in different climatic areas. Exposed serpentine outcrops with shallow and poorly developed soils mostly prevent taproot formation in forest terrains (Figs. 7 and 8).

Volcanism that started at the end of the Mesozoic and Early Tertiary was intensified during the Eocene and Neogene periods. Eocene volcanic rocks that are composed mainly of andesite, andesitic tuff, and agglomerate are common in the Köroğlu Mountains in the northwest and east of Turkey. Central volcanism that formed isolated or individual volcanic cones occurred mostly during the Upper Neogene and Quaternary periods. Most of east Turkey is covered by basaltic lavas, some of which are plateau basalts occurring in the mountains of Kargapazarı, Yalnızcam, and Allahüekber and Erzurum–Kars–Ardahan plateaus in the northeast of Turkey. The high mountains of Turkey such as the mountains of Erciyes, Hasan, Kara in central Turkey, and Ağrı (Mount Ararat), Süphan, Nemrut, and Tendürek in the north of Lake Van were formed during this period. The majority of the volcanic cones are strato-volcanoes that are composed of basalts, andesites, lahars, tuff, and sand/cinder layers. Almost all of east Turkey is covered by basaltic lava flow features (Fig. 9).

Andosol-like soils/Cambisols are common on the flat lands of the old basaltic surface in the west of Kula and the lowland edge of Karacadağ in southeast of Turkey. Volcanic Regosols or Psamments are found near the Cappadocia region, southeast of central Turkey.

1.4 Quaternary

During the Quaternary, Turkey was subjected to climatic changes/fluctuations. During the last Glacial period, the mountainous areas (higher than 2000 m) encircling the coastal terrains were occupied by glaciers (Fig. 10). Cold and arid climatic conditions prevailed in Turkey and the lake levels were raised due to the low evaporation in central Turkey. The level of the sea, encircling Turkey, was 125 m lower than the present (Atalay 1992a, 1995, 1998). Consequently, soil forming processes were almost nullified at that period. During the early Holocene, the melting of the glaciers in the highlands and the increase of precipitation led to catastrophic floods. The catastrophic features can be observed as the thick colluvial deposits on the edge of the mountains and sandy gravel deposits along the Euphrates and Tigris Rivers, especially in the vicinity of Diyarbakır city and the Birecik town. Deltas such as Çukurova (Cilician plain) and Asi in the south of the country, Eşen, Büyük Menderes, Küçük Menderes and Gediz in the Aegean Sea region, and Bafra and Çarşamba in the Black Sea region have started to form in the Quaternary according to the present-day sea level. The alluvial soils (Fluvisols) in these regions are found in the upper and central parts and hydromorphic alluvial soils (Gleysols) are common on the lower parts of the deltas. Sandy materials or sand dunes and Arenosols appear along the deltas near the sea (Fig. 11).

Fig. 7 Poorly weathered serpentine, north of Gölbaşı Town, Adıyaman Province

Fig. 8 Poor stands of red pine (*Pinus brutia*) on exposed serpentine in the vicinity of Kemer town, Antalya, west of Taurus Mountains

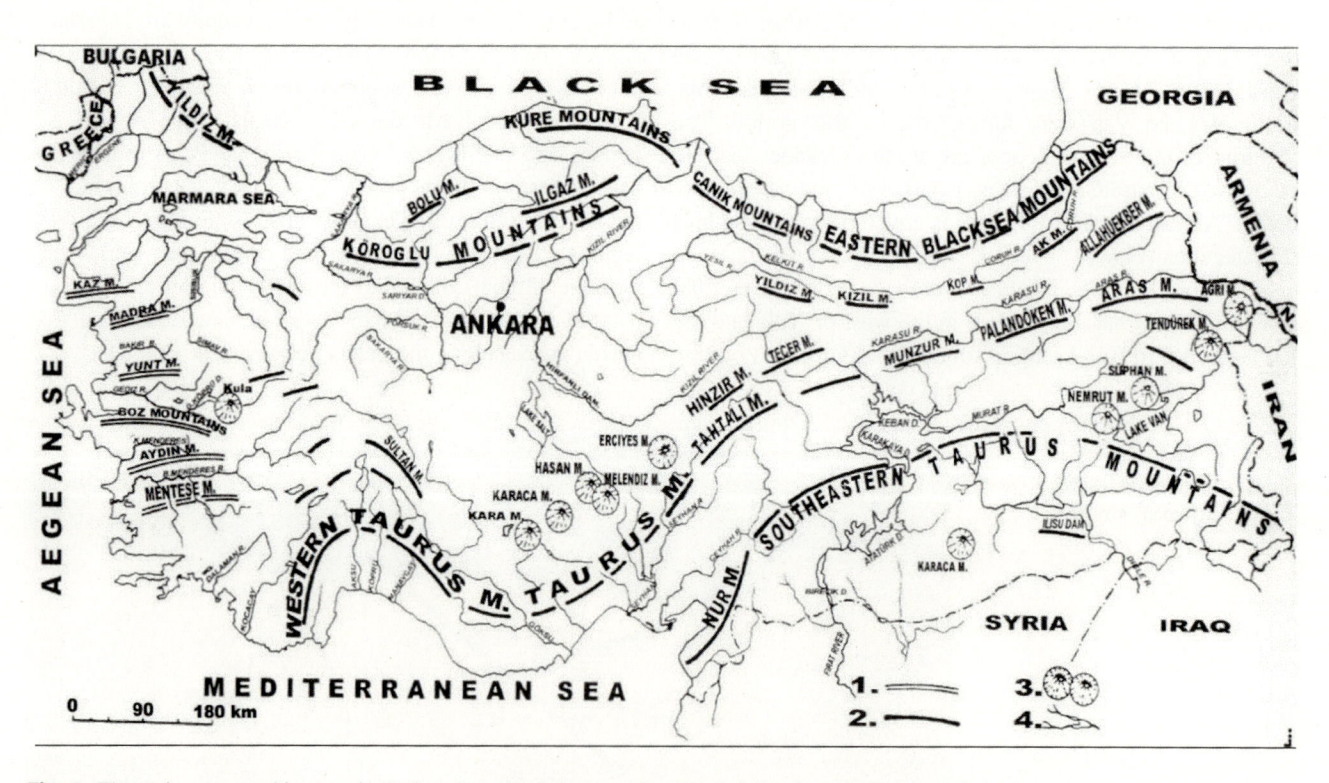

Fig. 9 The major geomorphic areas in Turkey, 1 epeirogenic areas, 2 orogenic belts, 3 volcanic cones, 4 rivers

2 Topography

Turkey is generally a highland country with rugged topography. Its mean elevation in Thrace is 1132 and 1162 m in Asia Minor. The elevation of Turkey increases from the west to the east, and from the coastal belt toward the orogenic range extending both in the north and south of Turkey.

According to the altitudinal belts, the area lying between 0 and 500 m covers an area of 17.5%, the 500–1000 m elevation area is about 26.6%, and the 1000–2000 m elevation belt accounts for 45.9% of the total land of Turkey (Table 1).

The topographic inclination of Turkey and especially of eastern Turkey is high due to the fact that mountainous areas

Fig. 10 Glaciated mountains during the last glacial period in Turkey (Atalay 1987)

Fig. 11 The Eşen delta and its sand dunes showing ripples (*arrow*) subjected to wind erosion (south Turkey)

Table 1 Altitudinal belts of Turkey (Tanoğlu 1947)

Altitudinal belt (m)	Area (km^2)	Total land area (%)
0–250	79,254	10.4
250–500	53,912	7.1
500–1000	201,999	26.6
1000–1500	230,775	30.4
1500–2000	118,284	15.5
More than 2000	75,754	10.0
Total	759,987	100.0

have been deeply dissected by the running water and/or streams. Indeed, the inclination with more than 40% covers an area of 46% of the total land area of Turkey, whereas the inclination more than 15% accounts for 80% of the total land area (Table 2).

The inclination in the mountainous areas is more than 40%. Less-inclined areas are generally found in the lowlands of Thrace and central part of central Turkey and the deltaic terrains (Oakes 1954).

2.1 Geomorphic Units of Turkey

Topographically Turkey can be divided into seven units as in the following:

2.1.1 Orogenic Belts

They extend both in the north and in the south. There are two main orogenic belts in Turkey, namely the Taurus and the north Turkey mountains (Fig. 9).

Taurus Mountains

These mountain ranges extend along the coast of the Mediterranean Sea and north of southeast Turkey. The Taurus Mountain range is the continuation of the Alp ranges in Europe. They start near the Kerme Gulf, continue in the direction of north–south in the Teke Peninsula, after the Antalya–Isparta conjunction they extend parallel to the Mediterranean coast and to the north– south direction in the Gulf of İskenderun (Nur Mountains). The southeast Taurus Mountains run concave to the north part of southeast Turkey.

The main mountains and their elevations in the direction of west to east are the Akdağ M. 3014 m, Bey M. 3069 m, Barla (Isparta) M. 2734 m, Dedegöl M. 2992 m, Geyik M 2890 m, Bolkar–Ala mountain range Aydos H. 3480 m, Medetsiz H. 3529 m, Demirkazık H. 3758 m, Bey M. 3075 m, Nurhak M. 3081 m, Bey M. (Malatya Mountains) 2608 m, Akçakara (Bingöl) M. 2940 m, Buzul (Cilo) M. 4116 m, and Sat M. 3794 m.

In the central part of Turkey, the main orogenic ranges are the north continuation of the Taurus mountains, these are the Tahtalı, Uzunyayla, Tecer, and Mercan (Munzur) (3293 m), Elçler, in the north of Muş plain, Karakaş (1748 m) in the Gaziantep plateau and the Raman Mountains (1260 m) in southeast of Turkey.

North Turkey Mountains

These mountains run parallel from Thrace to the coast of the Black Sea. The main mountains and their elevations are as follows: Yıldız (1031 m), Küre (2019 m), Canik (1194 m), and Giresun Mountains (3038 m) and the east Black Sea Mountain (Kaçkar H. 3932 m). The second ranges extending to the south part of the north Turkey Mountains are the Köroğlu (2499 m), Ilgaz (2546 m), Çamlıbel, Çimen, Mescit–Ala (2329 m), and Yalnızçam Mountains (3202 m).

2.1.2 Epeirogenic Areas

The epeirogenic areas are related to the vertical tectonic movements following the Alpine orogeny. Some part of the metamorphic rigid massif was uplifted and some was depressed along the fault lines, leading to the formation of the horst and graben topography due to rifting. The west part of Turkey was subjected to the vertical tectonic movements of the north–south extension (Atalay 1987). In west Turkey, the main horst extending in the north–south direction is called the mountains of Kaz, Manisa, Boz, Aydın, and Menteşe. The main grabens in which alluvial plains and rivers are found are Edremit, Bakırçay, Gediz, Small, and Büyük Menderes. The differences in elevations between the graben and horsts are evident in the Gediz graben which is about 100–200 m and over 2500 m in the Boz Mountains.

Rift systems are also found in the Lake Regions located southwest of Turkey. Here the depressed areas were occupied by the lakes such as Burdur, Acı, Beyşehir, and Eğirdir, and uplifted blocks fit the main mountain ranges in Söğüt, whereas the east part of Turkey underwent compressive movements bound to the northward movement of the

Table 2 Distribution of the inclination classes of Turkey (Oakes 1954)

Inclination class (%)	Area (km^2)	Total land area (%)
0–1	62,428	8.14
1–3	25,105	3.81
3–8	48,361	6.30
8–15	15,938	2.07
15–40	264,862	34.40
40>	351,813	45.70
Total	768,507	100.00

Arabian rigid plate. Here grabens containing plains are located along the east and north Turkey strike–slip fault lines.

The basins and/or plains of Muş, Malatya, Elazığ, Erzincan, and Erzurum were formed as the result of the collapse of the area along the fault lines. In north Turkey one can see tectonic basins extending along the north Turkey fault lines. The main basins are the Erzincan, Kelkit River corridor, Erbaa Niksar, Ladik, Çerkeş, Bolu, and Adapazarı, where the highlands encircling such depressions fit uplifted blocks.

2.1.3 Volcanic Areas

The Volcanic areas are due to the central volcanic activities and the accumulation of the lavas and pyroclastic materials such as the lahar, volcanic ashes, and sands to a relative altitude of more than 2000–3000 m. The main volcanic mountains in central Turkey are the Erciyes (3917 m), Melendiz (1898 m), Büyük Hasan Mountain (3268 m), and Karadağ (1819 m). The main volcanic mountains in east Turkey are Nemrut (2828 m), Süphan (4058 m), Tendürek (3660 m), Kısır (3197 m), and Ala Mountains (3138 m) in the vicinity of Lake Çıldır west and east, respectively, and Ağrı (Mount Ararat) (5137 m).

The volcanic Cappadocian area, one of the main touristic centers of Turkey, in the southeast part of central Anatolia displays different landforms such as the earth pillars and badland topography. The formation of the earth pillars is related to the erosion resistance of the volcanic materials. Namely, volcanic sand and tuff are subject to erosion, but lavas are resistant. The development and shape of the earth pillars capped by boulders, which serve to protect the main structure from erosion by rain, have been determined by the lithologic stratification properties, inclination, and crackiness of the layers. Indeed, the earth pillars have been developed in the soft volcanic deposits which are composed of ignimbrite, tuff, lahar, volcanic ash, and sand layers. Here, ash and sandy deposits are readily subjected to erosion, but ignimbrite and basalt layers that resist erosion remain on the ash and tuff layers as skull-caps leading to the development of the pillars (Figs. 12 and 13).

The thick and resistant lahar caps standing on the tuff layers protect the earth pillars from erosion and are responsible for the formation of larger pillars. Well-developed earth pillars are seen on the horizontal sequence of the pyroclastic material.

Conical isolated hills a few meters (generally 2–5 m) in height are common on the soft lahar and ignimbrite in the vicinity of Göreme town. These hills have been formed by the deepening and widening of the stream beds both on the erosional and pediplain surfaces—the old pediment surfaces similar to the River Kızılırmak valley. Debris flow is a common process on the steep slopes due to unconsolidated pyroclastic deposits. There is no vegetation cover or very

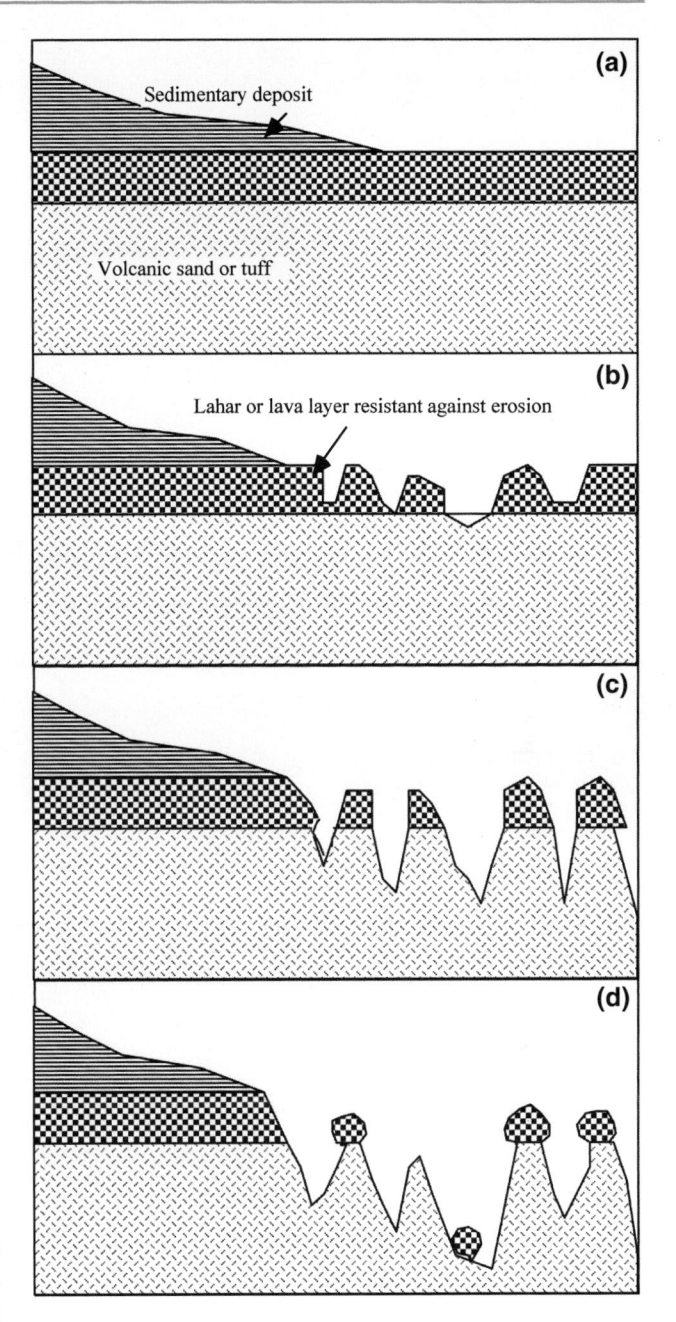

Fig. 12 The formation of earth pillars. **a** Typical stratification of volcanic land. **b** The beginning of erosion along cracks of lava or lahar. **c** Erosion of tuff and sand material. **d** Remnant of the massive lava layers resistant to erosion forming skull-caps on the volcanic sand and tuff (Atalay and Mortan 2011)

sparse herbaceous species grow within the area due to the continuous erosion.

The tuff deposit areas with a poor soil (Regosols) cover are the favorable sites for vineyard and potato cultivation, enabling easy penetration of water and air and in turn of the vine roots. Between Kula and Adala in the west part of Turkey there is a volcanic landform of 36 km length and 14 km width. Here volcanic plateaus, 68 smooth

Fig. 13 Earth pillar formed due to erosional differences between volcanic tuff and lava (basalt, *arrow*) in Cappadocia with dominant Regosols

Fig. 14 Young basaltic lava outcrops in the vicinity of Kula town, west of Turkey

volcanic cones, and young volcanic cones have developed due to the variable aged volcanic activities. Andosol-like soils have developed here on old basaltic formations. Young basalt flows and cones formed c. 8000 years before present are devoid of soil due to the unfavorable climatic conditions and the short time elapse in soil development (Erinç 1970; Atalay 2011b) (Figs. 14 and 15).

The orogenic mountain ranges of Turkey have been deeply dissected by the rivers causing the formation of the high inclinations of the valleys. Soil in the mountainous areas reflects the physical and chemical properties of rocks

Fig. 15 Young and old volcanic cones and basalt lava flow layers. Andosol-like soils are found both on the smooth surfaces of the old and young volcanic cones and the arable land formed on the basaltic lava layer

Fig. 16 Colluvial deposits in Cocak valley, in the south edge of the Taurus Mountains, north of Mersin

or parent materials due to intense erosion. Hard rocks such as granite, gneiss, and andesite are outcropping on the steep slopes of north and west Turkey. Cambisols/Inceptisols and Leptosols/Entisols are common on the steep slopes composed of sedimentary rocks. Rock avalanches and debris flows occur on the steep slopes of the high mountains. On the other hand, there are deep colluvial deposits on the lower slope of the mountains due to the intense material transportation coming from the upland areas as are in the Taurus and Ilgaz Mountains (Figs. 16 and 17).

The striking factor of soil formation and soil type is related to the direction of the mountain ranges, their aspect, and altitude in these areas. As a general rule, the mountain ranges extending to the east–west direction mostly prevent the wind fronts both coming from the south and north directions. For this reason, the north facing slopes of the coastal ranges of the north of Turkey Mountains receive higher precipitation than the south slopes due to the wind fronts over the Black Sea which are considerably prevented by their high north slopes. In contrast, the south slopes of the

Fig. 17 Rock avalanches (*arrow*) on the Ilgaz Mountains in the middle part of the Black Sea region

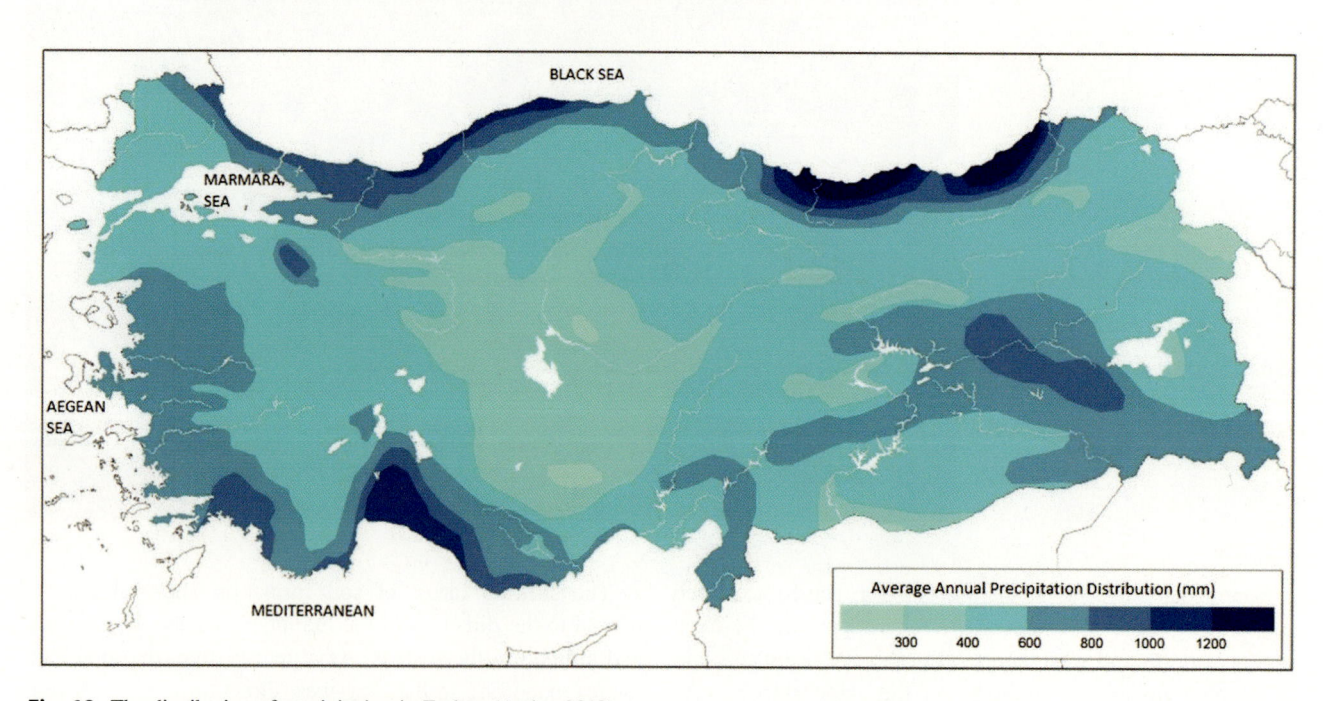

Fig. 18 The distribution of precipitation in Turkey (Atalay 2012)

Taurus Mountains receive abundant precipitation. Thus, the rainy areas of Turkey are the north slopes of the north and south slopes of the Taurus Mountains. But the depressions and inland parts of Turkey receive lower rainfall due to their locations within the rain shadow areas (Atalay 2011a, b, 2014) (Fig. 18).

In the map, rainy areas of Turkey are the north slopes of the north of Turkey Mountains and the south slopes of the Taurus Mountains.These conditions are determinant in soil formation and the types of soils to form via erosion susceptibility. In this context, mature soils, mainly acid brown forest soils (Luvisols/Alfisols), are found on the densely

Fig. 19 Granitic rock exposed on the steep slopes of the Barhal basin, south of Kaçkar Mountain, east of the Black Sea Region with Oriental spruce (*Picea orientalis*)

covered forest areas especially on the north facing slopes of the Black Sea coastal mountains which are humid-mild and humid cold climatic areas (Fig. 19), whereas calcareous brown forest soils, Leptosols/Entisols, and Cambisols/Inceptisols that formed under the semiarid climate are widespread on south facing slopes of these mountains (Figs. 20 and 21). Lithosols (Leptosols) are found on the steep slopes of the Çoruh valley (Fig. 21).

Soil types frequently change depending on parent material and topographic conditions in the mountainous areas. Mature and climatic soil types appear on the flat and slightly undulating areas, but Lithosols (Leptosols) are found on the steep slopes, Rendzinas are widespread on the Neogene deposits, and Red Mediterranean soils are common on the karstic land (Figs. 22, 23). Moreover, Podzols are found as patches over granite of Mount Uludağ in Marmara (24).

2.1.4 Plains and Basins

The plains of Turkey are composed of the deltaic deposits and accumulated sedimentary deposits in the depressions. Excluding the delta plains, almost all plains appear within the tectonic depressions and/or basins which have been formed by vertical tectonic movements. Here the rivers and streams with meanders are flat lands, dejection fans and cones which are common on the edges of the plains (Fig. 25). The lowlands of most basins of Turkey were occupied by lakes and swamps in the past. The plains of the country appear not only at the coastal areas but also at more than 2000 m elevations of highlands.

The main basins containing the plains are Muş, Van, Malatya, Elazığ, Erzurum, Erzincan, Horasan–Pasinler in east Turkey, Konya and Eskişehir in central Turkey, Adapazarı, Düzce, Erbaa Niksar in north Turkey, Bakırçay, Gediz, Büyük, and Küçük Meanders in west Turkey.

There are close relationships among the geomorphic units of the basins and deltas (Fig. 26). Alluvial soils, for example, appear on the flat land of plains and basins, and colluvial soils are found on dejection fans and colluvial deposits on the edge of the mountains (İnci 1991). Clayey soils occur on the back swamp deposits, sandy soils occur on the natural levees along the valleys, hydromorphic and hydromorphic alluvial soils are found where the ground water level is high (Fig. 26). Saline soils occur together with saline ground water tables and over irrigated fields of south east Turkey (Dinç et al. 1987, 1991a, b, 1993; Driessen and de Meester 1969; Driessen 1970) (Fig. 27).

On the other hand, terraces that formed by the lowering (incision) of the base level of the basins due to rejuvenation contribute to the formation of the different soil depths and soil types (Figs. 28 and 29). For instance, the depth of the Vertisols increases with the altitude of terraces that formed by the removal of the basin by fluvial erosion in the Muş basin (Atalay 1983).

There is a close relationship between the geomorphic surfaces and soil formation in the Harran Plain in the south east of Turkey. Vertisols formed on the clayey deposits transported from the upland limestone areas. Calcisols developed on the lowland part of the plain under fluctuation

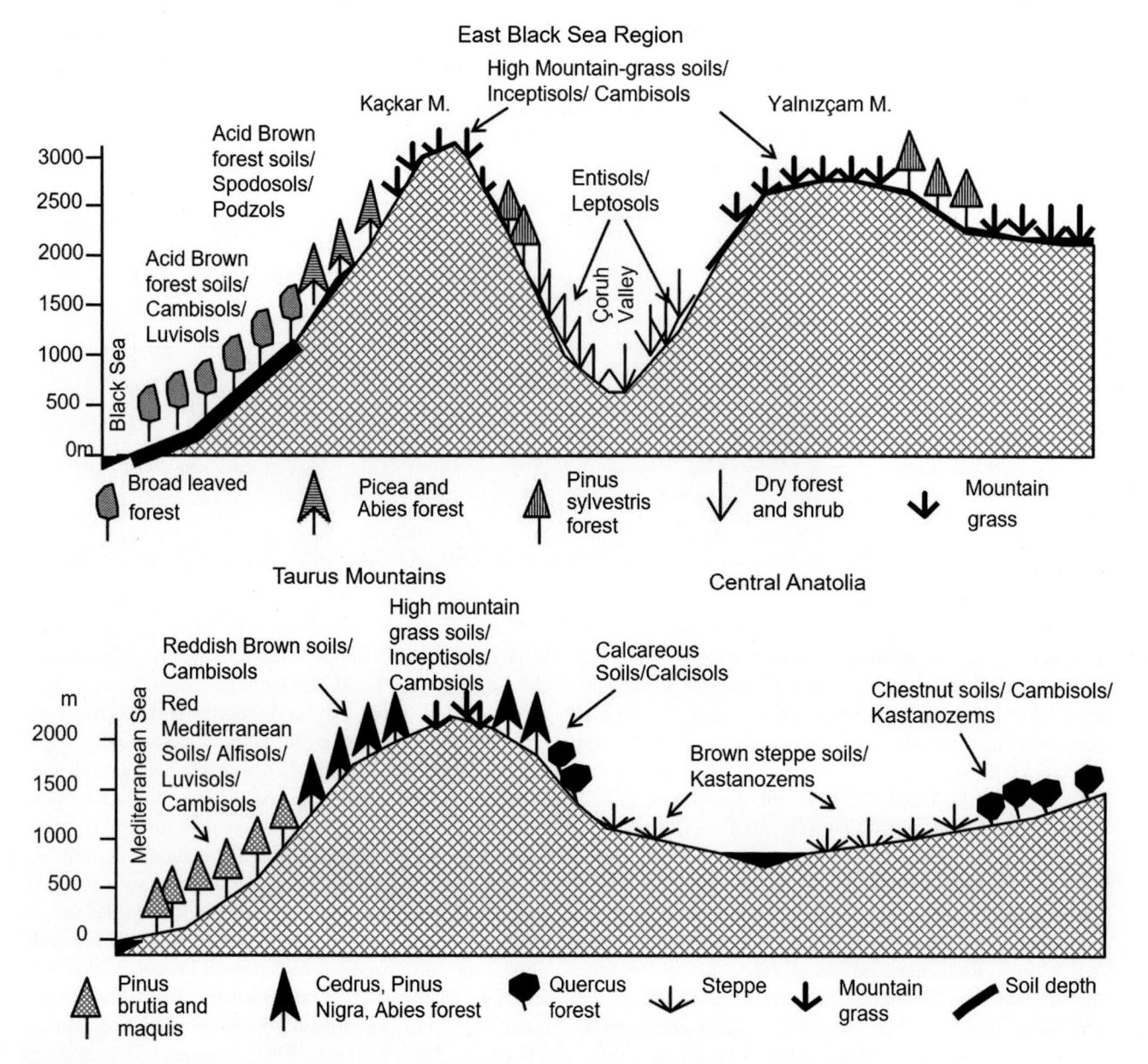

Fig. 20 Soil types on the south and north slopes of the north Turkey and Taurus Mountains (Atalay 2011b)

of hot and dry climatic conditions. The upland areas encircling the plain are the areas of Lithic Torrifluvents/Leptic Fluvisols (Dinç et al. 1987) (Fig. 30).

Saline soils mainly Solonchaks and alkaline soils are found in the playa areas occupying the lower levels of the Konya–Ereğli, Çumra plains in central Turkey. Histosols occur in the low part of the basin occupied by shallow lakes and swamps like in the east part of Erzurum, near Lake Köyceğiz, Düzce, and Yeniçağa depressions.

2.1.5 Plateaus

There are numerous plateaus in Turkey; they occur at sea level and continue to the highlands of 2000–2500 m elevation which developed due to the closing of the Neotethys in the east Mediterranean region since the late Cretaceous. This event resulted in the collision of the African and Arabian plates with the Eurasian plate along the Hellenic arc to the west and the Bitlis–Zagros suture zone to the east (Wilson and Bianchini 1999). Plateaus are found on the lateral Neogene sedimentary

Fig. 21 A general view from the Çoruh Valley passing to the east Black Sea Mountains

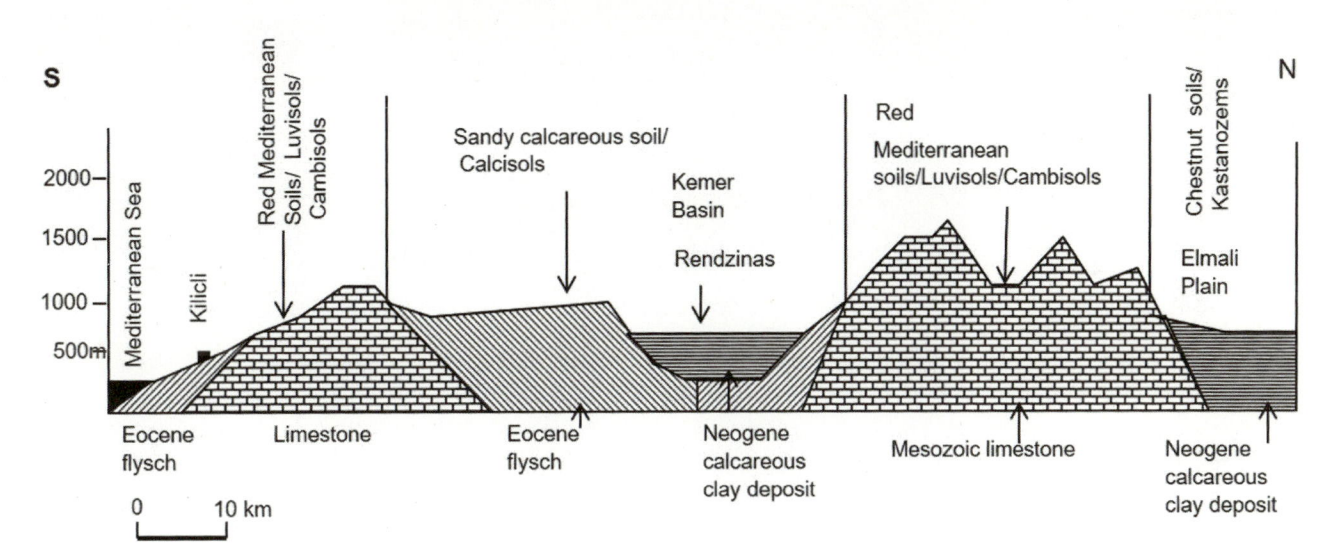

Fig. 22 Different types of soils formed on various parent materials/geologic formation on the west Taurus Mountains (Atalay 1987)

strata and basalt lavas were deeply cut by the streams. The Neogene sedimentary plateaus of Obruk, Cihanbeyli, Haymana, and Bozok are located in central Turkey (Fig. 31).

The Yazılıkaya plateau extends on the trachyte, volcanic sand, and tuff in the northwest of central Turkey underlying Andosol-like soils/Highland Grass Soils (local-Turkey designation). The high and flat plateaus are found on the basalts of the Erzurum–Kars–Ardahan, Yalnızçam, and Allahüekber mountains of north Turkey (Fig. 32).

There are many plateaus in the transitional area between the Aegean Region and central Turkey. One of them is the Ulubey–Banaz plateau, here the horizontal Neogene strata alternating with marl, clayey limestone, and sand stone were deeply dissected more than 150 m from place to place by the

Fig. 23 Steep slopes and rock outcrops of the Zap River with intense soil and parent material erosion, in southeast of Turkey

Fig. 24 Podzolic soil under fir (*Abies bornmulleriana*) and cold-humid climatic conditions on the undulating hills of Uludağ, Bursa, Marmara region

tributaries of the Büyük Menderes River (Atalay 2011a, b) (Fig. 33).

The Taşeli plateau which is located in mid-Taurus extends on the Neogene (Miocene) clayey and sandy limestone formations. In southeast Turkey, the leading plateaus developed on limestone are the Gaziantep, Şanlıurfa, and Adıyaman plateaus. The dry agricultural lands are widespread on the plateaus of the semiarid central Turkey with Cambisols/Mollisols (Brown steppe, Chestnut color, and Rendzina soils). Red Mediterranean soils (Luvisols and Calcisols) are formed under the dry Mediterranean climate of the southeast Turkey plateau.

Fig. 25 Küçük Menderes plain on a rift trough. Fluvisols/Fluvents are common in the alluvial material in the graben, and Leptosols are widespread on the steep slopes of the horst due to strong erosion

Fig. 26 Hydromorphic alluvial soil (Gleysol) along a valley, west Turkey

The transitional area between the basin and the plateaus has different types of soils due to the different types of geomorphic features. The best example for this can be given from the Gediz depression and the uplands. Here, the Fluvisols are found in the alluvial deposits of the Gediz basin, whereas Gleysols occur on the lowland where the ground water table is high. The maturity of the Luvisols increases from the lower surfaces of the basin to the plateau of the Menderes massive where soil development is on gneiss (Atalay et al. 1989) (Fig. 34).

Fig. 27 Saline soil formed by capillary rise in the Söke plain of the Büyük Menderes delta

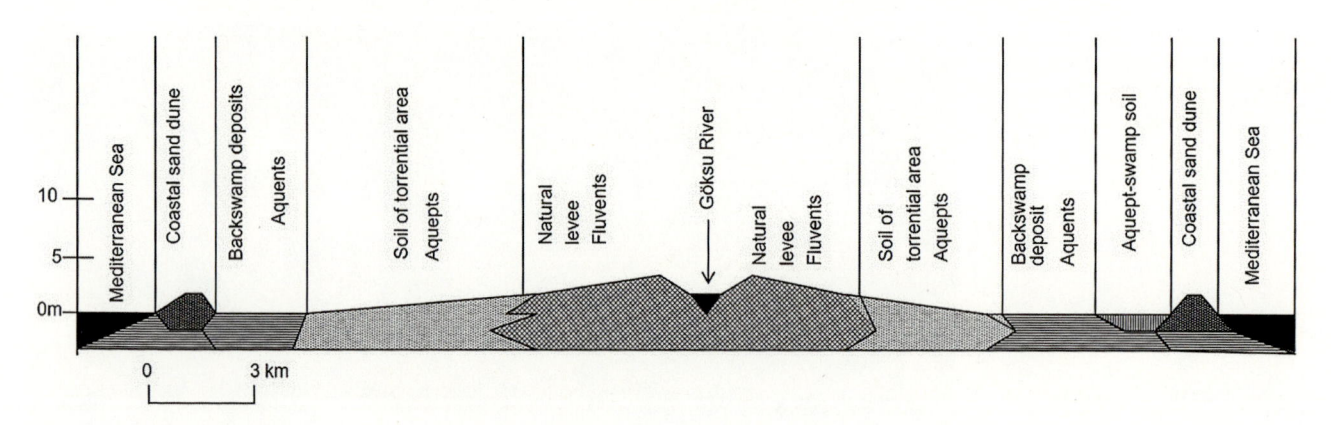

Fig. 28 Soil types of the Göksu Delta that formed via geomorphic properties in the mid-mediterranean region of Turkey (Modified from Özus 1988)

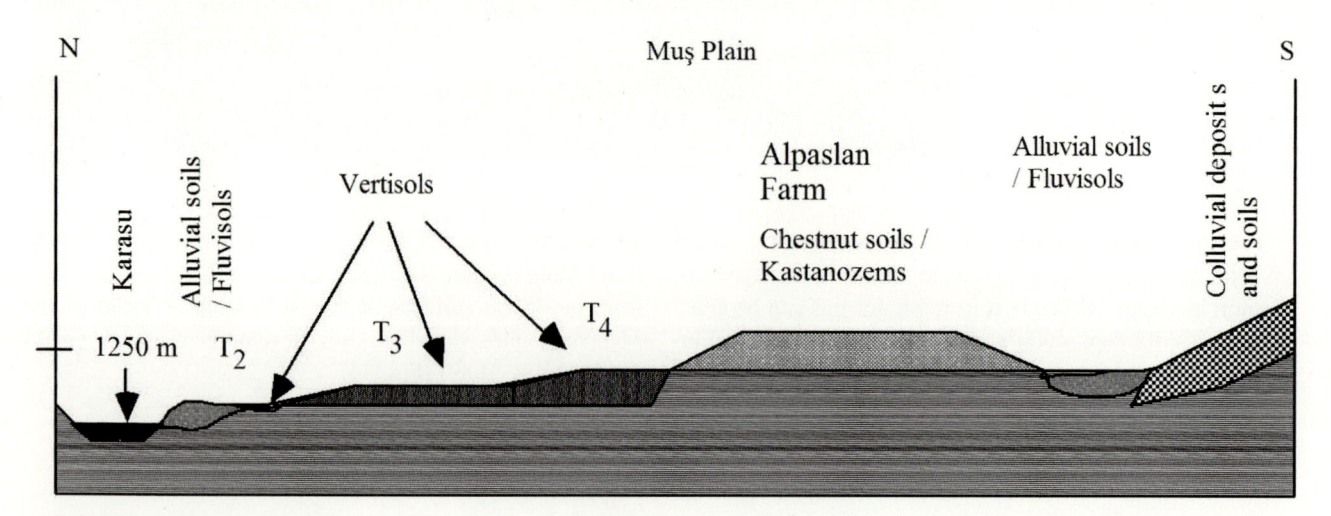

Fig. 29 Soil catena in the Muş Basin (Atalay 1983). Deep Vertisols are formed on the T_4 terraces, and chestnut colored soils (Cambisols) appear on the older surfaces of the basin

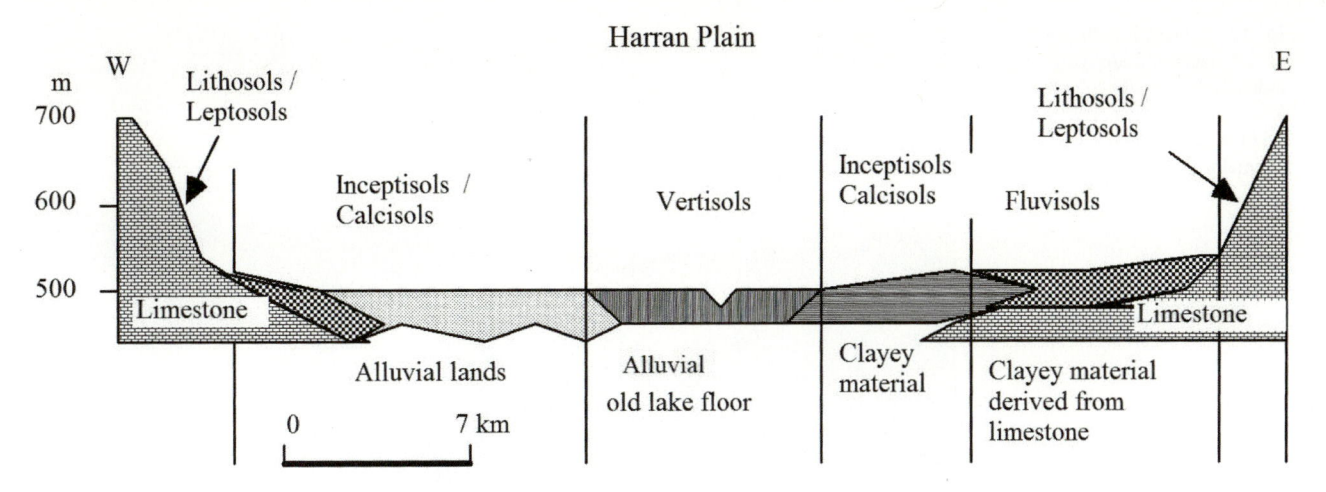

Fig. 30 Soils of the Harran Plain in southeast Turkey (modified from Dinç et al. 1987)

Fig. 31 The Obruk plateau and lake (sink-hole lake/karstic lake) formed by the dissolution of limestone in the south part of central Turkey

2.1.6 Karst Topography

Turkey has several types of karstic landforms containing lapies', caves, dolines, uvalas and poljes, canyon, and ground river valleys (Figs. 35, 36, 37 and 38). Karstification and karstic landforms are related not only to the depth and purity of the limestone, but also to the climate, elevation, and tectonic movements (Fig. 39).

The Karst mostly covers the middle and west parts of the Taurus Mountains and also occur in the southern parts both of central and the southeast Turkey region. The formation of the deep and large karstic depressions, especially poljes, is related to vertical tectonic movements. Large poljes are found within the tectonic basins and/or corridors. The intensity of the uplift of the Taurus Mountains has been led by the intensity of the karstification process leading to the increase of the karstic lands. Subsequent to the uplifts and intense karstification, the vast amount of stored water induced the dissolution of the limestones along the crack systems and fault lines. Consequently, the groundwater systems and caves developed and the polje areas formed, especially along the weak zones of the fault lines. Thus, the main poljes of Turkey appear within the grabens such as the Bucak, Seki, Çeltikçi, and Suğla in the Taurus Mountains.

Fig. 32 A plateau surface of basaltic material underlying Andosol-like soils/Highland Grass Soils with A and C horizons in the Yalnızçam Mountain, northeast Turkey

Fig. 33 The Ulubey–Banaz plateau and canyon valley formed by the upper tributary of the Büyük Menderes River

The lakes occupying the old poljes increase the intensity of the dissolution especially along the weak zones of the formations and finally convert sinkholes of the karstic depressions into groundwater systems. Some rivers draining through the karst have shifted downward following the uplift of the Taurus Mountains (Fig. 39). One can observe this phenomenon via the misfit valleys and wind gaps of the Taurus ranges. The 3-cave systems developed in vertical direction within the upper Cretaceous limestones of the south part of the Lake Suğla polje of the Taurus are well-known nature conservation sites in the country. Some dolines have collapsed and induced the dissolution of the limestones on the Mediterranean coastal belt of Silifke after the Taurus uplift. Some of the best examples for the cylindrical dolines or the karstic pits are Cennet (Paradise) and Cehennem (Hell) caves found in the Mersin area. These sinkholes and caves

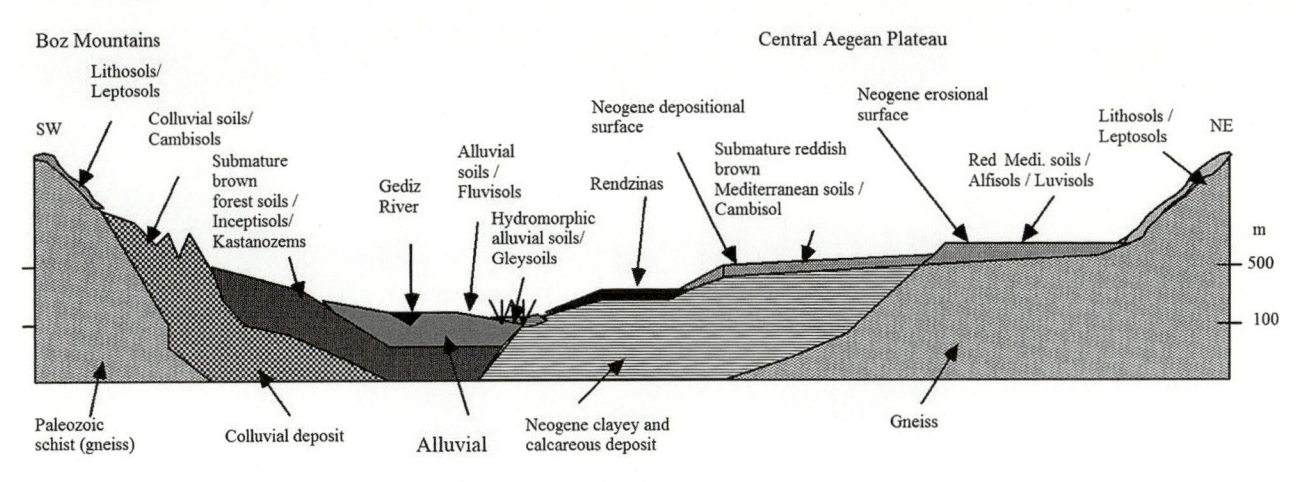

Fig. 34 Soil types between the Gediz graben and the plateau area with the changing parent materials and geomorphic surfaces (Atalay et al. 1989)

Fig. 35 A canyon valley near Kemer town, west of Antalya

containing stalagmites and stalactites follow the old dried up valleys and are the unique cave research areas of Turkey.

The uplifting of the north and south Turkish mountains has formed the deep and narrow canyon valleys. For example, the s of Eşen, Aksu, Köprü, Manavgat, Kadıncık, Seyhan, and Ceyhan flowing into the Mediterranean Sea, and Zap River, a main tributary of Tigris, have deep and long canyons. Other prominent canyon valleys are found in the Tortum and Oltu Rivers which are the prominent tributaries of the rivers Çoruh and Cide in the north part of Turkey.

The Damlataş (Alanya), Narlıkuyu, Koyungöbeği, Dim, Cennet, and Cehennem caves are located in the old river channels widespread in the karstic areas of the Taurus. Karstic accumulation features containing travertine deposits are also present in the Antalya and Denizli areas. The Antalya travertine deposits containing three main terraces are more than 100 m thick and take place on the south edge of the Taurus Mountains and continue toward the Mediterranean Sea. The world known Pamukkale (Hierapolis) is an active travertine and its waters are rich in terms of calcium bicarbonate and spread on the land of the flat terraces along the fault lines (Atalay 1987; Atalay et al. 2008; Dinç et al. 1991a, b). Karstic lands on sloping terrain are not suitable for agricultural practice, but good forest stands and natural regeneration occurs in good conditions due to the penetration of the tree roots toward the deeper parts of the

Fig. 36 Lapies' formed on the steep slopes of Mesozoic limestones in the Gidengelmez Mountain, west Taurus Mountains

Fig. 37 Dolines in the upper part of the Taşeli plateau, middle Taurus Mountains

limestones via the cracks. On the other hand, the seeds of the forest trees such as *Pinus brutia, P. nigra,* and especially *Cedrus libani* germinate and the roots of the seedlings follow the water seepage along the cracks. The roots along the cracks may attain a length of a meter during a vegetation period (Atalay 2014).

Luvisols allocated to agriculture occur on the karstic depressions and mostly on poljes, whereas deep Luvisols are common on the Seki polje in the west Taurus. Rendzinas are found on the poljes preoccupied by the ancient lakes as it is in the Bucak depression, N of Antalya. Marly or calcareous surfaces are widespread at ancient lake bottoms of the

Fig. 38 The Çeltikçi polje (karst plain) in north of Antalya, west part of the Taurus Mountains

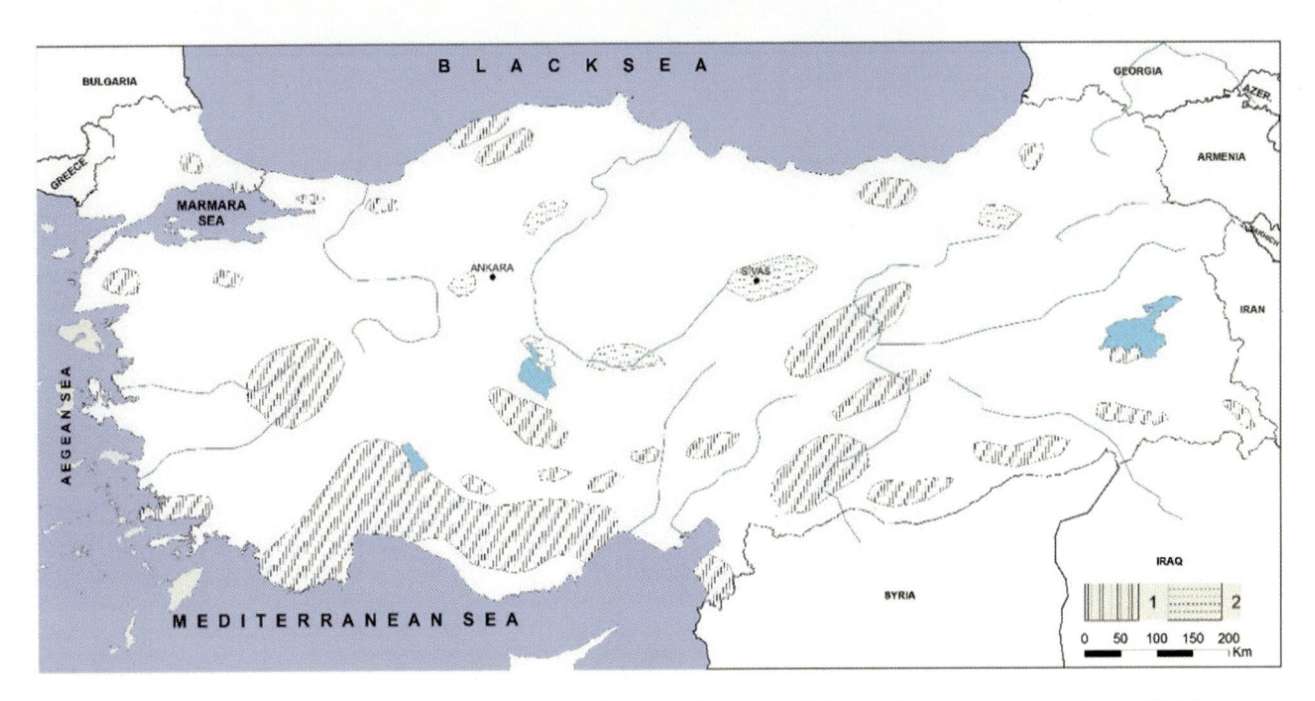

Fig. 39 The distribution of karstic lands of Turkey, (*1*) karstic features on limestone, (*2*) karstic features on gypsum (Atalay 2011a)

Holocene. The best examples of these surfaces are in the Baklan (Denizli province), Bucak, and Çeltikçi depressions.

Gypsic parent materials are also responsible for the dissolution of the topographic forms. The various depressions which have been formed by the chemical dissolution of the gypsum layers are encountered in the vicinity of the Sivas Province. The bottoms of the depressions are harvested for cereals like wheat and barley (Fig. 40).

2.1.7 Glacial Topography

During the Last Glacial Period the mountainous areas of more than 2000 m high in the coastal belts and central

Fig. 40 Doline formed by the dissolution of gypsum near Sivas town, central Turkey

Fig. 41 Cliffs of the high levels of the Last Glacial period in Lake Akşehir

Turkey were occupied by glaciers (Fig. 41). At that period, most parts of Turkey were occupied by desert steppe on which wind activities were dominant. The sea level was lowered down to 125 m from its present level. On the other hand, the lower part of Turkey, i.e., the Konya–Ereğli basin, was occupied by a pluvial lake and most of the present lake levels (Lake Tuz, Burdur, Akşehir, Eber) were raised due to the low evaporation of that period. The ancient cliffs along the lakes clearly reveal the earlier higher levels of the lakes (Atalay 1987, 1992a; Kapur et al. 1999a, b; Ekinci 2012) (Figs. 42 and 43).

During the Early Holocene most of the glaciers started to melt with the increase of temperature and precipitation. Subsequently, the "U"-shaped valleys and cirques mostly with lakes were on the glaciated mountainous areas. These glacial topographic landforms occur in both the east and north parts of Turkey and the Taurus Mountains and are the evidence of the lack or slightly active pedogenesis during the

Fig. 42 Glacial topographic forms, the "U"-shaped valley and cirque lakes in the east Black Sea Mountains

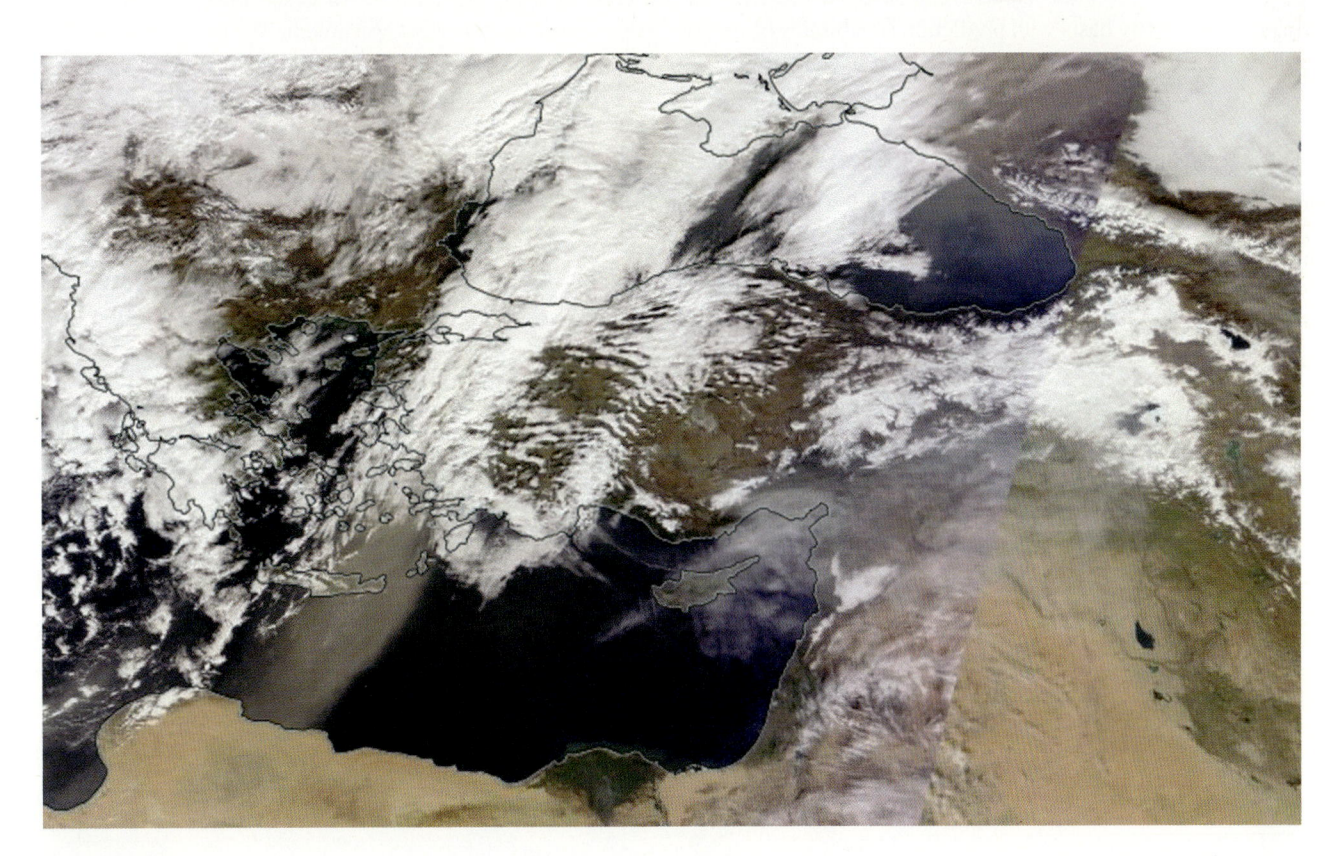

Fig. 43 The February 1, 2015 dust transport event as recorded by the MODIS Terra satellite (https://earthdata.nasa.gov/data/near-real-time-data/rapid-response)

early Pleistocene glacial climatic conditions. However, the High Mountain-Grass Soils/Cambisols/Inceptisols presently occurring on the past glaciated areas of the mountains may be the clues of Holocene soil formation. The old cliffs around the lakes of the Konya basin together with the deltaic deposits (the rounded pebbles) and the aeolian materials in the soils clearly reflect the existence of the glacial periods and late Pleistocene to early–mid-Holocene soil formation in Turkey (Dinç et al. 1991a, b).

3 Saharan Dust Over Turkey

Transport of Saharan Desert dust that affects Turkey is often activated by cyclonic depressions that develop over the Tyrrhenian Sea. The warm front, that is responsible for the uplift and transport of the desert dust, sweeps across the Sahara and carries away the clay size fraction of the desert soil. These episodic dust transport events can be predicted 3 days in advance through the dust transport models (Anonymous 2015).

The outputs of such models can also be reached via Internet on a daily basis with prediction capabilities of nearly 100% over the years since the model parameterization factors do not change at all over the desert regions. Model outputs can be verified via MODIS (Terra and Aqua) and SUOMI satellites that constantly supply images that clearly show evidence for the transport of Saharan dust. February 1, 2015 was one of the most spectacular dust transport events as predicted by the DREAM dust transport model (Fig. 43).

The transport of dust can easily be traced over the marine region since seawater acts as an excellent background medium. But over the ground, it is not easy to detect nominal size dust transport events even for an expert eye. The other obstacles that hinder the observation of dust are the clouds. Thus, so far the observation of dust is confined to specific regions within the synoptic-scale meteorological event due to the reasons stated above.

Another tool that illustrates the potential impact of desert dust is backward or forward air mass trajectory analysis that can easily be performed by using NOAA ARL HYSPLIT facilities (http://ready.arl.noaa.gov/HYSPLIT.php). With this tool it is possible to trace and illustrate the trajectories of any air mass for any time bracket up to 315 h that have the potential of transporting desert dust either forward or backward in time at different elevations. The 72 hourly air mass forward trajectories for the February 1, 2015 indicates the forward trajectories of air mass at 500 (red), 1500 (blue), and 3000 (green) meters above the mean sea level height. It can clearly be seen that the flow of desert dust and the air mass trajectories are in perfect agreement. The geographical extent of the air mass reached all the way to the interior of Siberia within 72 h and further illustrates the impact of Saharan Desert, as shown in Fig. 44.

Transport of the dust can easily be documented by using the various tools mentioned above. Learning the history of

Fig. 44 Seventy-two hour forward trajectory of air mass initiated from Africa on February 1, 2015

Pedogeomorphology

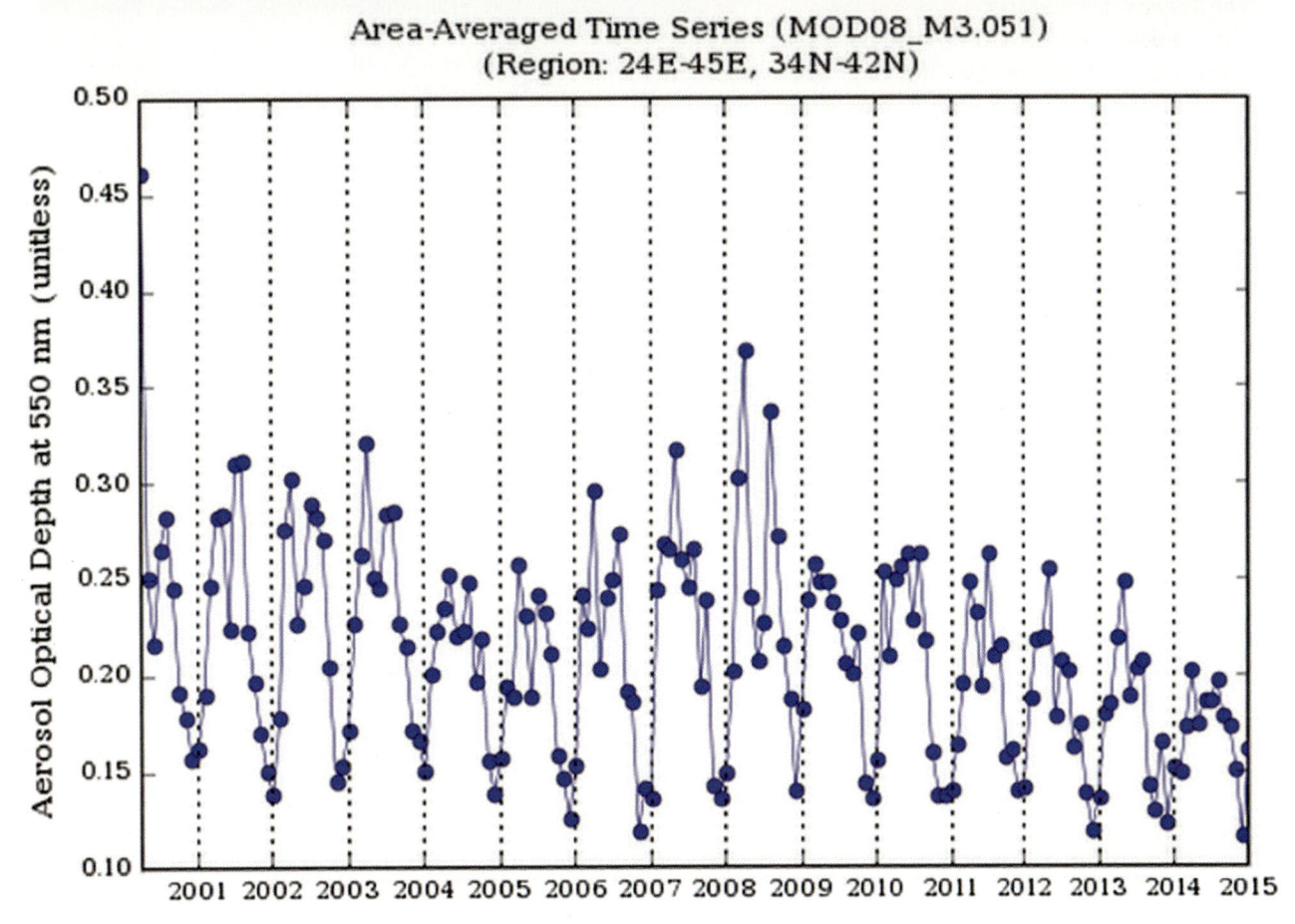

Fig. 45 Long-term variations of AOD over Turkey (2000 till 2015) (http://disc.sci.gsfc.nasa.gov/giovanni)

the dust transport is also useful in making rough estimates about its impact on the receiving body, i.e., the land. At present the Giovanni (http://disc.sci.gsfc.nasa.gov/giovanni) site offers excellent computing and visualization capabilities for satellite-derived Aerosol Optical Depth (AOD) values computed at the 550 nm visual band. This powerful computing facility gives us an opportunity to analyze the spatial and temporal distribution of the impact of desert dust.

The monthly average AOD data for the region that is shown in Fig. 45 is computed for the years 2000 till 2015 and shows a clear decreasing trend since 2008.

This monthly trend gives us a climatic perspective about the migration of desert dust over the region and is in conflict with the expected impacts of climate change for the region. Various GCM models predict an increase in temperature and decrease in precipitation for the Mediterranean and especially for the eastern Mediterranean as shown by Dubrovský et al. (2014). This would automatically raise the expectations of more dry hence dustier conditions over the region but the satellite-derived data shows just the opposite.

The 15-year dust transport history of the region can also be computed as cumulative deposition once again by using the powerful tool offered by Giovanni. Satellite-derived mass deposition based on a 15-year data stream reveals a mere 10–15 µg/cm^2 dust deposition over the topsoil of Turkey on an annual basis.

References

Anonymous (2015) SC-DREAM8b v2.0 Atmospheric dust forecast system. http://www.bsc.es/earth-sciences/mineral-dust-forecast-system/bsc-dream8b-forecast. Accessed on 6 Dec 2015

Atalay İ (1983) The geomorphology and soil geography of Muş Basin. Aegean University Faculty of Letter Publication No. 25. İzmir, Turkey. 154 p (in Turkish)

Atalay İ (1987) Introduction to geomorphology of Turkey, 2nd edn. Publication of the Aegean University Faculty of Literature, No. 9, İzmir, Turkey (in Turkish)

Atalay İ (1992a) The paleogeography of the Near East (From Late Pleistocene to Early Holocene) and human impact. Ege University Press, İzmir, Turkey (in Turkish)

Atalay İ (1992b) The ecology of beech (*Fagus orientalis* Lipsky) forests and their regioning in terms of seed transfer. Ministry of Forestry, Forest Trees and Seeds Amelioration Ins. Pub. No. 5, Ankara, 209 p

Atalay İ (1995) Effects of climatic changes on the vegetation in the Near East. Bulletin de la Sociêtê de Gêographie D'Êgypte 68:157–177

Atalay İ (1998) Paleoevironmental conditions of the Late Pleistocene and Early Holocene in Turkey. In: Alsharhan AS, Glennie KW, Whittle GL, Kendall CG StC, Balkema AA (eds) Quaternary deserts and climatic change. Rotterdam/Brookfield, pp 227–238

Atalay İ (2011a) Geography and geopolitics of Turkey, 7th edn. Meta Press, İzmir, Turkey (in Turkish)

Atalay İ (2011b) Soil formation, classification and geography, 4th edn. Meta Press, İzmir, Turkey

Atalay İ (2012) Applied climatology, 2nd edn. Meta Press, İzmir, Turkey (in Turkish)

Atalay İ (2014) Ecoregions of Turkey. Ministry of Forestry Publication No: 163. ISBN 975-8273-4-8, İzmir, Turkey (in Turkish)

Atalay İ, Mortan K (2011) Regional geography of Turkey, 5th edn. İnkilap Pub, İstanbul, Turkey (in Turkish)

Atalay İ, Sezer L, Işık Ş, Mutluer M (1989) The effect of climate, parent material and geomorphic factors on the formation of the soils in the Aegean region. Aegean Geogr J 5:32–34

Atalay İ, Efe R, Soykan A (2008) Mediterranean ecosystems of Turkey: ecology of Taurus Mountains. In: Efe R, Cravins G, Öztürk M, Atalay İ (eds) Natural environment and culture in the mediterranean region. Cambridge Scholar Publishing, England, pp 1–38

Cangir C, Kapur S, Ekinci H (1990) Clay mineralogy and related landslides of an Alfisol formed on the Oligocene marine deposits of the Tekirdağ Township. III. National clay symposium proceedings, Ankara, Turkey, pp 177–188

Dinç U, Yeşilsoy S, Sayın M, Şenol S, Kaya Z, Çolak AK, Sarı M, Özbek H, Yeğingil İ (1987) Soils of the Harran Plain. Project of the Turkish Scientific and Technological Research Council, Ankara, Turkey

Dinç U, Şenol S, Kapur S, Sarı M (1991a) Catenary Soil relationships in the Çukurova Region, Southern Turkey. Catena 18:185–196

Dinç U, Şenol S, Kapur S, Sarı M, Derici MR, Sayın M (1991b) Formation, distribution and chemical properties of saline and Alkaline soils of the Çukurova Region, Southern Turkey. Catena 18:173–183

Dinç U, Şenol S, Kapur S, Atalay İ, Cangir C (1993) Soils of Turkey. University of Çukurova Publication No. 12, Çukurova University Press, Adana (in Turkish)

Driessen PM (1970) Soil salinity and alkalinity in the Great Konya Basin, Turkey. Centre for Agricultural Publishing and Documentation, Agricultural research reports 743. Wageningen

Driessen PM, de Meester T (1969) The soils of the Çumra area, Turkey. Department of Tropical Soil Science, Agricultural University of Wageningen, Research reports 720, 105 p

Dubrovský M, Hayes M, Duce P, Trnka M, Svoboda M, Zara P (2014) Multi-GCM projections of future drought and climate variability indicators for the Mediterranean region. Reg Environ Change 14 (5):1907–1919

Ekinci D (2012) Glacial morphology of Turkey (Turkey). Proceedings of 7th Turkey-Romania geographical academic seminar. June 1–9, Antalya, Turkey. In: Atalay İ, Balteanu I, Efe R (eds) Potential and problems of natural environment in Turkey and Romania, Inkilap Publishing Company, İstanbul, Turkey. pp 159–180

Erinç S (1970) The young volcanic relief from Kula to Adala. J Geogr Inst, İstanbul University, 17:7–32

İnci U (1991) Miocene alluvial fan—alkaline playa lignite-trona bearing deposits from an inverted basin in Turkey: sedimentology and tectonic controls on deposition. Sed Geol 71:73–97

Kadir S (2007) Mineralogy, geochemistry and genesis of smectite in Pliocene volcaniclastic rocks of the Doğanbey Formation, Beyşehir basin, Konya, Turkey. Clays Clay Miner 55:402–422

Kapur S, Atalay I, Ernst F, Akça E, Yetiş C, İşler F, Öcal AD, Uzel İ, Şafak Ü (1999a) A review of the late quaternary history of Turkey. Special issue, Olba II, Mersin University Publications, pp 233–278 (in Turkish)

Kapur S, Saydam C, Akça E, Çavuşgil VS, Atalay İ, Karaman C, Özsoy T (1999b) Carbonate pools in soils of the mediterranean. A case study from Turkey. In: Kimble RJM, Eswaran H, Stewart BA (eds) Global climate change and pedogenic carbonates, Lewis Publications, London, New York, Washington, DC USA

Oakes H (1954) Soils of Turkey. Ministry of Agriculture, Soil Conservation and Farm Irrigation Division Publication No. 1, Ankara, Turkey

Özus A (1988) Genesis-classification and physical, chemical and mineralogical properties of the Silifke Plain soils. Publication of the Çukurova University Institute of Science (Unpublished Doctorate Thesis), Adana (in Turkish)

Tanoğlu A (1947) Elevation zones of Turkey. J Turk Geogr 9(10):37–55

Wilson M, Bianchini G (1999) Tertiary-quaternary magmatism within the mediterranean and surrounding regions. In: Durand B, Jolivet L, Horvath F, Séranne M (eds) The mediterranean basins: tertiary extension within the Alpine Orogen. Geological Society of London Special Publication, 156, pp 141–168

Soil Geography

Mehmet Ali Çullu, Hikmet Günal, Erhan Akça, and Selim Kapur

1 Introduction

This account of the 'Soils of Turkey' has developed from an initial effort to update the composition and content of the Soil Map of Turkey (Fig. 1) and it was undertaken in two phases. During the *first phase* a comprehensive set of data was assembled to update the map by means of existing data sources, such as: (i) the soil map of Turkey at a scale of 1:25,000 (previously prepared by the former General Directorate of Soil and Water (TOPRAKSU-Turkish acronym) of the Ministry of Agriculture) and the GIS-based dataset developed by the contemporary General Directorate of Agricultural Reform (GDAR-TRGM of the Ministry of Food, Agriculture and Livestock), (ii) the CORINE Land Cover map, at a scale of 1:100,000, (iii) the modified and simplified soil parent material map at the 1:1,000,000 scale (GDAR-TRGM 2013) prepared by an interdisciplinary team of pedologists, geologists, geomorphologists, and GIS experts, (iv) the Digital Terrain Model SRTM-30, obtained by resampling of the 90 m resolution SRTM, used to compute the average slope and aspect for each soil polygon.

These inputs proved to be very useful in extending the Pedotransfer Rules (PDR) and Pedotransfer Functions methodology, utilizing the experience and judgement of the interdisciplinary team mentioned above. The PDR adopted for this purpose are mainly qualitative, and they involve an assumption that the confidence level of individual inferred attributes can be assigned relative weightings (King et al. 1994; Van Ranst et al. 1995).

The *second phase* in the process of compiling the Soil Map of Turkey involved a program of field surveying and associated description of new profile pits. This initiative was undertaken by the authors of this book in order to check and, where necessary, to correct the soil classes allocated according to the criteria used by WRB (IUSS Working Group WRB 2015).

Decisions concerning the location of new soil-pits designed to provide supplementary data for this second phase were based on the compilation, integration, and synthesis of existing bodies of pedological information from all regions of Turkey, including the results of recent soil surveys undertaken by other teams, along with the detailed and reconnaissance soil surveys published by the Turkish Ministry of Food, Agriculture and Livestock (GDAR-TRGM 2013), the universities, TÜBİTAK (The Scientific and Technological Research Council of Turkey), and the General Directorate of State Agricultural Establishments (TİGEM). Additionally, the resulting database was integrated with satellite imagery, enabling correlations of soil type and related properties with geomorphological attributes, such as topographic gradient, land cover, land use, and climatic differences.

All these data compilations were subsequently analyzed in a GIS environment and the soils were classified according to the WRB (IUSS Working Group WRB 2015) procedures which are primarily intended to facilitate the correlation of national and local systems with international soil classification schemes. Furthermore, the local peculiarities of the indigenous soils of Turkey were noted in the field then named using the rich range of WRB (IUSS Working Group WRB 2015) qualifiers, thus complementing and enhancing the distinctive attributes for each soil class. The soil classification team faced various difficulties in this

M.A. Çullu (✉)
Department of Soil Science and Plant Nutrition,
University of Harran, Şanlıurfa, Turkey
e-mail: macullu@harran.edu.tr

H. Günal
Department of Soil Science and Plant Nutrition,
Gaziosmanpaşa University, Tokat, Turkey
e-mail: hikmet.gunal@gop.edu.tr

E. Akça
School of Technical Sciences, University of Adıyaman,
02040 Adıyaman, Turkey

S. Kapur
Department of Soil Science and Plant Nutrition,
University of Çukurova, 01330 Adana, Turkey

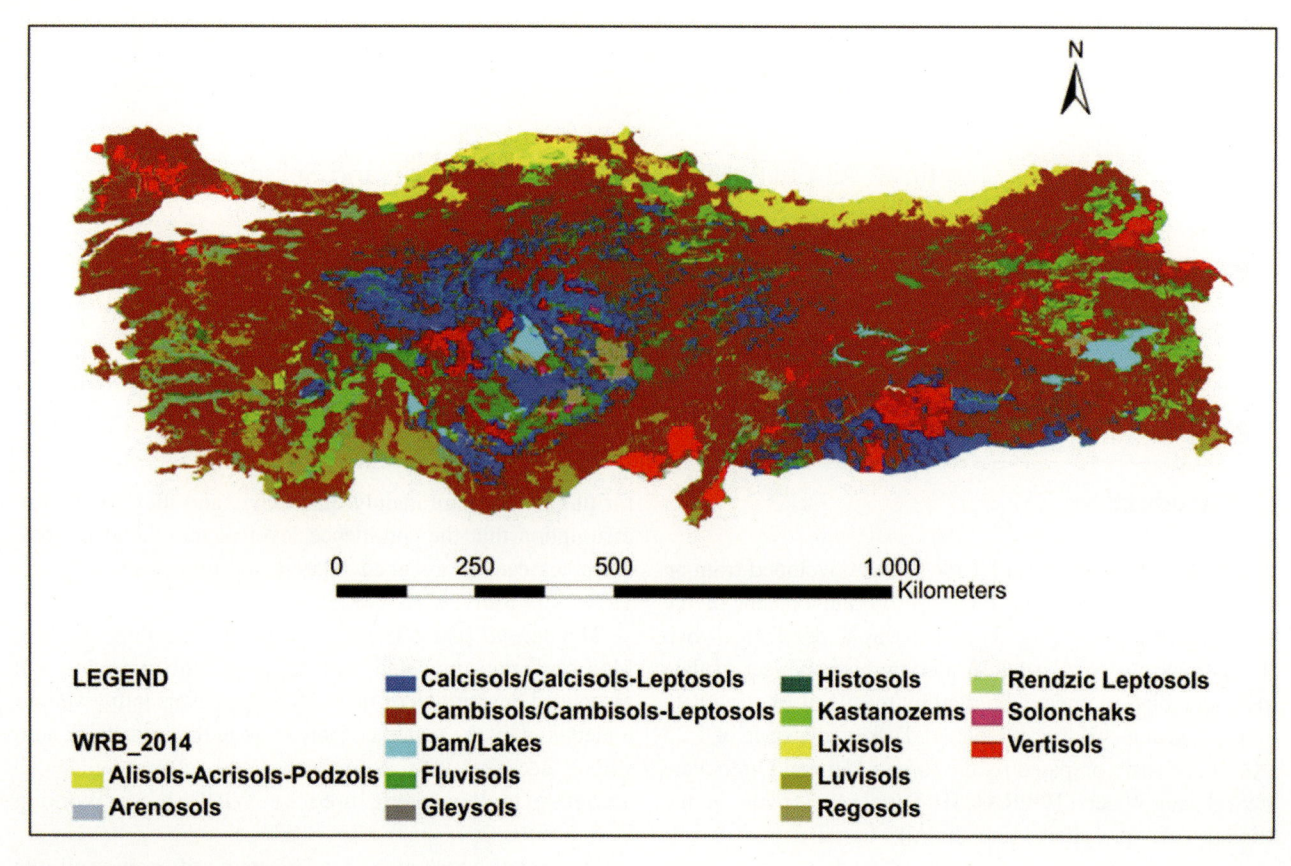

Fig. 1 WRB soil map of Turkey

process which they overcame by creating separate boxes for individual soils in the legend of the soil map. Most of these classificatory problems are attributable to the highly undulating topography and erodibility potential of soils within the semi-arid climatic areas of Turkey. As a consequence, during this study numerous soil complexes (e.g., Cambisols-Leptosols) were defined and identified, taking account of the sloping and undulating surfaces on which these soils were developed and the active tectonic topography, typified by graben-valley systems, that characterizes many parts of Turkey. Moreover, some of the soil groups of the WRB did not occur in Turkey and some were not mapped due to scarce distribution.

2 Mapped Soil Groups

The most widely distributed soils of Turkey, i.e., the existing and/or mapped WRB Soil Groups are the Cambisols/Cambisols-Leptosols (63.55%), Fluvisols (9.50%), Calcisols/Calcisols-Leptosols (9.39%), Vertisols (5.15%), Alisols-Acrisols-Podzols (3.25%), Kastanozems (2.85%) and Luvisols (2.05%) (Table 1). The Cambisols, Calcisols, Vertisols, and Fluvisols are the most appropriate and extensively managed soil ecosystems primarily for field crops and cereals. Further, the Calcisols, Luvisols, Acrisols-Alisols-Podsols, and Arenosols bear utmost significance in the production of particular crops such as the olives, carobs, apricots, hazelnuts, and stone pine. These soil ecosystems represent the long-standing management history of the country. On the other hand the Solonetz, Solonchaks, Gleysols, and Regosols are the most problematic soil ecosystems for agricultural production in need of supplementary activities for sustainable use (Table 1).

3 Unmapped Soil Groups

The dominant Mediterranean and continental climates of Turkey have not favored the formation of the highly weathered soils such as the Plinthisols, Nitisols, Durisols, and Ferralsols in any part of the country. As for the Cryosols, they may scarcely occur in east Turkey, namely at Mounts Süphan, Ağrı (Ararat), and Buzul (glacier) but were not mapped or defined in any study up to the present (Fig. 2). Retisols were also not mapped in the dominantly semi-arid Turkey due to the absence of the prominent

Table 1 Distribution of WRB groups in Turkey

Group name	Area (ha)	Distribution (%)
Cambisols and Leptosols	48,958,936.22	63.55
Fluvisols	7,338,590.54	9.50
Calcisols and Leptosols	7,266,094.83	9.39
Vertisols	3,957,069.74	5.15
Alisols-Acrisols-Podzols	2,498,952.52	3.25
Kastanozems	2,219,766.04	2.85
Luvisols	1,458,955.98	2.05
Regosols	989,711.55	1.62
Rendzic Leptosols	839,379.12	1.05
Solonchaks	750,346.09	0.98
Lixisols	385,311.14	0.50
Arenosols	74,870.34	0.09
Gleysols	23,761.09	0.03
Histosols	5,845.06	0.02
Total area	76,767,590.26	100.00

'albeluvic glossae' forming in the Argic horizon under boreal or temperate climatic conditions. Similarly, the Gypsisols are not mapped or mentioned in this book, due to their scarce and coalescent occurrence to Cambisols and Calcisols in central Turkey (300 mm annual precipitation), despite the favorable conditions to form the non-cemented secondary gypsum accumulation that meet the gypsic horizon criteria of the WRB (IUSS Working Group WRB 2015). Some of the other unmapped soil groups such as the Planosols and Stagnosols occur at unmappable small scales and at marginal areas disregarded for their low agricultural productivity and expensive amelioration to be performed for so-called income-generation which overlooks their crucial significance for the environment. The Umbrisols may occur along the high western slopes of the Black Sea region coasts of the country as patches. However, their inappropriate use for agriculture and scarcity along with the probable absence of some of the WRB (IUSS Working Group WRB 2015) qualifiers (especially an umbric horizon) are the major reasons for being unmapped. The Chernozems, Phaeozems, Anthrosols, and Technosols, however, besides their probable scarcity, may also not possess some of the qualifier properties of the relevant soil groups of the WRB (IUSS Working Group 2015) (Fig. 2) as explained below.

3.1 Chernozems and Phaeozems

The Chernozems and Phaeozems are the soils that are climatically possible to occur in eastern Turkey under a cold snowy forest climate with dry summers (humid micro-thermal Köppen class). However, they were not described in any of the soil mapping efforts in the country, nevertheless this led us to consider their occurrence and include them in the Kastanozems as scarce and coalescent/complex groups in this wider spread soil group bearing some similar properties. The dominant soils mapped as the Kastanozems in the north, west–central, and especially eastern parts of the country have relatively thinner and less dark humus-rich horizons together with higher carbonate accumulations in the subsurface horizons than the Chernozems and are not leached compared to Phaeozems which also have darker humus-rich horizons.

3.2 Anthrosols

Some soils of the Kızkalesi (Korykos) (the soils occurring behind the man-made wall terraces of the late Roman period in Southern Turkey) Calcisol Ecosystem/Anthroscape given in the Calcisol chapter (Section "The Kızkalesi (Korykos) Calcisol Ecosystem/Anthroscape") may likely be regarded as Anthrosols with degraded and/or partly lost Terric and Irragric qualifier properties due to the abandonment of the area after the Late Roman period (Fig. 2). The Terric qualifier could have stood/may stand for the soil materials filled by the late Roman societies behind polygonal rock-cut wall terraces and the Irragric for the irrigation performed via rock-cut channels and aqueducts from the karstic sources of the southern Taurid slopes for mostly olives, carobs and vineyards. Nevertheless, the abandoned Anthrosols behind the wall terraces have most likely encountered soil change during millennia. This change was/is probably due to the particular intrinsic structural development of the added (human transported) Calcisols (likely changed to Anthrosols

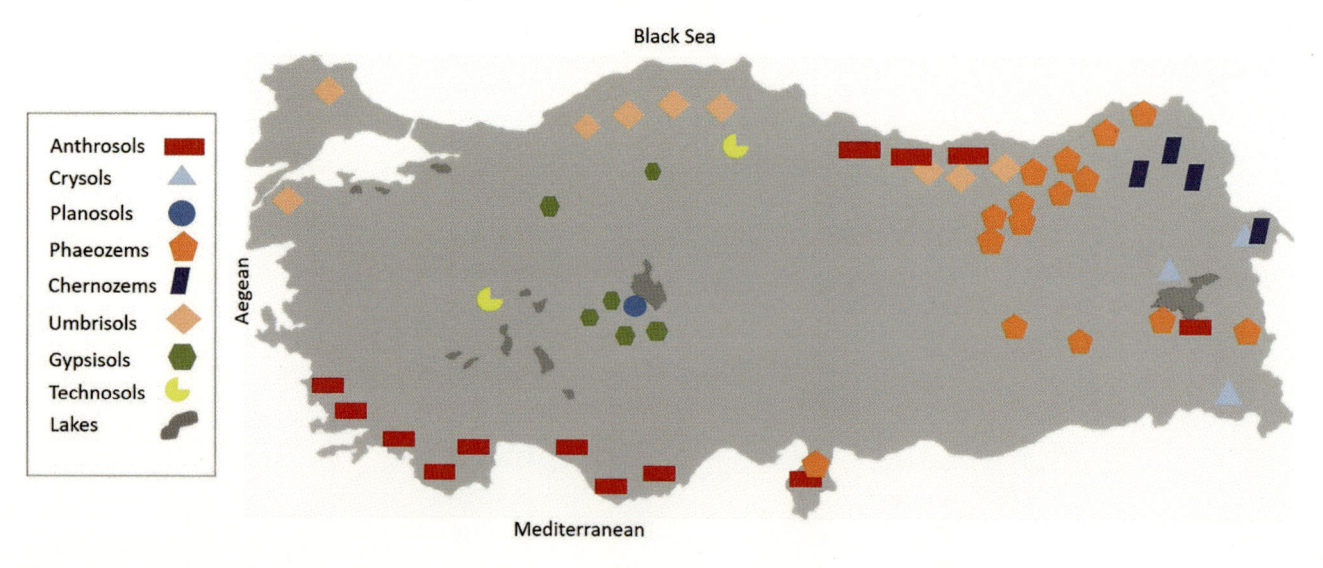

Fig. 2 The unmapped soil groups of Turkey

after intense cultivation and irrigation) enhanced by the rich smectitic clay contents together with the seasonal extremities of the Mediterranean climate and the secondary succession of the maquis-garrigue vegetation. The other probable Anthrosol area is in the north of Turkey (the Black Sea region) with annual mean rainfall of over 1000 mm (max. 2500 mm), which are included in the Acrisols-Alisols-Podsols ecosystem complex. These soils have been allocated to tea cultivation for at least a century and for hazelnuts since the historical periods. The Terric and the Hortic qualifiers are the probable Anthrosol indicators expected to occur in the soils of this complex. The transported-added soils over the Acrisols-Alisols-Podsols ecosystem of the steep to moderately steep slopes of the Black Sea Mountain ranges have been the terraces to shelter patches of Anthrosols which have not been described by the authors conducting soil mapping in the area. This is partly due to the scale of the map produced for this book and the random distribution of the probable Anthrosols among the Acrisols-Alisols-Podsols complex.

3.3 Technosols and Soils of Archaeological Sites

Technosols were not described by previous studies as the country possesses vast arable lands and in turn the absence of the will to utilize the contemporary or ancient land filled areas for cultivation. However, the landfills of the widespread marble quarries in the karstic areas filled with waste rocks and stones as well as the numerous archaeological sites with human reworked soils may shortly lead to the need for mapping Technosols. Technosols are prone to develop at traditional dumping areas of brick and pottery manufacturing zones (containing high amounts of clay and soil materials) mostly in central, western, and southern parts of the country.

In addition to the well-defined Technosols of the WRB IUSS Working Group (2015) an extract of the country's renowned archaeological site soils which may possess technic or relevant properties in need of the scholarly attention of the WRB IUSS Working Group (2015) is given herein. These studies concern the soils of the renowned historical sites of the country, such as the prestigious Bitinian (Graeco-Roman 2nd Cen. BC) tomb (Bursa, north western Turkey) underlying a Regosol with a Technic qualifier (a Colluvitechnic Calcaric Regosol) (Boyraz et al. 2011). This Regosol reflects the Late Holocene post-depositional weathering/soil formation of the deposit overlying the tomb. Another renowned site in south Turkey discovered in the early 1940s by the eminent Turkish archaeologist H. Çambel studied by Paksoy (2000) revealed the formation of probable initial technic qualifier properties at a Late Roman midden deposit/dump (containing garbage and earlier stone/construction materials of the midden tank) of the late Hittite site in Karatepe-Kadirli in south Turkey. Paksoy (2000) also revealed the formation of secondary rutile needles on quartz grains of the midden deposit indicating higher alkaline conditions at the time of formation. Further studies revealed soil (Technosol) formation in the Domuztepe site adjacent to Karatepe on fortress walls, around furnaces and in human transported basaltic soils by physical, chemical, mineralogical, and micromorphological analyses (Çambel and Knutstad 1997; Mermut et al. 2004). Isotope dating (ca. 2700 BP) results of the basaltic Technosols of the site confirmed the settlement age of the Late Hittite/Roman inhabitants determined by H. Çambel's earlier studies (Çambel and Knutstad 1997; Mermut et al. 2004). Akça et al. (2016)

analyzed the soil materials (Technosols/sediments) of the indoor floors of the Tell-Kurdu (6th millennium BC) settlement site in Antakya via micromorphology determining secondary phosphorus minerals (vivianite) in the presence of frequent bone fragments as butchery remnants. This study also sought to reveal indoor household activities via pedogenic processes concerning leaching (indoor water use) and formation of the microstructure within human made floor plaster and mats. Akyol and Demirci (2000) determined the habitation intensities in the Çatalhöyük Neolithic site referring to the distribution of the phosphorous fraction contents in the soils.

References

Akça E, Özbal R, Başkaya H, Kapur S (2016) The story of the sediments from 6th millennium BC Tell Kurdu under the microscope (submitted to Eurasian Soil Science)

Akyol AA, Demirci Ş (2000) Phosphorous analyses on the systematic soil samples of the Neolithic Çatalhöyük site. In: 15 archaeometry excavation meeting, pp 55–63

Boyraz D, Başkaya HS, Akşit İ, Arocena J, Polat S, Dingil M, Şahin M, Şahin D, Kaynak G, Akay SK, Yılmaz Ö, Akça E, Biçici M, Kapur S (2011) Preliminary submicroscopy of a vertebral bone fragment from a Bitinian tomb of 2nd century BC in Bursa, western Turkey. TÜBA-AR Turk Acad Sci J Archaeol 14:151–158

Çambel H, Knutstad JE (1997) The studies of Kadirli, Karatepe-Aslantaş and Domuztepe sites. In: XIX proceedings of the excavation studies meeting, Ankara, pp 483–496

GDAR-TRGM (2013) General directorate of agricultural reform, ministry of food, agriculture and livestock. Soil database for Turkey. In the GIS Media. Ankara

IUSS Working Group WRB (2015) World reference base for soil resources 2014, update 2015 international soil classification system for naming soils and creating legends for soil maps. World soil resources reports No. 106. FAO, Rome, 203 p

King D, Daroussin J, Tavernier R (1994) Development of a soil geographical database from the soil map of the European communities. Catena 21:37–56

Mermut A, Montanarella L, FitzPatrick EA, Eswaran H, Wilson M, Akça E, Serdem M, Kapur B, Öztürk A, Tamagnini T, Çullu MA, Kapur S (2004) Excursion book 20–26 Sept 2004. In: 12th international meeting on soil micromorphology. EC-JRC EUR 21275 EN/1. 58 p

Paksoy S (2000) The Physical, chemical and mineralogical properties of late hittite period soils in Domuztepe (Karatepe). MSc thesis in graduate school of sciences. Çukurova University, Adana. YÖK thesis number: 112332. 68 p

Van Ranst E, Thomasson AJ, Daroussin J, Hollis JM, Jones RJA, Jamagne M, King D, Vanmechelen L (1995) Elaboration of an extended knowledge database to interpret the 1:1,000,000 EU soil map for environmental purposes. In: European land information systems for agro-environmental monitoring. In: King D, Jones RJA, Thomasson AJ (eds). EUR 16232 EN. Office for Official Publications of the European Communities, Luxembourg, pp 71–84

Cambisols and Leptosols

Erhan Akça, Sevda Polat, Somayyeh Razzaghi, Nadia Vignozzi, Zülküf Kaya, and Selim Kapur

1 Introduction

Topography seems to be a major and consistent soil-forming factor that has modified in the past and/or modifies in the present the soils of Turkey. Thus, the common sloping topography together with the semiarid climate and moderate vegetative cover dominant are mediating the progress of soil formation in the country. In this context, the soils with moderate depth and A-Bw-C horizon sequences are classified as the Cambisol complexes with neighboring Leptosols. The Leptosols included are the eroded and shallow versions of the Cambisols with A–C horizon sequences as it is in the case of Calcisols–Leptosols and Luvisols–Leptosols. These complexes are located within tectonic-influenced topographies rich in graben valleys and sloping systems of the south, southeast, east, and west of Turkey.

The IUSS Working Group WRB (2015) defines Cambisols as soils having a cambic horizon within 50 cm of the soil surface and lower limit at 25 cm or deeper below the soil surface or an anthraquic, hydragric, irragric, plaggic or terric horizon, or a petroplinthic, pisoplinthic, plinthic, salic, thionic, or vertic horizon starting within 100 cm of the soil surface, or one or more layers with andic or vitric properties with a combined thickness of 15 cm or more within 100 cm of the soil surface. Cambisols may be correlated with inceptisols of the soil taxonomy (Soil Survey Staff 2014), 'brown soils' as 'Braunerden' (Germany), 'Sols bruns' (France), Brunisolic soil (Canada), and or 'Burizems' (Russia) (FAO/ISRIC 2006).

E. Akça (✉)
School of Technical Sciences, University of Adıyaman, 02040 Adıyaman, Turkey
e-mail: eakca@adiyaman.edu.tr

S. Polat
Ministry of Forestry and Water Affairs, Eastern Mediterranean Forestry Research Institute, Tarsus, Turkey

S. Razzaghi · Z. Kaya · S. Kapur
Department of Soil Science and Plant Nutrition, University of Çukurova, 01330 Balcalı, Adana, Turkey

N. Vignozzi
CRA Agricultural Research Council, Rome, Latium, Italy

© Springer International Publishing AG 2018
S. Kapur et al. (eds.), *The Soils of Turkey*, World Soils Book Series,
https://doi.org/10.1007/978-3-319-64392-2_8

2 Parent Materials

Cambisols in Turkey generally develop on readily weathered unconsolidated sedimentary rocks, ophiolites, marls, basalts, granites, and calcretes under a semiarid climate. However, the development is more common on basic rocks than acidic ones. Cambisols mostly overlie slightly sloping (0–2%) or undulating lands with sparse to moderate vegetation (Fig. 1). The pluvial and interpluvial periods of the Quaternary (Kapur et al. 1990) led to the weathering of relatively soft rocks where Cambisols developed. These climatic fluctuation periods of the Pleistocene were also responsible for the development of the widespread Cambisols of Turkey on the colluvials—the mass earth material movements—the mudflows (Erol 1999) (Fig. 2). Cambisols with weak B-horizon development are also found on river terraces that are relatively older than the late Holocene alluvial deposits (Erol 1999). Consequently, Cambisols may occur on a large variety of rock materials and topographies and thus are distributed almost in all the geographic regions of the country neighbored by Leptosols.

3 Environment

Although Cambisols are more dominant under the forest communities of the Tauride mountain ranges (south Turkey) than the Leptosols, they are mostly cultivated for agricultural crops due their wide and even distribution and higher fertility than the Leptosols along the easily accessible topographies of the country. Cambisols of Turkey, along with some of Leptosols, cover a land area of about 48,958,936.22 ha which is 63.55% of the whole agricultural land (Fig. 3).

Fig. 1 The sloping Cambisol–Leptosol topography in Ceyhan, South Turkey

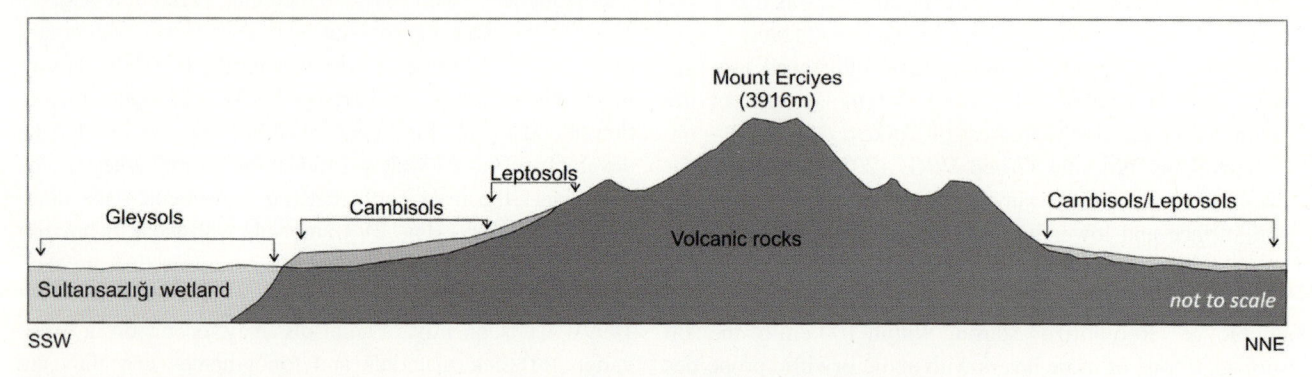

Fig. 2 A Cambisol–Leptosol toposequence along a colluvial continuum in central Turkey (Erol 1999)

4 Profile Development

The cambic horizon is the diagnostic horizon for the Cambisols. Organic matter (OM) contents of Cambisols other than the ones of the Black Sea and Marmara regions (Güzel 1980) are around 2% due to the moderate to sparse vegetation (Evrendilek et al. 2004). But OM contents are high varying from 3.6 to 7.2% under red Pine cover along the slopes of the Mediterranean and Aegean Taurides.

4.1 The Karataş Cambisol of Central Turkey

The Karataş series soil of central Turkey is a typical Cambisols located at the Konya–Karapınar area and developed from lacustrine-beach sediments deposited as flats in the former Pleistocene lake (Groneman 1968). The Karataş soils (Fig. 4) are commonly nearly level, in places gently sloping, mostly deep (60–100 cm), somewhat excessively drained, carbonatic, coarse and moderately coarse-textured soils. They occur westward of the abandoned village of Karataş.

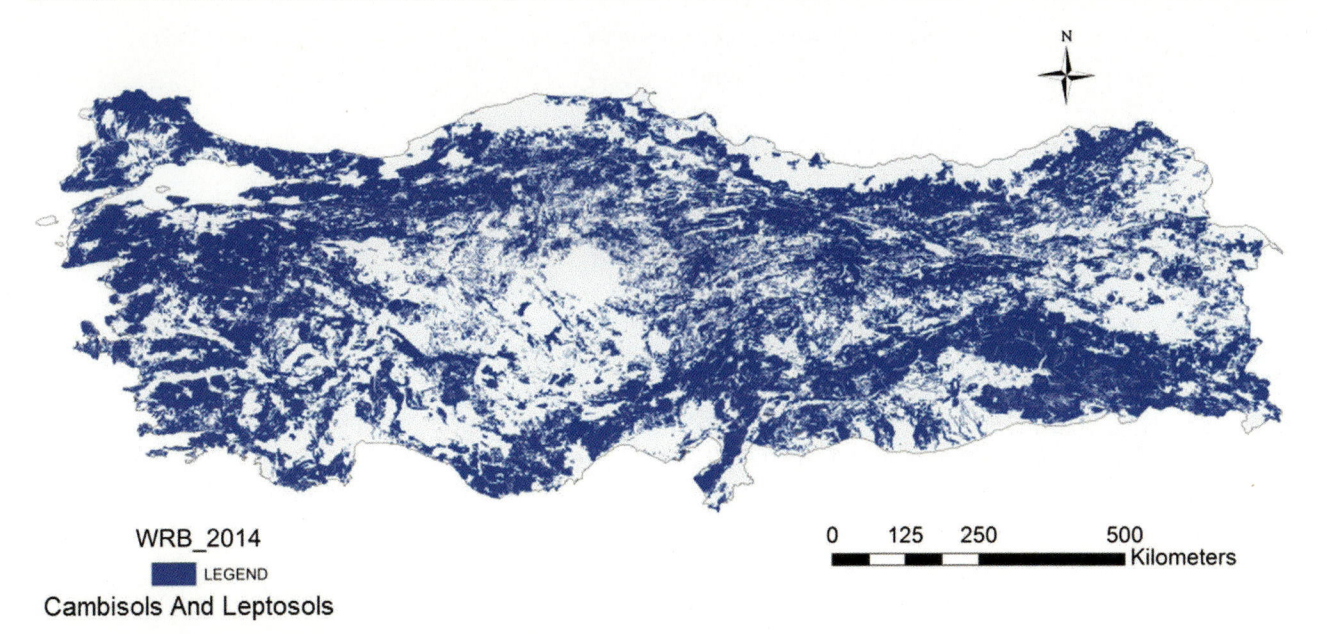

Fig. 3 Distribution of the Cambisols and Leptosols in Turkey

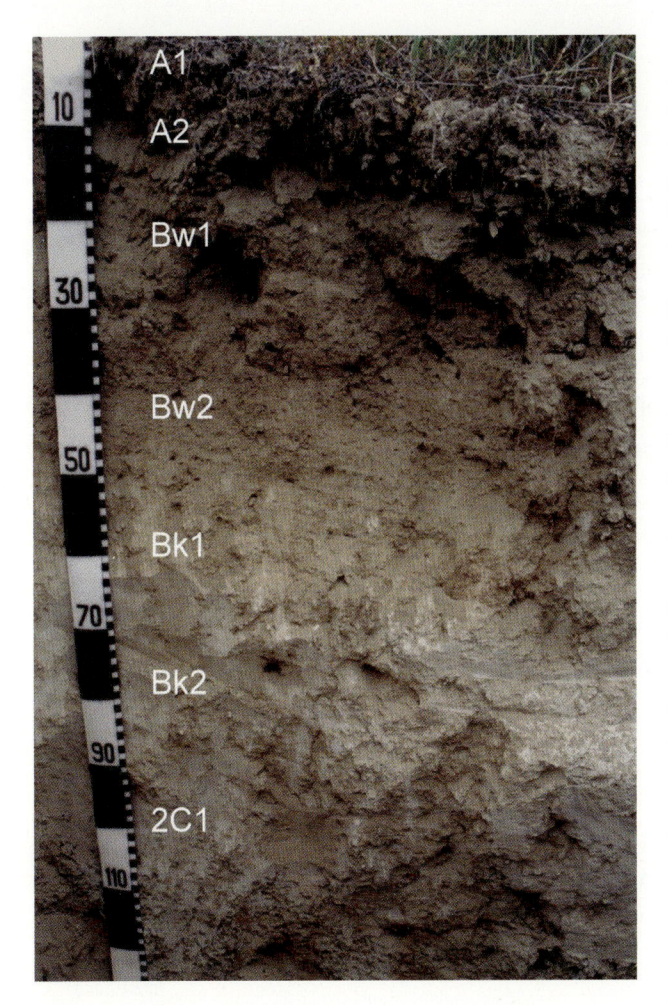

Fig. 4 The Karataş soil in Konya–Karapınar, central Turkey

The pH varies from 7.6 to 8.0 (Table 1) with low CEC values (8.7–16.5 cmol$_c$ kg^{-1}) due to the relatively low OM (0.5–1.0%) and clay mineral contents. The carbonate contents are high (from about 40–54%) due to the limestone parent materials and the micronutrient (Zn, Mg, and Fe) levels are low as ascertained in all the other soils of the area (Fig. 5).

Smectite is the dominant clay mineral in the Karataş soils followed by the low amounts of palygorskite and kaolinite together with some amorphous clay throughout the profile. Similar to many soils of the semiarid inland areas of Turkey, the widespread Cambisols of central Turkey are rich in biological activity. This may enhance organic matter accumulation and microstructural development and in turn increase percolation. The fine granular materials in the surface and subsurface horizons of the Karataş soils are most probably due to the presence of the earthworm pellets and insect larvae. The Karataş soils are dominated by calcite followed by lesser quartz and moderate amounts of gypsum (Fig. 6). Calcite and gypsum in the soils are inherited from the parent materials and from the ancient lake sediments, respectively.

4.2 Cambisols of East and Southeast Turkey

The widely distributed denuded Calcaric Cambisols of east and southeast Turkey—The long standing rainfed areas allocated for cereals—have developed on limestone and ophiolitic materials under continental and dry Mediterranean

Table 1 Site-profile description, and some physical, chemical properties of the Karataş soils in Konya–Karapınar, central Turkey

IUSS working group WRB (2015)	Calcaric Cambisol/Cambic Calcisol
Soil Survey Staff (2014)	Typic Calcixerept
Soil series	Karataş
Location	Karapınar erosion research farm, Ministry of rural affairs, Konya–Karapınar 600 northwest of Karataş village.
Coordinates	37° 42′ 17.29″N, 33° 30′ 34.43″E
Elevation	1009 m
Climate	Weakly Arid, transitional from Mediterranean
Vegetation	Very dense natural vegetation (Alhagi, Agropyron, Centaurea, Eryngium, Stipa, Achillea, Adonis, Festuca, Minuartia, Alyssum ve Poa Marribium parviflorum, Salvia cryptantha, Artemisia fragrans)
Parent material	Ancient lacustrine beach at the edge of limestone terrace
Geomorphic unit	Limestone terrace
Topography	Subnormal relief, gently sloping 1–2% S–N
Erosion	Wind erosion, weak
Water table	Deeper than 2 m
Drainage	Excessive
Infiltration	Rapid
Stoniness	No stones
Soil moisture regime	Xeric
Soil temperature regime	Mesic

Horizon	Depth (cm)	Description
A1	0–10	Light yellow (moist) 2.5Y 7/3, Fine sand, single grain, very friable, non-sticky non-plastic, thick mat of medium to fine roots, strongly effervescent, diffuse, smooth boundary
A2	10–22	Light yellow (moist) 2.5Y 7/3, fine sandy loam, single grain, very friable, non-sticky, non-plastic, few fine roots, strongly effervescent, few, fine, roots, minimal shell fragments, smooth gradual boundary
Bw1	22–45	Light yellow (moist) 2.5Y 7/3, fine sandy loam, weak subangular blocky breaking into single grain, very friable, non-sticky, non-plastic, no roots, 1–3 cm few rounded pebbles, thin and fine carbonate pendants, compact, strongly effervescent, no stones, diffuse gradual boundary
Bw2	45–60	Light gray (dry) 2.5Y 8/2, fine sandy loam, weak subangular blocky breaking into single grain, very friable, non-sticky, non-plastic, no roots, 1–3 cm few rounded pebbles, thin and fine carbonate pendants, compact, strongly effervescent, no stones, abrupt boundary
Bk1	60–85	Light gray (dry) 2.5Y 8/2, fine sandy loam, angular blocky breaking into subangular blocky, very friable, non-sticky, non-plastic, no roots, abrupt boundary
Bk2	85–105	Pale yellow (dry) 2.5Y 8/3, fine sandy loam, weak angular blocky breaking into subangular blocky, very friable, non-sticky, non-plastic, no roots, abrupt boundary
2C	105–130	Light yellow (dry) 2.5y 7/3, clay loam, massive breaking into fine angular blocky, compact, very firm when moist, slightly sticky and plastic, strongly effervescent

Horizon	A1	A2	Bw1	Bw2	Bk1	Bk2	2C
Depth (cm)	0–10	10–22	22–45	45–60	60–85	85–105	105–130
pH	7.8	7.8	7.7	8	7.8	7.8	7.6
EC (dS/m)	0.06	0.1	0.15	0.18	0.25	0.3	0.25
CEC (cmol$_c$ kg^{-1})	8.7	13.5	15.6	14.2	14.6	18.6	16.5
CaCO$_3$ (%)	43.5	39.6	44.6	44.2	53.6	52.7	43.8
Org. Carbon (%)	0.5	1	1.1	0.9	0.8	0.5	0.5

(continued)

Table 1 (continued)

Horizon		A1	A2	Bw1	Bw2	Bk1	Bk2	2C
P$_2$O$_5$ (kg/ha)		80	18	17	18	17	16	3
Particle Size	Sand	89.5	74.8	75.4	59.6	67.5	66.9	25.6
Distributions	Silt	5.7	10.04	16.8	33.5	13.7	13.6	38.6
%	Clay	4.8	15.2	7.8	6.9	18.8	19.5	35.8
Texture		S	SL	SL	SL	SL	SL	CL

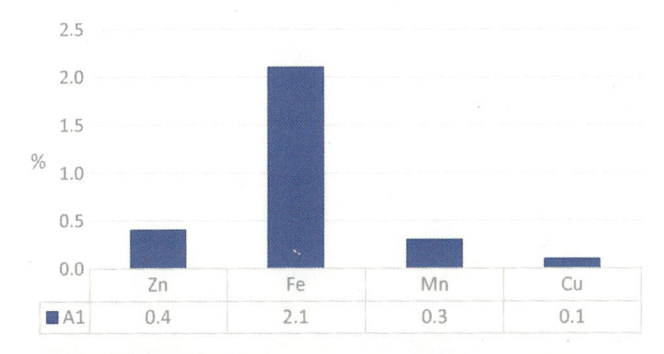

Fig. 5 The micronutrient level of the surface horizon (A1) of the Karataş soils (Akça 2001), Konya, Karapınar, central Turkey

	Zn	Fe	Mn	Cu
A1	0.4	2.1	0.3	0.1

climatic conditions and generally contain only 1.0% of organic matter due to overgrazing (Fig. 7) (Akça and Kapur 2014). The widely distributed Cambisol grazelands of the Van Province and its environs are fragile to erosion also due to extensive grazing. However, at depression sites rich in groundwater, the Cambisols–Vertisols of the region host several hundreds of natural medicinal and fodder species (Fig. 8) (Özgökçe and Özçelik 2004).

The rich composition of the natural vegetation is supported by the high nutrient capacity of the Cambisols developed on basaltic volcanic parent materials with high iron, calcium, and potassium contents in eastern Turkey particularly around Lake Van (Çimrin and Boysan 2006; Schmincke and Sumita 2014). Some of these nutrient-rich Cambisols in Muş, Van, and Ağrı regions are cultivated for fodder crops and particularly for vetch, barley, and alfalfa in relatively small fields due to the harsh topography of the region. Thus, labor-intensive cropping is common on limited land in the area which is ultimately abandoned by the young generation enhancing migration to the already highly populated cities.

4.3 The Cambisols of Southern Turkey, the İnköy, and Göksu Soils

4.3.1 The İnköy Soils

The other common Cambisols of the country are the İnköy soils, located at the central Tauride (south Turkey) forest slopes (at an altitude of about 1600 m) with annual rainfall

Fig. 6 Gypsum crystals in the 2C horizon of the Karataş soil (cross-polarized light)

Fig. 7 The denuded lands of east Turkey allocated to rainfed cereal production

Fig. 8 Grazeland in Cambisol—Vertisol landscape, Muş, east Turkey

Table 2 Site profile descriptions and some physical and chemical properties of the İnköy soils (south Turkey)

IUSS Working Group WRB (2015)	Chromic Cambisol (Humic)
Soil Survey Staff (2014)	Humic Lithic Haploxerept
Location	İnköy Valley, Tarsus, Mersin, Turkey
Coordinates	37° 17′ 38″–37° 17′ 52″NE, 34° 39′ 39″–34° 39′ 49″EW
Altitude	1548 m
Landform	High gradient mountain
Position	Middle slope
Aspect	north east
Slope	>40%
Drainage class	Excessive
Groundwater	N/A
Eff. soil depth	50 cm
Parent material	Colluvial
Climate	Mediterranean
Land use	Afforestation area, terraced
Vegetation	*Pinus nigra, Cedrus Libani*

Horizon	Depth (cm)	Description
A	0–12	Dark reddish brown 5YR3/3 (dry), sandy loam, fine medium
		Stable aggregates, no effervescence, dense hairy roots, wavy
		Boundary
Bw	12–27	Dark reddish brown 5YR4/3 (dry), loamy, fine to medium
		Moderate aggregates, slightly effervescent, stony, dense hairy
		Roots, wavy boundary
BC	27–42	Reddish brown 5YR4/6 (dry), loamy, stable medium size angular
		Blocky, no effervescence, sparse hairy and taproots, wavy
		Boundary

Horizon	A	Bw	BC	C
Depth (cm)	0–12	12–27	27–51	51
Sand (%)	54.4	50.2	50.8	50.2
Silt (%)	27.5	38.0	35.9	31.7
Clay (%)	18.1	11.8	13.3	18.1
Texture	SL	L	L	L
pH	7.83	7.91	7.88	7.89
EC (dS/m)	0.26	0.24	0.24	0.28
$CaCO_3$ (%)	0.94	1.21	0.68	0.81
Organic matter (%)	7.49	7.55	8.96	7.37
Carbon (C) (%)	4.34	4.38	5.19	4.27
Humin substances (%)	2.12	2.14	2.06	–
Humic acid/humin substances (%)	15.9	11.8	14.4	–
Fulvic acid/humin substances (%)	84.1	88.2	85.6	–
Humin substances/org. matter (%)	28.3	28.3	23.0	–
Nitrogen (N) (%)	0.29	0.29	0.34	0.31
C/N	15.1	15.1	15.3	13.8
Available P (mg/kg)	11.23	8.58	9.34	8.73
CEC ($cmol_c$ kg^{-1})	47.61	48.16	51.46	48.14

(continued)

Table 2 (continued)

Horizon	A	Bw	BC	C
Field capacity (%)	38.18	36.36	37.96	37.90
Wilting point (%)	26.45	24.79	23.69	26.40
Available water (%)	11.73	11.57	14.27	11.50
Water stable aggregates (%)	42.76	55.88	56.38	–

varying from 900 to 1000 mm (relatively higher than the average rainfall of the Mediterranean coast which is about 700 mm/year), and are the Chromic Cambisols. The calcium carbonate contents of the Chromic Cambisols are less than the neighboring Leptosols and the Calcaric Cambisols due to the higher decalcification caused by the higher rainfall. However, the illuviation of the clay size particles has not been sufficiently advanced to develop Luvisols in the area. The high precipitation, the moderately dense forest cover, and the relatively lower annual average temperature (9.5° C) than the coastal Mediterranean zone (annual average temp. 19.2° C) enhance the organic matter accumulation, i.e., the carbon sequestration of the Chromic Cambisols (Table 2). The İnköy soil described on the Taurus ranges is developed on crystalline limestone under cedar plantation with moderate depth and A-Bw-BC-C horizon sequences (Table 2, Fig. 9).

The İnköy Chromic Cambisol ecosystem also represents the widespread native and long enduring cedar *(Cedrus libani)* ecosystem with mixed tree communities of black pine *(Pinus Nigra)* and juniper *(Juniperus spp.)*. The root zones of the mixed tree communities and especially of the cedar (age 60-years) and the black pine enhance the development of uniform (size and distribution) microstructural units/aggregates. The cedar and black pine root zones also provide high pore spaces such as 33% for the former and 39% for the latter, thus enhancing the water holding capacity of the soils.

The development of the well-sorted and uniformly distributed granular microstructure and porosity is responsible for the high porosity/water holding capacity and the vigorous activities of the organic exudates (biofilms), organic acids and mycorrhiza of the root zones of the Chromic Cambisol ecosystems of the forest environments (Figs. 10 and 11). The biological quality of the İnköy soils under black pine/natural grass vegetation and cedar plantation is reflected by the similar distribution of the well-sorted and uniform sized granular MSUs (microstructural units)/aggregates throughout the thin sections of the horizons (Fig. 12). The cedar and black pine root zones also provide high pore spaces (determined by polarizing microscopy) estimated at 33% for the former and 39% for the latter, thus enhancing water holding capacities.

Fig. 9 The İnköy Chromic Cambisol under cedar canopy

4.3.2 The Göksu Soils

Further west from the İnköy site in south-central Turkey, the Calcaric Cambisol ecosystems (the Göksu soils) are the dominant forest communities of the Mediterranean and Aegean back coasts. However, the central and eastern Turkish Calcaric Cambisols underlie sparse steppe vegetation with a drier climate and lower sequestered organic carbon (Razzaghi 2014). The red pine *(Pinus brutia)* is the dominant tree species on the Calcaric Cambisol ecosystem

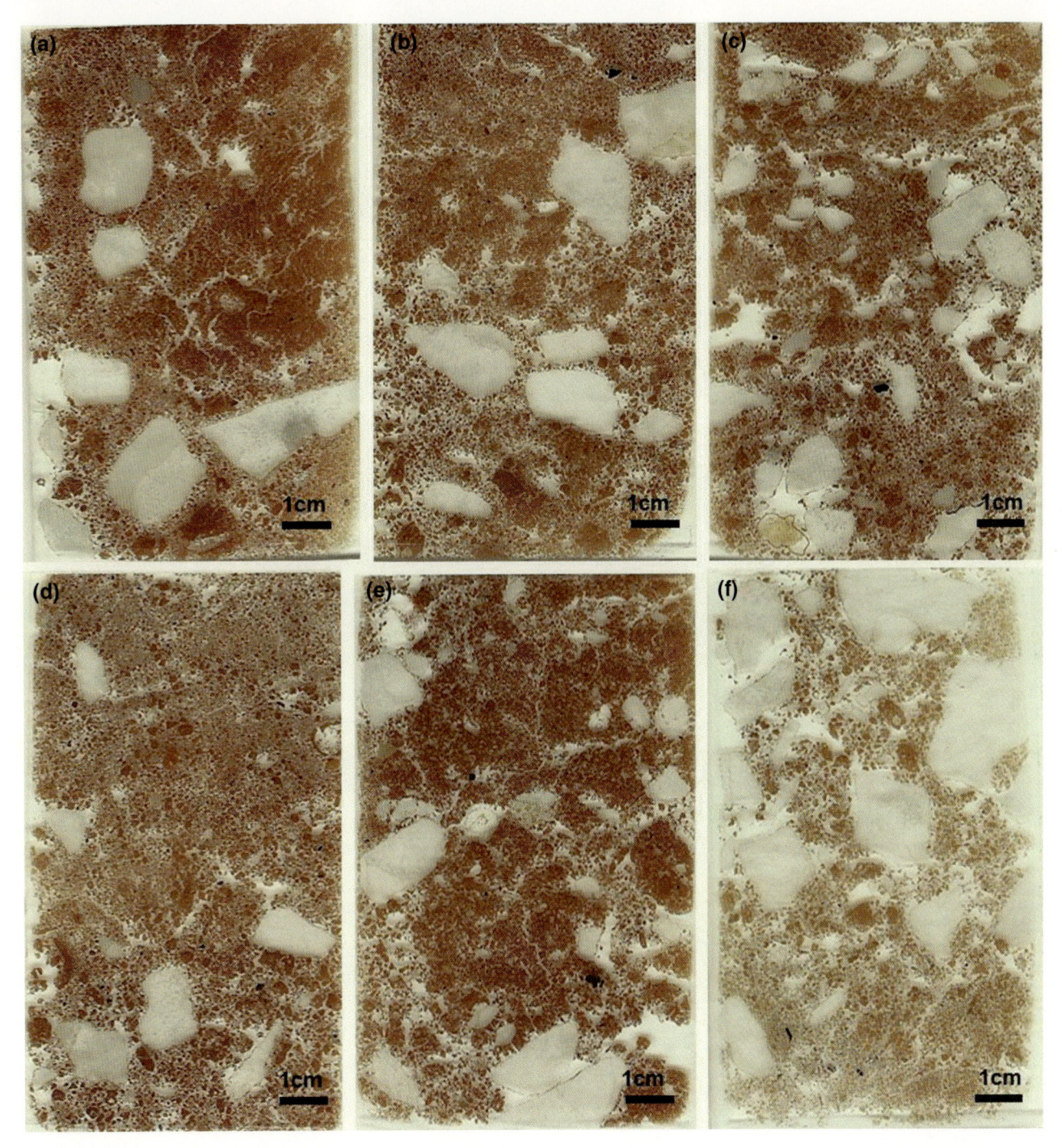

Fig. 10 The well-sorted and uniformly distributed MSUs/aggregates in thin sections of the different İnköy soil root zones. Horizons A (0–3 cm) and Bw1 (3–10 cm) of the İnköy soils under 13-year-old cedar plantation canopy (**a**), horizons A (0–5 cm) and Bw (5–10 cm) under 13-year-old black pine plantation canopy (**b**), horizons A (0–3 cm) and Bw/C (3–10 cm) under grass vegetation adjacent to 13-year-old black pine plantation (**c**), horizon A (0–10 cm) under 60-year-old natural cedar canopy (**d**), horizons A (0–3 cm) and C (3–10 cm) under grass vegetation adjacent to the natural cedar canopy (**e**), horizons A (0 −3 cm) and Bw (3–10 cm) under 13-year-old natural black pine canopy (**f**)

(Fig. 13) in the Mediterranean and Aegean associated with various oak species (Quercus spp) and black pine (*Pinus nigra*). This mixed tree community provides a relatively high microstructure development and a higher structural stability index. The abundant MSUs coated by biofilms of plant-root exudates determined by the SEM (Fig. 14) ensure the enhanced water storage capacity with nutrient supply higher than the central and eastern Anatolian Calcaric

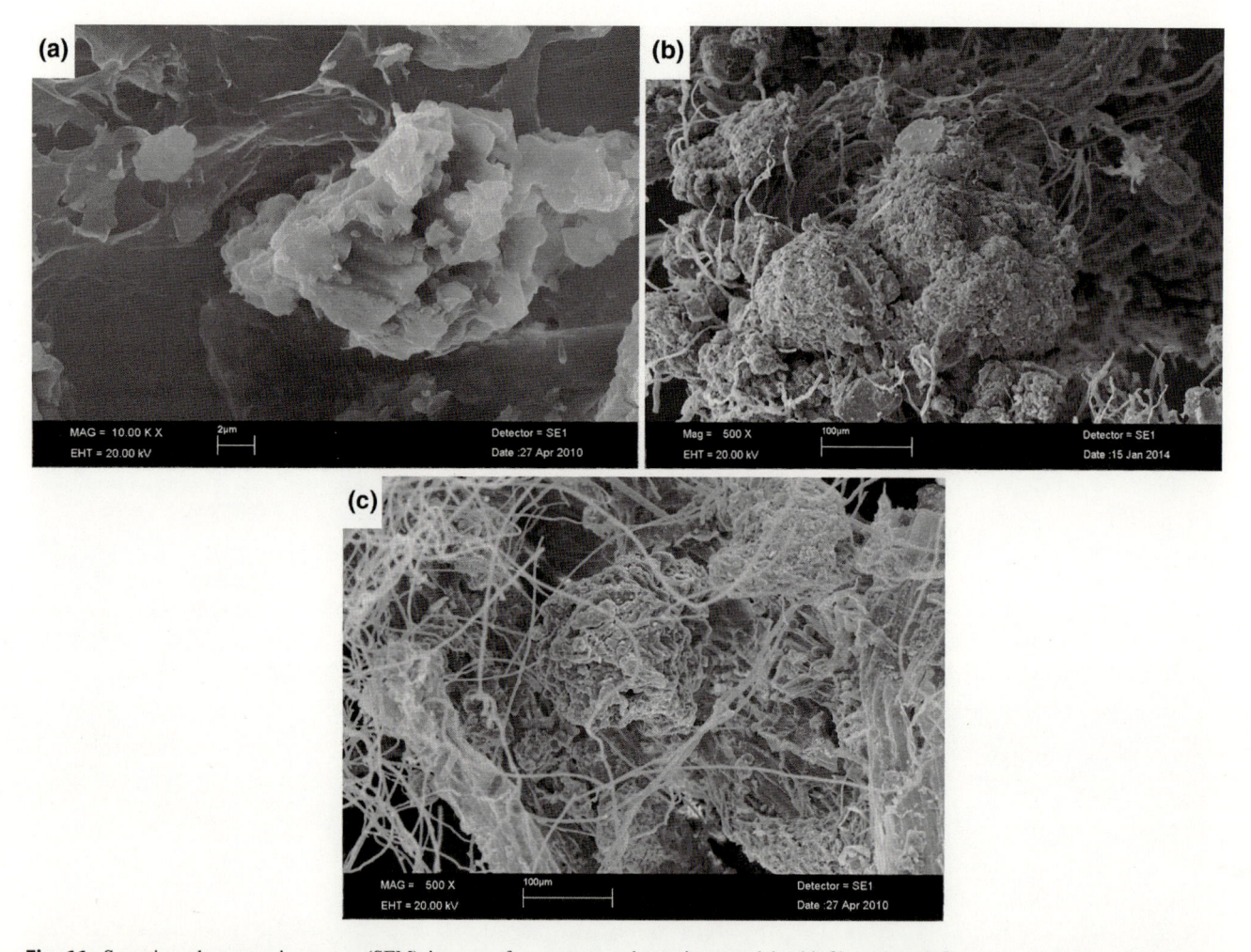

Fig. 11 Scanning electron microscope (SEM) images of aggregate and matrix coated by biofilm (**a**) and fine roots and mycorrhizalhyphae enhancing aggregate/MSU formation in the Ah (**b**) and Bw (**c**) horizons under the cedar canopy of the İnköy Chromic Cambisol

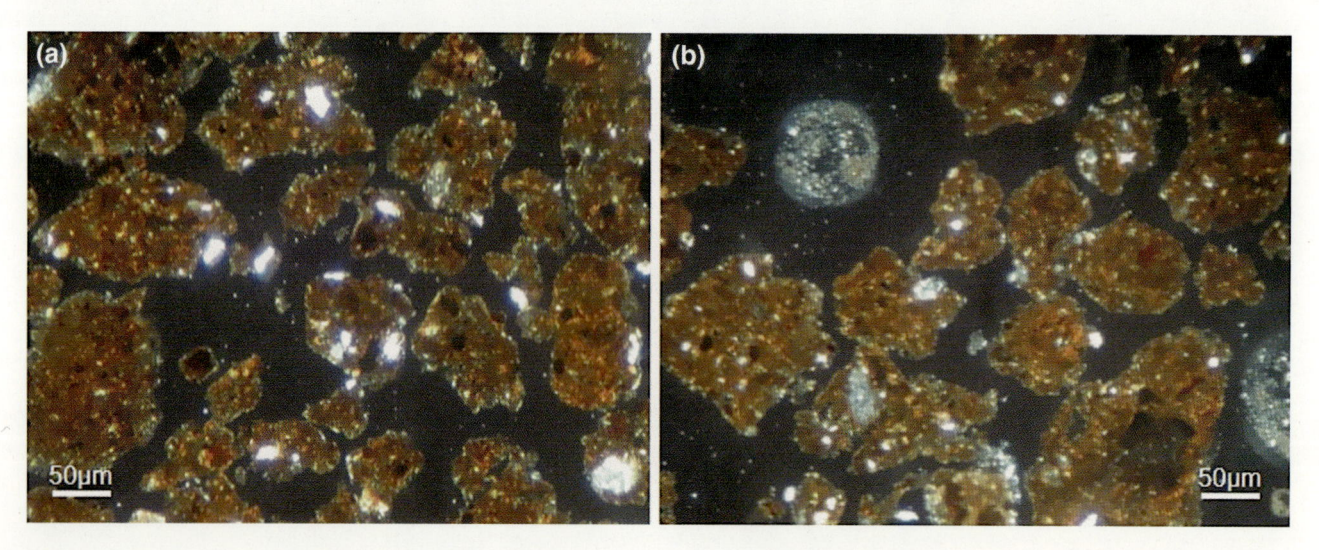

Fig. 12 Very fine-to-fine subangular to *oval* granular aggregates in horizon A (0–3 cm) of the İnköy soil under 13–year-old cedar plantation canopy (**a**), very fine-to-fine subangular to *oval* granular aggregates in horizon A (0–3 cm) under grass vegetation adjacent to 13–year-old black pine plantation (**b**), (cross-polarized light)

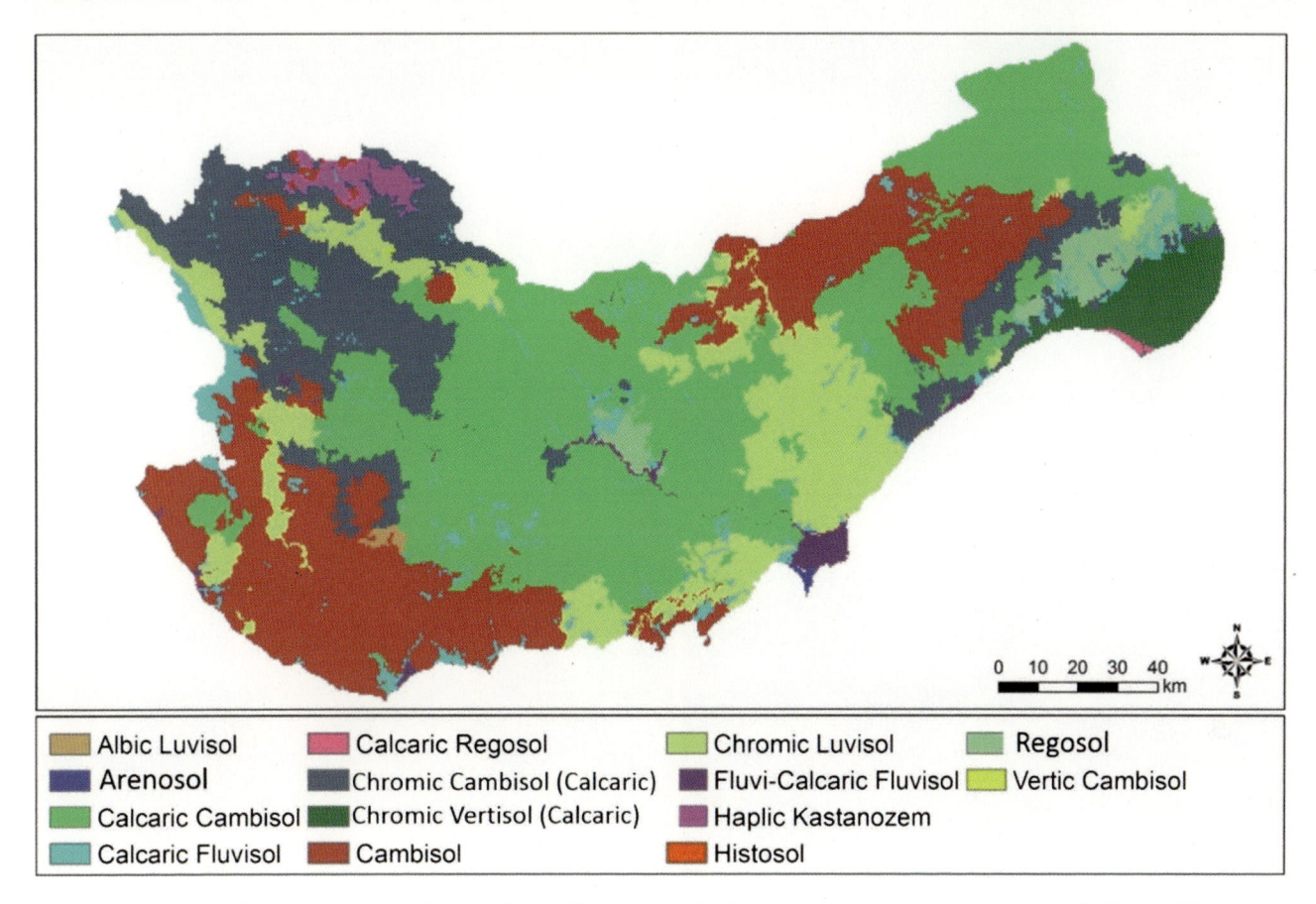

Fig. 13 Soils and the widely distributed Calcaric Cambisol ecosystems in the Göksu catchment of south-central Turkey (IUSS Working Group WRB 2015)

Cambisols. The pore characteristics of the Bw horizons of the Calcaric Cambisols under red pine and oak canopies studied by image analysis (Pagliai et al. 1983) revealed that all pore shapes, sizes, and the interaction of pore shapes x size classes were significantly different in both species (Razzaghi 2014). However, the elongated pores were determined to be higher within the 100–400 μm size limit, whereas the irregular pores were dominant from 500 to 1000 μm and larger in both red pine and oak root zones (Fig. 15) (Razzaghi 2014).

Elongated pores are known to be highly significant in easing the root penetration as well as in increasing the transmission of water and air in the soil. In any event, the irregular pore spaces are irrevocable of prime importance in storing the water, especially at the shallow to moderately deep soils of the semiarid mediterranean karstic ecosystem, i.e., the Cambisol, Luvisol, and Calcisol ecosystems.

Below is the description of physical and chemical properties of the Göksu soil (Table 3) formed on limestone parent material with an Ah, A2, Bw, and Bk/C horizon sequence underlying the canopy of red pine (*Pinus brutia*). There is a 5–cm-deep red pine litter over the soil surface.

The Calcaric Cambisols of the Mediterranean (in the Tauride mountain zone) are slightly alkaline with pH ranging from 7.4 to 7.8. The decomposition of the biomass and formation of the organic, citric, and oxalic acids as intermediate products (Blume et al. 2010) have most likely enhanced the microstructure/aggregate development of the forestal Calcaric Cambisols by the consequent development of the biofilms also coating and stabilizing the aggregates (Kantarcı 2000; Razzaghi 2014) (Fig. 16). Zinc deficiencies are occasionally observed in the susceptible field crops such as corn and beans cropped in the Mediterranean forest-agriculture mixed areas of the Calcaric Cambisols. The pH of the Calcaric Cambisols of the Taurides appears to be around 7 and may only slightly differ at each crop rhizosphere for nutrient uptake, namely in the Leptosols and Luvisols (Sağlıker and Darıcı 2007; Altun 2008). In general, the pH of the Calcaric Cambisols underlying a maquis vegetation and overlie limestone parent materials fluctuates from moderate to slightly alkaline.

The decomposition of the litter over a Calcaric Cambisol under the canopy of red pine is expected to provide high amounts of organic matter to the soil as it is in the case of the

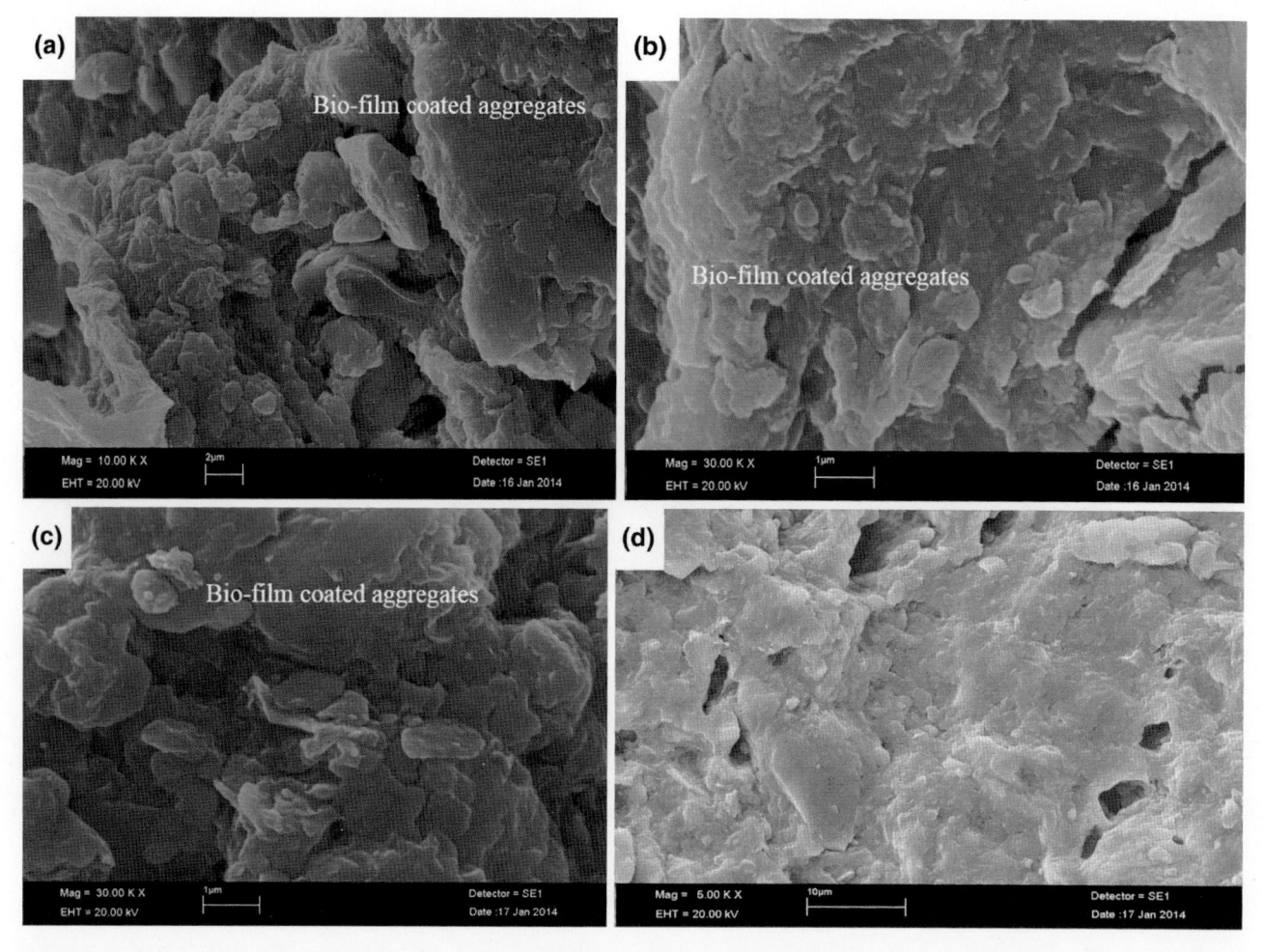

Fig. 14 Clay sheets/aggregates coated by biofilms (**a**, **b**, **c**) and surface of welded and coated MSUs (**d**) in the Bw horizons of the Göksu soils under red pine and oak canopies, respectively

organic matter contents of the Oe (41.3%) horizon of the Göksu Calcaric Cambisols (Table 4) (Razzaghi 2014). The lower organic matter content of the Oe horizon (humus horizon) of the Göksu soil is due to the rapid decomposition of the organic residues. Elsewhere, Polat (2012) has determined that the carbon contents of the Chromic Cambisols under the canopy of black pine (*Pinus nigra*) were higher in the Oe (50.1%) horizon than the adjacent Calcaric Cambisols studied by Razzaghi (2014). Moreover, the carbon contents of these soils under the canopy of *Cedrus libani* were determined to be 41.3% in the Oe horizon (Polat 2012).

The particle size distribution of the Göksu Calcaric Cambisol varied from 38 to 54% in sand and 39–44% in silt in the surface horizons most likely due to the deposition of the wind-blown materials (Razzaghi 2014) (Table 3). The widespread loamy Calcaric Cambisols of the Mediterranean Göksu catchment area (the Göksu soil/the Göksu Cambisol ecosystem) under red pine and oak canopies consist of various macro-pores (especially over 100 microns, Fig. 14) that provide easy and good aeration and in turn suitable conditions for plant production (Sağlıker and Darıcı 2007; Razzaghi 2014).

4.4 Soil Formation in Cambisols Under Forest and Pasture Vegetation in Northwest and South Turkey

4.4.1 The Uludağ Cambisol, Northwest Turkey

The highland (Uludağ: Mount Ulu) Cambisols/Inceptisols (Acidic Brown Forest Soils, pH 4–5) of northwest Turkey in Bursa are deep soils developed on biotite-rich granite under forest canopy at steep slopes (25–35%) and about 1700 m elevations. The Uludağ Cambisols are moderately deep to shallow under highland grass vegetation between 2000 and 2300 m elevations. Their high sand size fraction is composed of abundant quartz and feldspars followed by biotite (Güzel 1980). Chlorite is frequently neoformed from unweathered biotite in the fine sand fractions of the upper part of the solum as an initial soil formation (alteration of

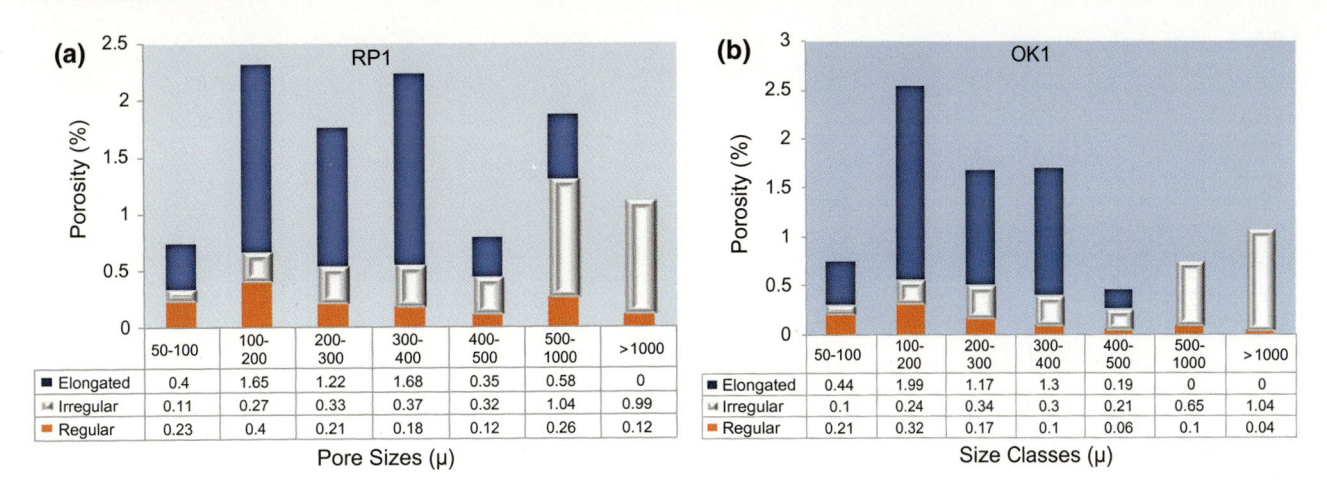

Fig. 15 The pore distribution of the Göksu Cambisol Bw horizons under red pine (**a**) and oak (**b**) in southern Turkey (Razzaghi 2014)

RP1

Pore Sizes (μ)	50-100	100-200	200-300	300-400	400-500	500-1000	>1000
Elongated	0.4	1.65	1.22	1.68	0.35	0.58	0
Irregular	0.11	0.27	0.33	0.37	0.32	1.04	0.99
Regular	0.23	0.4	0.21	0.18	0.12	0.26	0.12

OK1

Size Classes (μ)	50-100	100-200	200-300	300-400	400-500	500-1000	>1000
Elongated	0.44	1.99	1.17	1.3	0.19	0	0
Irregular	0.1	0.24	0.34	0.3	0.21	0.65	1.04
Regular	0.21	0.32	0.17	0.1	0.06	0.1	0.04

Table 3 Site profile descriptions and some physical and chemical properties of the Göksu soil (south Turkey) (Razzaghi 2014)

IUSS working group WRB (2015)	Calcaric Cambisol
Soil Survey Staff (2014)	Typic Calcixerepts
Location	On the Silifke-Mut highway-12 km to Silifke—on the right side
Coordinates	36° 40′ 28.14″N, 36° 57′ 45.91″E
Elevation	107 m
Climate	Mediterranean
Soil moisture regime	Xeric
Soil temperature Regime	Mesic
Vegetation	Natural Maquis vegetation with natural red pine (*Pinusbrutia*) and oak *(Quercus coccifera)* trees
Parent material	Limestone
Geomorphic unit	Mountain slope
Surface topography (relief)	Wavy, 10% slope
Slope direction	west to east
Erosion	Slight water erosion
Drainage	Good
Infiltration	Good
Stoniness	>15%

Horizon	Depth (cm)	Description
Oe	0–5	Fresh to highly decomposed tree leaves (litter). No stones, abrupt wavy boundary
Ah	5–10	Brown to dark brown 7.5YR2/3 (dry), silty loam, strong, very fine, rounded granular, friable when dry, firm when moist, non-plastic, non-sticky, no stones, many fine roots and high biological activity, moderately effervescent, clear wavy boundary
A2	10–19	Brown 7.5YR4/3 (dry), silty loam, fine, moderate, subrounded to rounded granular, friable when dry, firm when moist, slightly sticky and plastic, no stones, many, fine roots, moderately effervescent, clear wavy boundary
Bw	19–28	Reddish brown 7.5YR4/6 (wet), silty loam, weak, medium, subangular granular, friable when dry, firm when moist, slightly sticky and plastic, no stones, many, fine roots, moderately effervescent, clear wavy boundary
Bk/C	28–45	Light brown 7.5 YR 5/6 (wet), silty loam, weak, medium, subangular blocky, friable when dry, firm when moist, no stones, few main roots, strongly effervescent, rare carbonate mycelia and 1–2 mm carbonate nodules, clear wavy boundary

(continued)

Table 3 (continued)

Horizon	Ah	A2	Bw	Bk/C
Horizon	Ah	A2	Bw	Bk/C
Depth (cm)	5–10	10–19	19–28	28–45
pH	7.85	7.36	7.34	7.41
EC (dS/m)	0.56	0.28	0.23	0.19
CaCO$_3$ (%)	2.7	7.6	5.3	11.4
CEC (cmol$_c$ kg^{-1})	93.7	38.3	28.2	22.8
Organic Matter (%)	6.72	3.42	1.62	1.25

Horizon		Ah	A2	Bw	Bk/C
Depth (cm)		5–10 Hydrometer	10–19	19–28	28–45
Sand (%)		53.53	37.91	40.49	41.20
Silt (%)		38.74	44.27	39.91	39.14
Clay (%)		7.73	17.82	19.60	19.66
Texture		L Sedigraph	L	L	L
Very fine sand	(%)	6.73	7.66	9.1	10.26
Fine sand	(%)	5.93	7.34	7.89	7.57
Coarse sand	(%)	30.18	17.79	14.94	15.55
Total sand	(%)	42.85	32.79	31.93	33.39
Fine silt	(%)	29.35	32.35	32.93	35.55
Coarse silt	(%)	9.73	13.37	13.35	11.69
Total silt	(%)	39.08	45.73	46.28	47.23
Clay	(%)	18.07	21.48	21.79	19.38
Texture		L	L	L	L
Field capacity (%)		47.83	17.8	19.88	17.42
Permanent wilting point (%)		27.57	9.99	9.32	9.15
Available water (%)		20.26	7.81	10.55	8.27

primary minerals) product of a highland environment with 5.2° C mean annual temperature and about 1500 mm mean annual precipitation. Chlorite is also present in the clay size fraction in trace amounts alongside the dominant kaolinite (weathered from feldspars), smectite (weathered from biotite), vermiculite (weathered from biotite), and illite (weathered from biotite) indicating the initial incorporation to the soil matrix despite its frequent presence in the sand size fractions. The Cambisols of the Uludağ environment (the Cambisol ecosystems) are the benchmark soils of Turkey for thesequestration of organic carbon (4–5%) and total nitrogen (0.4–0.5%) in their surface horizons (Güzel 1980).

4.4.2 The Ulukışla Cambisol, South Turkey

The Ulukışla Cambisol of south Turkey is a clear reflection of the effect of the plant factor in soil formation. The Ulukışla Cambisol site is the experimental grazed and ungrazed plots of the highland Cambisols formed on the trachyandesites of the Taurides in Ulukışla (Adana, south Turkey) revealed significant results in soil maturation processes. The higher amounts of clay size fractions, clay minerals (smectite, illite, kaolinite), and related structural stability indices in contrast to the lower contents of organic matter and carbonates (Kapur et al. 1982) in the grazed plots indicated a more advanced soil development.

Fig. 16 Red pine canopy over the Göksu soil (**a**), the Göksu soil profile (**b**), aggregates of the horizons (**c**) (south Turkey). *Note* the strong effervescence in A2, Bw and Bk/C horizons (**c**)

However, the almost twofold higher organic matter contents in the ungrazed plots documented the significance of nature conservation for carbon sequestration and mitigation of climate change. The dominant presence/distribution of *Festuca ovina* and *Poa bulbosa*—the preferable fodder for the small ruminants—in the ungrazed plots (with dense fine root distributions in the upper 30–40 cm) was stated to be responsible for the decreased leaching of the rain water through the soil profile and in turn the lower weathering and/or the lower soil formation.

The dominant presence/distribution of *Festuca ovina* and *Poa bulbosa*—the preferable fodder for the small ruminants —in the ungrazed plots (with dense fine root distributions in the upper 30–40 cm) was stated to be responsible for the decreased leaching of the rain water through the soil profile and thus the lower weathering and/or the lower soil formation.

The clay mineral contents of a number of highland/forest soils, namely Cambisols developed on various parent rocks (dolerite, granodolerite, diorite, granite, shale, sandstone and

Table 4 The organic matter contents of the Göksu soils surface horizons (Calcaric Cambisols) (Razzaghi 2014)

Horizon	Depth (cm)	Air dry weight (g/0.25 m^2)	Dry matter (%)	Ash (%)	Organic matter (%)	Org. C (%)
Ol	0–2	233.50	88.66	6.71	93.29	54.11
Of	2–4	93.80	88.08	14.13	85.87	49.81
Oh	4–5	34.40	90.01	50.79	49.21	28.54

Fig. 17 Sour cherry orchards on Cambisols in the east Aegean region

shale, and biotite gneiss), have been found to be dominated by illite followed by kaolinite, vermiculite, and chlorite, whereas the Luvisols of the highland/forest areas developed on limestones contained abundant montmorillonite and illite (Mitchell and Irmak 1957). The overall physical quality of the highland/forest Cambisols is most likely influenced by the high smectite, illite, and vermiculite contents that are the major cause of their high water storage capacity and in turn probably the major factor in the increased amounts of sequestered carbon in their surface horizons.

5 Management

Cambisols are moderately deep soils on slightly to moderate sloping lands of Turkey that provide relatively better crop production opportunities than Leptosols. Cambisols are generally managed in rainfed conditions due to the lack of irrigation infrastructures on sloping topographies. Almond, apple, cherry, sour cherry, plum, and apricot productions are yield and quality-wise successful and common on central, eastern, east Aegean, and southeastern Turkish Cambisols (Fig. 17). The Calcaric Cambisol ecosystems/anthroscapes, including the vine and olive 'Terroirs' of the Mediterranean and Aegean (see Calcisols), are mostly allocated for traditional vine and olive productions (Fig. 18) with the increasing use of contemporary cultivation technologies that area using more computer controlled equipment in irrigation and fertilization as drip irrigation (Eswaran et al. 2013).

Smectite-rich Calcaric Cambisols, with high cation exchange and water holding capacities, can enhance nutrient uptake and the amount of the available water in the root zone soils of rainfed vineyards and olives (Polat 2012; Razzaghi 2014). Chromic Cambisols in the Black Sea and the north central Anatolian region are cultivated for kiwi, vineyard, and cherry production, whereas rainfed cereals (barley and wheat) are the major crops of Cambisols of the country, especially in eastern and central Turkey. However, the

Fig. 18 Traditional species and contemporary management in vineyard cultivation on the Calcaric Cambisols in Tarsus (south Turkey)

relatively high clay and low organic matter contents, sloping topographies and moderate soil depths are the limiting factors for agricultural management and higher crop yields on Cambisols. Thus, reduced tillage, input of organic-rich soil amendments and/or manure, legume and fallow rotations, and drip irrigation at appropriate sites can enhance the productivity of the crops grown on the widely distributed Cambisol ecosystems of Turkey. Cambisols are also the prime soil ecosystems for sloping forested land. The Tauride mountain ranges of the southern and western back coasts of Turkey comprise the best forest soil ecosystems for the dominant Cambisols underlying the red pine communities. Natural and recently planted red pine areas have been highly successful in protecting the soil from erosion.

References

Akça E (2001) Determination of the soil development in karapinar erosion control station following rehabilitation. Ph.D. thesis, University of Çukurova, Institute of Natural and Applied Sciences, Adana, Turkey

Akça E, Kapur S (2014) Soils. In: Güner A, Ekim T (eds) Illustrated flora of Turkey, İş bank cultural series, vol 1. pp 70–111

Altun N (2008) The geology and soil properties of Urla-Seferhisar (İzmir). Republic of Turkey Ministry of Environment and Forestry, Ege Forestry Research Institute, İzmir

Blume HP, Brümmer GW, Horn R, Kandeler E, Kögel-Knabner I, Kretzschmar R, Stahr K, Wilke BM (2010) Scheffer/Schachtschabel: Lehrbuch der Bodenkunde. Spektrum Akademischer Verlag, 520 p

Çimrin KM, Boysan S (2006) Nutrient status of van agricultural soils and their relationships with some soil properties. Yüzüncü Yıl Univ, J Agric Sci 16(2):105–111

Erol O (1999) A geomorphological study of the Sultansazlığı Lake Central Anatolia. Quaternary Sci Rev 18(4):647–657

Evrendilek F, Çelik İ, Kılıç S (2004) Changes in soil organic carbon and other physical soil properties along adjacent Mediterranean forest, grassland, and cropland ecosystems in Turkey. J Arid Environ 59(4):743–752

Eswaran H, Berberoğlu S, Cangir C, Poyraz D, Zucca C, Özevren E, Yazıcı E, Zdruli P, Dingil M, Dönmez C, Akça E, Çelik İ, Watanabe T, Koca KY, Montanarella L, Cherlet M, Kapur S (2013) The anthroscape approach in sustainable land use. In: Kapur S, Eswaran H, Blum WEH (eds) Sustainable land management, Springer, London/Berlin, pp 1–50

FAO/ISRIC (2006) IUSS (2006) World reference base for soil resources. A framework for international classification, correlation and communication. World soil resources reports, 103 p

Groneman AF (1968) The soils of the wind erosion control camp area, Karapınar, Turkey. Ph.D. thesis, Agricultural University, Wageningen, Netherlands, 160 p

Güzel N (1980) Influence of vegetation and topography on mineral weathering and formation in three soil profiles formed from granite in the Marmara region. University of Çukurova, Faculty of Agriculture Pub., no.140/32, Dilek Press, Adana, 40 p

IUSS Working Group WRB (2015) World reference base for soil resources 2014, update 2015 international soil classification system for naming soils and creating legends for soil maps. World soil resources reports No. 106. FAO, Rome, 203 p

Kantarcı MD (2000) Soil science. İstanbul University, Faculty of Forestry No. 462, İstanbul, p 420 (in Turkish)

Kapur S, Tükel T, Çavuşgil VS, Kaya Z, Yeşilsoy MŞ (1982) A comparison of clay minerals of grazed-ungrazed soil profiles with similar topography, parent rock and aspect in Ulukışla, Adana, Turkey. In: 1st National clay symposium of Turkey, Adana. University of Çukurova Press, Adana, pp 410–419

Kapur S, Çavuşgil VS, Şenol M, Gürel N, FitzPatrick EA (1990) Geomorphology and pedogenic evolution of quaternary calcretes in the northern Adana Basin of southern Turkey. Zeitschrift für Geomorphologie 34:49–59

Mitchell WA, Irmak A (1957) Turkish forest soils. J Soil Sci 8(2):184–192

Özgökçe F, Özçelik H (2004) Ethnobotanical aspects of some taxa in East Anatolia Turkey. Economic Botany 58(4):697–704

Pagliai M, Lamarca M, Lucamente G (1983) Micromorphometric and micromorphological investigations of a clay loam soil in viticulture under zero and conventional tillage. J Soil Sci 34:391–403

Polat S (2012) Physical, chemical, mineralogical and micromorphological properties of cedar (*Cedrus libani* a. rich) and black pine (*Pinus nigra* Arnold) soils in natural and planted karstic areas. PhD thesis, University of Çukurova, Institute of Natural and Applied Sciences, Adana, 297 p

Razzaghi SM (2014) Mineralogical and micromorphological Characteristics of red pine (*Pinus brutia*) and Oak (*Quercus coccifera*) Root Zone Soils in the Göksu Catchment. PhD thesis, University of Çukurova, Institute of Natural and Applied Sciences, Adana, 295 p

Sağlıker HA, Darıcı C (2007) Nutrient dynamics of Olea europaea L. Growing on soils derived from two different parent materials in the eastern mediterranean region (Turkey). Turk J Bot 29:255–262

Schmincke HU, Sumita M (2014) Impact of volcanism on the evolution of Lake Van (eastern Anatolia) III: periodic (Nemrut) vs. episodic (Süphan) explosive eruptions and climate forcing reflected in a tephra gap between ca. 14 ka and ca. 30 ka. J Volcanol Geoth Res 285:195–213

Soil Survey Staff (2014) Keys to soil taxonomy, 12th edn. USDA-Natural Resources Conservation Service, Washington, DC, p 372

Fluvisols

Hasan Özcan

1 Introduction

Fluvisols correlate with Fluvents and Fluvaquents of the USDA Soil Taxonomy (Soil Survey Staff 2014). The natural fertility of most Fluvisols and their attractive dwelling sites on river levees and higher parts in marine landscapes were recognized since the prehistoric times (IUSS Working Group WRB 2015). Fluvisols are found on alluvial plains, river fans, valleys, and tidal marshes on all continents and in all climate zones. Under natural conditions periodical flooding is fairly common. These soils have a clear evidence of sedimentary stratification. Soil horizons are weakly developed, but a distinct topsoil horizon may be present. Fluvisols commonly have A and C horizon sequences formed predominantly by fluvial activity. Fluvisols are classified as "Alluvial soils" (Russia, Australia), "Fluvents" (Soil Survey Staff 2014), "Auenböden" (Germany) and "Sols mineraux bruts d'apport alluvial ou colluvial" or "Sols peu évolués non climatiques d'apport alluvial ou colluvial" (France).

2 Parent Materials and Profile Development

Fluvisols predominantly form on the water borne sediments associated with rivers and flood plains as well as the shorelines of lakes and seas. Characteristics of Fluvisols vary greatly depending mainly on the composition and sequence of the sediments. Their so-called parent materials are recent (Holocene and Pleistocene) alluvial, colluvial, lacustrine, and marine sediments.

The Fluvisols have formed from aridic to humid climate and very different soil moisture and temperature regimes in Turkey. The most common Fluvisols have formed on the material transported by rivers in the upper Pleistocene and Holocene. Fluvisols of Turkey have A-C, A-AC-C horizon sequences. Due to the different transported materials, they have different textures and carbonate contents. The physical and chemical properties of the two widely distributed Fluvisols in Tarsus-Mersin (the Çukurova region, south Turkey) and Umurbey Plains (north west Turkey) respectively are given below (Tables 1 and 2). The former is a major delta in the country, dominant in Fluvisols and Vertisols that has been pioneering the historical and long-standing tradition of cotton growing and weaving culture developed in the area. This plain is worthy to meet the term standing for 'a human reshaped land unit, that may be named after a major crop and traditional technology, i.e., the "Cotton Anthroscape of Adana" (Çelmeoğlu 2011).

Fluvisols have also developed in mountainous east Turkey (Van Province) in river valleys associated to Cambisols and Leptosols (Çimrin et al. 2004). The acidic volcanic rocks gave rise to the development of relatively low pH Fluvisols with vertic properties due to abundant smectite contents. The acidic volcanic rocks gave rise to the development of relatively low pH Fluvisols with vertic properties due to abundant smectite clays.

3 Environment and Characteristics

The Fluvisols of the Çukurova, Bafra, Çarşamba, Büyük and Küçük Menderes, and Gediz Plains are fertile soils. The soil characteristics of the Umurbey plain highly change over the course of the floodplain development, where deposition builds up the geographic pattern of a river delta. The variable geographic patterns developing via the alternating retrogradation-progradation process cycles are responsible for the inherent differences of the soils of the deltas in Turkey (Bal et al. 2003) (Fig. 2). An appropriate example for this is the prehistoric-historic progress of the progradation-sedimentation process and related soil distribution in the Aydıncık delta (south Turkey) determined by Bal et al. (2003).

H. Özcan (✉)
Department of Soil Science and Plant Nutrition, Çanakkale Onsekiz Mart University, Çanakkale, Turkey
e-mail: hozcan@comu.edu.tr

© Springer International Publishing AG 2018
S. Kapur et al. (eds.), *The Soils of Turkey*, World Soils Book Series, https://doi.org/10.1007/978-3-319-64392-2_9

Table 1 Site-profile description and some physical and chemical properties of the Tarsus soil (south Turkey) Dinç et al. (1995)

IUSS Working Group WRB (2015)	Haplic fluvisols calcaric eutric
Soil Survey Staff (2014)	Typic Xerofluvent
Location	Çöplü Village-Tarsus-Mersin, south Turkey
Coordinates	36S, 686 787E, 4074 404N
Altitude	3 m
Landforms:	Young river terrace
Position	Lower part
Aspect	–
Slope	0–1%
Drainage class	Good
Groundwater	–
Eff. soil depth	+120 cm
Parent material	Fluvial
Köppen climate	Mediterranean
Land use	Agriculture
Vegetation	Cotton

Horizon	Color Moist	Roots	Structure	Clay	Silt	Sand
Ap	2.5Y4/4	Intense, fine	Granular	27	47	25
A2	2.5Y5/5	Intense, fine	Subangular blocky—granular	24	55	21
2A	10YR3/2	Sparse, fine	Subangular blocky	32	39	29
2AC	10YR3/3	Sparse, fine	Massive	22	43	35
2C1	2.5Y4/4	Sparse, fine	Massive	23	49	28
2C2	5Y5/3	Very rare, fine	Massive	35	52	13
2C3	2.5Y5/4	–	Massive	30	48	22

Horizon	Depth (cm)	Organic matter (%)	pH (H$_2$O)	EC (dS/m)	CaCO$_3$ (%)	CEC (cmol$_c$kg^{-1})
Ap	0–10	1.37	7.4	0.06	20.9	22
A2	10–39	1.17	7.5	0.08	21.5	20.4
2A	39–60	1.5	7.5	0.36	14.6	21.1
2AC	60–73	0.39	7.6	0.44	19.3	19
2C1	73–94	0.46	7.3	0.43	20.6	13.6
2C2	94–112	0.63	7.5	0.5	22.6	19.5
2C3	112–150	0.36	7.5	0.5	20.1	17.8

4 Regional Distribution

Fluvisols are widespread in the Adapazarı, Sakarya, Bafra, Çarşamba plains in the Black Sea Region, Ergene and Meriç basin in the Trace Region, Susurluk, Nilüfer basin and vicinity of Ulubat Lake in the Marmara Region, Büyük Menderes, (Aydın-Söke Plains), Küçük Menderes (Ödemiş Plain), Gediz (Menemen Plain), Bakırçay (Dikili Plain) basins in the Aegean Region, Aksu-Serik plain in Antalya, Silifke delta, Ceyhan, Berdan, Yüreğir, and Seyhan plains in the Çukurova, Kahramanmaraş-Türkoğlu, and Hatay-Amik plains of the Mediterranean Region, the Konya closed basin and vicinity of the salt lake in central Turkey, Iğdır, Malazgirt, Erzincan, Erzurum, and Pasinler-Horasan plains

Table 2 Site-profile description and some physical and chemical properties of the Umurbey soil (northwest Turkey) (Yiğini 2006) (Fig. 1)

IUSS Working Group WRB (2015)	Haplic fluvisols eutric
Soil Survey Staff (2014)	Typic Xerofluvent
Location	Umurbey-Çanakkale, northwest Turkey
Coordinates	35T, 466 092E, 4 456 667N
Altitude	17 m
Landforms	Flood plain, river
Position	Lower part
Aspect	–
Slope	0.1%
Drainage class	Good
Groundwater	–
Eff. soil depth	+120 cm
Parent material F	Fluvial
Köppen climate	Mediterranean
Land use	Agriculture, peach orchards

Horizon	Colour dry	Moist	Mottles	Roots	Structure	Texture (%) Clay	Slit	Sand
Ap	2.5Y5/3	2.5Y4/2	None	Intense, fine	Subangular	15.39	19.94	64.67
A2	2.5Y5/3	10YR4/2	None	Intense, fine	Subangular blocky	17.85	17.85	68.85
C1	2.5Y5/3	10YR4/2	None	Medium, fine	Massive	15.32	6.18	78.51
2Ad	2.5Y5/3	10YR4/2	None	Medium, fine	Subangular blocky	13.99	25.66	60.36
2C	–	–	None	–	Single grained	–	–	–

Horizon	Depth (cm)	Organic matter (%)	pH 1:1	EC (dSm^{-1})	CaCO$_3$	CEC (cmol$_c$kg^{-1})
Ap	0–20	0.84	7.46	0.14	1.19	22.53
A2	20–36	0.48	7.60	0.14	0.85	22.14
C	36–75	0.19	7.65	0.14	1.46	22.29
2Ad	75–87	0.43	7.59	0.21	2.92	20.02
2C	87–150	–	–	–	–	–

in east Turkey (Dinç et al. 2001), Karamenderes plain (Çanakkale) (Everest and Özcan 2015), Diyarbakır (Ergani plain) (Fig. 3) in southeast Turkey and as patches in the river valleys throughout the country (Çimrin et al. 2004) (Fig. 4). Fluvisols of Turkey cover a land area of about 7,338,590.54 ha, which is 9.50% of the whole agricultural land (Fig. 3).

5 Management

Fluvisols are mainly allocated to the major and economically strategic crops of the country, namely to the non-irrigated cereals and irrigated to non-irrigated horticultural crops, as Turkey was and generally has been a rainfed ecosystem. The widely distributed rainfed. Fluvisols allocated for cereals in

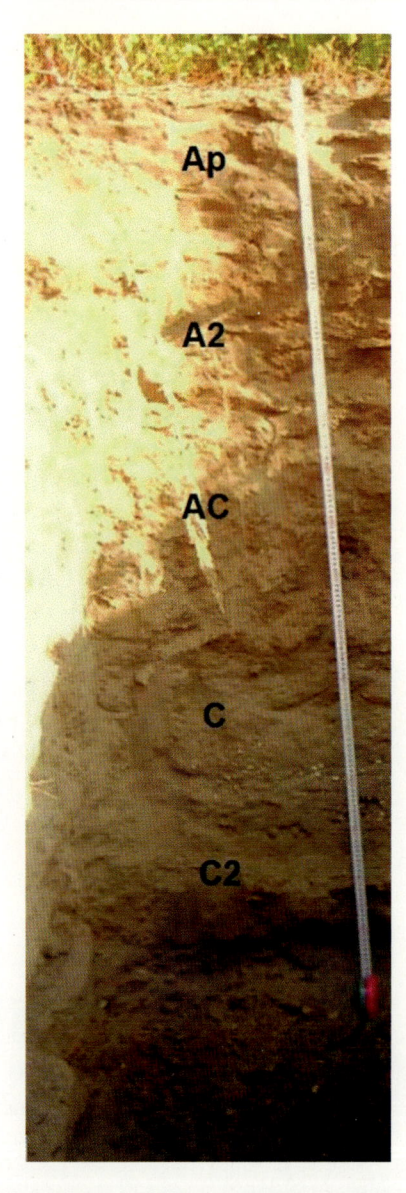

Fig. 1 Fluvisol profile in Umurbey–Çanakkale (northwest Turkey) (Yiğini 2006)

Turkey have been the well-established wheat ecosystems/terroirs since millennia, following historical deforestation. Their low organic carbon contents are somewhat uniform throughout their pedons documenting their stable and preserved state and potential to increase with renovated technologies and sustainable cropping and irrigation. The soil organic matter contents of the rainfed. Fluvisols may increase up to 3% in the surface horizons via sustainable vegetable cropping and irrigation as it is in the case of the long-cultivated rainfed cereal areas of the Troy Historical National Park in northwest Turkey (Çanakkale) (Everest 2015).

Most of the irrigated citrus, apricot, and vegetable orchards as well as cotton, maize, and sunflower fields are also widespread on the earlier rainfed. Fluvisol environments in southern and southeastern Turkey. Monitoring studies are underway in both areas especially in determining the C-sequestration and soil change expected to occur after the implementation of the gigantic irrigation scheme that concerns 1.8 M ha land in southeast Turkey. The Fluvisols of southern and southeastern Turkey bear a prime advantage for cotton and vegetable growing with their high K-retention capacity based on their moderate contents of kaolinite together with higher contents of smectite (montmorillonite) and illite. The illite and interstratified clay mineral contents of some Fluvisols were determined to be responsible for their higher exchangeable K contents which bear a significant advantage for crop production (Özbek et al. 1979). The high-magnesium clay contents of the Fluvisols in the Acipayam basin in western Turkey would most probably enhance the chemical quality of these soils as well as increase their capacity in developing a better microstructure via the pedogenic palygorskite determined to form in the soil environment rich in magnesium and alkali solutions (Güzel and Wilson 1985). Güzel and Kapur (1976) reported the presence of well-managed Fluvisols on the early to late Holocene terraces of the Ceyhan River alternating with Vertisols (similar to the Black cotton soils of India developed from transported-deposited volcanic dark colored soil materials). These deep soils of the Ceyhan Basin similar to the Seyhan Basin Fluvisols and Vertisols are allocated for rainfed and partly irrigated high yielding cotton in southern Turkey as is elsewhere in similar cropping ecosystems within a wheat rotation (Fig. 5). Some of the Vertisols and Fluvisols of the Holocene chrono-sequential catenas of the Mediterranean soil ecosystems consist of buried profiles with B horizons reflecting the difference in the magnitude of the soil formation conditions/processes of the buried materials compared to the overlying contemporary soil consisting of only A-C horizon sequences (Dinç et al. 1991a).

Many of the Fluvisols in Turkey have been degraded by soil sealing, namely by inappropriate land use and

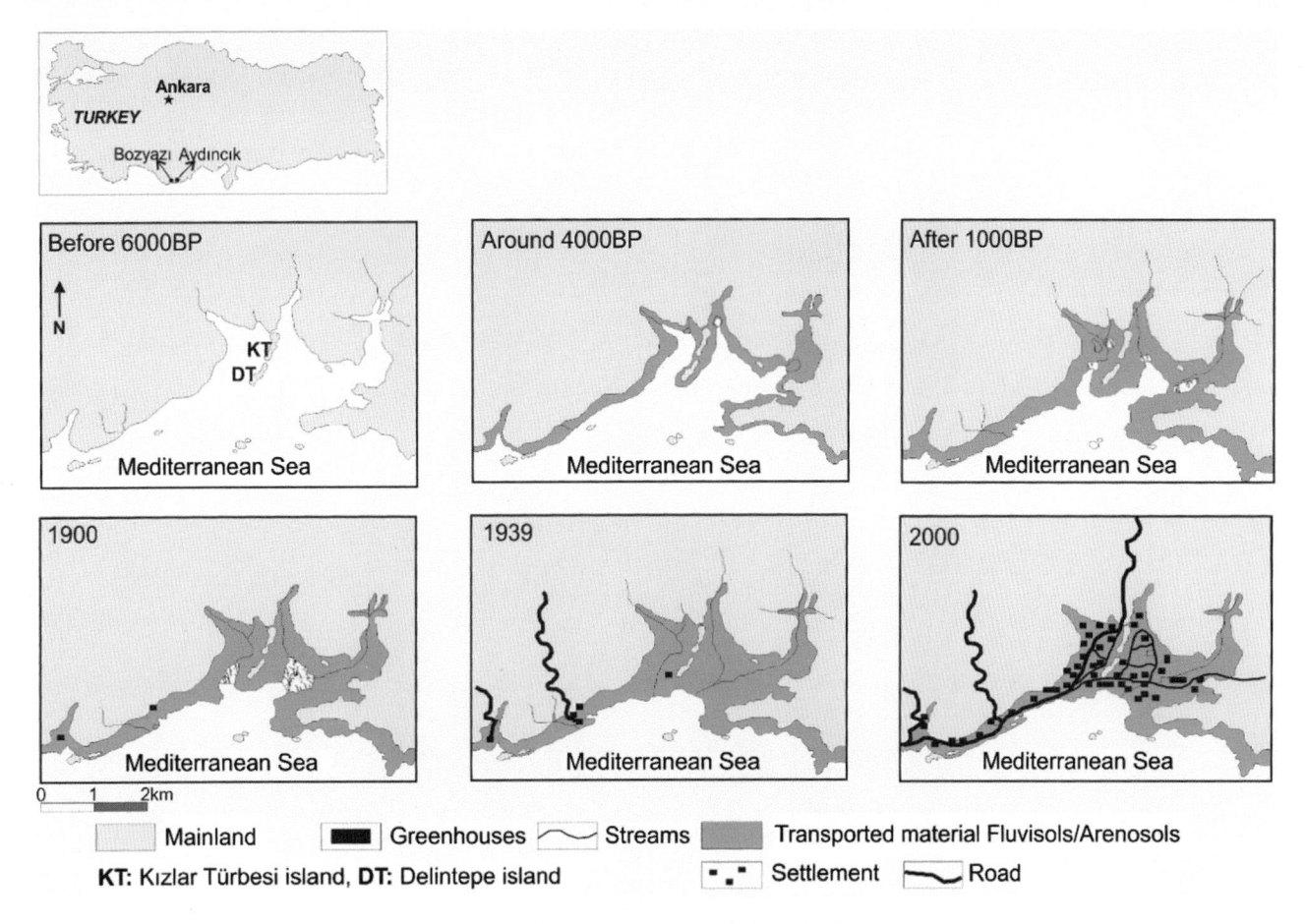

Fig. 2 Fluvisol–Arenosol development in the Aydıncık Delta in south Turkey (Bal et al. 2003)

urbanization, industrialization, highway constructions, tourism facilities, and secondary house buildings especially at the coastal areas of Adana, Mersin, Bursa, Thrace, Adapazarı, İzmir, and Antalya as well as at inland regions of Ankara, Eskişehir and Konya. The fertile Fluvisols deposited at the surrounding areas of the major river of central Anatolia- the Kızılırmak- are being extracted as clay/soil sources for making pottery and bricks since the Early Roman period. Today this area is one of the largest historical soil-bound industry sites of Turkey (Gürtan and Munsuz 1969) highly degraded by clay/soil extraction. Studies were undertaken in order to assess the technological properties of the presently used and the potential raw material/clay sources by Ünver et al. (1987) seeking to determine the more appropriate raw material sites for the tile-brick industry and in turn the inappropriate lands for cultivation (the non-agricultural lands).

Some Fluvisols in the country suffer from salinity, alkalinity, and drainage problems due to high or frequently fluctuating groundwater levels bound to inappropriate and excess water extraction and irrigation. The malpractice of wetland restoration actually seeking conversion to cultivated land is also responsible for the unpredictable and unexpectedly fluctuating groundwater levels (Fig. 6). This misuse of the soil creates highly inappropriate conditions for the sustainable management of groundwater use and salinity management in the extensive central Turkey ecosystem of the Konya Basin (Dinç et al. 1991b). Similar problems concerning crop and halophyte productions based on salinity management also prevail in the irrigated soils of the Seyhan

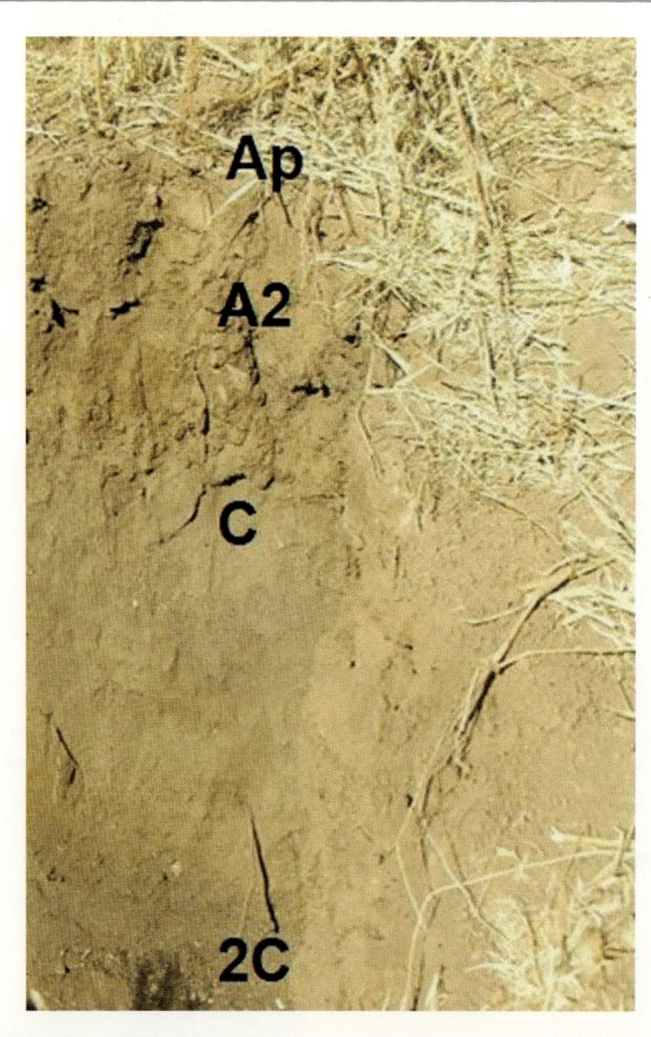

Fig. 3 Fluvisol profile in Ergani-Diyarbakır (southeast Turkey)

Basin due to the excess use of the irrigation water and inappropriate extraction of the groundwater sources (Akça et al. 2017). Studies are underway in establishing income generating local halophyte and appropriate groundwater management programs seeking sustainable land management via water use and crop production associations especially for the Arenosol and Fluvisol areas.

Vast areas are now partly irrigated within immense irrigation projects in central and especially southeastern Turkey seeking income generation for rural populations. The recently ongoing southeastern Anatolian Irrigation Project seeks to irrigate 1.8 M ha fertile rainfed land (partly accomplished) until the near future. Additionally, the traditionally cultivated rainfed olive and pistachio orchards in the southeast and the wheat crops in central Turkey on some Fluvisols are now partly irrigated for higher yields by deep ground water wells of about 300–400 m depths. However, indispensably, the degradation of the long-established soil pore system and salinity problems, (as is the case for the Calcisols and the Cambisols of the area) are increasing at the expected rate as determined by the recent studies carried out by the University of Harran in Şanlıurfa (southeast Turkey) (see Solonchaks). Studies are also conducted in these areas and throughout Turkey in order to establish a holistic sustainable land management plan of which is also foreseen by the National Action Plan of Desertification of the country (NAP-D, MEF 2006). Moreover, as a further requirement of the UNCCD targeting the protection of the soils and the environment, the 10-year strategic program of the NAP-D of Turkey is accomplished by the General Directorate of Combatting Desertification and Erosion of the Ministry of Forestry and Water Affairs (UNCCD, 10-YEAR strategy alignment, MWF 2014) in coordination with the Soil Science Society of Turkey. The Ministry of Food, Agriculture and Livestock is also highly active in developing SLM programs over consolidated land since 1967 with enhanced project progress since 2011.

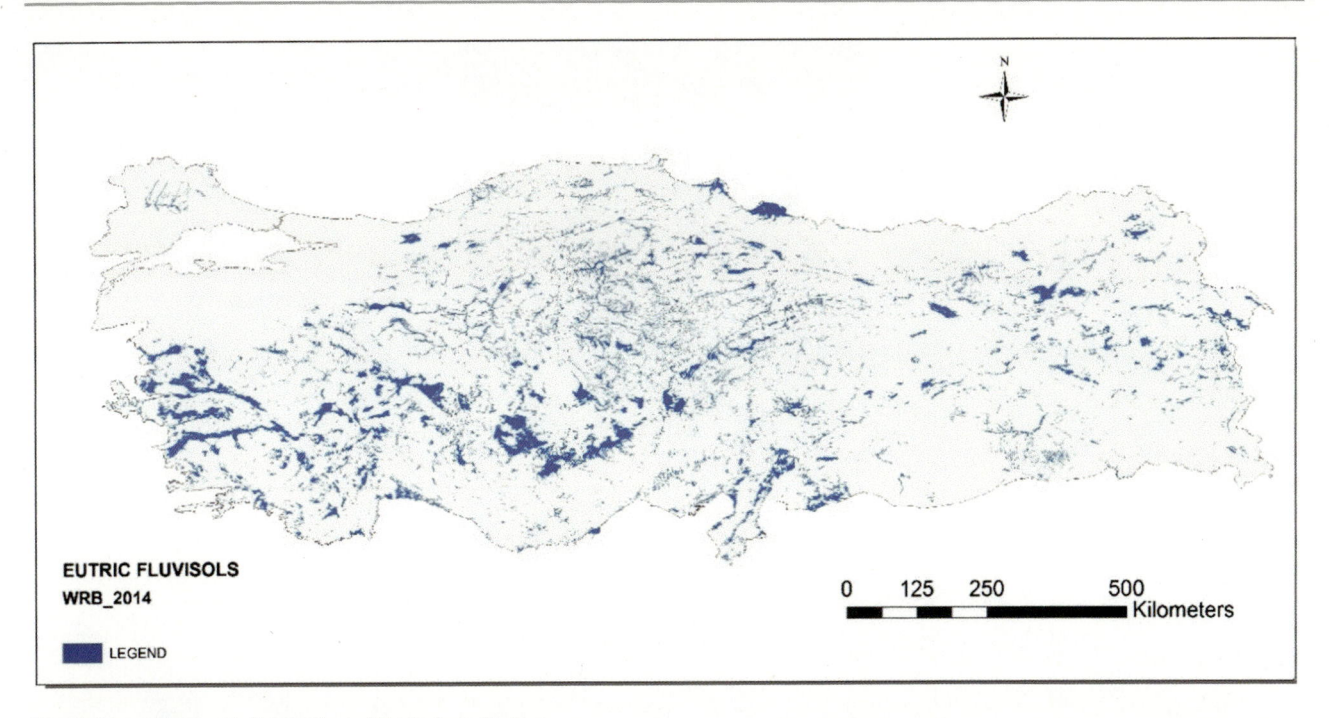

Fig. 4 The widespread distribution of fluvisols in Turkey

Fig. 5 The cultivated Fluvisol–Vertisol landscape on the banks of the Ceyhan River (south Turkey)

Fig. 6 A drained Fluvisol wetland shifted to grazeland in central Turkey

References

Akça E, Kume T, Nagano T, Donma S, Watanabe T, Kapur S (2017) Salinity management for the foreseen climate change conditions in the highly productive agricultural areas of Turkey. ICCAP projects book, Springer, Hexagon (in prep.)

Bal Y, Kelling G, Kapur S, Akça E, Çetin H, Erol O (2003) An improved method for determination of holocene coastline changes around two ancient settlements in Southern Anatolia: a geoarchaeological approach to historical land degradation studies. Land Degrad Dev 14(4):363–376

Çelmeoğlu N (2011) The historical anthroscape of Adana and the fertile lands. In: Kapur S, Eswaran H, Blum WH (eds) Sustainable Land Management. Springer, Berlin, pp 259–285

Çimrin KM, Akça E, Şenol M, Büyük G, Kapur S (2004) Potassium potential of the soils of the Gevaş region in eastern Anatolia. Turk J Agric Forest 28:259–266

Dinç U, Şenol S, Kapur S, Sarı M (1991a) Catenary soil relationships in the Çukurova region, southern Turkey. Catena 18:185–196

Dinç U, Şenol S, Kapur S, Sarı M, Derici MR, Sayın M (1991b) Formation, distribution and chemical properties of saline and alkaline soils of the Çukurova region, southern Turkey. Catena 18:173–183

Dinç U, Sarı M, Şenol, S. Kapur S. Sayın, M. Derici MR. Çavuşgil V, Gök M, Aydın M, Ekinci H, Ağca N, Schlichting E (1995) Soils of Çukurova Region, University of Çukurova, Faculty of Agriculture, No. 26, Adana, Turkey. (in Turkish)

Dinç U, Şenol S, Kapur S, Cangir C, Atalay İ (2001) Soils of Turkey. Univ Çukurova Fac Agric Pub No 51 (in Turkish)

Everest T (2015) The soils of the Troy Historical National Park and their management. Onsekiz Mart University, Graduate School of Natural Sciences, Doctoral Thesis, Çanakkale, (unpublished) (in Turkish)

Everest T, Özcan H (2015) Classification of Karamenderes' Plain, Troy Region Right Cost Fluvial Soils. Soil Water J 4(2):p21:29

Gürtan N, Munsuz N (1969). The Avanos pottery and some technological properties of the soils used in production. Publication of the Ankara University Faculty of Agriculture No: 404–254, Ankara, p 75 (in Turkish)

Güzel N, Wilson J (1985) High-Magnesium clays from alluvial soils of the Acıpayam plain, southern Turkey. Proceedings of the 5th meeting of the European clay groups. Charles University, Prague, pp 117–128

Güzel N, Kapur S (1976) Soil-Geomorphology relationships on the terraces of the Ceyhan River. Ann Fac Agric Univ Çukurova 2:117–143

IUSS Working Group WRB (2015) World reference base for soil resources 2014, update 2015 International soil classification system for naming soils and creating legends for soil maps. World Soil Res Rep No 106 FAO Rome, 203 p

MWF (Min. of Water and Forestry) (2014) Alignment of Turkey's NAP with UNCCD 10- Year Strategy and Reporting Process. FAOProject, GCP/TUR/060/GFF

MEF (Min. of Environment and Forestry) (2006) The Desertification National Action Program of Turkey (NAP-D). In: Düzgün M, Kapur S, Cangir C, Akça E, Boyraz D, Gülşen N (eds) Ministry of environment and forestry publication No 250. ISBN 975-7347-51-5. State Meteorology Institute Press. 110 p

Özbek H, Kaya Z, Derici MR, Kapur S (1979) Potassium availability of some soils of S. Turkey as related to clay mineralogy. In: soils of Mediterranean type of climates and their yield potential, Proceedings of the 14th Colloquium of the International Potash Institute held in Bern

Soil Survey Staff (2014) Keys to Soil Taxonomy, 12th edn. USDA-Natural Resources Conservation Service, Washington, DC, 372p

Ünver İ, Kapur S, Çanga M, Çavuşgil VS (1987) Relationships between technological properties and clay minerals of the present and potential raw materials for the brick- tile industry at some regions in Turkey. In: Proceedings of the 3. Nature Clay Symposium, 21–27 Sept, METU, Ankara, pp 203–216 (in Turkish)

Yiğini Y (2006) Detailed soil survey and mapping of Çanakkale Umurbey Plain. Çanakkale Onsekiz Mart University, Graduate School of Natural Sciences, Master Thesis, Çanakkale, Turkey (unpublished) (in Turkish)

Calcisols and Leptosols

Erhan Akça, Salih Aydemir, Selahattin Kadir, Muhsin Eren, Claudio Zucca, Hikmet Günal, Franco Previtali, Pandi Zdruli, Ahmet Çilek, Mesut Budak, Ahmet Karakeçe, Selim Kapur, and Ewart Adsil FitzPatrick

1 Introduction

Calcisols are soils with substantial accumulation of secondary lime and the name of Calcisols derives from the Latin calx which means lime. They are described as "*Soils having a calcic or petrocalcic horizon within 100 cm of the surface and no diagnostic horizons other than an ochric or cambic horizon, an argic horizon which is calcareous, a vertic horizon, or a gypsic horizon underlying a petrocalcic horizon*" (IUSS Working Group WRB 2015). Calcisols are common in/on highly calcareous parent materials and widespread in arid and semiarid environments. Formerly used soil names for many Calcisols included Desert soils and Takyrs and in the US Soil Taxonomy, most of them belong to the Calcids (Soil Survey Staff 2014).

Calcisols of Turkey widely occur especially in association with Leptosols at highly sloping and eroded highlands and at some locations with Cambisols. They also associate with Gleysols, Solonchaks, Gypsisols, and Vertisols in depressions and valley bottoms/lowlands and with Luvisols and Cambisols in the Turkish-Mediterranean highlands and sloping skirts of coastal mountains (Dinç et al. 2001). The Cambic Vertic Calcisols are the most widespread Calcisols in Turkey. Calcisols in general develop on level to hilly land of arid and semiarid environments. The natural vegetation is most often sparse and dominated by xerophytic shrubs (maquis) and trees and/or ephemeral grasses.

Typic Calcisol profiles have a pale brown surface horizon and substantial secondary accumulation of lime (calcium carbonate) that occurs within 100 cm of the soil surface. Calcretes (caliche)-Petrocalcic horizons—are the major carbonate deposits—pools—generally located on the Pleistocene-Early Holocene landscapes of the present semi-arid to arid and subtropical regions. Calcrete surfaces are widespread in almost all parts of Turkey except the north (Fig. 1).

E. Akça (✉)
School of Technical Sciences, University of Adıyaman, 02040 Adıyaman, Turkey
e-mail: erakca@gmail.com

S. Aydemir
Department of Soil Science and Plant Nutrition, University of Harran, Şanlıurfa, Turkey

S. Kadir
Department of Geological Engineering, Osmangazi University Eskişehir, Eskişehir, Turkey

M. Eren
Department of Geological Engineering, Mersin University, Mersin, Turkey

C. Zucca
International Center for Agricultural Research in the Dry Areas, ICARDA, Amman, Jordan

H. Günal
Deprtment of Soil Science and Plant Nutrition, University of Gaziosmanpasa, Tokat, Turkey
e-mail: hikmet.gunal@gop.edu.tr

F. Previtali
Department of Geosciences, University of Milan (Bicocca), Milan, Italy

P. Zdruli
Mediterranean Agronomic Institute of Bari, CIHEAM, Bari, Italy

A. Çilek
Department of Landscape Architecture, University of Çukurova, Adana, Turkey

M. Budak
Department of Soil Science and Plant Nutrition, Siirt University, Siirt, Turkey

A. Karakeçe
Ministry of Food, Agriculture and Livestock, Dir. of the Koçaş Agricultural Establishment, Aksaray, Turkey

S. Kapur
Department of Soil Science and Plant Nutrition, University of Çukurova, 01330 Adana, Turkey

E.A. FitzPatrick
Department of Plant and Soil Science, University of Aberdeen, Aberdeen, UK

© Springer International Publishing AG 2018
S. Kapur et al. (eds.), *The Soils of Turkey*, World Soils Book Series,
https://doi.org/10.1007/978-3-319-64392-2_10

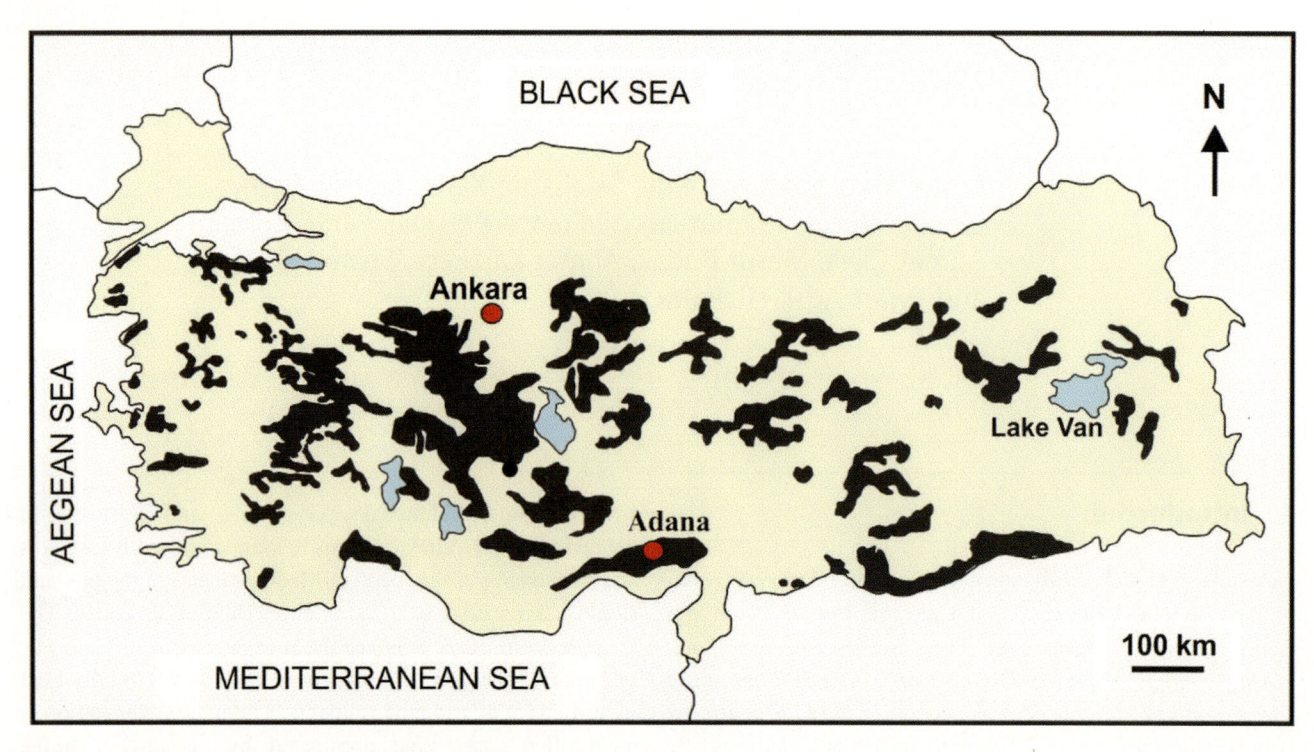

Fig. 1 Distribution of calcretes in Turkey (Erol 1984)

Development of the calcrete depends on the magnitude of the past and present carbonate sources together with the climatic fluctuations of the Quaternary (Şenol et al. 1991, 1993) (Fig. 2).

Calcretes, presently, occur on the preserved surfaces of the Pleistocene of arid to semiarid climatic regions as well as subhumid and subtropical environments as buried paleocalcretes (Reeves 1970; Goudie 1975). Calcretes in Turkey have mainly formed in or on clay-stones, limestones, shales, especially in or on marls, fine Neogene sediments, mud-flows, and shallow marine-lacustrine environments (Marshall 1983; Kapur et al. 1986, 1987, 1993; Gürbüz and Kelling 1993).

Fig. 2 General pedo-stratigraphic section of the Mediterranean region of Turkey (Şenol et al. 1991)

2 Distributions, Geomorphic Evolution, and Formation of Calcisols and Calcretes

Calcisols (especially Cambic Vertic Calcisols) are widespread in southeastern, southern, western (Aegean), eastern, and central regions of Turkey where calcium carbonate-rich materials are transported-deposited by alluvial and colluvial processes (mud-flows) along catenary sequences (Tanju 1977; Dinç et al. 1998) (Fig. 3). Calcisols of Turkey, along with the Leptosols, cover a land area of about 726,609,483 ha which is 9.39% of the whole Agricultural land (Fig. 3).

2.1 The Calcisol/Calcrete Terrace Systems, Southern Turkey

Calcisol formation was reported on old alluvial terrace systems in the south, central-west, and Aegean Regions of Turkey including a sequence of early Pleistocene to early Holocene deposits (Erol 1984; Kapur et al. 1984, 1990, 1993, 2000; Yetiş et al. 1995; Mermut et al. 2004). Kapur et al. (1990, 2000) modeled this as a sequence of mudflow and alluvial terraces where soil formation was correlated with morphological positions and ages (Fig. 4). One of the best examples for this terrace system is the Balcalı, Adana Pleistocene-Holocene toposequence (Fig. 5), widespread in the Mediterranean belt of Turkey, and similar to numerous sites along north Africa (especially in Tunisia) and southern Europe (especially in central and southern Spain). The uppermost terrace (TH1-early Pleistocene calcrete surface on Pliocene clay) is the oldest surface of the toposequence and comprises the thickest pedogenic calcrete described by Kapur et al. (1987) as consisting of completely calichified conglomerates ("calcrete-conglomerate").

The calcrete underlies a highly rubified (10R) and completely decalcified soil (Luvisol/Alfisol) of an Ap/Bt/Ckm horizon sequence formed in/on the calichified conglomerate/calcrete conglomerate. The calcrete of the sequence overlies a column (see Sect. 3) horizon, the upper part of which has been cemented to form the calcrete (caliche) (Fig. 5). The TH2 surface (early-mid Pleistocene) has a calcrete conglomerate (caliche with columns) with higher amounts of reddish-brown (2.5YR) soil material in between the parent rock and a probably autochthonous and partly allochthonous soil cover (A/Ckm1/Ckm2/Bwk) (Fig. 6).

The column or 2Bwk horizon of this section is occasionally deeper compared to the TH1 surface and intergrading to the moderately weathered Pliocene clay with vertical calcite columns (Figs. 5 and 6). The slope of TH2 (middle Pleistocene) has a more weathered and deeper reddish-brown (2.5YR) soil (Rhodic Luvisol) of Ap/Bt/Ckm, which is most probably autochthonous on the underlying deep calcified conglomerate. The calcified conglomerate of this section is not as massive as the calcrete conglomerate of the previous profile although having a crust formed on its surface at the soil boundary and continuing as patches along the section. Absence of a continuous calcrete along the toposequence may indicate removal of autochthonous soil, followed by dissolution and partial removal

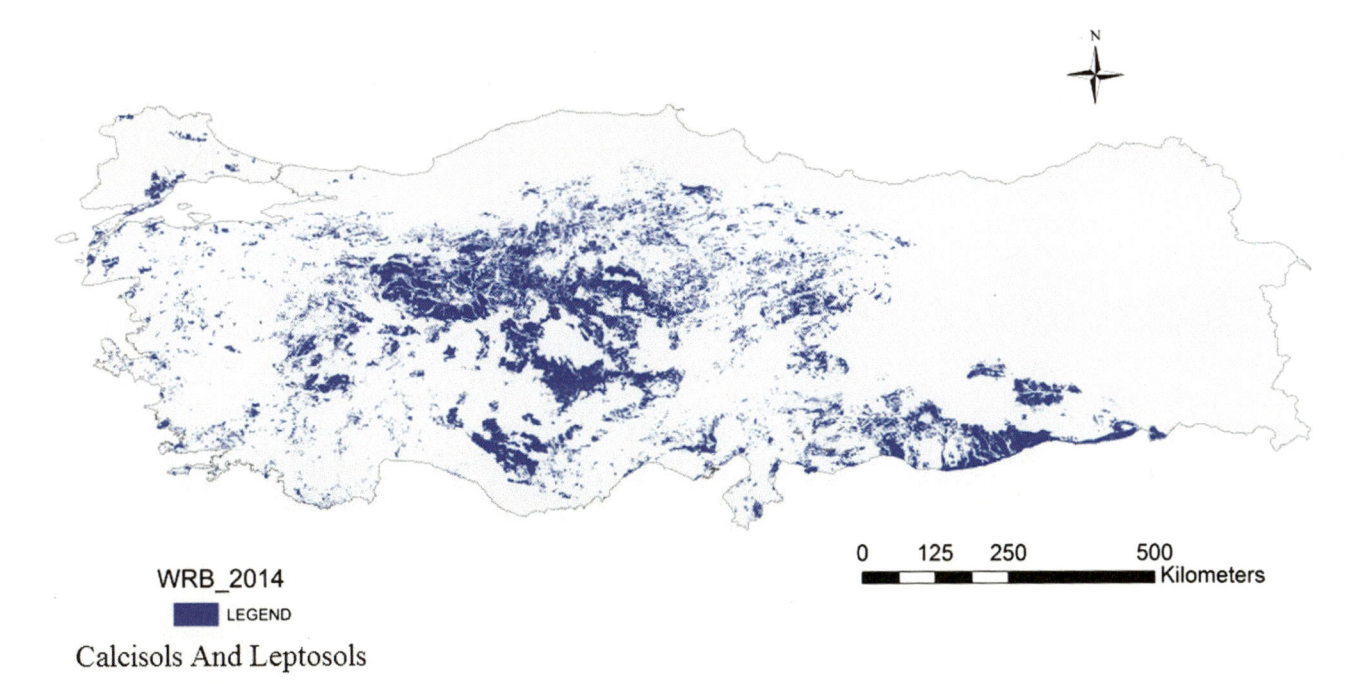

Fig. 3 Distribution of Calcisols and Leptosols in Turkey

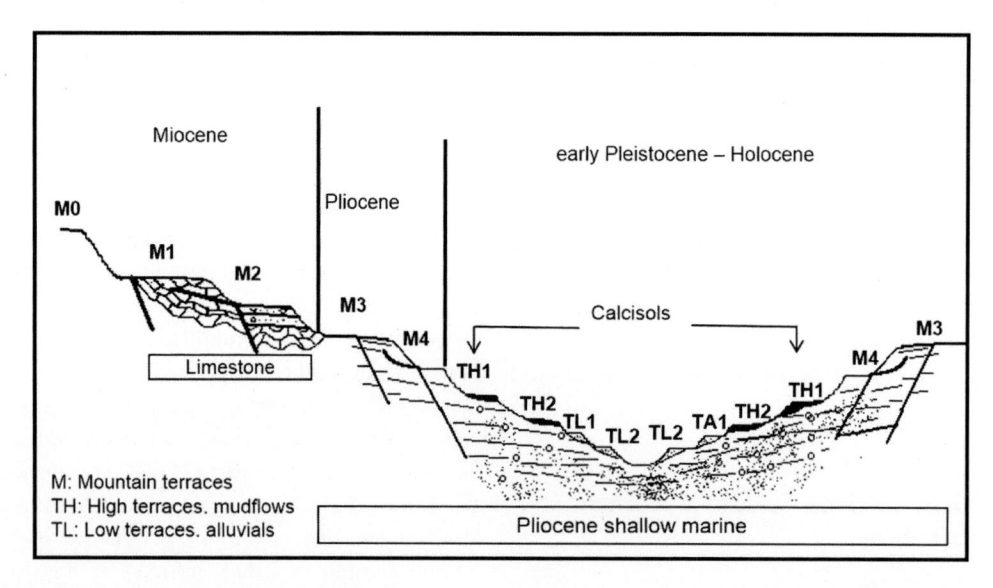

Fig. 4 Mountain-Colluvial-Alluvial Terraces and Calcisol Formation on TH surfaces (Kapur et al. 1990 and 2000, modified from Erol 1984)

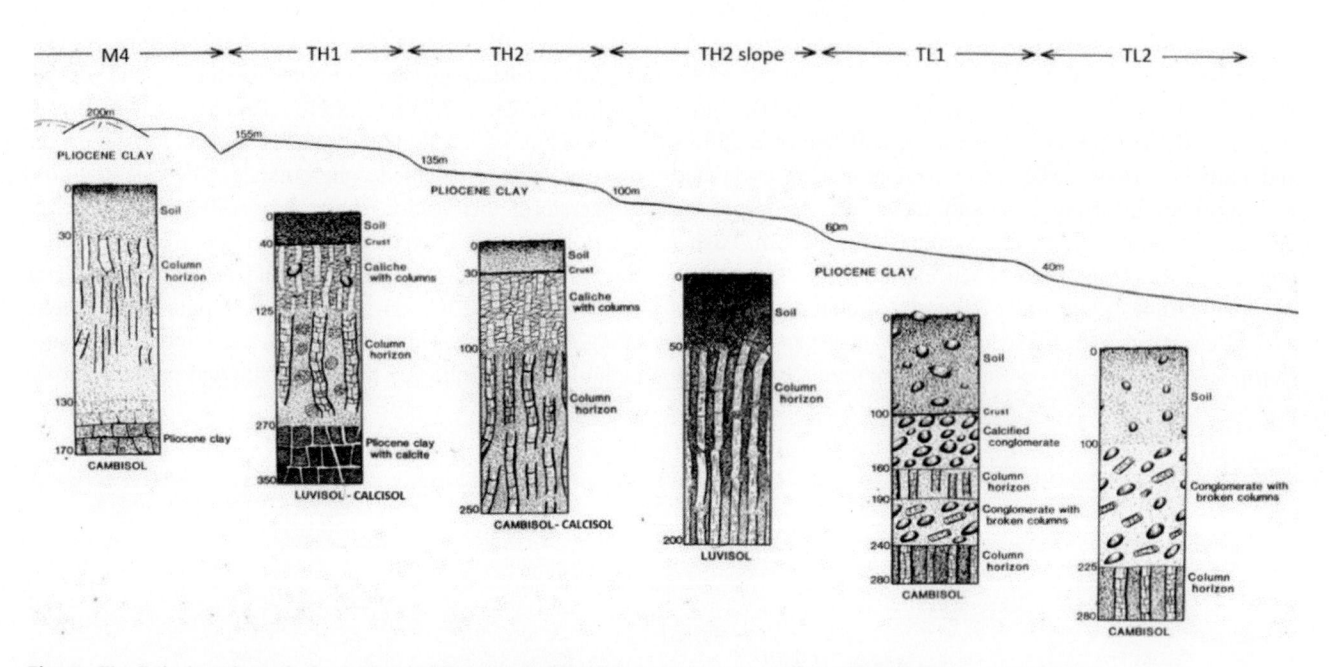

Fig. 5 The Balcalı, Adana glacis, the Luvisol-Calcisol-Cambisol toposequence in southern Turkey (Kapur et al. 1990)

of the pre-existing calcrete and consequently, burial by soil from an upper surface. This assumption would most likely account for the similar mineralogical paragenesis and properties determined in the soils along the studied transect (Figs. 5 and 6).

The TL1 surface (mid Pleistocene-early Holocene) is a thin layer of reddish-brown soil (Cambisol) with frequent rock fragments—conglomerate rubble—overlying a deep rubified reddish-brown (2.5YR) weathered column horizon. The reddish-brown soil (Cambisol) at the surface of this section with an A/Bw/Ckm horizon sequence is separated by a discontinuous crust with the conglomerate as well as a discontinuous calcrete conglomerate, indicating the transportation of the material. Presence of glauconite (determined by XRD) in the column horizon of this section indicates a probable transgression–regression between the lowermost calcified layer (continental) and the column layer (presence of glauconite of marine origin during deposition followed by continental arid conditions indicated by column formation associated with abundant palygorskite precipitation). The TL2 (Holocene) of this toposequence has a shallow Calcisol/Cambisol at the surface that is developed in transported material over a column horizon which is the actual dipped continuation of the column layers of TL1 as well as

Fig. 6 Petric Calcisol, the Sarıçam soil on TH1-TH2 surfaces—Adana, south Turkey (**a**), roots along clay coatings in Ckm1 horizon (**b**), column horizon of Sarıçam soil (2Bwk horizon) underlying massive calcrete (**c**), abundant clay coatings in the column or 2Bwk horizon (**d**)

the Pliocene clay deposit with vertical calcite columns of TH2 (Fig. 5).

2.2 The Calcisols of Southern Turkey, the Sarıçam Soil

The Sarıçam soil (Petric Calcisol) is widely distributed on the TH1 and TH2 surfaces located in the north of Adana and Mersin provinces. It is frequently overlain by a massive/desiccated calcrete that has preserved the features of the vigorous pedogenic processes undergone during the Pleistocene period. The most prominent pedogenic feature of the Sarıçam profile and relevant sites is the development of the column horizon(s)/layer(s) (the Bwk and Ckm horizons) (term coined by Kapur et al. 1991) below the massive calcrete. The column horizon with calcite columns manifests the most significant micro-pedogenic process by the abundant distribution of the red clay coatings on the surfaces of the microfractures between the calcite columns (Figs. 6 and 7). These may indicate the occurrence of a sequence of processes initiating the leaching of the mobile to moderately mobile Si and alkali elements and the concentration of the immobile elements of Mg, Al+Fe. This in turn, most likely, causes the precipitation of Fe (oxyhydr) oxide phases and Fe (liberated following the alteration of the precursor minerals) in the crystal structures of palygorskite and especially of the consequently leached/illuviated Fe bearing smectite, which was determined to be Ferrogenious smectite by Özkan and Ross (1979) in similar and nearby soils to the Sarıçam (Fig. 6). In situ development (precipitation) of intermatted and protruding as well as bridging palygorskite between the calcite crystals in the columns most likely enhances aggregate development (Kapur et al. 1991) and most likely increases water retention via the increasing stable aggregates in the similar soils of southeastern Turkey (Kettaş et al. 1991) (Fig. 7) (work is underway to determine porosity characteristics and structural stability of the palygorskite—calcite aggregates of these soils). Palygorskite is the only pedogenic mineral formed during the Pleistocene and its presence can be accepted as an evidence of the dominant arid climatic conditions when most Calcisols of Turkey were formed (Kapur et al. 2000; Küçükuysal et al. 2013; Kaplan et al. 2013).

The higher palygorskite contents compared to smectite in the Ckm1 and Ckm2 horizons was different in the A-horizons that may indicate the contemporary aeolian smectite contribution (most probably Holocene) by Saharan input (Kubilay et al. 1997, 2000; Andreucci et al. 2012) and/or the probable less drier pedo-environmental conditions that would not favor

Fig. 7 Clay coatings developing in 2Bwk of TH1 surface of the Sarıçam soil (Kapur 2000) (*cross polarized light*) (**a**), pure micritic matrix with occasional *thick* to moderate clay coatings (CC) and palygorskite coatings (P) repeated calcification destroying clay coatings followed by the development of acicular pseudomycelia (**b**), palygorskite bundles forming between rhombohedric calcites of horizon 2Bwk (SEM-scanning electron microscope) (**c**), intermatted palygorskite fibers (SEM) (**d**), palygorskite fibers protruding from matrix (TEM–transmission electron microscope) (**e**)

the transformations/weathering or neoformations of smectite to palygorskite in the surface horizon (see Luvisols, Fig. 9). The unusually low contents of the clay size fraction (6.1%) within the 2Bwk horizon of the Sarıçam soil is most likely due to the high amount of sand and silt-sized aggregates (stable to partly stable) composed of clay size minerals and especially of palygorskite and calcite aggregates mentioned above (Table 1; Fig. 7). These partly stable aggregates might be responsible for the higher water holding capacity (Akça et al. 2006). The high humic acid contents of the similar horizons of the Balcalı soil (Calcisol on Pleistocene calcrete terraces, located in the University of Çukurova campus, Adana, Turkey) determined by Sokolovska (2002) are most likely the other reason for the higher water storage of the similar horizons of the Calcisols in the Mediterranean calcrete/Calcisol ecosystem. This consequently documents the need for appropriate soil management provided these soils are allocated to crop cultivation. Deeper phases of the Sarıçam soil, cultivated for cotton under dry-farming, overlying the massive calcrete (Ckm) of the Sarıçam profile in Table 1 consisting of an Ap/A12/B21t/B22t/Ckm horizon sequence, were described and studied by Cangir (1982). The Ap, A12, and B21t horizons of this Calcisol were dominant in palygorskite/illite and kaolinite, whereas smectite was dominant in the B22t and Ckm horizons followed by palygorskite/illite. The increase of smectite together with the increases in the clay fraction and Fe_2O_3 and MgO contents most likely indicated clay transportation–deposition (illuviation) via leaching (Cangir 1982).

2.3 The Calcisols of Central Turkey, the Ankara, Niğde, and Kap Soils

The Calcisols of central Turkey (Ankara) mostly developed on the calcretes of the Middle Pleistocene are similar in chemical and clay mineral (smectite and palygorskite contents) compositions to the other calcretes throughout Turkey (Küçükuysal and Kapur 2014). The calcretes of Ankara have developed on/in the flat to slightly undulating consecutive shallow marine-lacustrine and fluvial deposits/or red paleosols rather than being developed on/in mudflow-glacis systems like in the Mediterranean areas (south and southeast) of Turkey (Fig. 8). These consecutive calcrete layers are the potential providers of proxy data for reconstructing the Quaternary paleoclimates (Küçükuysal et al. 2013).

Other widespread deeper Calcisols from the Niğde province overlie shallow marine-lacustrine sediments and comprise pronounced Bw horizons (Fig. 9). Although these soils are prone to salinisation and are frequently moderately saline (Table 2), they are the extensive ancient lacustrine Calcisol/pasture ecosystems of central Turkey allocated for the grazing of small ruminants as well as being the wheat basket of the country. The higher P, Na contents, and the sand fraction of the Bw1 horizon may reflect the occurrence of an older surface and contemporary burial by the present horizon A (Table 2).

The most significant Calcisol-Leptosol site/ecosystem in central Turkey is the Kap (particular Calcisol initially named upon the widespread south eastern shallow Calcisols overlying calcretes) soil ecosystem developed on/in Mio-Pliocene marls and conglomerates, and Pleistocene lacustrine marls. The widely distributed Kap soils in central Turkey overlying Plio-Pleistocene lacustrine marls (TH1-TH2 surfaces) are composed of A/Ap-Bw-Ckm horizon sequences with varying colors which frequently are brown (10–7.5YR) and rarely reddish-brown (5YR). The presence of the repeated highly crystalline calcrete-thick Ckm, despite the absence of the column horizon (e.g., in Koçaş, Konya lacustrine Kap soil ecosystems), points to a similar paleo-climate with dry and wet fluctuations, but different in magnitude than the one prevailing in the Adana and western regions of Turkey (Burdur and İzmir) (Fig. 10).

The Kap soils (especially of Koçaş, Altınova and Niğde, central Turkey) developed on autochthonous lacustrine land surfaces by precipitation-crystallization of the carbonate-rich solutions, occurring throughout the Pleistocene climatic fluctuations. The weakly developed and infrequent column horizons underlying massive calcretes with diffuse-friable calcite columns indicate occurrence of a rapid and dry period followed by a wet (pluvial) one impeding development of deep clayey layers similar to the upper levels of the mud-flows. Fluvial terraces-TL1-TL2- indicated by Erol (1981) (Fig. 4) are the playa and steppe marls of this section (as coined by de Meester 1970) (Fig. 11). The TH surfaces are bound by the Neogene uplifted structural terraces (de Meester 1970). These are the M1 and M2 land surfaces in Erol (1981, 1991) (Fig. 11).

2.4 The Calcisols of Southeastern Turkey

2.4.1 The Nizip and Kap Soils

Two Calcisol sequences are widespread in southeastern Turkey with horizon sequences of Ap-Bw/Bkm-Ckm and Ap-Ckm (Fig. 12; Tables 2, 3 and 4). The former is the Nizip Calcisol extensively allocated for traditional pistachio production (a pistachio terroir for millennia, since the early Greco-Roman periods) and the latter is the most widespread Calcisol which is shallower with a more massive underlying calcrete extensively allocated for wheat production. This is the Kap soil sequence in SE Turkey representing the TH surfaces that developed on/in the indurated/calcified mud-flow deposits transported from the Eocene-Oligocene and Oligocene-Miocene uplifted limestone formations of the two fault sides (the Tektek and Fatik ranges) of the Şanlıurfa graben (Figs. 13 and 14; Table 4). Associations of

Table 1 Site-profile descriptions and some physical, chemical and mineralogical properties of the Sarıçam soil, Adana, south Turkey

IUSS Working Group WRB (2015)	Petric Calcisol
Soil Survey Staff (2014)	Petrocalcic Calcixerepts
Location	Sarıçam area, 18 km northeast of the Adana province, south Turkey
Coordinates	N37°06′ 38″ E35°31′ 35″
Altitude	250 m
Landforms	Colluvial
Aspect	South-West
Slope	1%
Drainage class	Well-drained
Groundwater	–
Eff. Soil depth	50 cm
Parent material	Plio-quaternary carbonate-rich sediments
Köppen Climate	Mediterranean
Land use	Reforested maquis area
Vegetation	Evergreen forest

Horizon	Depth (cm)	Description
A	0–20	Dark brown (7.5YR 3/4), clay loam, brown (7.5YR 4/4) dry, weak to moderate medium subangular blocky to granular structure, firm to friable, common very fine to medium roots, slightly effervescent, abrupt smooth boundary
Ckm1	20–80	Light brownish gray (5YR 7/1), very strongly effervescent, diffuse wavy boundary
Ckm2	80–120	Light brownish gray (5YR 7/2), very strongly effervescent, clear wavy boundary
2Bwk	120–150	Pale orange (5YR 8/2) matrix of cemented clay with carbonates, many distinct red (10R 5/6–5/8) clay films/coatings and dendritic diffuse manganese dioxide concretions, strong coarse prismatic to angular blocky structure, firm, strongly effervescent

Horizon	Depth (cm)	Organic C	$CaCO_3$ (%)	pH 1:1	EC (dSm^{-1})	Sand	Silt (%)	Clay	CEC* (cmol$_c$ kg^{-1})
A	0–20	1.0	23	7.4	0.39	46.3	26.1	27.6	31.2
Ckm1	20–80	0.2	83	8.2	0.66	9.2	22.6	68.2	12.0
Ckm2	80–120	–	69	8.3	0.97	23.9	43.7	32.4	25.1
2Bwk	120–150	0.3	27	8.0	0.87	44.6	49.3	6.1	

Horizon	P Olsen (mg kg^{-1})	Ca (cmol$_c$ kg^{-1})	Mg (cmol$_c$ kg^{-1})	Na (cmol$_c$ kg^{-1})	K (cmol$_c$ kg^{-1})	BS (%)	Clay suites
A	7.3	46.6	3.3	0.1	1.3	100	Smectite, palygorskite, kaolinite
Ckm1	4.1	66.1	6.3	0.2	0.3	100	Palygorskite, smectite, kaolinite
Ckm2	2.1	67.7	6.1	0.2	0.3	100	Palygorskite, smectite, kaolinite
2Bwk	3.4	40.8	0.7	0.2	0.1	100	Smectite, palygorskite, kaolinite

EC electrical conductivity, *BS* base saturation, *CEC* cation exchange capacity
* Extractable Ca may contain Ca from calcium carbonate or gypsum. CEC base saturation set to 100

authigenic illite to palygorskite and smectite to palygorskite and kaolinite in this profile are possibly related to the high contents of aluminum and potassium (Mermut et al. 2004) which are regulated by weathering, alteration, and neoformation. Soil and calcrete samples collected from the Kap soil of southeast and central Turkey have different $CaCO_3$ contents within their profiles and calcretes. These differences may point to variability in climatic fluctuations and in turn decalcification as well as addition of wind-blown materials soils to the surface horizons (recalcification) during the Pleistocene, throughout the Holocene and today (Figs. 15 and 16).

The increase in sand and silt fractions along with the Ca, Mg, and K cations and decrease in clay fraction along with

Fig. 8 The consecutive calcrete layers (*arrows*) developed on/in the shallow marine-lacustrine to fluvial deposits in Ankara, central Turkey (Küçükuysal et al. 2013)

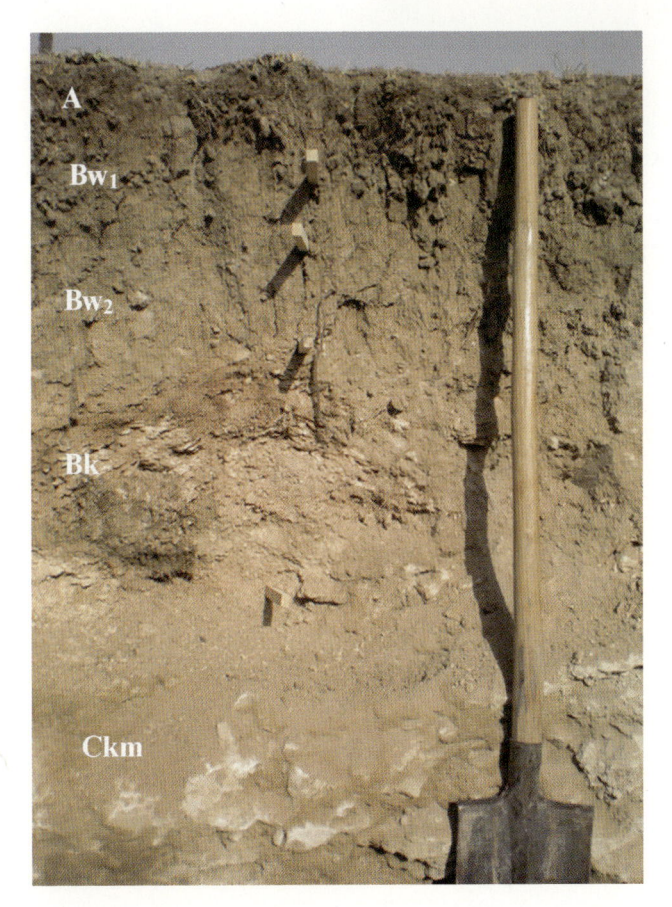

Fig. 9 Calcisol formed over shallow marine sediments in Bor town of the Niğde province in central Turkey

the CEC most likely indicate an aeolian input of the coarser grain sizes to the surface horizon. The abrupt increase in $CaCO_3$ of the Bkm horizon (49%) compared to the Ap (22%) and Bw (23%) horizons indicates a moderate to strong decalcification process that might have occurred during pedogenesis.

The decrease in clay and silt size fractions parallel to the decrease in $CaCO_3$ content and CEC values from the Bw to the Ap horizon in contrast to the increase in organic matter may point out to the formation of the surface horizon from the underlying calcrete coupled with leaching reflected by the slight increase of $CaCO_3$ and the Ca+Mg cations in the Bw horizon. The semi-quantitative increase of kaolinite at the surface horizon may also show the increased surface weathering and/or aeolian addition during pedogenesis.

Pedogenesis in the Aftermath of Calcrete Formation in the Kap and Akçakale Soils

Clay size particles of the Kap Calcisols (southeast Turkey) were most likely illuviated into the massive calcrete following the multiple calcification of the massive calcrete (Fig. 17).

This multiple calcification seems to have initiated the formation of the frequent 'ovalith' nodules determined in the massive calcretes of this area. Ovalith features most likely formed following the cohesion of the clay present in the matrix with the secondary pedogenic micritic calcites that developed via the multiple calcification in the Ckm horizons of the Kap soils. Rounded to oval ovaliths (sometimes coated with illuviated clays) from the Kap soils were occasionally determined in the Akçakale Solonchaks likely indicating their transportation from the Kap soil areas following their final separation from the physically weathered calcrete and/or Ckm horizons. The transportation of the ovaliths from the north (The Kap soils) to the south (the Akçakale soils) of Şanlıurfa has most probably occurred during the Pleistocene

Table 2 Site-profile descriptions and some physical and chemical properties of the Bor Calcisol, central Turkey (Budak 2012)

USS Working Group WRB (2015)	Petric Calcisol
Soil Survey Staff (2014)	Petrocalcic Calcixerepts
Location	16 km northwest of Bor town of Niğde province, central Turkey
Coordinates	37°4' 76" N–34°21' 54" E
Altitude	1155 m
Landforms	Shallow marine deposits
Aspect	
Slope	0–1%
Drainage class	Well-drained
Groundwater	–
Eff. Soil depth	80 cm
Parent Material	Shallow marine deposits
Köppen Climate	Mediterranean
Land use	Pasture
Vegetation	Major plant cover is formed from Taraxacum *farinosum*, Panderia *pilosa*, Onopordum *davisii* and Halimione *verrucifera*

Horizon	Depth	Description
A	0–16	Dull yellowish brown (10YR 5/3) dry and (10YR 5/4) moist, clay, moderate medium subangular block and strong medium prismatic structure, very firm, sticky and plastic, moderate fine and rare tap roots, strongly effervescent, salt crystals on ped surfaces, abrupt smooth boundary
Bw	16–26	Dull yellow orange (10YR 6/4) dry and (10YR 6/3) moist, moderate medium prismatic structure, silty clay, very firm when dry, sticky and plastic, strongly effervescent, moderate fine and tap roots, clear smooth boundary
Bk1	26–41	Dull yellow orange (10YR 6/3) dry, carbonate accumulation surface grayish yellow (2.5 Y 7/3) dry, clay, strong coarse prismatic structure, slightly firm when dry and moist, strongly effervescent, calcium carbonate on ped surfaces, distinct wavy boundary
Bk2	41–80	Light gray (5Y 8/2) dry and dull grayish yellow (2.5 Y 6/3)moist, clay, slightly firm when dry and friable when moist, very strongly effervescent, rare fine and tap roots, surface soil brought by animals, distinct wavy boundary
Ckm	80–112+	Light gray (5Y 8/2) dry and moist, very firm, strongly effervescent

Horizon	P Olsen (mg kg^{-1})	Ca	Mg (cmol$_c$kg^{-1})	Na (cmol$_c$kg^{-1})	K	BS (%)	Boron (mg kg^{-1})
A	1.01	22.31	5.12	6.60	3.37	100	23.41
Bw	4.05	17.29	4.07	14.49	3.26	100	21.07
Bk1	1.74	17.86	3.84	9.01	1.96	100	14.14
Bk2	15.36	16.01	2.68	7.82	1.91	100	9.68
Ckm	14.42	17.03	2.90	1.31	1.23	100	2.12

EC electrical conductivity, *BS* base saturation, *CEC* cation exchange capacity

climatic fluctuations with mass movements (mud or earth flows) (Fig. 18) (Kapur et al. 1991).

2.4.2 The Sırrın Soils

The Sırrın soils are deep, well-drained, moderately alkaline and occur on the broad alluvial plains of the Holocene to Pleistocene. This soil occurs on level to nearly level (0–1%) land at elevations of about 550 m. The plough layer of these soils is strong brown (7.5YR 4/6) and clayey with a surface granular mulch over weak medium subangular blocky structure. Clayey textures were found in all horizons, and color was uniform throughout (7.5YR 4/6) except for a 7.5YR 5/6 color at 167 cm. Beginning at a depth of 140 cm, a few black Fe–Mn stains were noted. Calcium carbonate segregations were noted in sufficient quantity to meet criteria for a calcic horizon at 77 cm and extended to 200 cm. All horizons of the soil were moderately alkaline and violently effervescent suggesting both a substantial quantity of calcite and a relatively high surface reactivity for the mineral. This soil did not have slickensides but evidence of shrink-swell activity was

relatively weathered Ca- and K-feldspars and chlorite (Fig. 19) in the SEM micrographs (Aydemir 2001).

Clay fraction: Quantification of the total clay minerals of the soil based upon peak areas is presented in Fig. 20 and this soil shows a variable distribution of minerals with depth. The smectite content (52–41%) decreases from the surface to the subsoil. Likewise, kaolinite (19–16%) slightly decreases from the surface to the Bk2 horizon. Chlorite (9-18%) increases with depth, while palygorskite and illite show little quantity change with depth. The total clay mineral distribution order of this soil might be as smectite > kaolinite > chlorite > palygorskite = illite (Aydemir 2001).

The TEM images of the clay size particles of the deeper horizons (Bk2) show little weathering and fibers have a thickness range of 7.7–9 nm and lengths up to 1.7 μm. Smectite forms on the edges of the palygorskite fibers (Fig. 21b) (Aydemir 2001).

Smectite may have neoformed on the edges of the palygorskite fibers as seen in Fig. 21b pointing out to a regional shift to a milder (slightly wetter) climate of the late Pleistocene-early Holocene. This deformation of palygorskite and reformation of smectite is a significant paleo-climatic proxy in the context of semi- arid to arid calcretisation.

3 Evolution of Calcisols Associated to Calcretes

Calcisols/Cambisols in southern Turkey are mainly overlying calcretes (Kapur et al. 2000). Thus, the authors described the evolutionary sequence of pedogenic calcretes via the stages of column horizon developing on fine textured mudflow or in situ sediments with respect to toposequences as follows:

Stage 1: Deposition of the Pliocene clays,
Stage 2: Cracking of clayey material and formation of the large cuboidal structural units,
Stage 3: Leaching proceeding in the Pleistocene pluvials and calcification occurring along vertical continuations of the structural units,
Stage 4: Development of vertical calcite columns-columnar horizon (in the inter-pluvial of the Pleistocene) and formation of the moderately calcareous soil overlying,
Stage 5: Intense pluvials/inter-pluvials increased decalcification of soil and calcification along with rubification of the column horizon resulted in the formation of a massive calcrete with change—modification of shapes and sizes of columns due to dissolution—recrystallization of calcite.

Fig. 10 Calcisol over massive calcrete under cereal cultivation, Burdur, south Turkey

noted in many pressure faces throughout the lower Bk horizons (77–200 cm) (Aydemir 2001) (Tables 5 and 6).

Mineralogy of the Sırrın Soils

Sand and silt fractions: The sand (2–0.05 mm) and silt (50–2 μm) fractions of the Bk2 horizon have mixed mineralogy. The major minerals are quartz, calcite, and feldspars and the relatively minor minerals are chlorite, kaolinite and mica in the silt fraction. The sand fraction has the same major minerals but without chlorite, kaolinite and mica.

The morphologies of the silt-size minerals from the Bk2 horizon of the Sırrın soil also revealed the presence of

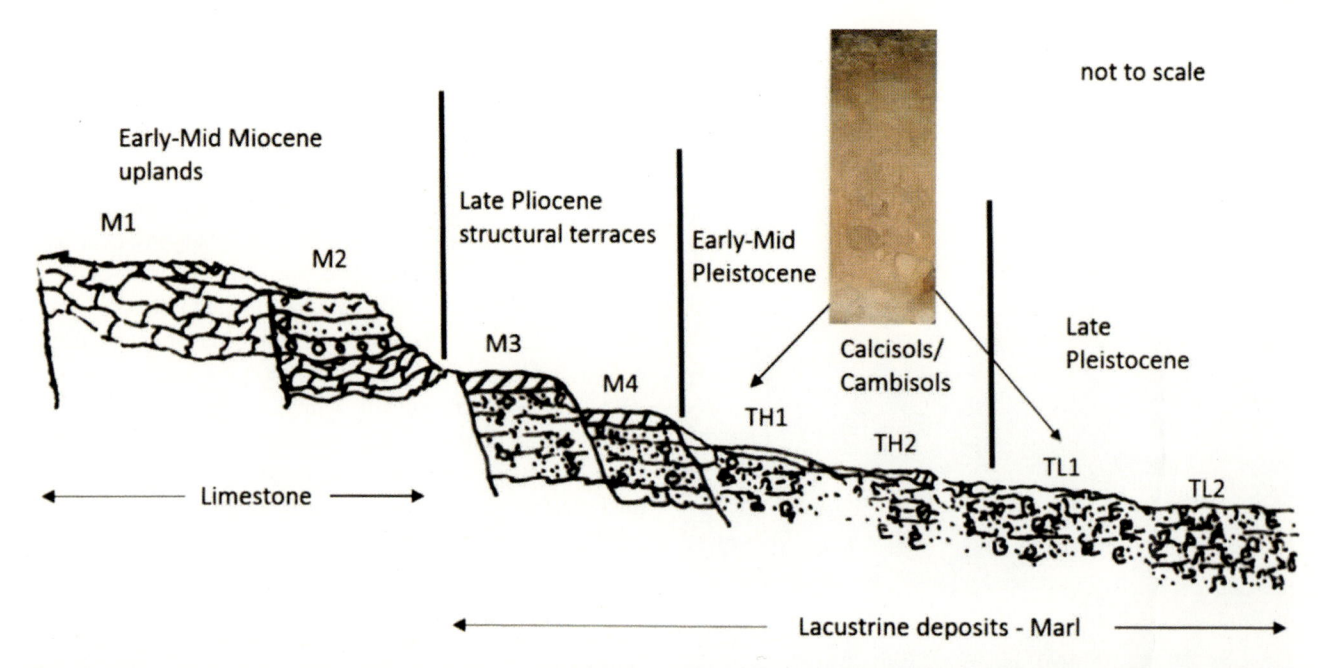

Fig. 11 Generalized toposequence of the Kap soils in central Turkey (modified by Kapur et al. 2000 from De Meester 1970)

Fig. 12 A Calcisol underlying pistachio trees in Nizip, Gaziantep, south Turkey

Table 3 Site-profile descriptions and some physical, chemical and mineralogical properties of the Nizip Calcisol, south Turkey

IUSS Working Group WRB (2015)	Petric calcisol chromic
Soil Survey Staff (2014)	Petrocalcic Calcixerepts
Location	18 km northeast of Nizip town, south Turkey
Coordinates	N36°58′ 48″ E37°44′ 39″
Altitude	540 m
Landforms	Colluvial
Aspect	North–South
Slope	1%
Drainage class	Well-drained
Groundwater	–
Eff. Soil depth	60 cm
Parent material	Plio-quaternary carbonate-rich sediments
Köppen Climate	Mediterranean
Land use	Cultivated
Vegetation	Pistachio nut

Horizon	Depth	Description
Ap	0–22	Dull reddish-brown (5YR 4/4), silty clay loam, reddish-brown (5YR 4/6) dry, weak to moderate medium subangular blocky to granular structure, firm to friable, moderate fine roots, strongly effervescent, surface stoniness of 30–40% (a cobbly surface horizon), abrupt smooth boundary
Bw	22–40	Reddish-brown (5YR 4/3) dry, moderate medium subangular blocky, few fine roots, strongly effervescent, firm, sticky and plastic, very strongly effervescent, clear smooth boundary
Bkm	40–64	Dull reddish-brown (5YR 5/4 dry), clay loam, bright reddish-brown (5YR 5/8), moderate to strong medium prismatic structure, friable to firm, strongly effervescent, clear smooth boundary
Ckm	64–95	Dull orange (5YR 6/4 dry), very strongly effervescent, mammilated pendants of carbonate, diffuse wavy boundary
2Bk	95–120	Reddish-brown (2.5YR 4/6), clay loam, bright brown (2.5YR 5/6) dry, moderate to strong medium prismatic structure, friable to firm, strongly effervescent

Horizon	Depth (cm)	Organic matter (%)	CaCO$_3$	pH 1:1	EC (dSm^{-1})	Sand	Silt (%)	Clay	CEC* (cmol$_c$kg^{-1})
Ap	0–22	1.4	22.0	8.1	0.26	34.4	38.7	26.9	27.0
Bw	22–40	1.2	23.0	8.2	0.26	9.2	22.6	68.2	38.0
Bkm	40–64	0.8	49.0	8.0	0.50	23.9	16.9	32.4	32.1
Ckm	64–95	0.6	52.0	8.0	0.42	38.9	11.2		16.0
2Bk	95–120	1.1	29.0	7.9	0.38	32.5	56.3		8.4

Horizon	P Olsen (mg kg^{-1})	Ca	Mg (cmol$_c$ kg^{-1})	Na	K	BS (%)
Ap	7.3	63.7	2.2	0.1	0.9	100
Bw	4.1	50.6	0.6	0.1	0.4	100
Bkm	2.1	60.8	1.4	0.1	0.6	100
Ckm	1.8	72.9	2.3	0.1	0.9	100
2Bk	3.2	73.7	2.7	0.1	1.0	100

EC electrical conductivity, *BS* base saturation, *CEC* cation exchange capacity
* Extractable Ca may contain Ca from calcium carbonate or gypsum. CEC base saturation set to 100

The sources of carbonate enrichment were probably the aeolian additions of calcite during the interpluvials and the surface waters during the pluvials.

Stage 6: Continued rubification to red hues of 2.5YR and 10R, with a hard, thin crust (Ø: 1–5 mm) forming over the massive calcrete which preserves the geomorphological surface that calcrete formed in/on (Fig. 22).

4 Management of the Calcisol Ecosystems/Anthroscapes

Calcisols developed on slightly undulating, flat and plain lands are mainly used in agricultural production. The moderately deep Calcisols with favorable chemical and physical characteristics support plant growth (Tables 1, 2, 3 and 4).

Table 4 Site-profile descriptions and some physical, chemical and mineralogical properties of the Kap soil, Şanlıurfa, south Turkey

IUSS Working Group WRB (2015)	Petric calcisol chromic
Soil Survey Staff (2014)	Petrocalcic Calcixerepts
Location	9.5 km west of Akçakale town center (south Turkey)
Coordinates	N36°43′ 44″ E38°51′ 46″
Altitude	383 m
Landforms	Colluvial
Aspect	East–West
Slope	2–4%
Drainage class	Well-drained
Groundwater	–
Eff. Soil depth	40 cm
Parent material	Miocene-oligocene limestone
Köppen climate	Mediterranean
Land use	Cultivated
Vegetation	Pistachio nut

Horizon	Depth (cm)	Description
Ap	0–17	Dark brown (7.5YR 5/6 moist, dry), clay, subangular blocky to fine granular structure, firm when dry, friable when wet, stick and plastic, moderate fine roots, strongly effervescent, abrupt smooth boundary
Bw	17–39	Reddish brown (5YR 4/3) dry, medium subangular blocky to fine granular blocky, firm when dry, friable when wet, sticky and plastic, abundant carbonate nodules, few fine roots, strongly effervescent, clear wavy boundary
Bkm	39–57	Dark reddish brown (7.5YR 5/6 moist), clay, medium subangular, firm when dry, friable when wet, sticky and plastic, calcrete development, strongly effervescent, few fine roots, clear wavy boundary

Horizon	Depth (cm)	Organic matter (%)	CaCO$_3$	pH (1:1)	EC (dSm^{-1})	Sand	Silt (%)	Clay	CEC (cmol$_c$ kg^{-1})
Ap	0–17	1.2	38.0	7.3	0.04	24..3	33.8	41.9	26.8
Bw	17–39	0.7	39.0	7.4	0.03	14.4	36.4	49.2	28.9
Bkm	39–57	0.5	44.0	7.8	0.04	20.7	30.2	49.1	25.7

Horizon	P Olsen (mg kg^{-1})	Ca+Mg (cmol$_c$kg^{-1})	Na	K	BS (%)	Clay suite
Ap	4.6	24.4	1.7	0.7	100	S, K, I+P
Bw	3.8	26.5	1.7	0.7	100	S, I+P, K
Bkm	1.9	23.3	1.8	0.6	100	–

EC electrical conductivity, *BS* base saturation, *CEC* cation exchange capacity, *S* smectite, *I* illite, *P* palygorskite, *K* kaolinite

However, intensive cultivation induces wind and water erosion and decreases organic carbon due to the rapid mineralization of soil organic matter. Fruit production is a primary management on Calcisols such as olive, carob, pistachio nut, vineyard, fig, apricot, peach, and apple in the Mediterranean, Aegean, and Thrace regions. Excess and inappropriate irrigation and tillage practices may decrease the quality of Calcisols by building up salts and deteriorating the stability of aggregates. This in turn may decrease the sustainable water supply provided by palygorskite and palygorskite- calcite rich aggregates for the rainfed olives, carobs, vineyard, and figs grown in the Calcisol ecosystems abundant in the west, south and southeast of Turkey.

The calcrete layers and column horizons laden with palygorskite-calcite rich aggregates (micro-structural units) (Fig. 7) are the essence in retention of water that is appropriately and sustainably made available to roots in the rhizospheres of the traditional olive and pistachio trees. These aggregates (the probable sand and silt-sized aggregates of the Sarıçam soil) are the outcomes of the long-enduring orchard cultivation that reshaped the root-zones of traditional Mediterranean crops (Kapur et al. 1991). They provide a balanced water supply to the olive and pistachio roots via their pore size and pore distribution (Akça et al. 2006) (Fig. 7). Moreover, the high surface and internal crystalline areas of the palygorskite crystals creating dehydration,

Fig. 13 The Kap soil (Leptosol/Calcisol) in Şanlıurfa (southeast Turkey)

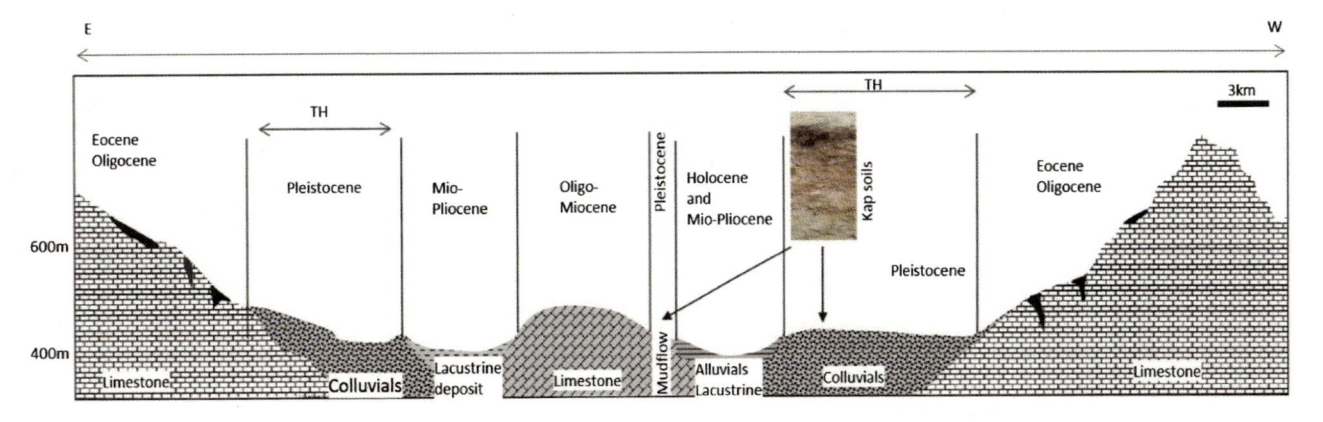

Fig. 14 The Kap soil of the TH surfaces in southeast Turkey formed in between the two graben ranges of Şanlıurfa (modified from Dinç and Kapur 1991)

zeolitic, and coordinated H_2O (Gionis et al. 2006) are most likely the subtle water sources for olive and pistachio trees in the Mediterranean rainfed conditions. Wheat, barley, and triticale cultivations are common in central and eastern Turkish Calcisols. Natural vegetation on Calcisols provides grazing to ruminants but overgrazing is a main constrain for sustainable use of central and eastern Turkish graze lands (Fig. 23).

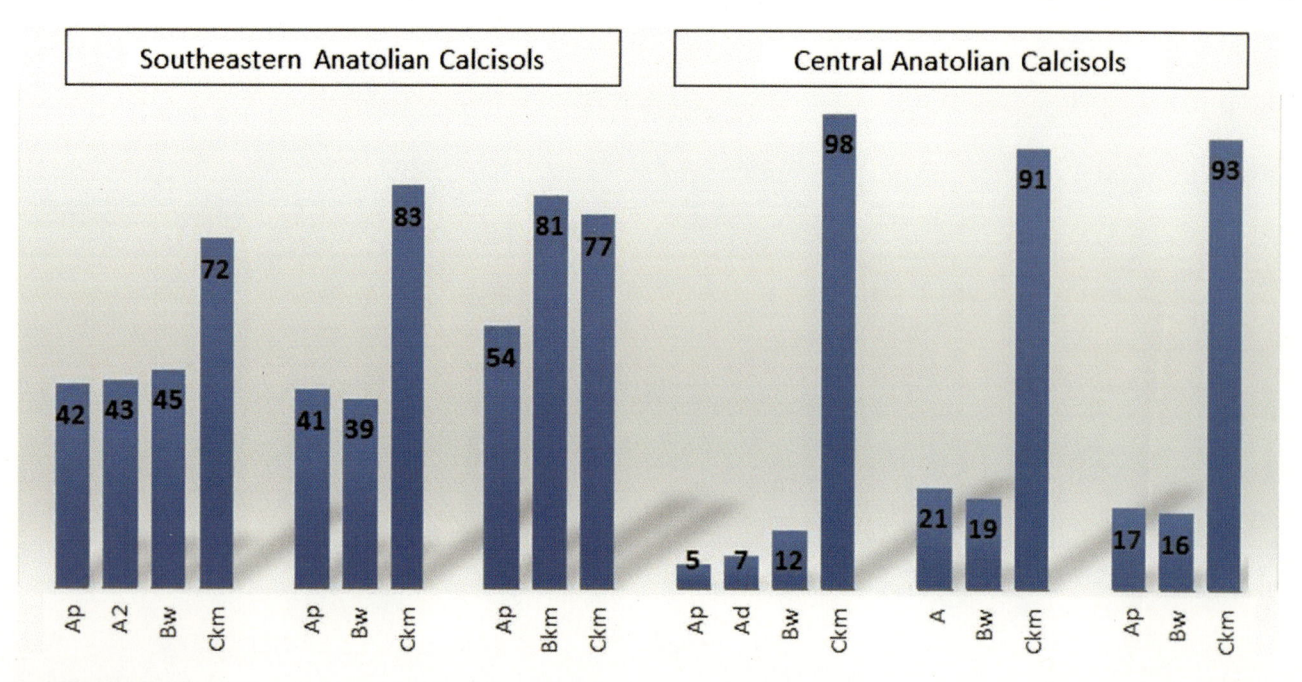

Fig. 15 CaCO$_3$ (%) contents of some southeast and central Turkish Kap soils (Calcisols)

4.1 The Tarsus Calcisol Ecosystem/Anthroscape

Calcisols in Turkey are among the most productive soils particularly in the southern and central parts of the country. The productivity of Calcisols led to the establishment of anthroscapes, i.e., human-shaped landscapes (Eswaran et al. 2013). For example, the Tarsus Anthroscape includes widespread indigenous land use types for the Mediterranean coastal areas of Turkey with high yield capacity. It covers the landscape units of the coastal strips with the highlands of the Taurus Mountains (Eocene-Miocene limestones), colluvials (Calcisols, Cambisols and Leptosols), Tertiary sediments and the TH-TL terrace soils (calcretes and the Luvisols, Cambisols, Calcisols, Leptosols) of the Pleistocene and the valley bottom soils, the Arenosols (Fig. 24). The slopes of the calcretes have been allocated to vineyards for the production of the well-known 'Tarsus White' grapes of a long-standing background for table consumption (Tangolar et al. 2009). The valley bottoms, i.e., the southernmost geomorphic units are devoted to citrus which was introduced by eastern trade in the late 19th century (Fig. 24). The eternal olive tree of the Pleistocene TH-TL terraces has been present in the Levant, Cyprus, and Turkey ever since humanity has recorded history.

4.2 The Kızkalesi (Korykos) Calcisol Ecosystem/Anthroscape

Some other particular Mediterranean Anthroscapes located in Kızkalesi (Korykos)-Erdemli, Mersin (south Turkey-the east part of the Mediterranean coast of Turkey) comprise the well-known polygonal cut rock terraces of the Late Roman—Byzantine period over the Leptosol—Cambisol—Calcisol—Luvisol-Technosol ecosystem. This human intervened agroecosystem comprises the ancient and long-established olive/carob/vineyard Calcisol/Cambisol ecosystems together with the extensive goat/sheep husbandry pathways of grazing (Fig. 25). The man-made terraces acted as water harvesting structures along with numerous cisterns (MEF 2006) which were constructed on the land highly adapted to the natural topography of the site (Fig. 24). The site of Kızkalesi (Korykos) (south Turkey) is one of the best examples for the Anatolian/Byzantine-Ottoman concepts of sustainable land and water management system (dating back to the Bronze Age) (Akça and Kapur 2014).

Similar man-made wall-terraces to the Kızkalesi (Korykos) are widespread in the countries of the Mediterranean basin (e.g., Greece, Italy, Malta, Israel, Spain and north Africa), constructed on karstic surfaces that were conserved for cultivation of the specific Mediterranean crops and also were useful for water harvesting besides increasing water infiltration and storage in the karstic fissure systems (Fig. 26) (Kapur and Akça 2004; MEF 2006).

4.3 The Nizip Calcisol Ecosystem/Anthroscape

The Nizip anthroscape (Fig. 27), which is another Calcisol-dominated landscape, is a dry Mediterranean and semiarid environment, widespread throughout the Middle East (esp. in Syria), abundant in north Africa (esp. in

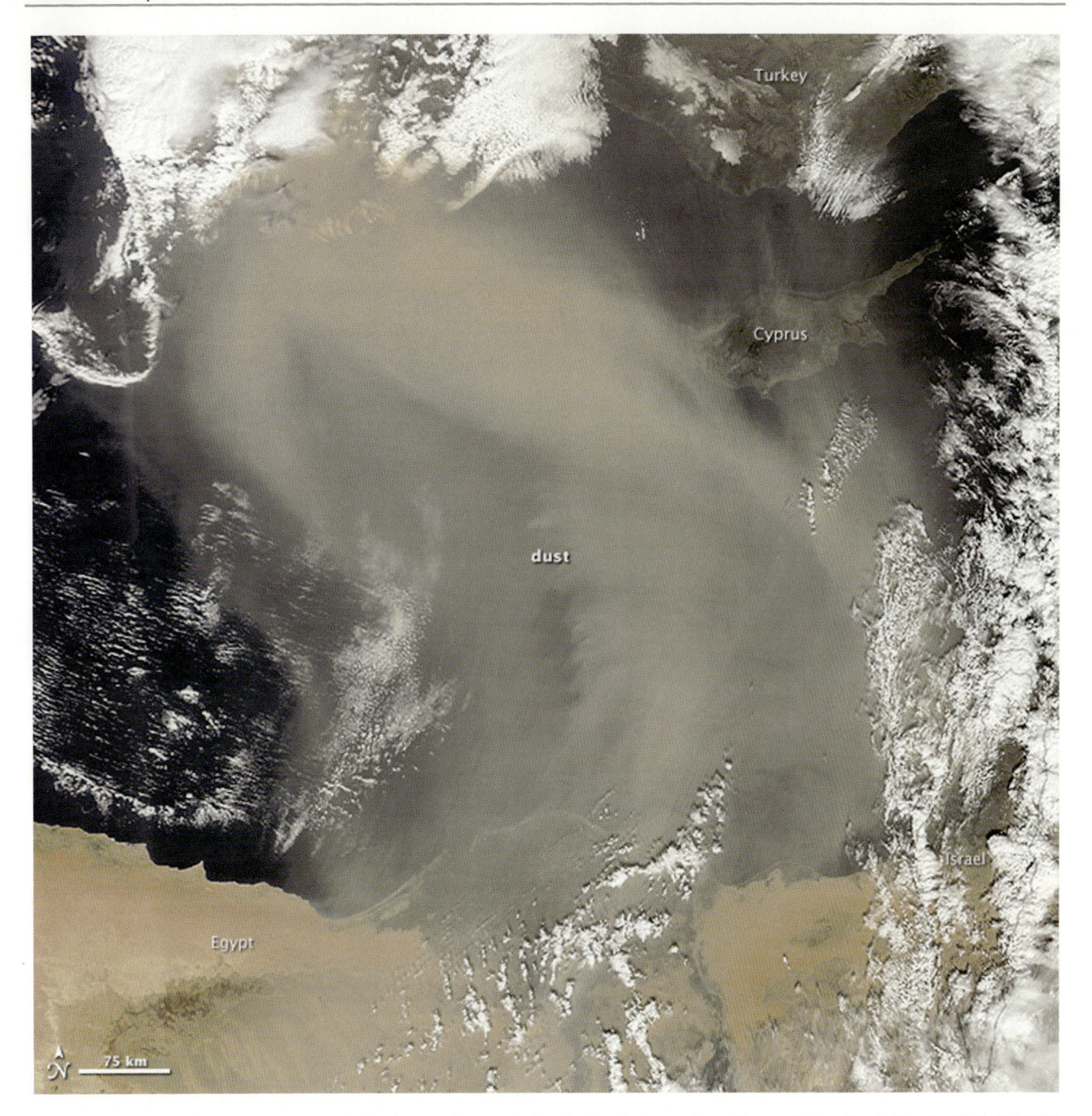

Fig. 16 Dust storm event of January 15, 2014 over Cyprus and Turkey. (http://earthobservatory.nasa.gov/NaturalHazards/view.php?id=83268 (We acknowledge the use of data products or imagery from the Land, Atmosphere Near real-time Capability for EOS (LANCE) system operated by the NASA/GSFC/Earth Science Data and Information System (ESDIS) with funding provided by NASA/HQ)

Tunisia) and frequent in southern Europe (esp. in Spain). The anthroscape comprises dominant Calcisols/Luvisols/Cambisols overlying calcretes and Tertiary crystalline limestone formations, with slightly undulating surfaces. The olives, pistachio and vine are indigenous species which have been present in the area for thousands of years. Vineyards are planted between the wide rows of olives and pistachio as pioneer plants for ready income-generation. The stony and shallow Kap soils located eastwards of the Nizip anthroscape are allocated for rainfed wheat production in southeast Turkey similar to the wheat terroirs of central Turkey.

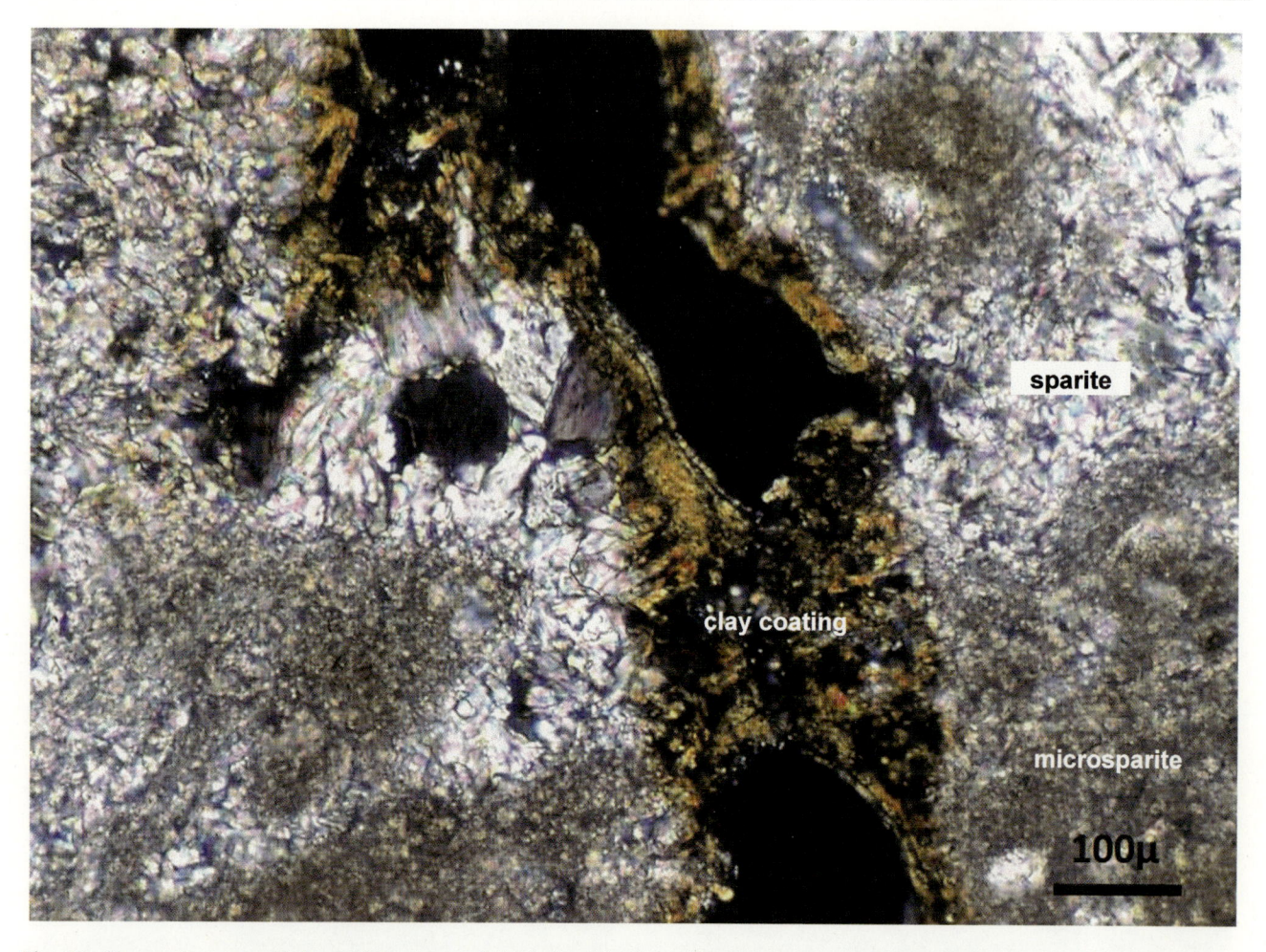

Fig. 17 Clay coatings in multiple calcified calcrete of Calcisols of the Kap soil, south Turkey (*cross polarized light*)

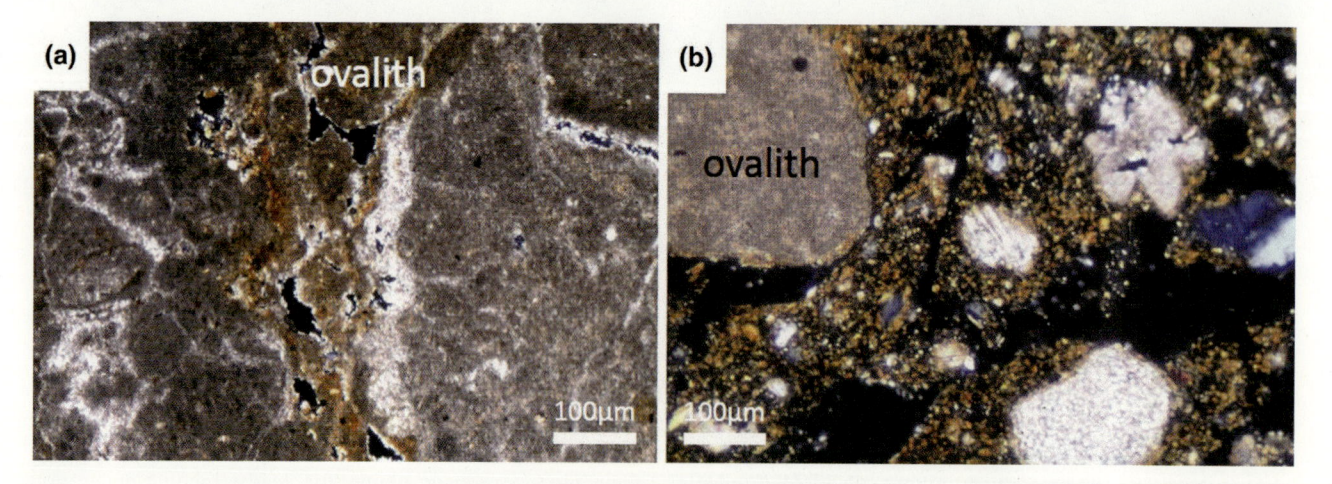

Fig. 18 Ovalith developing in calcrete matrix of the Kap soil (**a**), ovalith fragment in Akçakale soil, Şanlıurfa (south Turkey) (**b**)

4.4 The Adana Calcisol Ecosystem/Anthroscape

The Adana anthroscape, that has been established since the sixteenth century, to cope with the increasing fiber demand of Turkey by cotton production, faces destruction by the recently introduced maize (cash crop). Moreover, a few decades ago cotton production was shifted to the southeast of Turkey due to the vast amounts of water resources

Table 5 Selected morphological features of the Sırrın soils, Şanlıurfa, southeast Turkey (Aydemir 2001)

Horizon	Depth	Color	Texture	Structure	Consistency	Boundary
Ap	0–20	7.5YR 4/6	C	wk, m sbk	Very hard	Clear smooth
A	20–40	7.5YR 4/6	C	md m sbk	Very hard	Clear smooth
Bw	40–77	7.5YR 4/6	C	md m abk	Firm	Gradual wavy
Bk1	77–110	7.5YR 4/6	C	wk m pr/st m abk*	Firm	Gradual wavy
Bk2	110–140	7.5YR 4/6	C	md m pr/st m abk	Very firm	Gradual wavy
Bk3	140–167	7.5YR 4/6	C	st m abk	Very firm	Gradual wavy
Bk4	167–200	7.5YR 5/6	C	md m abk	Very firm	–

*The slash means that the first structure is parting to the second structure. Structure abbreviations: *st* strong, *md* moderate, *wk* weak, *co* coarse, *m* medium, *f* fine, *abk* angular blocky, *sbk* subangular blocky, *pr* prismatic, *pl* platy

Table 6 Site-profile description of the Sırrın soil (Aydemir 2001)

IUSS Working Group WRB (Group 2015)	Cambic Calcisol
Soil Survey Staff (2014)	Typic Calcixerepts
Location	Approximately 12 km southeast of Şanlıurfa
Landform	Bajada
Parent materials	Alluvium
Elevation	Approximately 550 m
Slope	<1%
Topography	Nearly level
Formation	Holocene alluvium
Drainage	Well-drained
Land use	Fallow
Sample date	August 8, 1997
Sampled by	C.T. Hallmark, S. Aydemir and M.A. Çullu
Remark	No water table was noted

supplied from the new gigantic dam network, low-cost labor and governmental subsidies. However, trends in shifting to the earlier cotton production lands today, i.e., to the more appropriate soil-cotton ecosystems in Adana, seems to increase due to the recently granted governmental subsidies. Rainfed olive groves and citrus orchards with drip irrigation are widespread on the calcrete surfaces of the area with Calcisols and highly productive Luvisols (Fig. 28).

All the studied calcrete surfaces (from M4 to the TL) in southern, southeastern, and western Turkey were most probably covered by plant species of the Mediterranean at least since 80,000 BP. These were the Mediterranean bio-indicators such as the Pistacia vera, Olea europea (olives), Pistacia terebinthus (rootstock for pistachio nut trees) and Quercus species predominating western and southern Turkey (Davis and Burrows 1994; Yılmaz 1996; Kapur et al. 1999). However, with the beginning of the human impact around 10,000 BP (the Neolithic), the natural vegetation of the Mediterranean region of Turkey was widely transformed, to generate the contemporary olive, vineyard and orchard cultural plant cover mosaic (Naveh 1984; Yılmaz 1996). The increasing population pressure, the Bronze and Iron Ages clearance of Quercus and Olea communities for Agriculture and fuel (for silver and gold smelting) brought about the use of the indigenous wheat species—*Triticum aestivum*—with goat and sheep grazing following the harvest. Today, rainfed cotton and cereal production together with vegetable fields and overgrazing of the harvest are occupying part of the TH and TL surfaces in south and southeast Turkey. Some of the TH surfaces in southern Turkey (Adana, Mersin) contain almost a bare calcrete crust which may be broken and incorporated to a

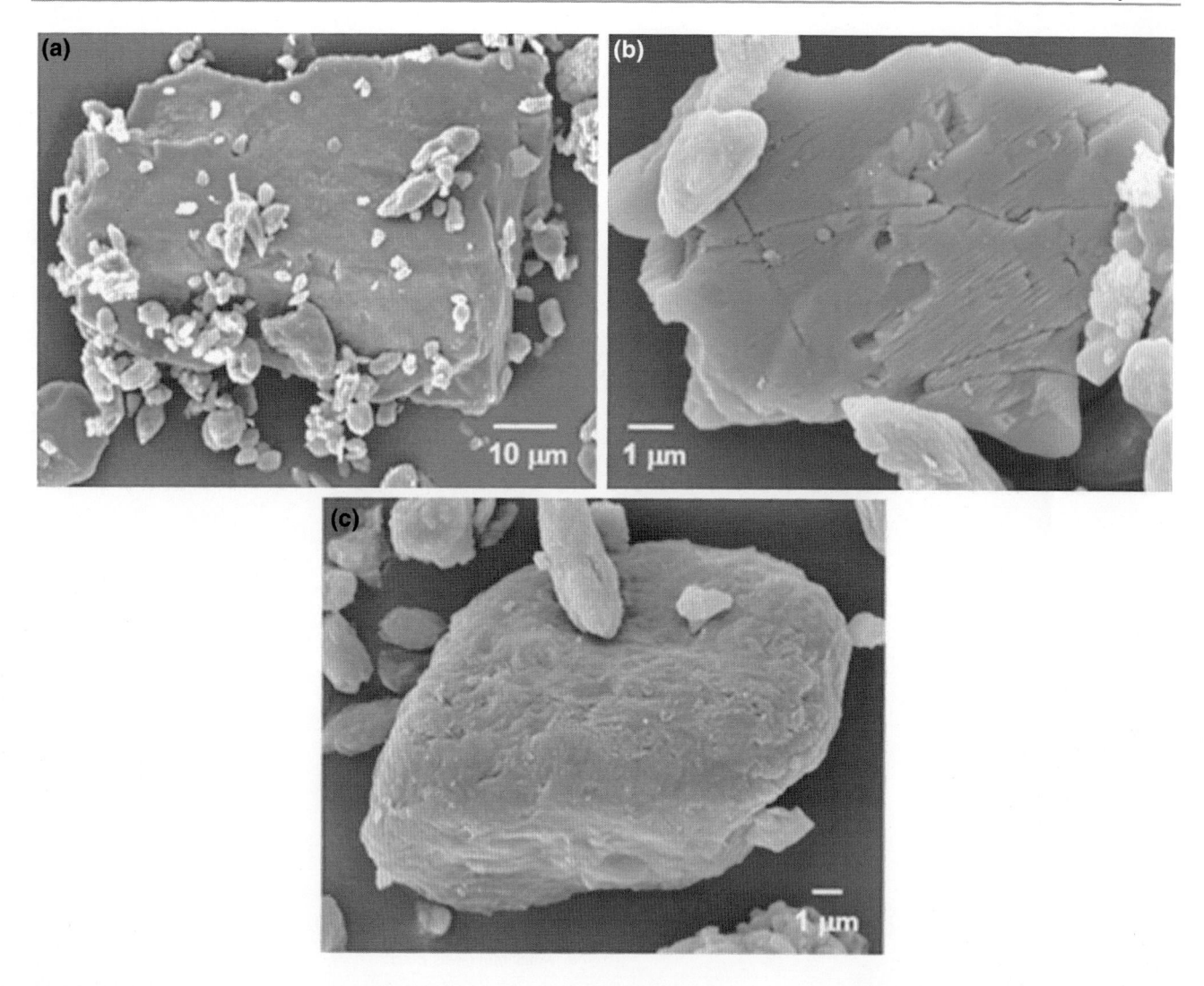

Fig. 19 Scanning electron micrographs of the silt fraction from the Bk2 horizon of the Sırrın soil. **a** Ca-feldspar **b** K-feldspar, and **c** chlorite (Aydemir 2001)

very shallow Leptosol by cultivation integrating to the underlying column horizon (Fig. 22, Stage 6). The TL surfaces in especially Mersin and Tarsus are covered with high yielding citrus trees which have adapted themselves to the moderately deep Luvisols and Calcisols overlying palygorskite-rich calichified conglomerates.

4.5 The Konya Calcisol Ecosystem/Anthroscape

The Konya anthroscape (Fig. 29) mainly overlies lacustrine sediments of the Great Konya Basin and has been an ancient cereal/small ruminant Anthroscape since c. 8000 BP—one of the sites where the 'agricultural revolution' of the Neolithic took place. Some of the lacustrine deposits have naturally dried during the late Holocene minor climatic fluctuations and some went through the so-called reclamation process—a highly popular practice of the 1940s and 1950s undertaken for gaining land for cultivation which was actually a major factor of environmental degradation, i.e., the destruction of the hydrologic cycle. Irrigation of the alluvia and Calcisols of the Great Konya Basin for the cash crops has been causing salinity build up due to geomorphic implications. Thus, irrigated cultivation is sought to be recommended as to be partly abandoned for the Konya area/anthroscape in the context of SLM in the recently revised proposal of the NAP-D (National Action Plan for Desertification of the UNCCD) of Turkey along the lines of

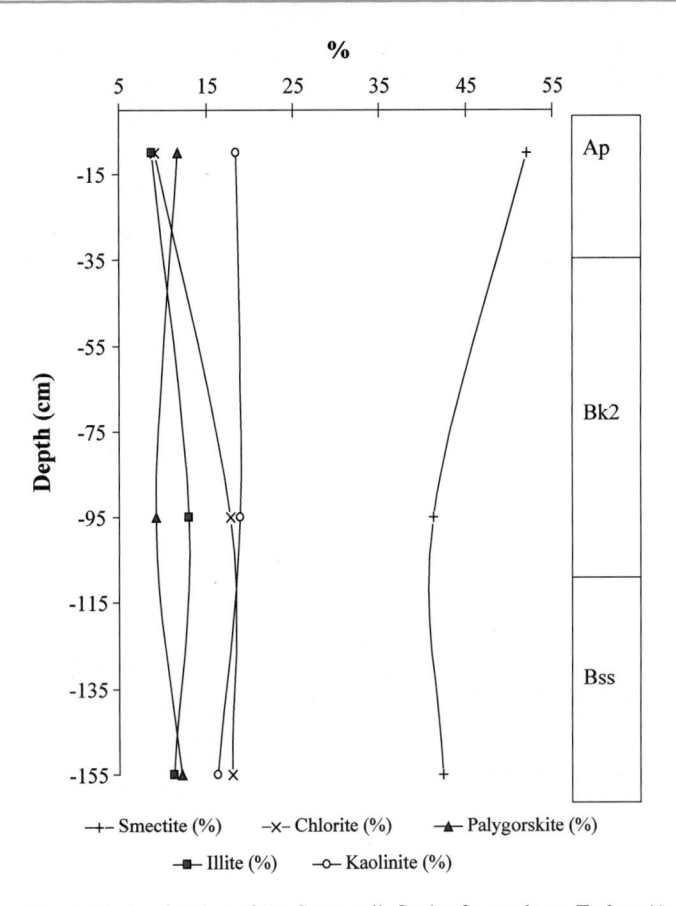

Fig. 20 Clay mineral distribution of the total clay fraction of the Sırrın soil, Şanlıurfa, southeast Turkey (Aydemir 2001)

the 10-year strategy plan. The recommendation concerns the preservation and improvement of the traditional grassland (for small ruminants) and wheat anthroscapes (MEF 2006) of the country. Moreover, it should be extended to cover the need to combat the unplanned overexploitation of the groundwater of central Turkey sustaining the wheat/Calcisol ecosystems via the anticipated SLM in the revised plan. Nevertheless, the wheat and small ruminant Calcisol ecosystems of central Turkey have irrevocably been the economic power of the country and source of the capital for industrial investment until today since the nineteenth century and even earlier.

4.6 The Van Calcisol Ecosystem/Anthroscape

The Cambisols and the Calcisols of the lacustrine and fluvial terraces (natural terraces) of the late Pleistocene and Holocene are the common geomorphic features in many parts of Lake Van (Valeton 1978), i.e., the Van anthroscape (Fig. 30).

These terrace soils, the Cambisols/Calcisols, have been reworked/cultivated earlier by the lost kingdom of Van, the Urartu culture (1200–800 BC) which has been documented to be highly advanced in sustainable land use (Çilingiroğlu 1994). The man-made terraces constructed on the natural Cambisol/Calcisol ecosystem of the Lake Van region have most probably been the productive farming areas and part of the Van anthroscape (i.e., the indigenous landscape/crop models introduced by Eswaran et al. 2005). The terraces were allocated for vineyards and orchards (walnuts and almonds) in the past. Presently, they are used as natural pastures and cultivated for sugar beet and cereals (Akça et al. 2008). Akça et al. (2008) have also stated that the Cambisol/Calcisols of the terraces were the soil ecosystems of high carbon sequestration potential by studying their ancient buried surfaces with high-carbon contents. They also mentioned that there was a possibility in increasing

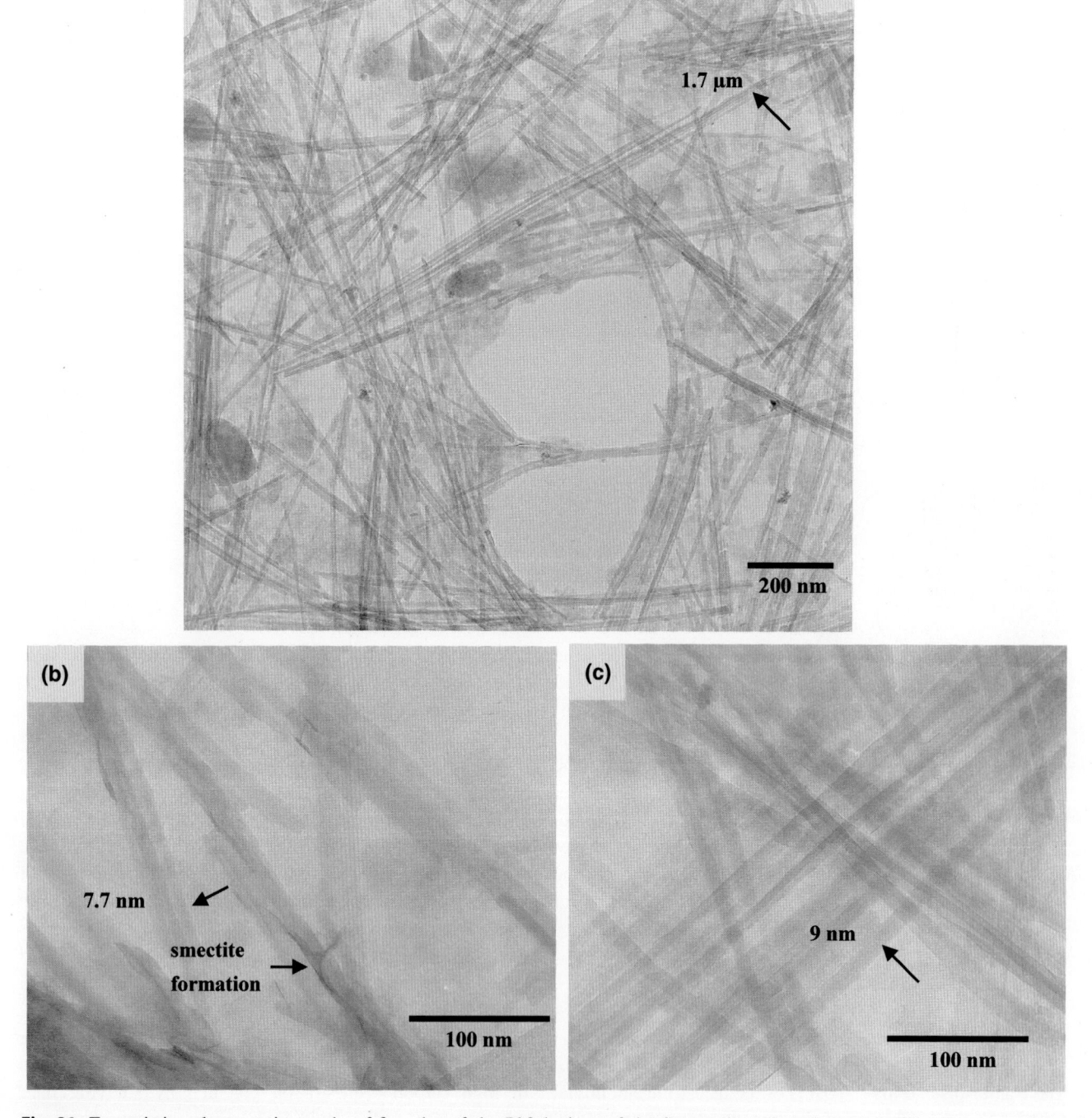

Fig. 21 Transmission electron micrographs of fine clay of the Bk2 horizon of the Sırrın soil. **a**, **b** and **c** palygorskite fibers in different magnifications and their widths (Aydemir 2001)

Calcisols and Leptosols

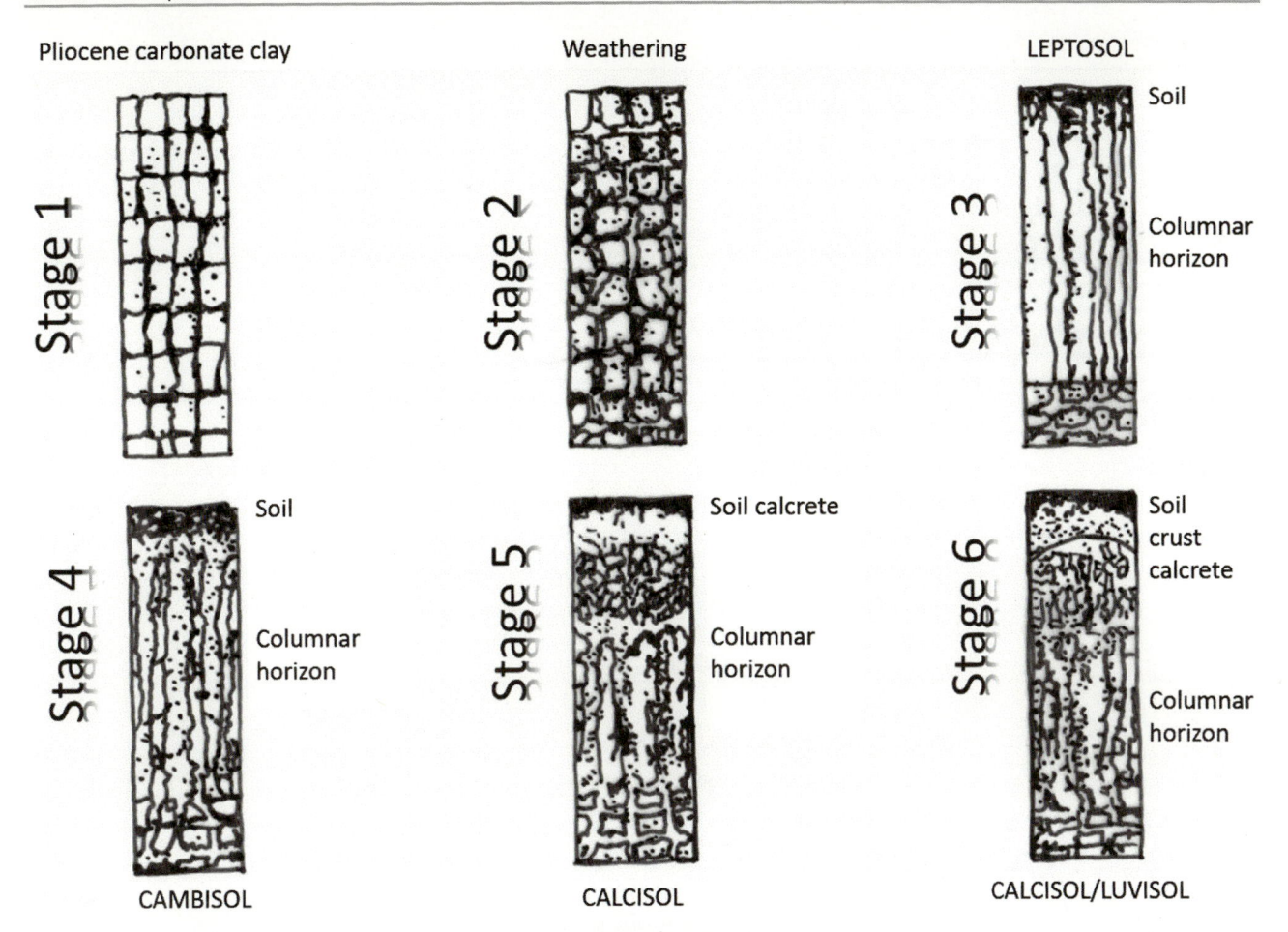

Fig. 22 Evolutionary sequence of pedogenic calcretes in south and west Turkey (Kapur et al. 1990)

Fig. 23 Overgrazing is one of the major causes of degradation of Calcisols in central Turkey

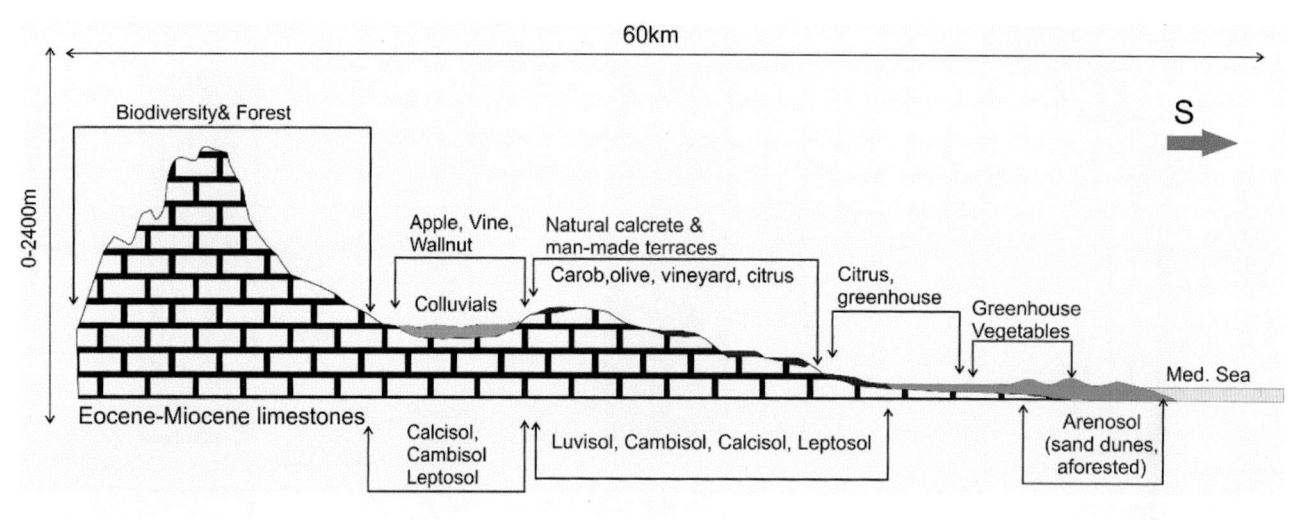

Fig. 24 The Tarsus Anthroscape southern Turkey

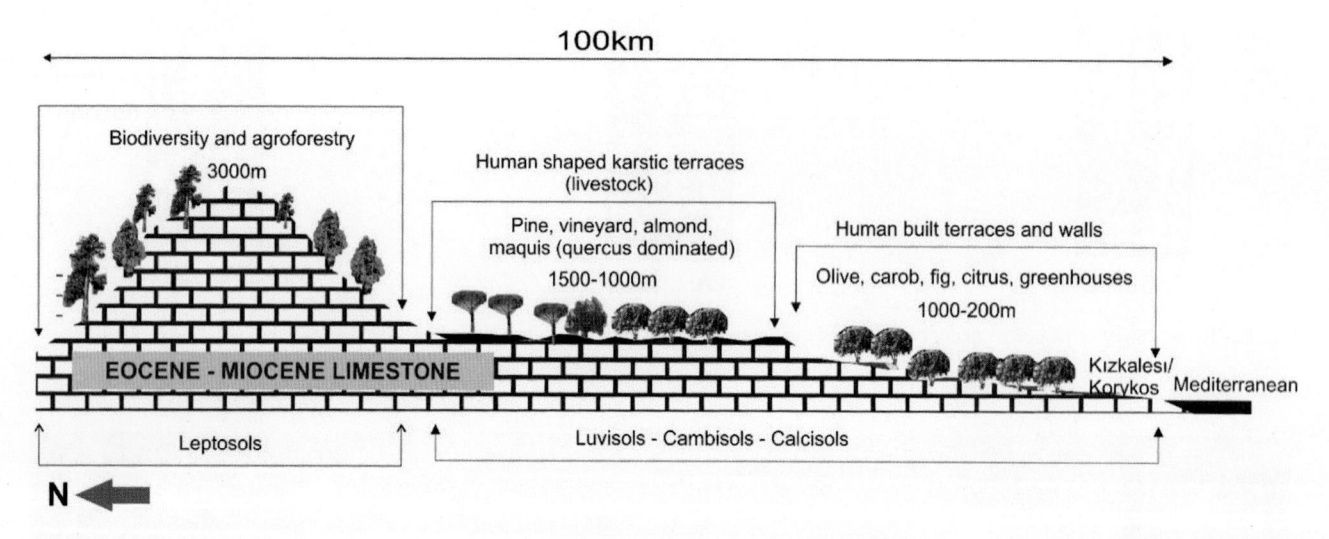

Fig. 25 The Kızkalesi (Korykos)-Erdemli, Mersin Anthroscape, south Turkey

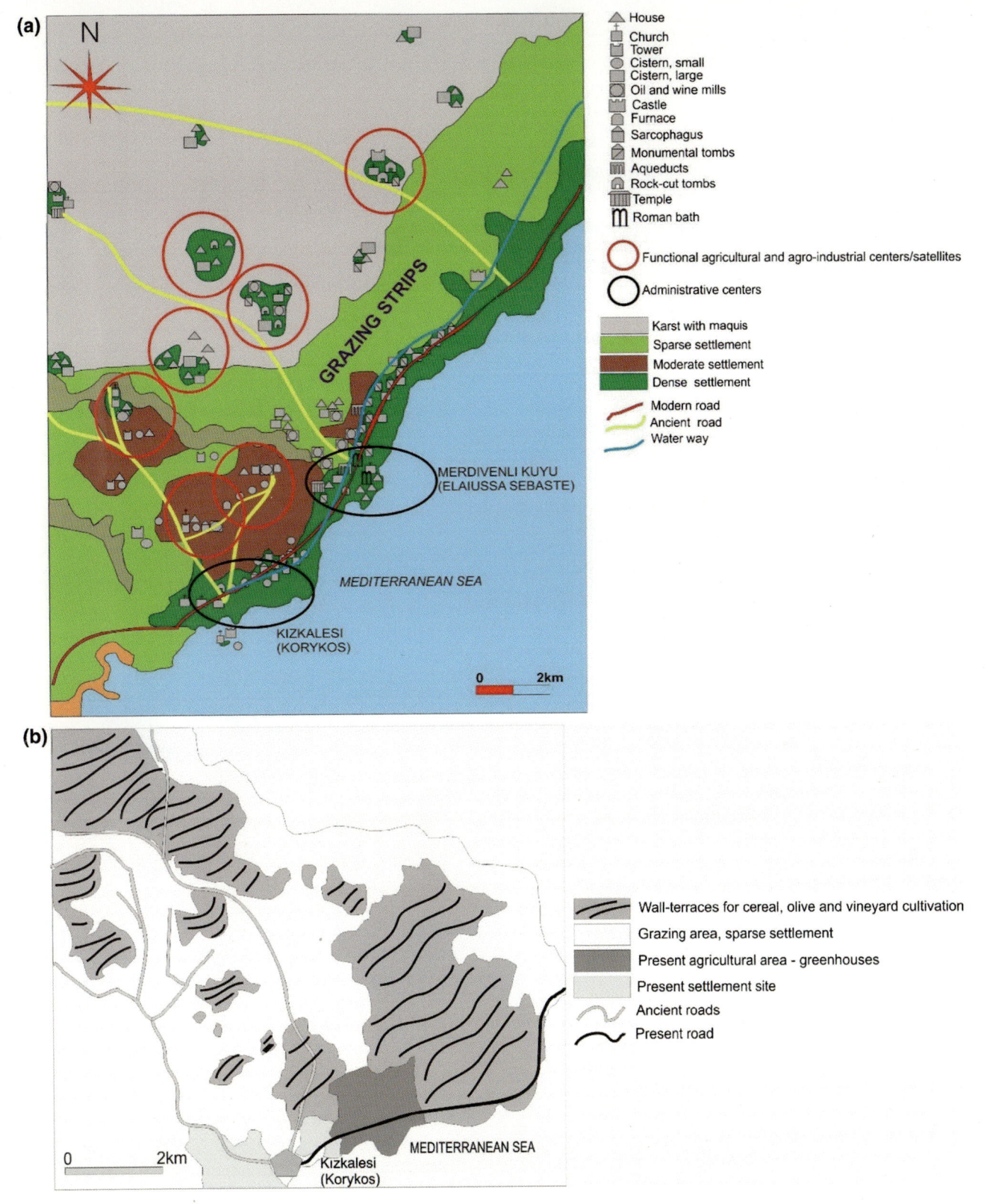

Fig. 26 The ancient Kızkalesi (Korykos)-Erdemli, Mersin settlement-management site (MEF 2006) **a** the man-made terraces **b** running parallel along the coastal borders of the ancient functional agricultural and agro-industrial centers and grazelands of Kızkalesi (south Turkey) (Kapur et al. 1999)

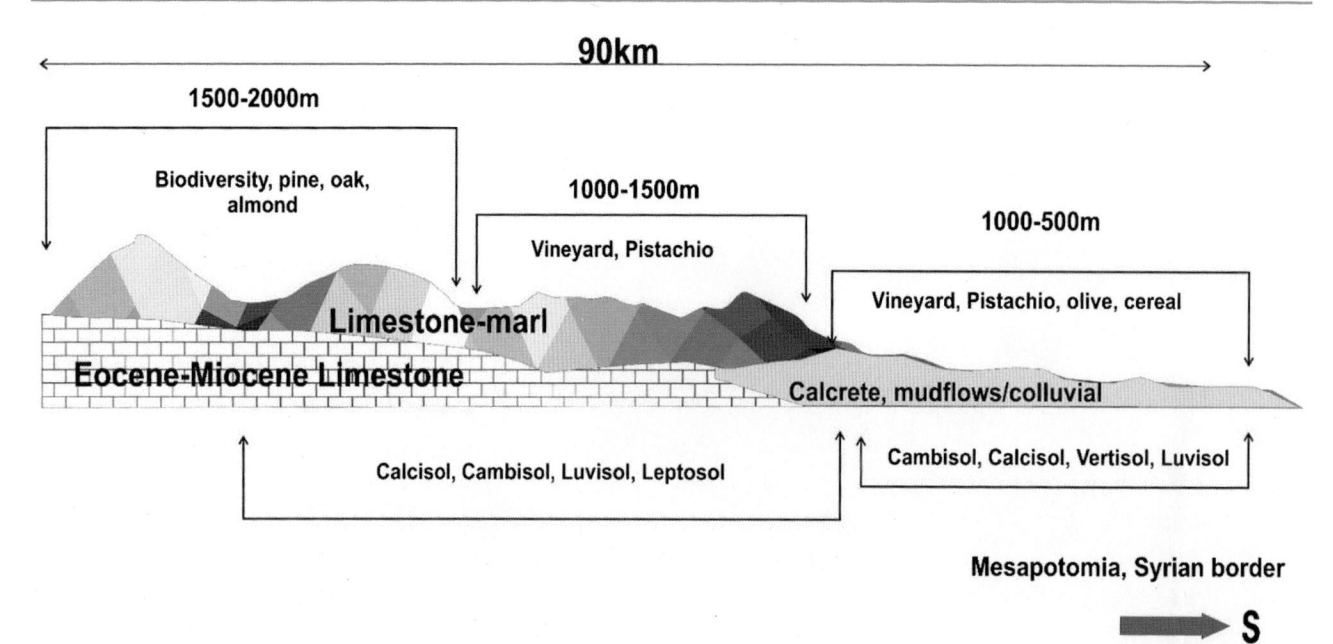

Fig. 27 The Nizip anthroscape (southeast Turkey)

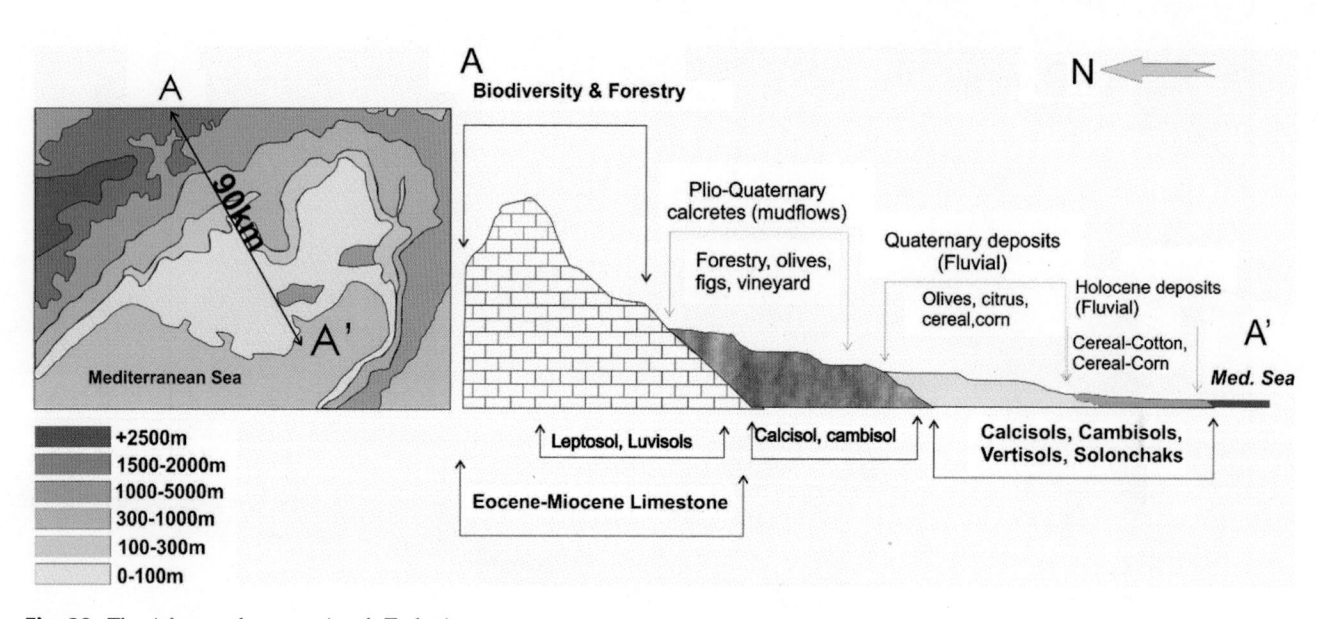

Fig. 28 The Adana anthroscape (south Turkey)

Calcisols and Leptosols

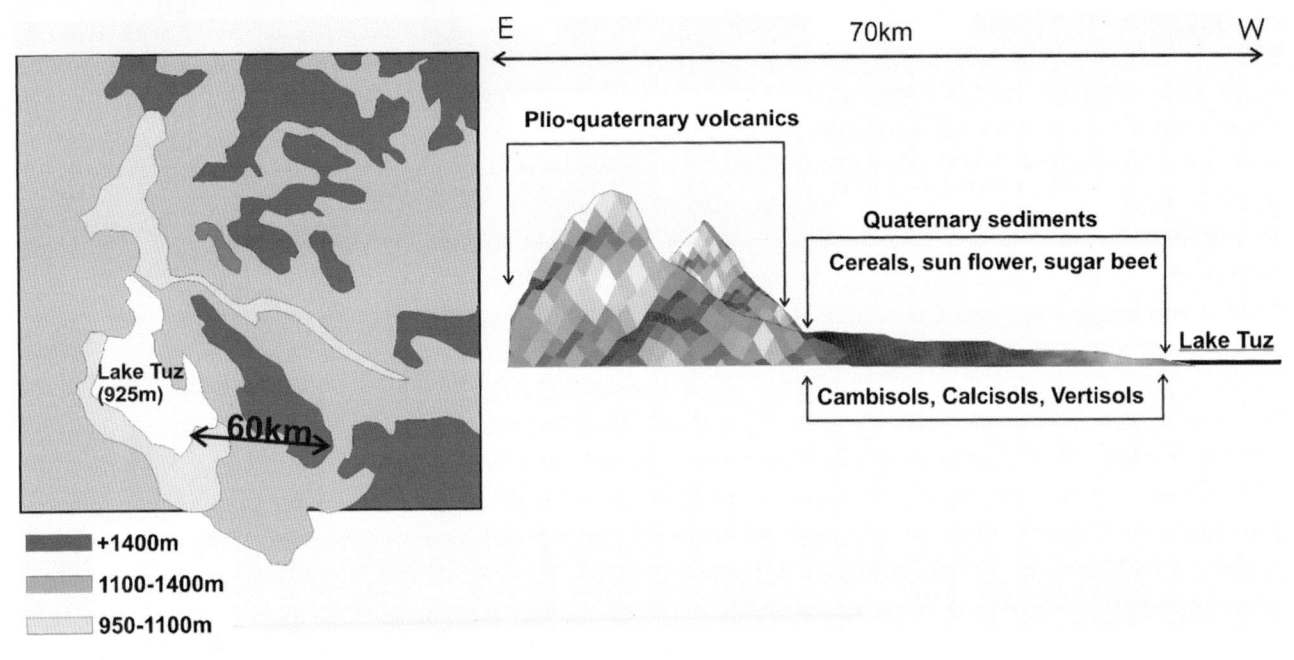

Fig. 29 The Konya anthroscape (central Turkey)

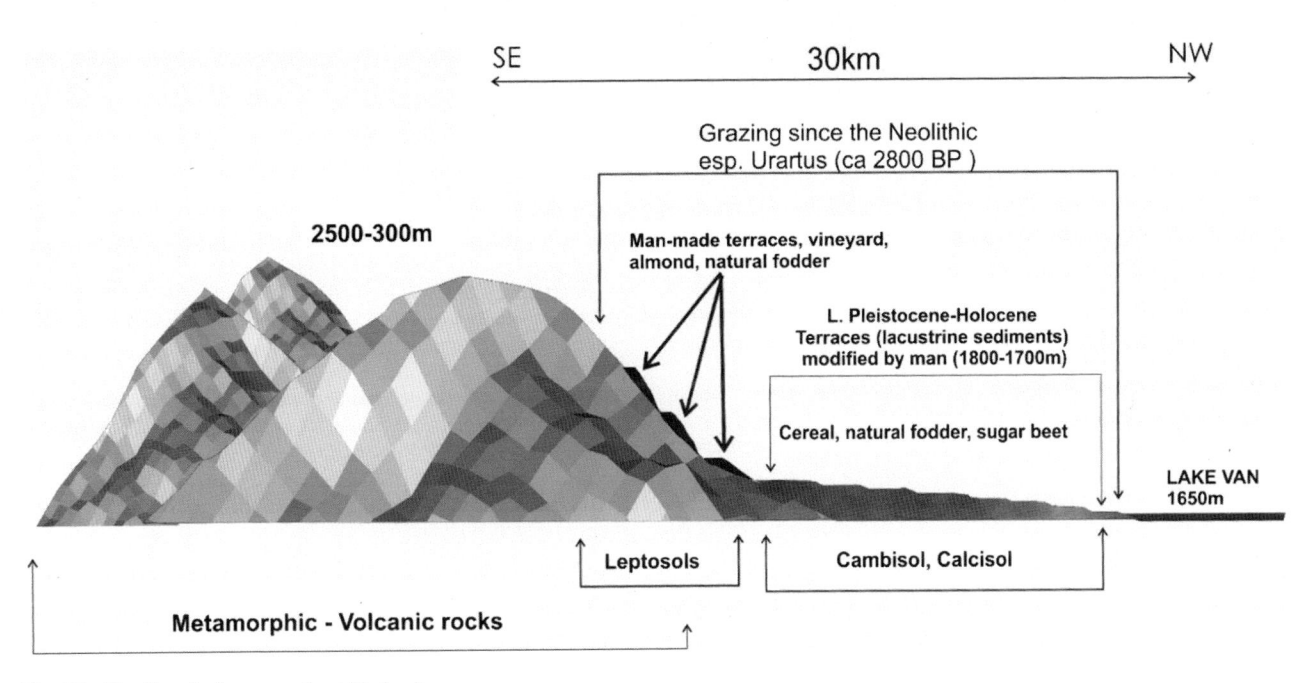

Fig. 30 The Van Anthroscape (east Turkey)

their contemporary carbon sequestration by developing sustainable carbon management programs suitable to the continental climate of the area and similar to the ancient (traditional) also seeking the conservation of the terraces.

Acknowledgements Professor Salih Aydemir extends his sincere gratitude to Profs JB Dixon (Texas A&M University, USA) and CT Hallmark (Texas A&M University, USA) for their invaluable efforts in the SEM laboratory and interpreting the images of the Sırrın soil.

References

Akça E, Çimrin KM, Ryan J, Nagano T, Topaksu M, Kapur S (2008) Differentiating the natural and man-made terraces of Lake Van, Eastern Anatolia, utilizing earth science methods. Lakes & Reservoirs: Research and Management 13(1):83–93

Akça E, Kapur S (2014) The anatolian soil concept of the past and today. In: Churchman GJ, Landa ER (eds) the soil underfoot: infinite possibilities for a finite resource, CRC Press, pp 175–184

Akça E, Saydam C, Steduto P, Zdruli P, Sarıyev A, Çelik İ, Kapur S (2006) Inherited agroscapes of the Mediterranean: carbon pools for sustainable natural resource management. In: Zdruli P, Steduto P, Kapur S, Akça E (eds) Workshop proceedings of ecosystem—based assessment of soil degradation to facilitate land users' and land owners' prompt actions. Adana, Turkey, 2–7 June 2003. Medcoastland Publication 1. IAM Bari, Italy, pp 395–401

Andreucci S, Bateman MD, Zucca C, Kapur S, Akşit İ, Dujanko A, Pascucci V (2012) Evidence of Saharan dust in upper Pleistocene reworked paleosols of northwest Sardinia, Italy: paleoenvironmental implications. Sedimentology 59:917–998

Aydemir S (2001) Properties of palygorskite-influenced vertisols and vertic-like soils in the harran plain of Southeastern Turkey. PhD Dissertation. Texas A&M University, College Station, TX, USA

Budak M (2012) Genesis and classification of saline alkaline soils and mapping with both classical and geostatistical techniques. PhD Dissertation Gaziosmanpasa University, Tokat, Ins. of Science, No: 322692

Cangir C (1982) The morphology and genesis of brown, reddish brown, terra rossa, rendzina and grumusolic soils formed on calcareous materials. Habilitation Thesis, the Faculty of Agriculture, University of Ankara, 135 p (in Turkish)

Çilingiroğlu A (1994) The history of the Urartians. Ege University, Faculty of Arts, 77, İzmir, Turkey, 75 p (in Turkish with English Abstract)

Davis FW, Burrows DA (1994) Spatial simulation of fire regime in Mediterranean-climate landscapes. The role of fire in Mediterranean-type ecosystems. Springer, New York, pp 117–139

De Meester T (1970) Soils of the Great Konya Basin. Centrum voor and bouwpublikaties en Landbouwdocumentatie (Netherlands), Pudoc, Wageningen, 302 p

Dinç U, Kapur S (eds) (1991) Soils of the Harran plain. Turkish Scientific and Research Council, Project No 534, Ankara, 57 p

Dinç U, Şenol S, Kapur S, Cangir C, Atalay İ (2001). Soils of Turkey. Çukurova University, Faculty of Agriculture Publication No: 51, Adana, 233 p (in Turkish)

Dinç U, Kapur S, Şenol S, Karaman C, Çullu MA, Aksoy E, Öztürk N, Öztekin E, Dingil M, Akça E, Çelik İ, Kılıç Ş, Ölmez A, Karadeniz MS, Vural H, Ergün H, Gündoğan R, Cangir C, Altınbaş Ü, Kurucu Y, Bolca M, Güler F, Özdemir İ, Durulmuş G, Durak A, Kılıç K, Dengiz O (1998) Detailed soil survey and mapping of Polatlı agricultural farm soils. TIGEM No: 25, Ankara (in Turkish)

Erol O (1981) Quaternary pluvial and interpluvial conditions in Anatolia and environmental changes in south-central Anatolia since the last glaciation. In: Prey W, Uarpmann HP (eds) Contributions to the environmental history of Southwest Asia. Beihefte zum Tübinger Atlas des Vorderen Orients, vol 8, pp 101–109

Erol O (1984) Neogene and quaternary continental formation and their significance for soil formation. In: Proceedings of the 1st National Clay Symposium. University of Çukurova, Turkey, pp 24–28 (in Turkish)

Eswaran H, Kapur S, Akça E, Reich P, Mahmoodi S, Vearasilp T (2005) Anthroscapes: a landscape unit for assessment of human impact on land systems. In: Yang JE, Sa TM, Kim JJ (eds) Application of the emerging soil research to the conservation of Agricultural Ecosystems. Publication of the Korean Society of Soil Science and Fertilizers, Seoul, Korea, pp 175–192

Eswaran H, Berberoğlu S, Cangir C, Boyraz D, Zucca C, Özevren E, Yazıcı E, Zdruli P, Dingil M, Dönmez C, Akça E, Çelik İ, Watanabe T, Koca KY, Montanarella L, Cherlet M, Kapur S (2013) The anthroscape approach in sustainable land use. In: Kapur S, Eswaran H, Blum WEH (eds) Sustainable land management. Springer, London/Berlin, Berlin, pp 1–50

Gionis V, Kacandes I, Kastridis ID, Chryssikos GD (2006) On the structure of palygorskite by mid- and near-infrared spectroscopy. Am Miner 91:1125–1133

Goudie AS (1975) Petrographic characteristics of calcretes (caliches): modern analogues of ancient corestones. Colloque sur les Croutes Calcaires, Strasbourg, France, pp 3–6

Gürbüz K, Kelling G (1993) Provenance of Miocene Submarine Fans in the northern Adana Basin: a test of discriminant function analysis. Geol J 28:277–295

IUSS Working Group WRB (2015) World reference base for soil resources 2014, update 2015 international soil classification system for naming soils and creating legends for soil maps. World Soil Resources Reports No. 106. FAO, Rome, 203 p

Kaplan MY, Eren M, Kadir S, Kapur S (2013) Mineralogical, geochemical and isotopic characteristics of Quaternary calcretes in the Adana region, southern Turkey: implications on their origin. Catena 101:164–177

Kapur S, Gökçen SL, Yaman S (1984) Caliche formation in the late Tertiary Adana basin, Turkey. International association of sedimentology. In: 5th European regional meeting of sedimentology, transactions, Marseilles, pp 230–231

Kapur S, Yaman S, Gökçen SL, Çavuşgil VS, Yazıcı H (1986) Occurrence, mineralogy and sedimentology of the carbonate formations in the Misis area, southern Seyhan basin Turkey. Geosound. 1(35):36

Kapur S, Çavuşgil VS, FitzPatrick EA (1987) Soil-calcrete (caliche) relationship on a quaternary surface of the Çukurova region, SeyhanPlain (Turkey). In: Fedoroff N, Bresson LM, Courty MA (eds) Soil micromorphology, Ässociation Francaise pour L' Etude du sol, Paris, pp 597–603

Kapur S, Çavuşgil VS, Şenol M, Gürel N, Fitzpatrick EA (1990) Geomorphology and pedogenic evolution of quaternary calcretes in the Northern Adana Basin of Southern Turkey. Zeitschrift für Geomorphologie NF Bd Heft 1:49–59

Kapur S, Sayın M, Gülüt KY, Şahan S, Çavuşgil VS, Yılmaz K, Karaman C (1991) Mineralogical and micromorphological properties of widely distributed soil series in the Harran plain. In: Dinç U, Kapur S (eds) TOAG-534 project special issue. TÜBITAK. Turkish Scientific and Technical Research Council, Publicatioin, pp 11–20

Kapur S, Yaman S, Gökçen SL, Yetiş C (1993) Soil stratigraphy and quaternary caliche in the Misis area of the Adana basin, southern Turkey. Catena 20:431–445

Kapur S, Atalay İ, Ernst F, Akça E, Yetiş C, İşler F, Öcal AD, Uzel İ, Şafak Ü (1999) A review of the late quaternary history of anatolia. Special issue, Olba II, Mersin University Publication, pp 233–278 (in Turkish)

Kapur S, Saydam C, Akça E, Çavuşgil VS, Karaman C, Atalay İ, Özsoy T (2000) Carbonate pools in soil of the Mediterranean: a case study from Anatolia. In: Lal R, Kimble JM, Eswaran H, Stewart BA (eds) Global climate change and pedogenic carbonates. Lewis Publishers, Boca Raton, Florida, pp 187–212.

Kapur S, Akça E (2004) Technologies, environmentally friendly indigenous. Encyclopedia of Soil Science. doi:10.1081/E-ESS120006648. Marcel Dekker

Kettaş F, Berkman A, Yeşilsoy MŞ, Aydın M (1991) Determination of soil-water retention characteristics of widely distributed soil series in the Harran plain. In: Dinç U, Kapur S (eds) Soils of the Harran plain. TÜBITAK TOAG 534. Ankara, pp 21–30

Kubilay NN, Saydam C, Yemenicioğlu S, Kelling G, Kapur S, Karaman C, Akça E (1997) Seasonal chemical and mineralogical variability of atmospheric particles in the coastal region of the Northeast Mediterranean. Catena 28:313–328

Kubilay N, Nickovic S, Moulin C, Dulac F (2000) An illustration of the transport and deposition of mineral dust onto the eastern Mediterranean. Atmos Environ 34:1293–1303

Küçükuysal C, Türkmenoğlu AG, Kapur S (2013) Multiproxy evidence of Mid-Pleistocene dry climates observed in calcretes in central Turkey. Turkish J Earth Sci 22:469–483

Küçükuysal C, Kapur S (2014) Mineralogical, geochemical and micromorphological evaluation of the plio-quaternary paleosols and calcretes from Karahamzali, Ankara (central Turkey). Geol Carpath 65(3):241–253

Marshall JF (1983) Lithology and diagenesis of the carbonate foundations of modern reefs in the southern Great Barrier Reef. BMR. J Aust Geol 8:253–265

MEF (2006) The desertification national action program of Turkey (NAP-D). In: Düzgün M, Kapur S, Cangir C, Akça E, Boyraz D, Gülşen N (eds) Ministry of environment and forestry Publication. No. 250. ISBN 975-7347-51-5. State Meteorology Institute Press, 110 p

Mermut AR, Montanarella L, FitzPatrick EA, Eswaran H, Wilson M, Akça E, Serdem M, Kapur, B, Öztürk A, Tamagnini T, Çullu MA, Kapur S (2004) Excursion book. In: 12th international meeting on soil micromorphology. Adana 20–26 Sept 2004. European Commission, EUR 21275 EN/1, Joint Research Center, Ispra, Italy, 56 p

Naveh Z (1984) The vegetation of the Carmel and Nahal Sefunim and the evolution of the cultural landscape. Sefunim Prehistoric Sites in Mount Carmel, Israel. BAR Int Ser 230:23–63

Özkan İ, Ross GJ (1979) Ferrogenious beidellites in Turkish soils. Soil Sci Am J 43(6):1242–1248

Reeves CC (1970) Origin, classification and geologic history of caliche on the southern High Plains, Texas and Eastern New Mexico. J. Geology 78:352–362

Soil Survey Staff (2014) Keys to Soil Taxonomy, 12th edn. USDA-Natural Resources Conservation Service, Washington, DC, p 372P

Sokolovska, MG, Bech J, Petkova L, Lansac A (2002) Comparative studies on the processes of humus formation in soils from Turkey, Greece, Bulgaria and Spain. In: 7th internationa meeting on soils with Mediterranean type of climate (selected papers). Zdruli P, Steduto P, Kapur S (eds) Options mediterraneennes, seria a: Mediterranean seminars, No: 50, pp 149–156

Şenol M, Kapur S, Gökçen SL (1991) Litho-stratigraphic and sedomentologic study of the tertiary-quaternary of the Adana-Balcalı area, Turkey. Geosound 18:79–91

Şenol M, Kapur S, Şahin S (1993) The quaternary of the Adana basin. In: Şengör AMC (ed) The quaternary of Turkey, ITU, Faculty of Mining, Ist (In Turkish)

Tangolar GS, Soydam S, Bakır M, Karaağaç E, Tangolar S, Ergül A (2009) Genetic analysis of grapevine cultivars from the Eastern Mediterranean region of Turkey, based on SSR markers. J Agr Sci 15(1):1–8

Tanju Ö (1977) Catenary soil relationships in Polatlı State agricultural farm, central Turkey. Publication of the Ankara University Press, pp 26–3, 524–541, Ankara (in Turkish)

Valeton I (1978) A morphological and petrological study of the terraces around lake Van, Turkey. In: Degens ET, Kurtman F (eds) The geology of Lake Van. The Mineral Research and Exploration Institute of Turkey (MTA) Publication No: 169. Ankara, p 158

Yılmaz KT (1996) Natural plants of the Mediterranean. Çukurova University Faculty of Agriculture Publication No. 141, 179 p

Yetiş C, Kelling G, Gökçen SL, Baroz F (1995) A revised stratigraphic framework for Later Cenozoic sequences in the northeastern Mediterranean region. Geol Rundsch 84:794–812

Vertisols

Hasan Özcan, Salih Aydemir, Mehmet Ali Çullu, Hikmet Günal, Muhsin Eren, Selahattin Kadir, Hüseyin Ekinci, Timuçin Everest, Ali Sungur, and Ewart Adsil FitzPatrick

1 Introduction

Vertisols are defined as "heavy clay soils with a high proportion of swelling clays. These soils form deep wide cracks from the surface downward when they dry out, which happens in most years" (IUSS Working Group WRB 2015). The repeated shrink–swell action due to the high content of smectitic clays causes pedoturbation. In this context, soil material consistently mixes within the soil profile and forms Vertisols that have *either* slickensides *or* wedge-shaped aggregates/peds within 100 cm of the soil surface (Ahmad 1983; Ahmad and Mermut 1996). Wedge-shaped peds have long axes tilted 10°–60° from the horizontal. The shrink–swell behavior of the high expansive clays causes gilgai microrelief to form (Figs. 1 and 2) along with periodic cracks and slickensides (Fig. 2) (SSSA 2008; IUSS Working Group WRB 2015; Soil Survey Staff 2014). The degree and frequency of wet and dry conditions in Vertisols are the most important parameters controlling the cracking intensity (Mermut et al. 1990), formation of slickensides, and eventually gilgai microtopography.

2 Parent Materials

Vertisols are usually very dark in color, with widely variable organic matter contents (1–6%). They form in calcium (Ca) and magnesium (Mg) rich materials such as limestone, basalt, or in areas of topographic depressions enriched with these elements. The materials that form Vertisols can be either allochtonous or autochtonous in origin. Vertisols are mostly found in lower landscape positions such as dry lake bottoms, river basins, lower river terraces, and other lowlands that are periodically wet in their natural states (IUSS Working Group WRB 2015). The landscapes of the Vertisols called the "black cotton soils" in the Thrace region of Turkey vary from lowlands to peneplains. Vertisols in Turkey have formed on different parent materials varying from fluvial (river terraces) and colluvial sediments to lacustrine and marine deposits containing marl, shale, claystone, flysch, limestone, and basalts. The ages of the parent materials vary from Cenozoic-Eocene to Quaternary-Holocene. The ages of basalts in the Ergani (Diyarbakır, southeast Turkey, Fig. 1), Hilvan, Harran, and Muş Plains are Eocene and Neogene, Eocene in Samsun-Bafra and Denizli-Acıpayam, Holocene in the Ceyhan and Seyhan plains, Neogene in the Çanakkale-Biga Plains and in the Thrace region (Tanju 1981; Atalay 1983; Dinç et al. 2001) (Fig. 3).

3 Environment

Although Vertisols cover only a small area of the land surface of the world, and only a subdominant portion even of any geographical zone, they bear a significant role for specific crops like cotton in semiarid dryland agriculture.

Fig. 1 Gilgai relief from (**a**) Ergani-Diyarbakır (southeast Turkey), cracks from (**b**) Kazova-Tokat, (north Turkey), (**c**) Biga-Çanakkale (northwest Turkey), (**d**) Malazgirt-Muş (east Turkey)

Vertisols are among the most productive soils in such environments. The major factor contributing to the productivity of Vertisols in semiarid environments is high water-holding capacity. Vertisols are mostly formed in warm, subhumid, or semiarid climates, where the natural vegetation is predominantly grass, savanna, open forests, or desert shrub. Vertisols are also extensively found in the tropics, subtropics, and warm temperate zones and known as *Dark Clays, Black Earths, Black Cotton Soils, Dark Cracking Soils, Grumusols, Regurs, Vlei Soils* (South Africa), and *Margalites* (Indonesia) in other classification systems (Dudal 1965).

Vertisols cover a total of 311 million ha or 2.4% of the global land area and an estimated 150 million ha is potential

Fig. 2 Vertisol profile with deep cracks (**a**, **b**) and slickensides (**a**, **b**, **c**) in Ergani, Diyarbakır, southeast Turkey and (**d**) in Alaçam, Bafra, Samsun province, north Turkey

crop land (Driessen and Dudal 1989). Vertisols of Turkey, mainly Chromic Calcaric Vertisols, cover a land area of about 3,957,069.74 ha which is 5.15% of the whole agricultural land (Fig. 3). Out of this, approximately 393.550 ha (23.84%) of Vertisols are present in the Thrace region. Vertisols in Turkey are distributed throughout the country and found mainly in southeast Turkey (Harran, Adıyaman, Gaziantep, Ceylanpınar, Mardin, Ergani, Cizre and Suruç Plains), the Çukurova region (east Mediterranean Region—Adana, Mersin, Kahramanmaraş and Antakya Plains), the West Mediterranean Region (Antalya, Burdur, Mersin Plains), the Aegean region (Bornova and Menemen Plains in İzmir), the central Black Sea region (Samsun-Bafra and Çarşamba Plains, Kazova Plain, Tokat), the Thrace region (Ergene Basin), Bursa and Çanakkale Plains, Central Anatolia (Konya Basin), and east Anatolia (Muş Plain) regions (Dinç et al. 2001).

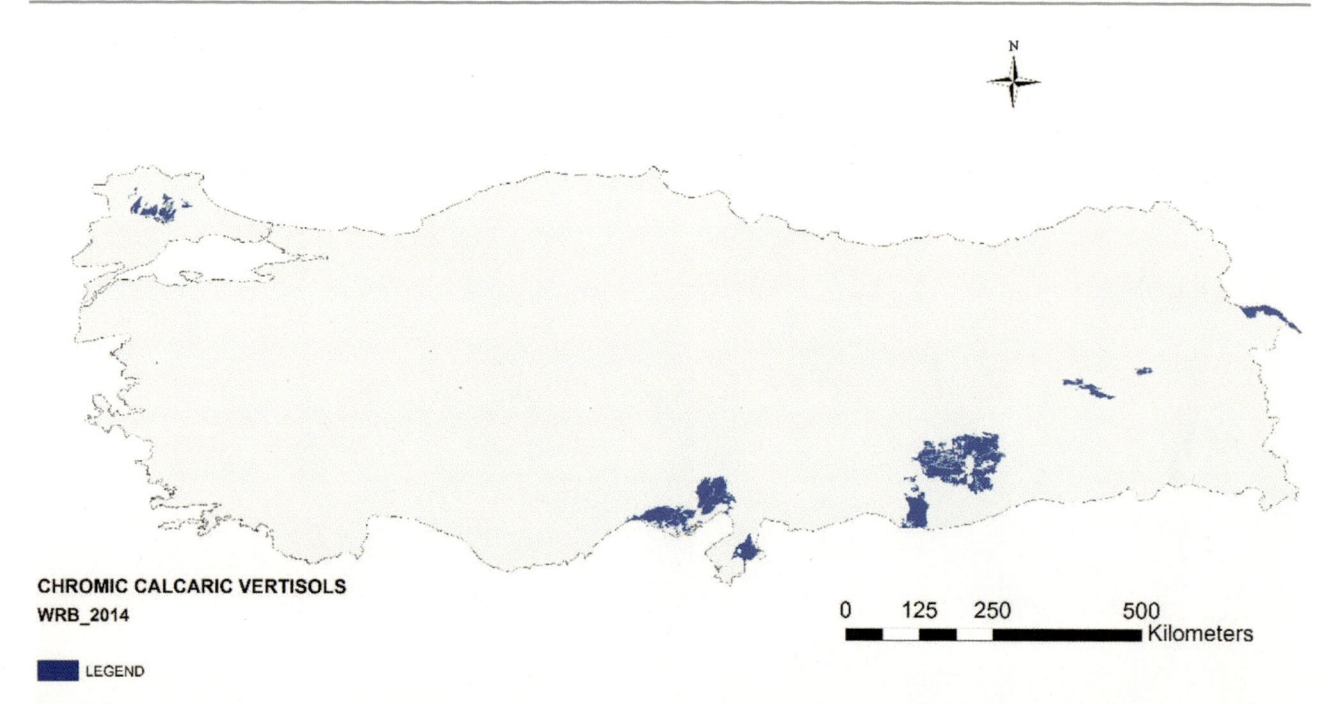

Fig. 3 The distribution of the Vertisols in Turkey

4 Profile Development

Vertisols in Turkey generally consist of A, C and A, B and C horizons, where the subsurface horizons are designated as the Ass in the former and Bss in the latter horizon sequences due to the slickenside formation (Akalan 1968; Dinç et al. 2001) and wedge-shaped structural aggregates. Despite the negative effect of the shrink–swell process occurring during the profile development of Vertisols, the fine sand-size heavy minerals (zircon, rutile, opaque and especially coated and/or weathered minerals with patches of clay development) were determined at the lower boundaries of the plough pans in southeast Turkey revealing the upper boundary of churning/pedoturbation (İnce and Kapur 1985). This may also likely point out to the development of a partly nutrient-rich weathered mineral layer with clay mineral formation on the fine sand-size grains. Some Turkish soil scientists have described and studied Vertisols with B horizons (developing under the surface horizons and/or plough pans) (Mermut et al. 1990, 2006; Ekinci 1990; Kapur et al. 1997; Dinç et al. 2001; Ekinci et al. 2004; Özsoy and Aksoy 2007; Çakmaklı 2008; Yakupoğlu et al. 2010; Dengiz et al. 2012) even at areas with excess irrigated cultivation causing extremes in their shrink–swell potential. These are mostly the Vertisols developed on older parent materials or the Vertisols formed on basalts containing dominant smectite and some palygorskite with kaolinite (Fig. 4).

Fig. 4 Basaltic Vertisol in Karacadağ, Siverek, Şanlıurfa, southeast Turkey

Table 1 Site profile description and some physical and chemical properties of a Vertisol (Harran 1 soil) from the Harran Plain, Şanlıurfa, southeast Turkey (Dinç et al. 1991; Mermut et al. 2006)

IUSS Working Group WRB (2015)	Chromic Vertisols
Soil Survey Staff (2014)	Chromic Calcixerert
Soil series	Harran 1
Location	32 km south of Şanlıurfa city on the Akçakale road (southeast Turkey)
Coordinates	37° 04′ 59″N, 38° 51′ 27″E
Elevation	390 m
Climate	Semiarid, transitional Mediterranean
Vegetation	Annual cropping (cereal, cotton, maize)
Parent material	Colluvial
Geomorphic unit	Mudflow
Topography	Subnormal relief, gently sloping 1–2% north to south
Erosion	Weak surface water erosion
Water table	Deeper than 140 cm
Drainage	Moderate
Infiltration	Moderate
Stoniness	No stones
Soil moisture regime	Xeric
Soil temperature regime	Mesic

Horizon	Depth (cm)	Description
Ap	0–14	Brown (5YR 5/4) dry and moist, clay, medium strong angular blocky, hard when dry, firm when moist, very sticky when wet, very plastic, strongly effervescent with disseminated lime, undulating horizon boundary
Ad	14–25	Brown (5YR 5/4) moist, clay, medium strong angular blocky and granular, very hard when dry, firm when wet, very sticky when wet, very plastic, strongly effervescent with disseminated lime, few vertical roots, undulating horizon boundary
Bss1	25–55	Brown (5YR 4/4) moist, clay, strong prismatic and medium strong angular blocky, very hard when dry, firm when wet, very sticky when wet, very plastic, strongly effervescent, few vertical roots, many slicken sides, undulating horizon boundary
Bss2	55–87	Brown (5YR 4/4) moist, clay, strong prismatic and medium strong angular blocky, very hard when dry, firm when wet, very sticky when wet, very plastic, strongly effervescent with disseminated lime, few vertical roots, many slicken sides, undulating horizon boundary
Bss3k	87–135	Brown (5YR 4/4) moist, clay, strong prismatic and medium strong angular blocky, very hard when dry, firm when wet, very sticky when wet, very plastic, strongly effervescent with disseminated lime, few vertical roots, medium secondary carbonate accumulation, many slicken sides, undulating horizon boundary

Horizon	Depth (cm)	EC (dS m^{-1})	pH 1:1	CEC cmol$_c$kg^{-1}	Na	K	Ca + Mg
Ap	0–14	0.77	7.3	35	2.1	1.6	32.3
Ad	14–25	0.71	7.4	35.9	2	1.3	32.6
Bss1	25–55	0.88	7.4	42.2	1.7	1.0	39.5
Bss2	55–87	0.96	7.4	43.1	2.1	0.9	40.1
Bss3k	87–135	0.83	7.5	54.1	2.2	1.0	50.9

Horizon	CaCO$_3$ (%)	Organic Mat. (%)	Sand (%)	Silt (%)	Clay (%)	Texture class
Ap	30.2	0.9	14.3	32.4	53.3	C
Ad	29.3	0.9	13.4	33.4	53.2	C
Bss1	29.5	0.7	13.2	26.3	60.5	C
Bss2	28.7	0.4	10.3	24.4	65.4	C
Bss3k	32.2	0.4	11.3	4.8	83.9	C

4.1 The Harran Basin Vertisols, the Harran Soils of Southeast Turkey

Dinç et al. (1991) and Mermut et al. (2006) described the most widely distributed matured Vertisols (mapped as the Harran 1 soil series) with a well-developed Bss-horizon (over 50% smectite in clay fraction) in southeast Turkey formed on/within the mudflow deposits of the Şanlıurfa graben (Table 1). Following irrigation via the southeastern Anatolian Project in 1995, these soils have been used for cotton and maize production (Fig. 5).

The other widely distributed Vertisols of Şanlıurfa (southeast Turkey), mapped as the Harran 2 and 3 soil series, contain Ass/Bwss and Bwk/Bwkss horizons, respectively. These soils also formed on the carbonate-rich mudflow transported colluvial materials of the Şanlıurfa graben (Figs. 6 and 7; Tables 2 and 3).

The high clay and smectite contents (Akalan 1968; Munsuz 1974; Dinç et al. 1991), and vigorous faunal activity of the Harran 2 soil are responsible for the development of the particular Vertisol microstructure, i.e., the aggregates/microstructural units of high structural stability (İnce and Kapur 1985; Mermut et al. 2004). This reflects the indispensable need for the sustainable use of water in irrigation and subsequently, the conservation of the microstructure and porosity as studied in thin sections (Figs. 8, 9, 10 and 11). The coarse to fine silt (50–20 µm) size aggregates and their intra-aggregate porosity are responsible for the water storage in the profile (Mermut et al. 2004) (Table 3).

4.1.1 Development of Microstructure in the Vertisols of Southeast Turkey (the Harran 2 Soil)

Studying the microstructure of Harran soils has been of prime significance due to the widespread irrigation activities taking place that are expected to cause change in the soil physical quality. The thin section description of the Harran 2 soil is presented below,

Ap Horizon: Granular structure composed of unaccommodated to partly accommodated irregular aggregates with frequent interaggregate pore space. Initial seasonal formation of incomplete structure with frequently developing fine sinuous to linear boundaries (Fig. 8). Occasional to frequent subspherical fecal pellets most probably developed at earlier periods coalescing with the granular matrix developed of contemporary arthropod activity (Fig. 8). Compaction features of fine to very fine sinuous, parallel crack patterns destroyed by faunal activity (Fig. 8).

A2/Ad Horizon: Frequent to occasional partly accommodated irregular aggregate development together with frequent patches of initial development of peds with rare incomplete oval to spheroidal peds (Fig. 15). Highly developed/complete versions of such features were determined by Kapur et al. (1997) in adjacent Vertisols of Beğdeş —Harran (Turkey) and Akko (Israel). Abundant faunal activity in channels with uniformly distributed fecal material (Fig. 9).

Fine-textured subsurface horizon with abundant faunal activity and destruction of stress-induced microstructure and seasonal peds earlier developed by tillage. Oval to irregular

Fig. 5 Vertisol terrain in the Harran plain, southeast Turkey and irrigation network

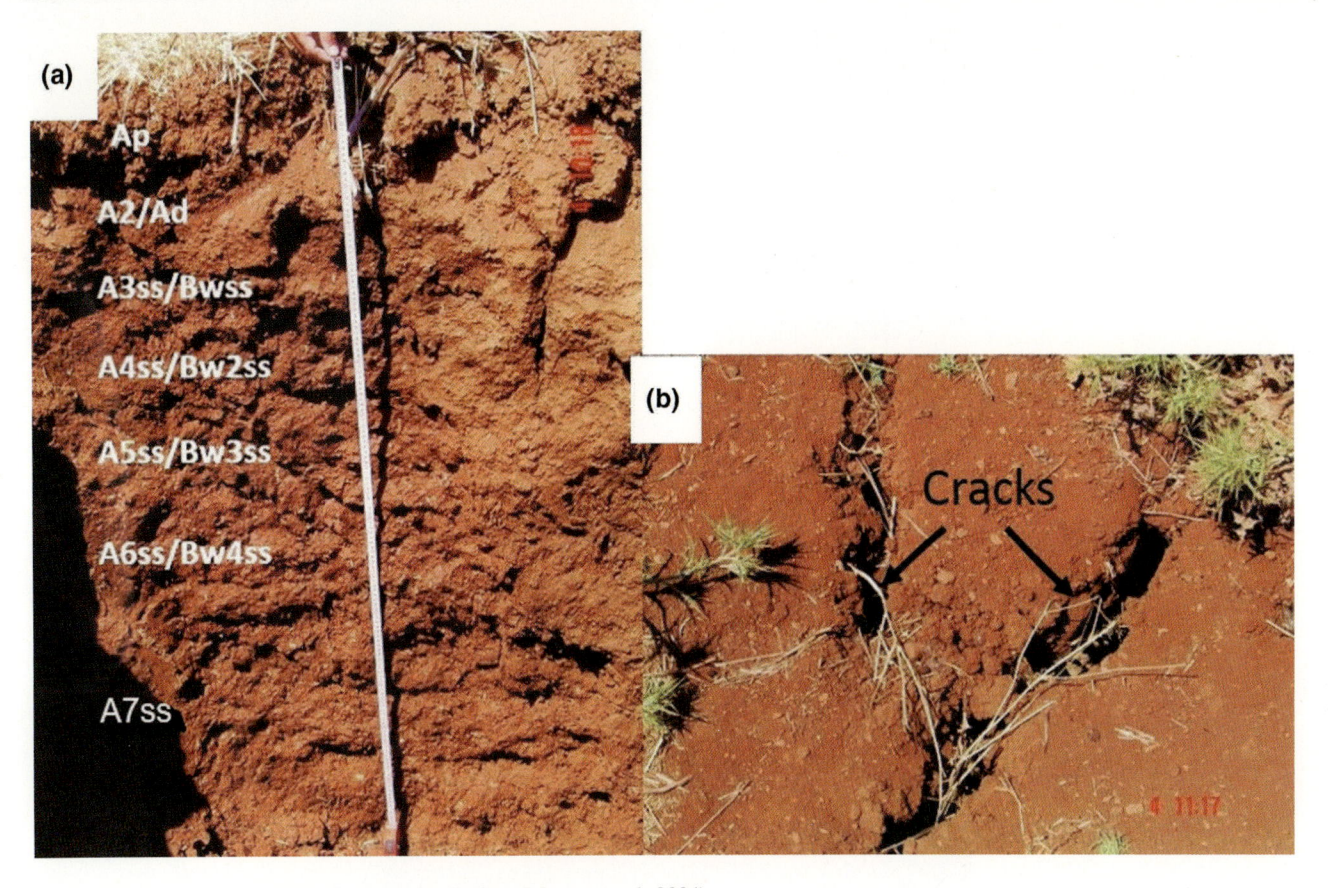

Fig. 6 The Harran 2 soil, Şanlıurfa, southeast Turkey (Mermut et al. 2004)

fine to very fine (50–100 μm) calcite nodules and frequent to occasional root remnants in matrix (Fig. 9).

A3ss Horizon: Initial development of occasional bifurcating very fine to medium incomplete/unaccommodated subcuboidal (wedge formation)–subangular peds (50–100 μm) 14 with thick (0.5–1 mm) to fine (10–20 μm) linear to sinuous outlines in massive matrix (Fig. 10). Rarely to occasionally continuous pore space between peds with discrete pores in rare rounded aggregates (Fig. 10). Fine-textured middle horizon with high amounts of expanding lattice clays common to Vertisols. Occasional rounded irregular micritic calcite nodules (Fig. 10) within matrix of faunal excrements.

Shrink and swell cycles creating high pressures during high water absorption causing the release of one portion of the soil slip over the other, ultimately forming slickenslides upon drying, thus yielding characteristic pore pattern.

A4ss Horizon: Continuous pore spaces (cracks, 0.25–1.00 mm) mainly linear or sinuous forming wedge-shaped soil areas with perpendicular intersections (45–60°) (Fig. 11). Unaccommodated rounded/spheroidal peds in incomplete very fine (<5 mm) wedges (aggregate) (Fig. 11). Fine-textured middle horizon with high amounts of expending lattice clays common to Vertisols. Shrinkage and swelling phenomena are the same as above in inducing development of slickenslides upon drying.

4.2 The Soils of the Harran Basin, the Basaltic Vertisols, the Karacadağ Soils of Southeast Turkey

Vertisols developed on basalts and basaltic materials are widespread in the southeast of Turkey (Fig. 4). The most prominent geomorphic terraces/plateaus formed after the eruptions of Karacadağ located between Şanlıurfa and Diyarbakır are overlain by the basaltic Cambisols and Vertisols (Figs. 12, 13, 14, 15 and 16; Table 4).

Fig. 7 The Harran 3 soil in the Harran plain, Şanlıurfa (southeast Turkey) (Çakmaklı 2008)

4.2.1 Neoformed Clay Minerals in Basaltic Vertisols (Southeast Turkey)

There are particular features that have developed via specific processes in the smectite-rich (Hocaoğlu 1973; Ergene 1977) basaltic Vertisols of southeast Turkey. These namely are the smectitic nodules (Kapur et al. 1991) developing in the amygdules of the basaltic parent materials of the basaltic Vertisols that were frequently neoformed and/or transformed to palygorskite under stagnant basic conditions controlled via arid and semiarid climates. Montmorillonite may neoform as in-fills with/in calcrete via precipitation in the magnesium and silica-rich basaltic fracture/crack/pore environment (Fig. 13) of southeast Turkey. Whevellite-Ca $(C_2O_4)\cdot(H_2O)$ also forms in pores and surfaces of basalts of the Harran plain via oxalic acid excreted from lichen roots (southeast Turkey) enriched with calcium from Saharan dust (Arocena et al. 2007).

Possible transformation and/or neoformation of smectite to palygorskite may indicate a shift to a drier climate, whereas the neoformed kaolinite may stand for the prevalence of a wetter climatic fluctuation taking place in south Turkey during the late Quaternary. Both neoformed clay minerals (smectite and kaolinite forming in amygdules of basalts) are most likely incorporated and weathered in the pedons of the basaltic Vertisols at consecutive dry 18 wet seasons increasing the clay contents and changing the

Table 2 Site profile description of Vertisol (Harran 2 soil) in Şanlıurfa, southeast Turkey (Mermut et al. 2004)

IUSS Working Group WRB (2015)	Chromic Vertisols
Soil Survey Staff (2014)	Haploxererts
Soil series	Harran 2
Location	12 km south of Şanlıurfa city on the Akçakale road (southeast Turkey)
Coordinates	36° 51′ 50.40″N, 38° 58′ 50.95″E
Elevation	360 m
Climate	Weakly arid, transitional from Mediterranean
Vegetation	Annual cropping (cereal, cotton, maize)
Parent material	Colluvial
Geomorphic unit	Mud flow
Topography	Subnormal relief, gently sloping 1–2% N–S
Erosion	Wind erosion, weak
Water table	Deeper than 2 m
Drainage	Moderate
Infiltration	Moderate
Stoniness	No stones
Soil Moisture Regime	Xeric
Soil Temperature Regime:	Mesic

Horizon	Depth (cm)	Description
Ap	0–10	Dull reddish brown (5YR 4/4), clay, reddish brown (5YR 4/6) dry, moderate to strong fine granular structure, firm, moderate biological activity, many fine roots, strongly effervescent, gradual wavy boundary
A2/Ad	10–30	Dark reddish brown (5YR 3/4), clay, reddish brown (5YR 4/6) dry, strong very coarse blocky structure, very firm, moderate biological activity, very few fine roots, moderate irregular fine to medium rarely diffuse nodules mainly of carbonate origin, strongly effervescent, gradual wavy boundary
A3ss/Bwss	30–53	Dark reddish brown (5YR 3/6), clay, reddish brown (5YR 4/6) dry, strong coarse prismatic to angular blocky structure, very firm, moderate irregular fine to medium and rarely diffuse carbonate nodules, moderate biological activity, very few fine roots, strongly effervescent, gradual wavy boundary
A4ss/Bw2ss	53–74	Dark reddish brown (5YR 3/6), clay, reddish brown (5YR 4/6) dry, strong coarse to very coarse angular blocky structure, very firm, strongly effervescent, moderate irregular fine to medium rarely diffuse carbonate nodules, gradual wavy boundary
A5ss/Bw3ss	74–100	Dull reddish brown (5YR 4/4), clay, reddish brown (5YR 4/6) dry, strong medium prismatic to angular blocky structure, very firm, moderate irregular fine to medium nodules with rare diffuse carbonate accumulations, strongly effervescent, gradual wavy boundary
A6ss/Bw4ss	100–150	Dark reddish brown (5YR 4/8), clay, reddish brown (5YR 4/6) dry, strong medium prismatic to angular blocky structure, very firm, rare irregular fine to medium modules, strongly effervescent, clear smooth boundary
A7ss	150–180	Reddish brown (5YR 4/8), clay, reddish brown (5YR 4/6) dry, strong medium angular blocky structure, very firm, strongly effervescent

Horizon	Depth (cm)	pH 1:1	CaCO$_3$ (%)	EC (dS/m)	Phosphorous (mg kg^{-1})	Org. Mat. (%)
Ap	0–10	7.8	22.0	0.64	25.05	1.0
A2/Ad	10–30	7.9	20.0	0.36	21.39	0.8
A3ss/Bwss	30–53	7.7	21.0	0.43	20.10	0.7
A4ss/Bw2ss	53–74	7.8	22.0	0.52	18.35	0.5
A5ss/Bw3ss	74–100	7.7	24.0	0.58	18.29	0.4
A6ss/Bw4ss	100–150	7.8	26.0	0.63	17.24	0.4
A7ss	150–180	7.8	28.0	0.71	15.81	0.4

(continued)

Table 2 (continued)

Horizon	Exchangeable cations					Particle size classes			Clay suite
	Ca	Mg	Na	K	CEC*	Clay	Silt (%)	Sand	
	cmol$_c$kg^{-1}								
Ap	75.8	5.9	0.1	1.8	45.7	54.2	37.9	7.9	S, K, V, M
A2/Ad	74.6	5.9	0.3	1.5	46.1	57.6	36.7	5.7	S, K, V, M
A3ss/Bwss	74.1	6.9	0.3	1.3	45.9	58.6	36.4	5.0	S, K, V, M
A4ss/Bw2ss	73.8	7.7	0.2	1.2	44.7	58.2	36.2	5.6	S, K, V, M
A5ss/Bw3ss	73.2	8.1	0.2	1.2	43.4	56.5	37.5	6.0	S, K, V, M
A6ss/Bw4ss	69.3	8.3	0.2	1.1	41.9	57.4	37.1	5.5	S, K, V, M
A7ss	69.2	8.8	0.3	1.1	40.2	55.2	39.6	5.2	S, K, V, M

S Smectite, *K* Kaolinite, *V* Vermiculite, *M* Mica
* Extractable Ca may contain Ca from calcium carbonate or gypsum. CEC base saturation set to 100

quality of the soils (Fig. 14). Crystallized calcite in basaltic vesicles or in the pores of the basaltic soils as well as the development of calcrete (frequently incorporated to the soil matrix via the vigorous shrink–swell action due to the high contents of smectite) in the matrices of the basaltic soils indicate the recalcification by aeolian inputs of the Saharan dust (Figs. 14, 15, and 16).

4.2.2 The Permanent Microstructural Units in Basaltic Vertisols (Southeast Turkey)

Carbonate and smectite induce the formation of the permanent microstructural units in the B horizons of the Southeast Anatolian Vertisols (Kapur et al. 1997; Stoops 2003) (Fig. 17). Some of the permanent microstructural units that are present in the ABss-horizons of the Vertisols, Vertic Inceptisols, and Vertic Cambisols formed on different parent materials in Turkey (Aslanpınarı and Beğdeş soil series) and Israel (Akko soil series) (Mermut et al. 2006) are spheroidal. Presence of the spheroidal microstructural units and increase of palygorskite and vermiculite in the ABss horizon of the Beğdeş soil, which is in contrast with increasing smectite in the Aslanpınarı and Akko soils, link with an increase in the structural stability index and an increase in the hydraulic conductivity (Kapur et al. 1997) (Fig. 18).

The most prominent soil forming process in the basaltic Vertisols is the development of the individual oval to spheroidal permanent microstructural units (MSUs)/aggregates surrounded by thick to moderate stress coatings (about 2–10 μm). The stress features around the permanent MSUs/aggregates develop via repeated shrink–swell action (Kapur et al. 1997) and may be regarded as a measure of maturity in the development of the basaltic Vertisols in Turkey (Fig. 6).

4.3 Soils of the Harran Basin, the Cepkenli and Kısas Soils of Southeast Turkey

The Cepkenli and Kısas Vertisols of the Harran basin are not as widespread as the Harran Vertisols in the southeast of Turkey, but are highly significant in documenting the neoformation of palygorskite and the probable transformation of palygorskite to smectite. This seems to contrast the neoformation of palygorskite and/or probable transformation of palygorskite from smectite in the Calcisols of south Turkey overlying calcretes (see Calcisol chapter in this book). But, nevertheless, the pedogenic features determined in the clay fraction and thin sections of the Kısas Vertisol (Sects. 4.3.3.2 and 4.3.3.3) point to an integration to a Calcisol and may be regarded as the paleo-climatic proxies of the Pleistocene to early Holocene in the southeast of Turkey (Tables 5 and 6).

Table 3 Site profile description of Vertisol (Harran 3 Soil) in Şanlıurfa, southeast Turkey (Çakmaklı 2008)

IUSS Working Group WRB (2015)	Calcaric Chromic Vertisols
Soil Survey Staff (2014)	Chromic Calcixerert
Soil series	Harran 3
Location	32 km south of Şanlıurfa city on the Akçakale road, (southeast Turkey)
Coordinates	37° 04′ 59″N, 38° 51′ 27″E
Elevation	390 m
Climate	Semiarid, transitional from Mediterranean
Vegetation	Annual cropping (cereal, cotton, maize)
Parent material	Colluvial
Geomorphic unit	Mudflow
Topography	Subnormal relief, gently sloping 1–2% N–S
Erosion	Weak surface water erosion
Water table	Deeper than 140 cm
Drainage	Moderate
Infiltration	Moderate
Stoniness	No stones
Soil Moisture Regime	Xeric
Soil Temperature Regime	Mesic

Horizon	Depth (cm)	Description
Ap1	0–12	Brown (7.5YR 4/4) moist, clay, moderate granular structure, firm, moderate biological activity, many fine roots, strongly effervescent, gradual wavy boundary
Ap2	12–30	Brown (7.5YR 4/4) moist, clay, strong coarse blocky structure, very firm, moderate biological activity, very few fine roots, strongly effervescent, gradual wavy boundary
Bwk1	30–66	Brown (7.5YR 3/4) moist, clay, medium angular blocky structure, firm, moderate irregular fine to medium and rarely diffuse carbonate nodules, moderate biological activity, very few fine roots, strongly effervescent, gradual wavy boundary
Bwk2ss	66–112	Brown (7.5YR 4/4) moist, clay, medium angular blocky structure, firm, strongly effervescent, moderate irregular fine to medium rarely diffuse carbonate nodules: strong slickensides
Bwk3	112–140	Brown (7.5YR 4/4) moist, clay, medium angular blocky structure, firm, moderate irregular fine to medium nodules with rare diffuse carbonate accumulations, strongly effervescent, gradual wavy boundary

Horizon	Depth (cm)	pH 1:1	CaCO$_3$ (%)	EC (dS/m)	Org. Mat. (%)
Ap1	0–12	8.40	26.5	0.92	1.11
Ad2	12–30	8.20	24.94	1.08	1.29
Bwk1	30–66	7.77	31.18	0.24	0.82
Bwk2ss	66–112	8.05	29.62	0.91	0.65
Bwk3	112–140	8.10	30.4	1.0	0.59

Horizon	Exchangeable cations (cmol$_c$kg^{-1})				Clay (%)	Silt (%)	Sand (%)
	Ca + Mg	Na	K	CEC*			
Ap1	42.61	3.48	7.43	40.96	53.56	36.32	10.12
Ad2	39.44	7.47	6.24	33.46	55.18	36.55	8.27
Bwk1	61.82	10.56	6.33	46.82	55.38	38.55	6.07
Bwk2ss	36.36	11.55	7.30	35.15	58.41	37.40	4.19
Bwk3	47.04	12.86	5.82	42.19	56.79	39.32	3.89

*Extractable Ca may contain Ca from calcium carbonate or gypsum. CEC base saturation set to 100

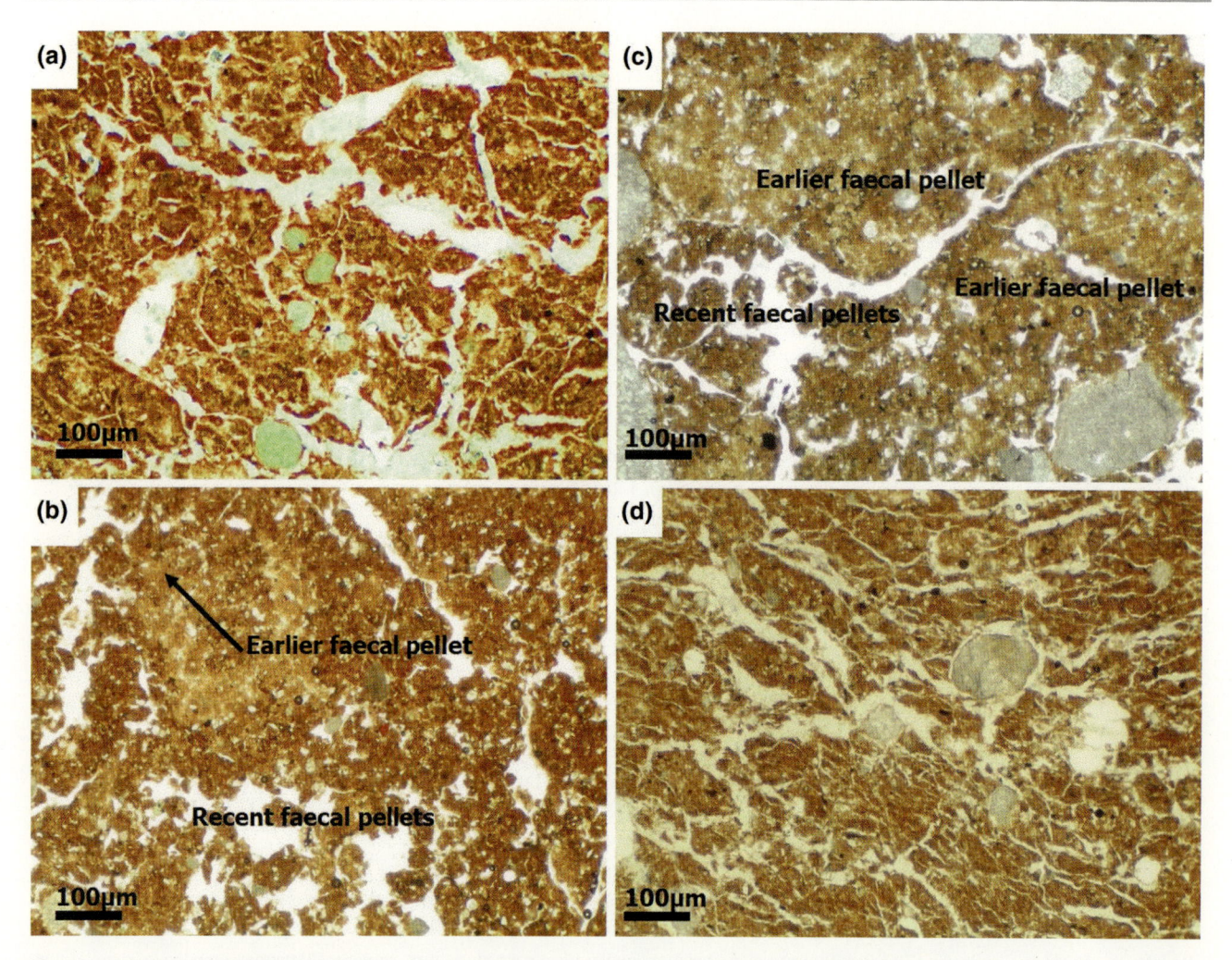

Fig. 8 Development of granular aggregates with root channels (**a**), Incorporation of probable earlier developed subspherical to rounded fecal pellets in contemporary finer excrements (**b**, **c**), Compaction features of parallel crack patterns (**d**) (plane polarized light)

4.3.1 Mineralogy and Micromorphology of the Cepkenli and Kısas Soils

Mineralogy of the Cepkenli Soil

Sand and Silt Fractions
The randomly oriented powder mount of sand (2–0.05 mm) and silt (50–2 μm) fractions of the selected horizons (Bw1 and By2) have mixed mineralogy. The major mineral phases include quartz, calcite, and feldspars (microcline, anorthite, and albite) and relatively minor minerals are kaolinite, palygorskite, and a 1.4 nm peak mineral likely to be chlorite or smectite in the silt fraction. The sand fractions have the same major minerals but lack peaks indicative of kaolinite, palygorskite, or the 1.4 nm clay mineral. The SEM images show the morphologies of chlorite and calcite in the silt fraction in Fig. 19. At higher magnification, the characteristic fiber morphology of palygorskite is determined between the cemented grains of calcite.

Clay Fraction
The X-ray diffraction (XRD) patterns indicate that the clay fraction of the Cepkenli soil includes smectite, chlorite, palygorskite, illite, kaolinite, quartz, and calcite.

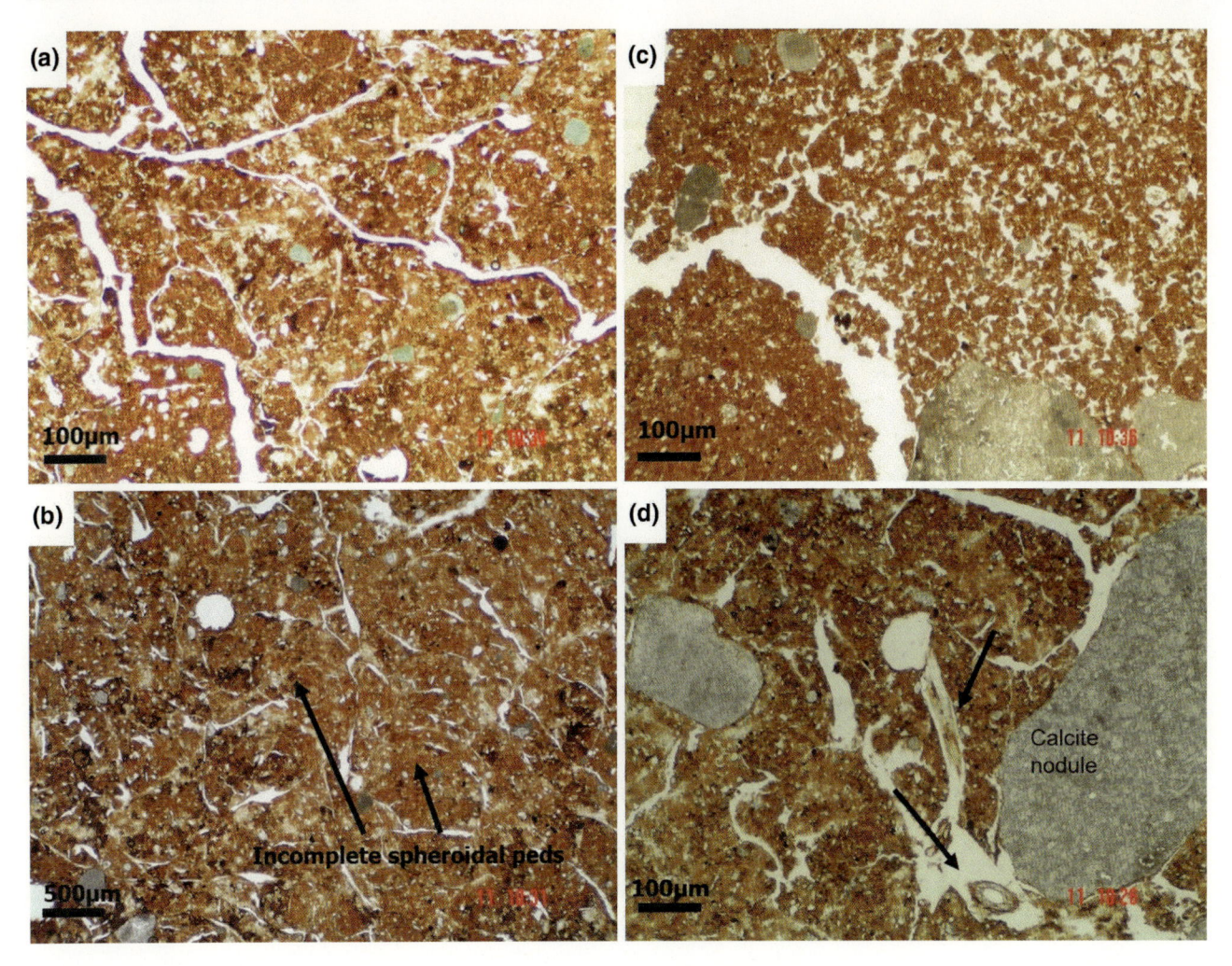

Fig. 9 Irregular partly accommodated aggregate development (**a**), incomplete spheroidal peds (**b**), Channels completely filled with uniformly distributed fecal material (**c**), Calcite nodules and root remnants (*arrows*) in channel (**d**) (plane polarized light)

Quantification of the total clay minerals of the Cepkenli soil using peak area is presented in Fig. 20. Data indicate that smectite tends to decrease from the surface (42%) to the subsoil (28%) with depth but again increases in the By2 horizon. Kaolinite shows a tendency to increase slightly (19–26%) with depth, reaching a maximum content in the Bss horizon. Chlorite, palygorskite, and illite each constitutes less than 20% of the clay fraction. Ranges for each are 13–20% for chlorite, 13–17% for palygorskite, and 8–15% of illite. Both palygorskite and illite have the lowest quantities in the surface (Ap1) horizon. In the total clay fraction, clay minerals might be ordered from the highest to lowest quantity as smectite, kaolinite, chlorite, palygorskite, and illite. The palygorskite content is higher than chlorite below a depth of 1 m.

Mineralogy of the Kısas Soil

Sand and Silt Fractions

The randomly oriented powder mounts of the sand (2–0.05 mm) and silt (50–2 µm) fractions of the Bw and Bk2 horizons have mixed mineralogy. Quartz, calcite, and

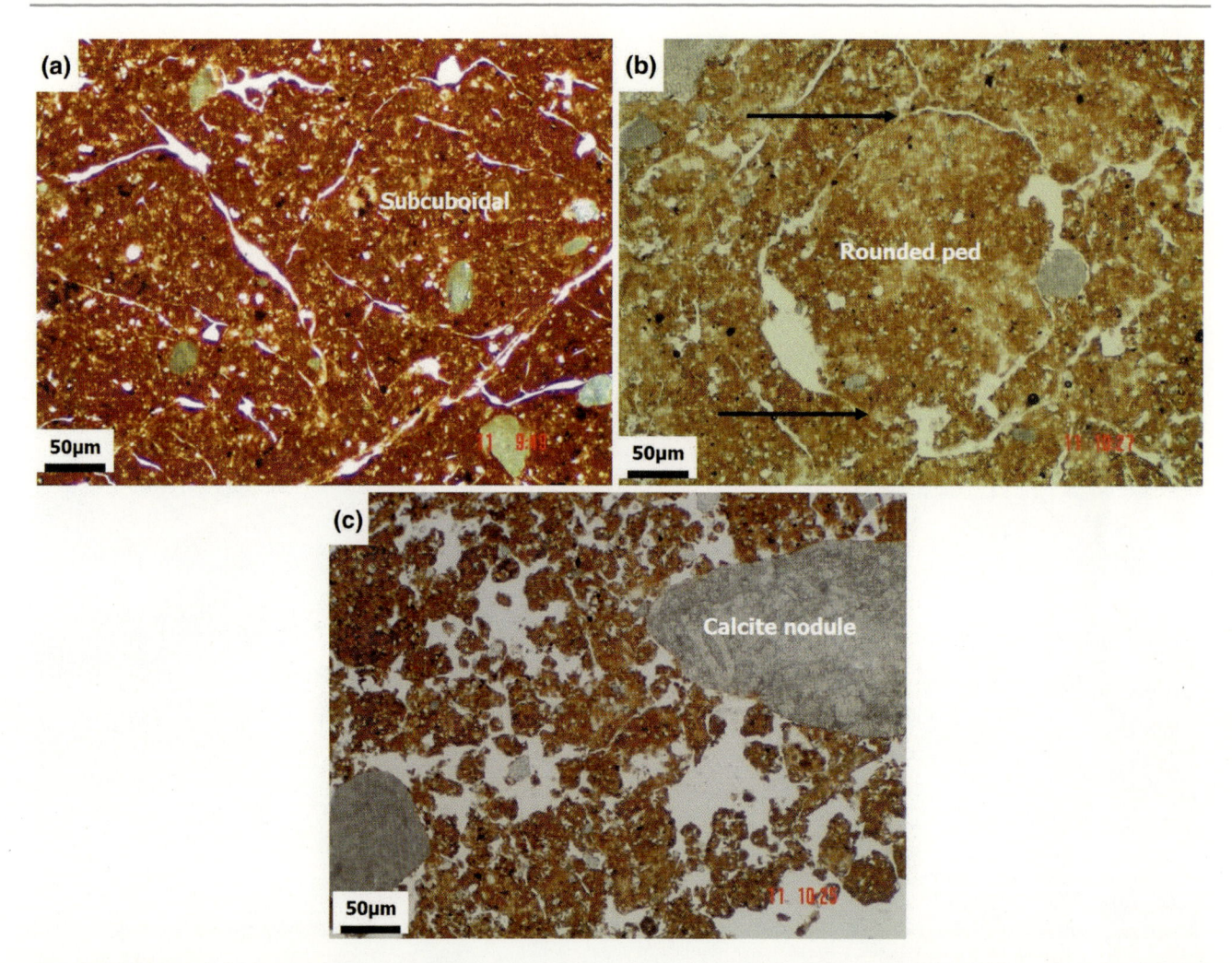

Fig. 10 Bifurcating incomplete subcuboidal (with very fine wedge formation)—subangular peds (**a**), Development of occasional rounded ped with wedge-shaped areas (*arrows*) (**b**), Calcite nodules of various marine origin within matrix of faunal excrements (**c**) (plane polarized light)

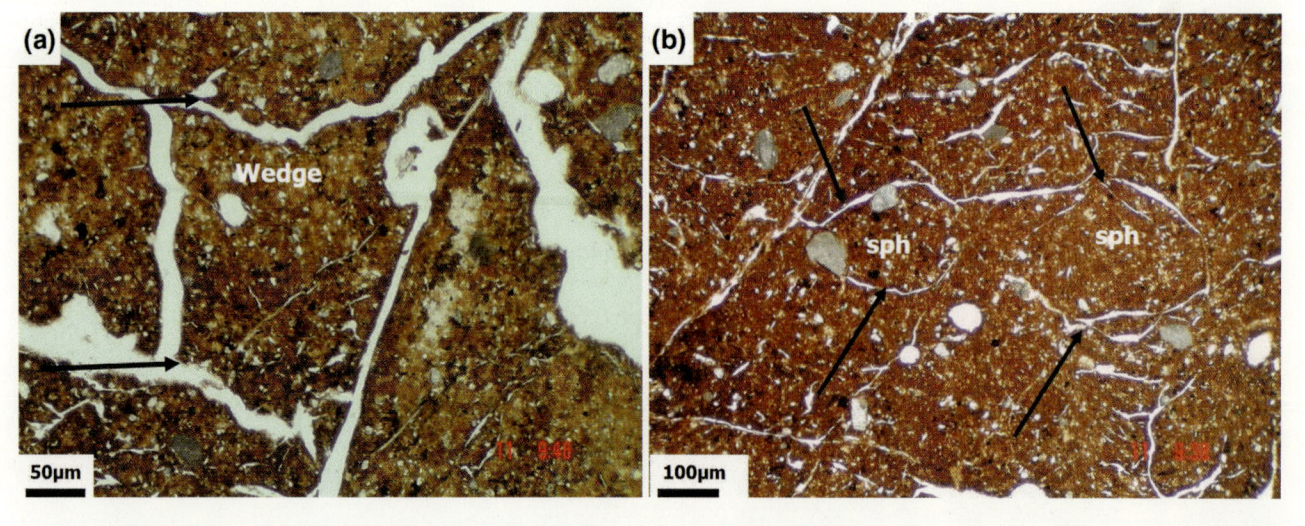

Fig. 11 Wedge-shaped soil areas (wedge with *arrows*) with perpendicular intersections/continuous pore spaces (45°–60°) (**a**), Unaccommodated rounded/spheroidal peds (*sphere* and *arrows*) in incomplete very fine (<5 mm) wedges (aggregate) (**b**) (plane polarized light)

Vertisols

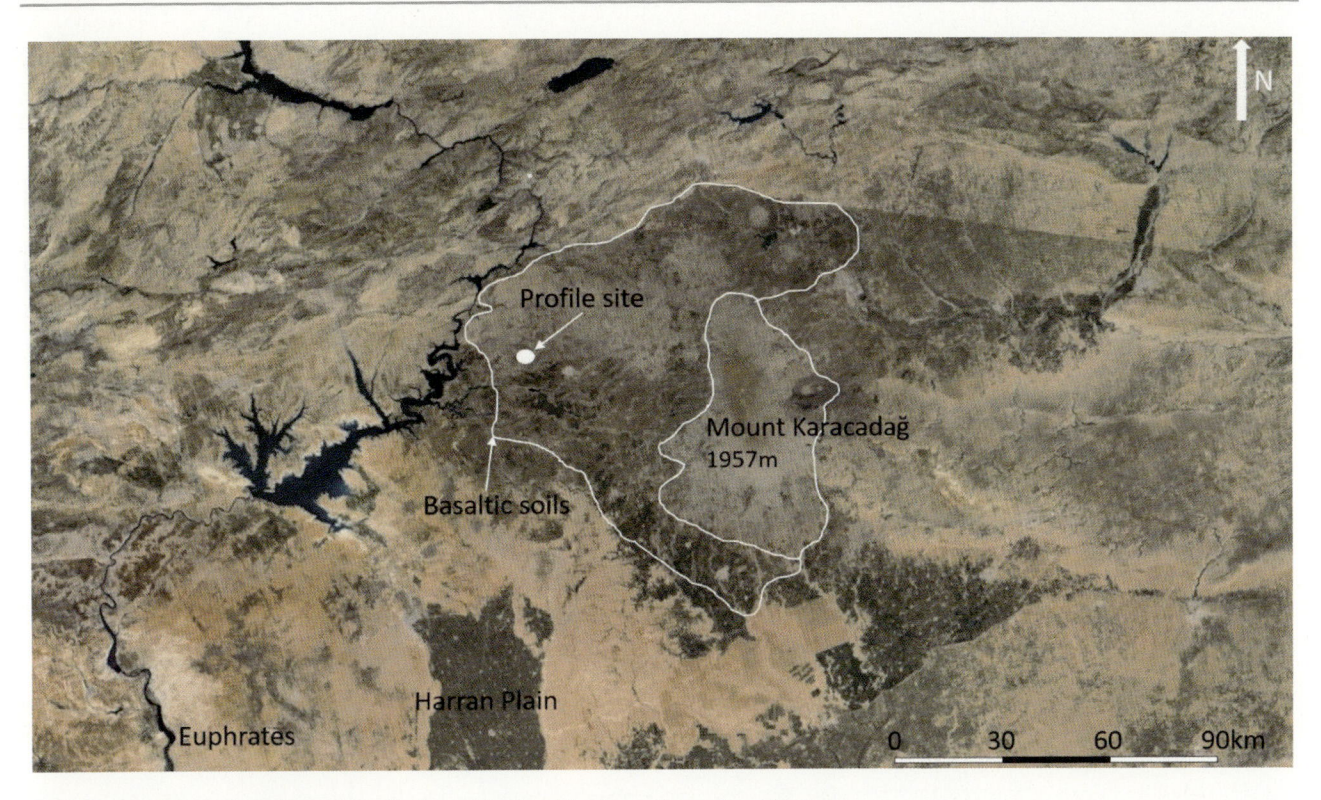

Fig. 12 Karacadağ soil profile and distribution of the basaltic plateau in southeast Turkey

Fig. 13 Montmorillonite in-fills of basalt cracks in Şanlıurfa, southeast Turkey

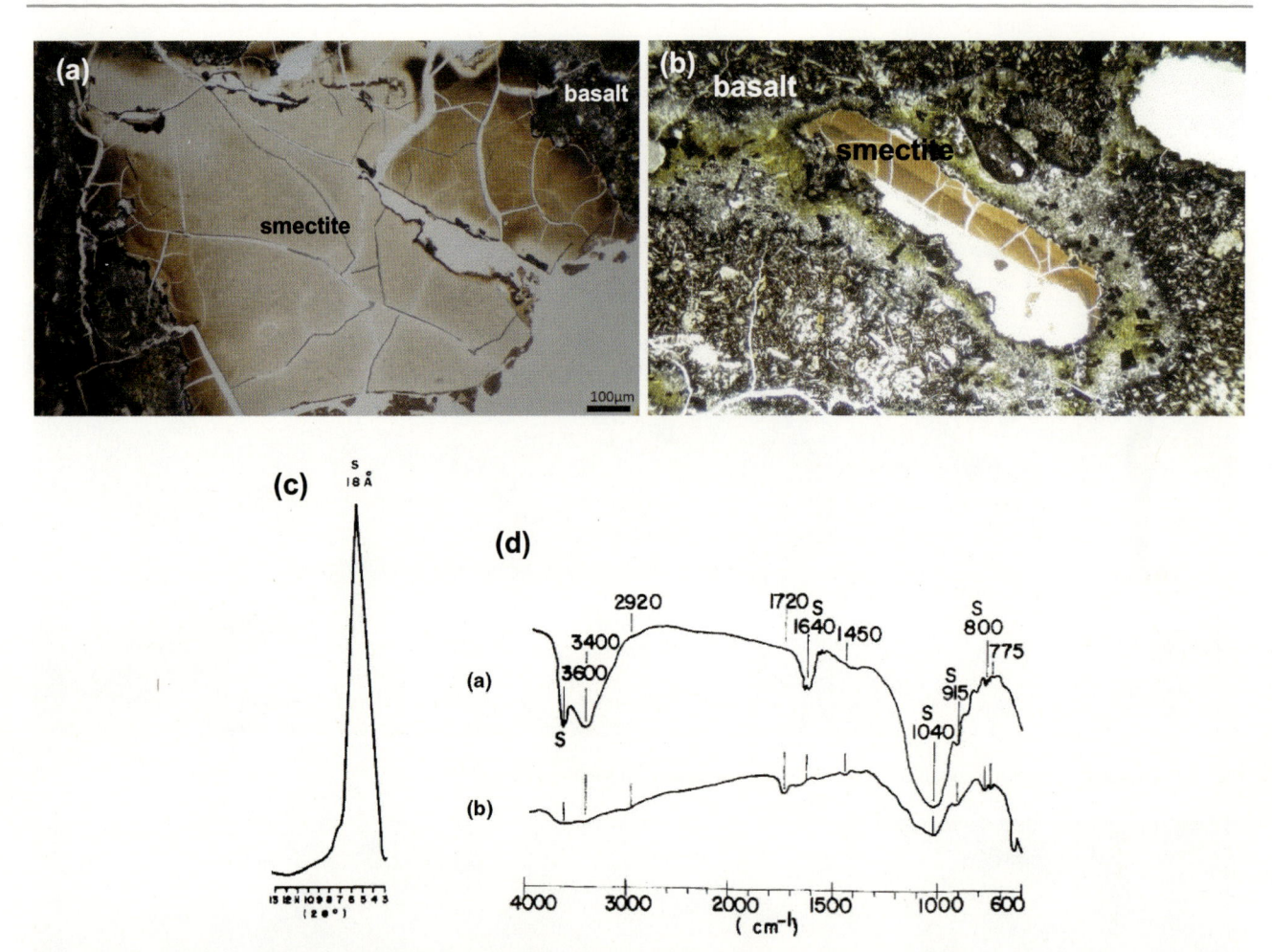

Fig. 14 The smectitic neoformations-nodules (**a, b**) in thin section (plane polarized light), smectite peak obtained by X-ray diffraction from the neoformed smectitic nodule (**c**) and di-octahedral smectite bands obtained by infrared spectroscopy in comparison to Mössbauer clay (Germany) (*upper line*) from the neoformed clay nodule peeled from the amygdules of the basaltic fragments incorporated in the basaltic Vertisols (*lower line*) (**d**)

feldspars are the major minerals, whereas chlorite, palygorskite, and kaolinite are the minor minerals of the silt fraction. The sand fractions of the Bw and Bk2 horizons show the same major mineral composition but the Bk2 horizon has a higher peak intensity for palygorskite than the Bw horizon. The sand fractions do not show significant chlorite and kaolinite peaks (Aydemir 2001).

The SEM of silt-size minerals (quartz, calcite, and palygorskite) and sand-size minerals (calcite and palygorskite) in the Bk2 horizon are shown in Figs. 21 and 22, respectively. Figure 21a shows a quartz grain, which has a layer-like morphology (supported with EDS data showing very high amount of Si in its composition). Figure 22a shows a very interesting sand grain identified as calcite with palygorskite fibers (Fig. 22b, c). This explains the palygorskite XRD peak in the sand fraction of Bk2 horizon (Aydemir 2001).

Clay Fraction

The clay fraction (<2 μm) XRD data of the Kısas soil revealed the presence of smectite, chlorite, palygorskite,

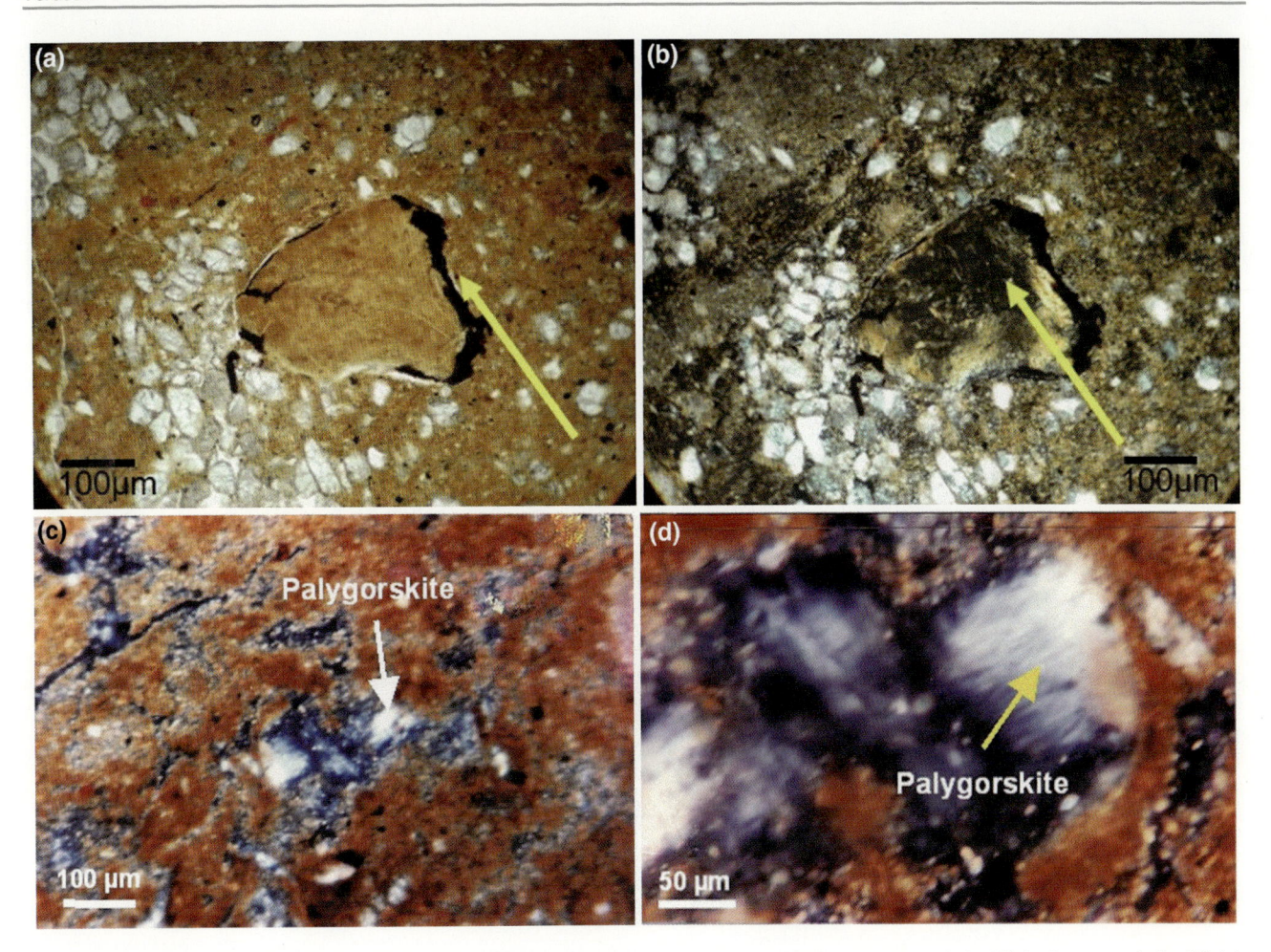

Fig. 15 Weathering smectite nodule (*arrow*) incorporated from amygdule by shrink–swell action (**a** plane polarized light, **b** cross-polarized light), alteration of smectite nodule to palygorskite (**c** and **d** cross-polarized light)

illite, kaolinite, feldspar, quartz. and calcite contents (Fig. 23). Data indicate that smectite is the highest in quantity though erratic in distribution, being greatest in the surface (67%) and decreasing erratically with depth. Kaolinite and chlorite tend to decrease with depth, though not markedly. Palygorskite has its highest content (40%) in the Bk2 horizon. Above the Bk2 horizon palygorskite was <12%. Illite is low in all horizons and was absent in the deepest horizon. In the total clay fraction, the clay minerals might be ordered as smectite > kaolinite > chlorite > palygorskite > illite, however, palygorskite is the second most abundant clay mineral in horizons deeper than 120 cm (Aydemir 2001).

Figures 24 and 25 show the morphology of fine clay particles under transmission electron microscopy (TEM) from selected horizons (Ap1, Bw, and Bk2). Figure 24a depicts the morphology of the fine clay-size particles of deformed palygorskite fibers and probably reformed nondescript thin flakes of smectite in the Ap1 horizon. Figure 24b, d, e show the deformed (palygorskite) and reformed (smectite) minerals in higher detail. Figure 24c shows the hexagonal shaped kaolinite particles. Figure 25 shows dissolution of palygorskite and formation of smectite in the Bw horizon. Figure 25a, b, and d show the fibers more common and less deformed than those in the Fig. 24a, b, d, e. Smectite is identified by the lattice fringes that have a

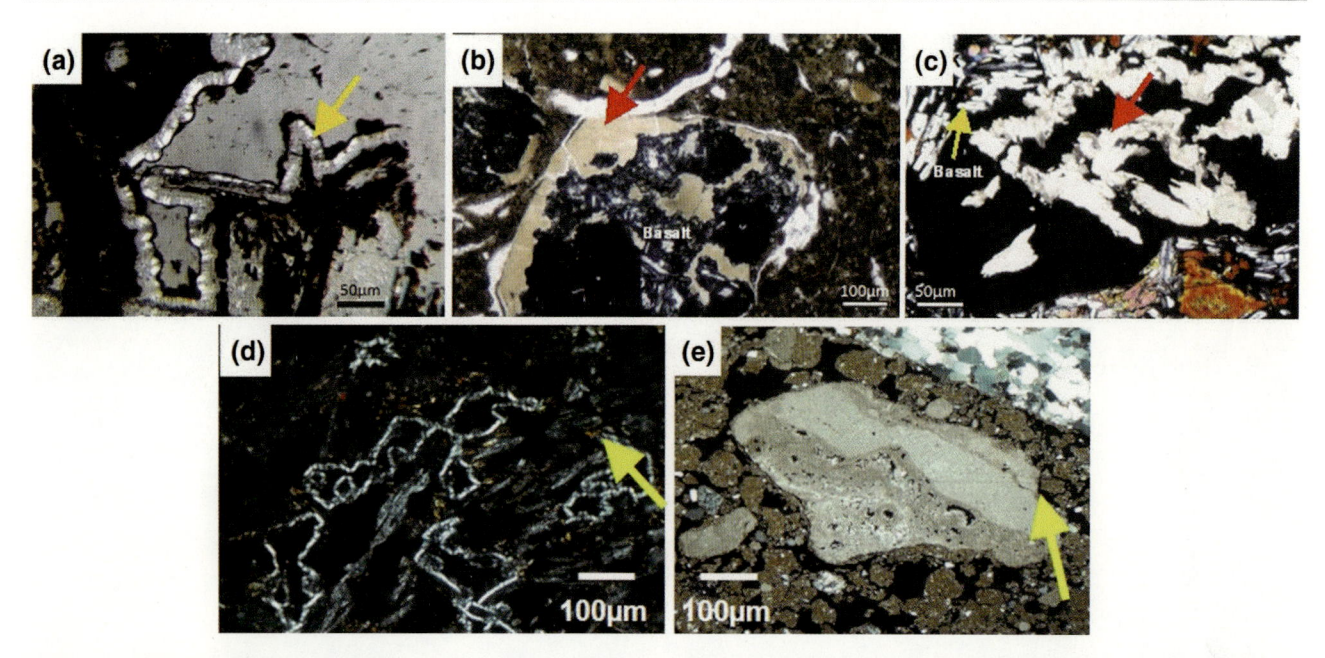

Fig. 16 Secondary vermicular kaolinite (*arrow*) forming in amygdule of basalt fragment in a basaltic Vertisol of Kilis (**a**) and Siverek (**b**), southeast Turkey, basalt fragment with neoformed smectite (*arrow*) nodule in matrix (**c**), secondary calcite (*red arrow*) forming in pore of basaltic soil (**d**), calcrete with crust on fragment surface incorporated in the matrix of a basaltic Vertisol from Siverek, southeast Turkey (**e**)

Table 4 Site profile description of the Karacadağ soil in Şanlıurfa, southeast Turkey (Kapur 1981)

IUSS Working Group WRB (2015)	Chromic Ochric Vertisol
Soil Survey Staff (2014)	Typic Chromoxerert
Soil series	Karacadağ
Location	37° 04′ 59″N, 38° 51′ 27″E
Elevation	1100 m
Climate	Semiarid, transitional from Mediterranean
Vegetation	Natural shrub vegetation
Parent material	Basalt
Geomorphic unit	Basaltic plateau,
Topography	Gently sloping 1–2%
Erosion	Weak surface water erosion
Water table	Deeper than 140 cm
Drainage	Good
Infiltration	Good
Stoniness	95% surface stoniness, basalt rock fragments (d: 90 cm)

Horizon	Depth (cm)	Description
A	0–30	Brown (7.5YR 4/2) moist and (7.5YR 4/4) dry, clay, coarse angular blocky structure, sticky and plastic, firm and massive, no effervescence, cracks of 1–5 cm width, gradual wavy boundary
Bwss1	30–50	Dark reddish brown (5YR 3/2) moist) and reddish brown (5YR 4/4) dry, clay, coarse angular blocky structure, sticky and plastic, firm when wet, no effervescence, cracks of 1–5 cm width, gradual wavy boundary
Bwss2	50–70	Dark reddish brown (5YR 3/2) moist and reddish brown (5YR 4/4) dry, clay, medium angular blocky structure, sticky and plastic, firm when wet, no effervescence, cracks of 1–5 cm width, slicken sides, gradual wavy boundary
Cr	70+	Basaltic rock

(continued)

Table 4 (continued)

Horizon	Depth (cm)	pH 1:1	CaCO$_3$ (%)	Salinity (%)	Org. Mat. (%)
A	0–30	7.60	2.20	0.09	2.40
Bwss1	30–50	7.70	3.60	0.08	1.70
Bwss2	50–70	7.80	3.40	0.09	1.50
Cr	70+	–	–	–	–

Horizon	Exchangeable cations (cmol$_c$ kg^{-1})				Clay (%)	Silt (%)	Sand (%)	
	Ca + Mg	Na	K	CEC				
A	46.0	2.4	4.1	51.7	58.5	28.9	12.6	S, K, I
Bwss1	49.3	4.1	0.05	54.9	56.7	31.0	12.4	S, K
Bwss2	48.2	2.7	0.03	53.3	62.5	16.4	21.2	S, K
Cr								

S Smectite, *K* Kaolinite, *I* Illite

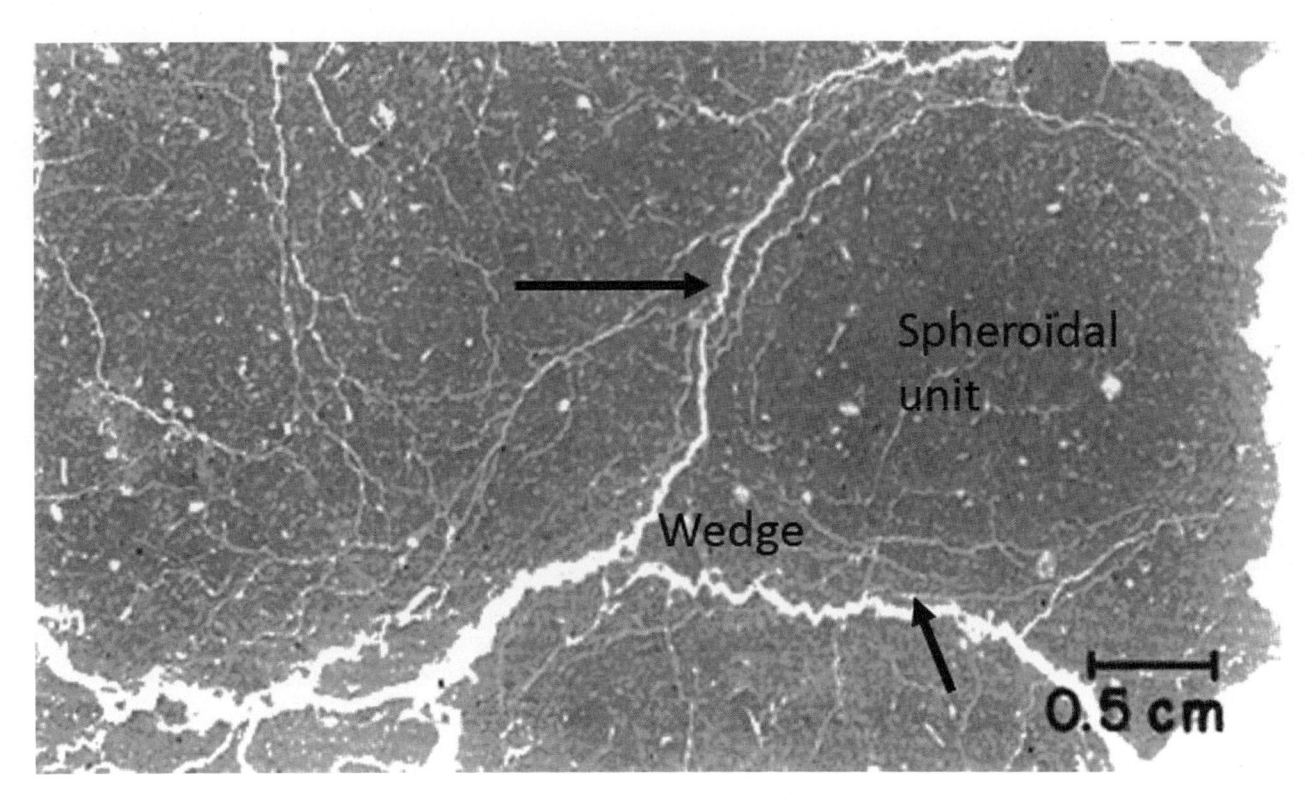

Fig. 17 Spheroidal microstructure in Beğdeş Vertisol southeast Turkey (Kapur et al. 1997)

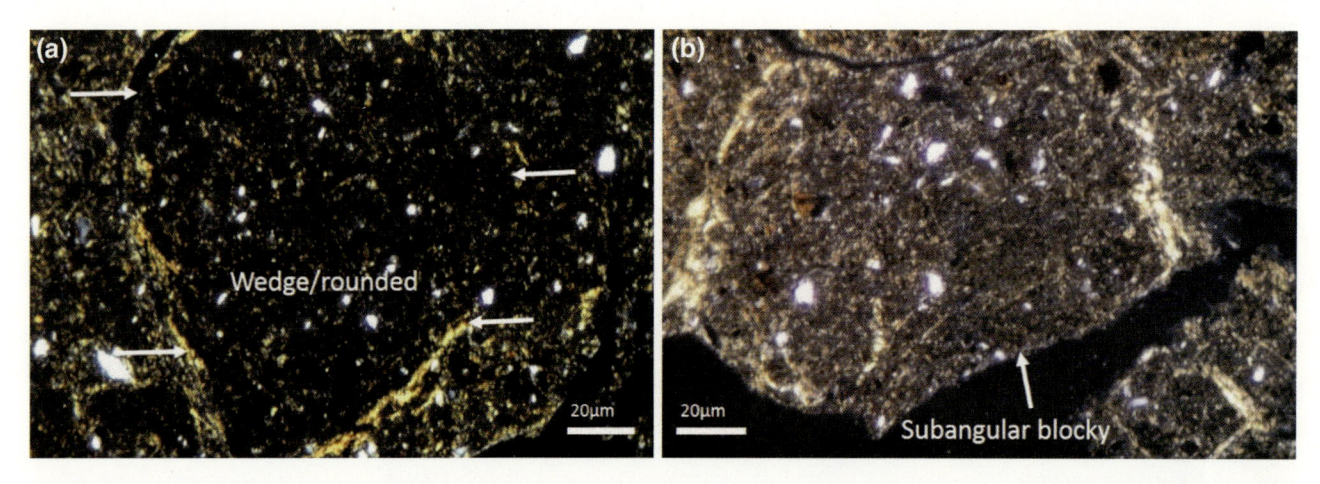

Fig. 18 Permanent microstructural units (MSUs)/aggregates surrounded by stress features/stress coatings, wedge–rounded (**a**), rounded (**b**) (Kapur et al. 1997) (cross-polarized light)

Table 5 Site description of and physical and chemical properties of the Cepkenli soils, Şanlıurfa, southeast Turkey (Aydemir 2001)

IUSS Working Group WRB (2015)	Gypsic Vertisol
Soil Survey Staff (2014)	Typic Calcixerert
Soil series	Cepkenli
Location	Approximately 6 km northeast of Şanlıurfa
Landform	Bajada
Elevation	400 m
Climate	Mediterranean
Vegetation	Fallow
Parent material	Alluvium
Geomorphic unit	Holocene alluvium
Topography	Nearly level, 0–1% slope
Water table	Deeper than 2 m
Drainage	Moderately well drained
Infiltration	Fine
Stoniness:	None

Remarks: There was ground water at 190 cm in pit after opening. The EC value of the ground water was 30 dS/m. There was a good indication of soluble salts in moist horizons as they are friable, EC values for horizons beginning at 58 cm were 0.7, 0.7, 1.7, 5.2, and 6.5 dS/m, respectively, to a depth of 184 cm. Soil EC values were on field extracted saturated pastes

Horizon	Depth (cm)	Description
Ap1	0–22	Brown (7.5YR 4/4) wet, very pale brown (10YR 7/4) dry, silty clay loam, moderate coarse subangular blocky parting to moderate fine and medium subangular blocky structure, very hard, common fine pores, common fine roots, moderately alkaline, violently effervescent, clear smooth boundary
Ap2	22–38	Strong brown (7.5YR 4/6) wet, light yellowish brown (10YR 6/4) dry, silty clay loam, structureless massive, extremely hard, few fine and medium pores, few fine roots, the Ap1 material with subangular blocky structure has fallen into cracks between massive materials, moderately alkaline, violently effervescent, clear smooth boundary
Bw1	38–58	Strong brown (7.5YR 4/6), silty clay, structureless massive, very hard, few fine pores, few fine roots, moderately alkaline, violently effervescent, gradual smooth boundary
Bw2	58–85	Strong brown (7.5YR 4/6), clay, weak coarse prismatic parting to weak

(continued)

Table 5 (continued)

Horizon	Depth (cm)	Description
		fine and medium angular blocky structure, very hard, few fine pores, few fine roots, moderately alkaline, violently effervescent, gradual smoothboundary
Bss	85–114	Strong brown (7.5YR 4/6), clay, weak coarse prismatic parting to weak fine and medium angular blocky structure, very firm, few fine pores, very few fine and medium roots, few small slickensides, moderately alkaline, violently effervescent, gradual smooth boundary
Bw3	114–143	Strong brown (7.5YR 4/6), clay, weak fine and medium subangular blocky structure, firm, few fine pores, very few medium roots, few (1%)fine (3 mm) limestone pebbles, moderately alkaline, violently effervescent, gradual smooth boundary
Bw4	143–160	Strong brown (7.5YR 4/6), clay, moderate fine and medium subangular blocky structure, friable, few fine pores, very few fine and medium roots, moderately alkaline, violently effervescent, gradual smooth boundary
By1	160–184	Strong brown (7.5YR 4/6), clay, moderate fine and medium subangular blocky structure, friable, few fine pores, very few fine and medium roots, few fine Fe–Mn stains, many fine to coarse selenite crystals, about 30%gypsum, moderately alkaline, violently effervescent, gradual smoothboundary
By2	184–210	Strong brown (7.5YR 4/6), clay loam, moderate fine and medium subangular blocky structure, friable, few fine pores, very few medium roots, few fine Fe–Mn stains, common fine to coarse selenite crystals, about 15% gypsum, moderately alkaline, violently effervescent

Horizon	Depth (cm)	Particle size distribution (%)										Texture class
		Sand						Silt		Clay		
		VC	Co	M	F	VF	Total	F	Total	F	Total	
Ap1	0–22	0.3	0.2	0.3	1.2	2.5	4.5	51.7	60.5	2.3	35.0	SiCL
Ap2	22–38	0.3	0.2	0.3	1.1	1.9	3.8	49.0	56.7	3.5	39.5	SiCL
Bw1	38–58	0.4	0.3	0.4	1.1	1.8	4.0	41.0	46.8	5.9	49.2	SiC
Bw2	58–85	0.1	0.3	0.3	1.1	2.2	4.0	37.3	43.4	10.0	52.6	SiC
Bss	85–114	0.3	0.2	0.3	1.3	2.5	4.6	35.6	43.9	13.7	51.5	SiC
Bw3	114–143	0.3	0.3	0.3	2.0	3.5	6.4	35.2	42.9	19.1	50.7	SiC
Bw4	143–160	0.3	0.4	0.7	2.3	3.7	7.4	34.3	42.3	18.2	50.3	SiC
By1	160–184	0.5	0.6	0.4	1.6	2.9	6.0	29.8	41.5	20.8	52.8	SiC
By2	184–210	1.3	0.9	0.6	1.0	2.2	6.0	30.0	36.7	21.6	57.3	C

	OC %	pH (H$_2$O) 1:1	NH$_4$OAc extractable bases (cmolc kg^{-1})					NaOAc CEC*	BS %	ESP %	SAR	Cal %	Dol %	CCE %	Gyp %
			Ca	Mg	Na	K	Total								
Ap1	1.15	8	58.2	5.7	0.7	1.9	66.5	37.3	100	2	1	20.5	1.8	22.3	0
Ap2	0.79	8.1	57.6	7.6	0.5	1.8	67.5	37.4	100	1	1	21.5	1.3	22.9	0
Bw1	0.62	8	58.1	9.6	0.5	1.3	69.5	34.1	100	1	1	23.7	0.5	24.2	0
Bw2	0.38	8.1	57.9	11.5	0.7	0.9	71	33.4	100	2	1	26.2	0.4	26.6	0
Bss	0.43	8.2	57.6	13.3	1.4	0.8	73.1	31.4	100	4	3	25.6	2	27.8	0
Bw3	0.34	8.2	46.1	9.5	3.9	0.7	60.2	28.1	100	11	9	28.3	1.8	30.2	0
Bw4	0.35	7.9	46.1	11.4	7.6	0.7	65.8	28.3	100	15	10	25.9	2.1	28.2	0.2
By1	0.39	8.1	146	15.6	12.7	0.7	175	27.7	100	22	16	23	1.2	24.3	8.9
By2	0.35	8.2	106	17.4	18.6	0.8	142	31	100	27	20	24.3	0.7	25.1	4.7

Table 5 (continued)

	Saturated paste extract						Bulk density		COLE (cm/cm)	WC 0.33 bar (%)
	EC	WC	Ca	Mg	Na	K	0.33 Bar	Oven dry		
	dS/m	%	meq/l				g/cm^3			
Ap1	0.8	56	5	1	2.3	0.2	1.28	1.45	0.042	26.8
Ap2	0.6	56	5	0.9	1.5	0.2	1.33	1.54	0.050	29.2
Bw1	0.7	59	5	1.2	1.4	0.1	1.42	1.7	0.062	25.2
Bw2	0.8	60	2	1.1	1.8	0.1	1.41	1.72	0.068	26.7
Bss	0.5	65	1.2	0.8	3.3	0	1.49	1.8	0.065	25.2
Bw3	1.6	71	3	1.7	13	0.1	1.42	1.74	0.070	27.0
Bw4	7.4	74	25	13.2	44.8	0.1	1.29	1.54	0.061	28.6
By1	10.5	87	22	23.8	77.4	0.2	1.29	1.54	0.061	33.1
By2	12.4	95	22	26.3	97.4	0.2	1.19	1.58	0.099	41.3

Note Fine silt = 0.02–0.002 mm, Total silt = 0.05–0.002 mm, Fine clay = <0.0002 mm, Total clay = <0.002 mm
VC very coarse (2.0–1.0 mm), *Co* coarse (1.0–0.5 mm), *M* medium (0.5–0.25 mm), *F* fine (0.25–0.10), *VF* very fine (0.10–0.05 mm), *SiCL* silty clay loam, *SiC* silty clay, *C* clay, *OC* organic carbon, *CEC* cation exchange capacity, *BS* base saturation, *ESP* exchangeable Na percentage, *SAR* sodium adsorption ratio, *Cal* calcite, *Dol* dolomite, *CCE* calcium carbonate equivalent, *Gyp* gypsum, *EC* electrical conductivity, *COLE* coefficient of linear extensibility, *WC* water content
* Extractable Ca may contain Ca from calcium carbonate or gypsum. CEC base saturation set to 100

Table 6 Site profile description and physical and chemical properties of the Kısas soils, Şanlıurfa, southeast Turkey (Aydemir 2001)

IUSS Working Group WRB (2015)	Chromic Vertisol Calcaric
Soil Survey Staff (2014)	Chromic Calcixererts
Soil series	Kısas
Location	Approximately 10 km south of Şanlıurfa
Landform	Bajada
Elevation	650 m
Climate	Mediterranean
Vegetation	Fallow
Parent material	Alluvium
Geomorphic unit	Deposit
Topography	Nearly level, 0–1% slope
Water table	Deeper than 2 m
Drainage	Well drained
Infiltration	Fine
Stoniness	No stone

Horizon	Depth (cm)	Description
Ap1	0–13	Strong brown (7.5YR 4/6), brown (7.5YR 5/4) dry, clay, weak medium subangular blocky parting to weak fine granular structure, hard, many medium and coarse pores, fine few roots, 5% coarse fragments, moderately alkaline, strongly effervescent, clear smooth boundary
Ap2	13–30	Reddish brown (5YR 4/4), strong brown (7.5YR 4/6) dry, clay, weak medium subangular blocky structure, very hard, common medium and coarse pores, few fine roots, 5% coarse fragments, moderately alkaline, strongly effervescent, clear smooth boundary
Bw	30–50	Yellowish red (5YR 4/6), clay, moderate fine and medium angular blocky structure, firm, common fine and medium pores, very few fine roots, 5% coarse fragments, moderately alkaline, strongly effervescent, gradual wavy boundary

(continued)

Table 6 (continued)

Horizon	Depth (cm)	Description
Bk	50–70	Yellowish red (5YR 4/6), clay, moderate fine and medium angular blocky structure, firm, few fine pores, very few fine roots, about 5% of horizon is secondary carbonate nodules which seem to be limestone ghosts, 5% coarse fragments, moderately alkaline, strongly effervescent, gradual wavy boundary
Bssk1	70–107	Yellowish red (5YR 4/6), clay, weak medium prismatic parting to moderate medium angular blocky structure, firm, few fine pores, very few fine roots, common medium slickensides, about 15% secondary carbonate nodules which seem to be limestone ghosts, 5% coarse fragments, moderately alkaline, strongly effervescent, clear smooth boundary
Bssk2	107–136	Light olive brown (2.5Y 5/3), moist, sandy clay loam, gradual wavy boundary, single grain, very friable, medium effervescent
Bk1	136–161	White (10YR 8/2), gravelly, clay, weak fine and medium angular blocky structure, extremely hard, no roots, about 40% of horizon is very hard white nodular calcite, concentrated near top of horizon, at intervals of about 10 to 15 cm are tongues of 5YR 4/6 material which has about 30% secondary carbonate nodules, moderately alkaline, strongly effervescent, clear wavy boundary
Bk2	162–200	Light yellowish brown (10YR 6/4), clay, weak fine and medium angular blocky structure, firm, no roots, few fine Fe–Mn stains, many continuous calcium carbonate segregations, about 10% of horizon is tongues and pockets of 5YR 4/6 clay, total of about 60% of horizon is secondary carbonate, moderately alkaline, violently effervescent

Horizon	Depth	Particle size distribution (%)									Texture
		Sand						Silt		Clay		
		VC	Co	M	F	VF	Total	F	Total	F	Total	Class
Ap1	0–13	1	1	1.1	1.3	1.9	6.3	38.8	46.1	2.6	47.6	SiC
Ap2	13–30	1.1	1	1.1	1	1.5	5.7	33.4	38.8	5.3	55.5	C
Bw	30–50	1.1	1	0.8	0.9	1.4	5.2	30.7	34.5	14.3	60.3	C
Bk	50–70	1.4	1.2	1	0.9	1.4	5.9	29	34.2	14.9	59.9	C
Bssk1	70–107	1.3	1.8	1.3	0.9	1.2	6.5	30.9	33.7	19.3	59.8	C
Bssk2	107–136	1.7	1.7	1.3	0.9	1	6.6	29.9	33.2	17.7	60.2	C
Bk1	136–161	4.4	3.5	3	2.6	2.1	15.6	39.6	41.7	15.5	42.7	SiC
Bk2	161–200	1.4	0.6	0.5	0.9	2.1	5.2	54.8	60	14.1	34.8	SiCL

	OC (%)	pH (H$_2$O) 1:1	NH$_4$OAc extractable bases (cmolc kg^{-1})					NaOAc			SAR	Cal (%)	Dol (%)	CCE (%)	Gyp (%)
			Ca	Mg	Na	K	Total	CEC* (cmolc kg^{-1})	BS (%)	ESP (%)					
Ap1	0.89	7.8	67.6	3.7	0.1	2	73.4	41.2	100	0	0	22.9	2.2	25.3	0
Ap2	0.79	7.8	68	3.7	0.1	1.5	73.3	40.6	100	0	0	24.5	1	25.6	0
Bw	0.32	7.8	69.3	5.7	0.2	1	76.2	39.9	100	0	0	28.2	0.9	29.3	0
Bk	0.66	7.8	68.8	5.6	0.2	1	75.6	40.3	100	0	0	26.4	1.7	28.2	0
Bssk1	0.27	7.8	57.2	5.7	0.2	0.9	64	38.4	100	0	0	30.3	2.2	32.7	0
Bssk2	0.46	7.9	58.6	5.9	0.2	0.9	65.6	36.4	100	0	0	32.9	0.8	33.9	0
Bk1	0.39	8	45.5	3.7	0.1	0.6	49.9	25.2	100	0	0	57.1	0.2	57.3	0
Bk2	0.39	8.1	34.1	2.8	0.1	0.3	37.3	17.2	100	0	0	63.8	1.4	65.4	0

		Saturated paste extract					Bulk density		COLE (cm/cm)	WC
		EC	WC	Ca	Mg	Na	K	0.33 bar	Oven dry		0.33 Bar %
		dS/m	%	meq/l				g/cm^3			
Ap1		0.4	64	4.5	0.4	0.2	0.2	1.19	1.58	0.099	0.099	36.8
Ap2		0.3	65	3	0.3	0.3	0.1	1.29	1.70	0.096	0.096	32.0
Bw		0.3	73	3	0.4	0.3	0.1	1.32	1.76	0.101	0.101	33.6

(continued)

Table 6 (continued)

		Saturated paste extract						Bulk density		COLE (cm/cm)	WC
		EC	WC	Ca	Mg	Na	K	0.33 bar	Oven dry		0.33 Bar %
		dS/m	%	meq/l				g/cm³			
Bk	0.3	74	3	0.4	0.3	0	1.35	1.76	0.092	0.092	31.7
Bssk1	0.3	75	3	0.4	0.4	0	1.35	1.80	0.101	0.101	32.7
Bssk2	0.3	76	3	0.5	0.4	0	1.35	1.76	0.092	0.092	32.1
Bk1	0.4	61	4.5	0.6	0.5	0	1.45	1.65	0.044	0.044	23.5
Bk2	0.3	67	3	0.6	0.4	0	1.47	1.60	0.029	0.029	24.2

Note Fine silt = 0.02–0.002 mm, total silt = 0.05–0.002 mm, fine clay ≤ 0.0002 mm, total clay ≤ 0.002 mm
VC very coarse (2.0–1.0 mm), *Co* coarse (1.0–0.5 mm), *M* medium (0.5–0.25 mm), *F* fine (0.25–0.10), *VF* very fine (0.10–0.05 mm), *SiCL* silty clay loam, *SiC* silty clay, *C* clay, *OC* organic carbon, *CEC* cation exchange capacity, *BS* base saturation, *ESP* exchangeable Na percentage, *SAR* sodium adsorption ratio, *Cal* calcite, *Dol* dolomite, *CCE* calcium carbonate equivalent, *Gyp* gypsum, *EC* electrical conductivity, *COLE* coefficient of linear extensibility, *WC* water content
* Extractable Ca may contain Ca from calcium carbonate or gypsum. CEC base saturation set to 100

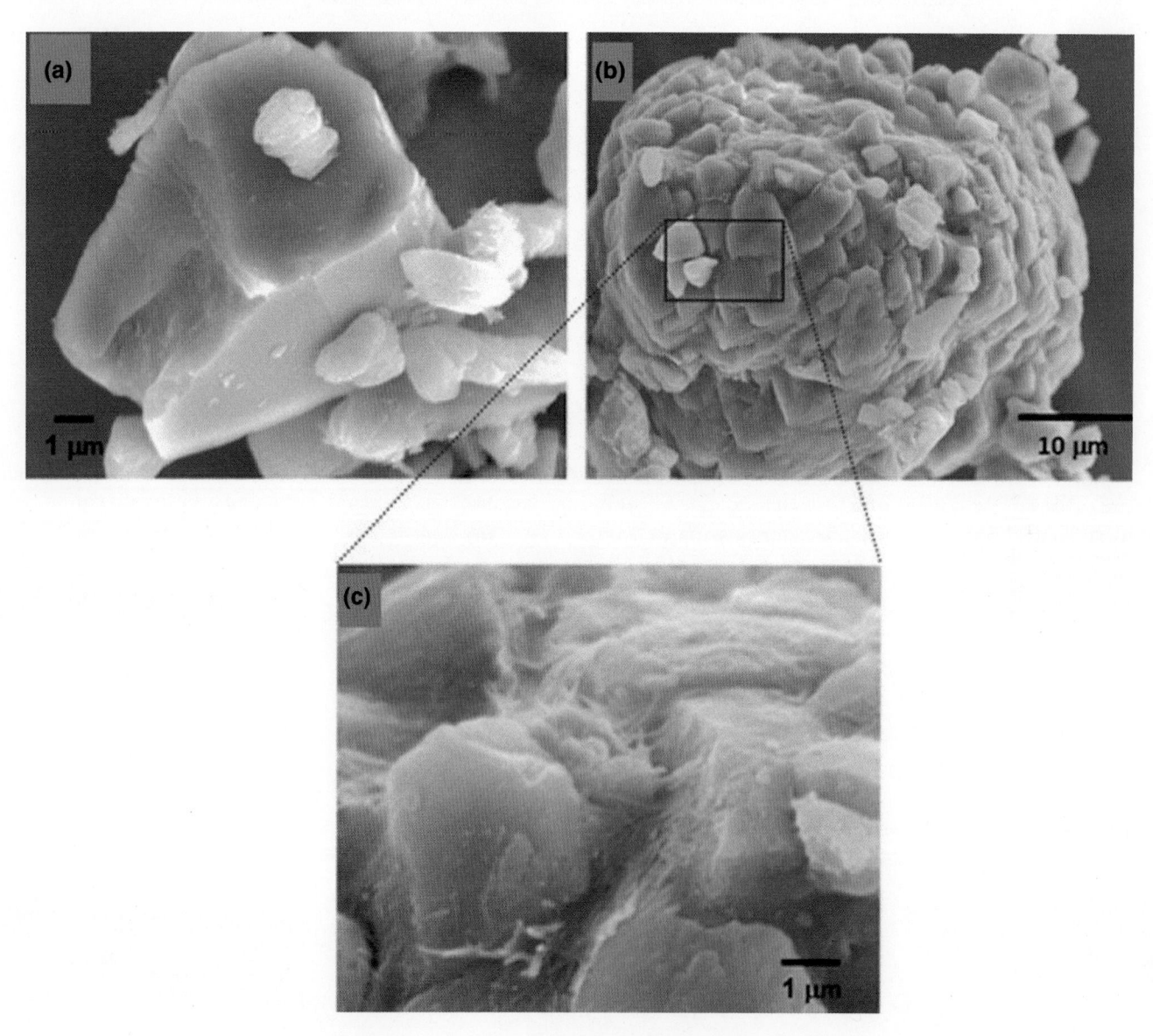

Fig. 19 Scanning electron micrographs of the silt fraction of the By2 horizon of the Cepkenli soil. **a** Chlorite, **b** Cemented grains of calcite, and **c** Greater magnification of (**b**) showing calcite grains and palygorskite fibers (Aydemir 2001)

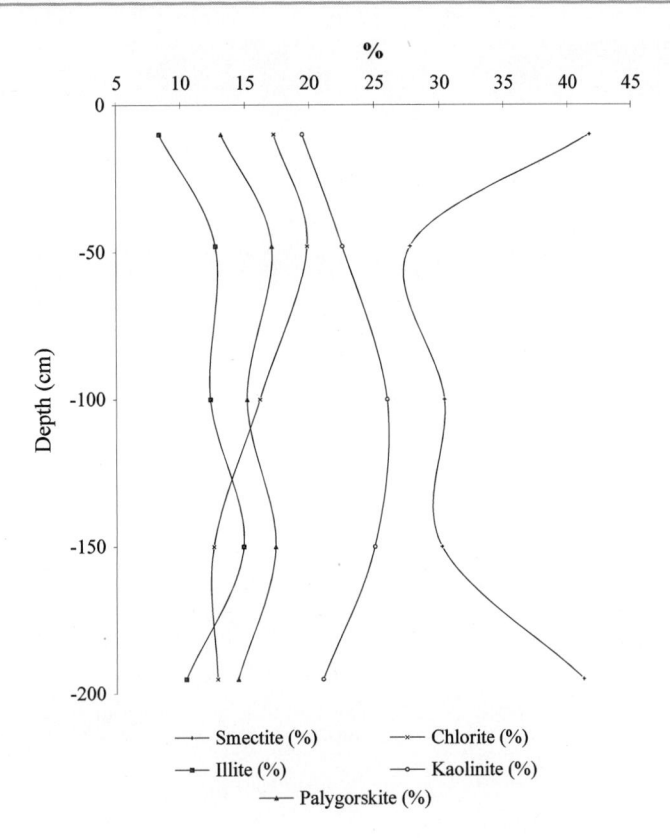

Fig. 20 The clay mineral distribution in the total clay fractions for the Cepkenli soil (Aydemir 2001)

1.36 nm interlayer distance indicating the first order of montmorillonite (Fig. 25c) (Aydemir 2001).

Micromorphology of the Cepkenli and Kısas Soils

The carbonate nodules/ovaliths determined in the matrix of the Ap and Bw2 horizons of the Cepkenli soils were most likely transported from the calcrete-bearing layers of the northern part of the plain (Fig. 26). In contrast, the sparitic channel in-fills in the Bk horizon of the Kısas soil point out to an ongoing process of initial calcretization and integration to a Calcisol (Fig. 27) (Aydemir 2001).

4.4 Vertisols from Bursa and Biga, Northwest Turkey, Çarşamba and Bafra, North Turkey

Özsoy and Aksoy (2007) studied the Vertisols formed on Neogene clay-lime deposits in the Bursa Province (Turkey) with prominent deep cracks of 1–4 cm width and 50–100 cm depth alongside gilgai formation at the surface (Table 7).

Yakupoğlu et al. (2010) studied the morphological characteristics of two Vertisols developed in the Çarşamba (Table 8; Fig. 28) and Bafra plains of the Black Sea region of northern Turkey. They classified the soils of the Çarşamba plain as Chromic Endoaquerts and Typic Haplusterts and Chromic and Eutric Vertisols according to the Soil Taxonomy (Soil Survey Staff 2014) and WRB (2015) respectively.

The Bafra Plain Vertisols were classified as Sodic Haplusterts, Typic Calciaquerts and Sodic Calciusterts (Soil Survey Staff 2014), and as Sodic Vertisols and Calcic Vertisols according to the IUSS Working Group WRB (2015).

The Vertisols in the Çanakkale province cover about 24,458.00 ha and the majority of them are located in Biga

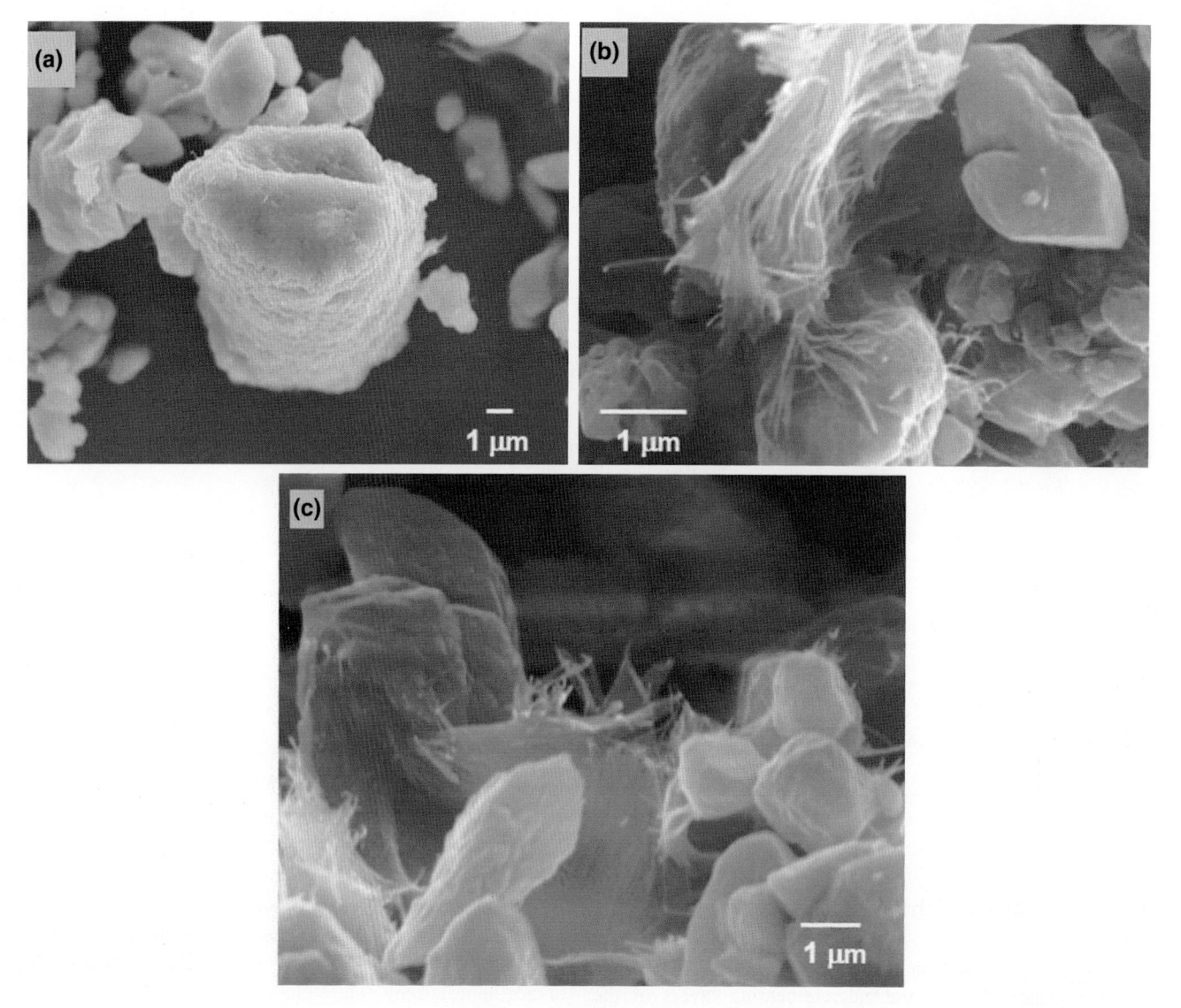

Fig. 21 Scanning electron micrographs of the silt fraction from Bk2 horizon of the Kısas soil. **a** Quartz grain, **b**, **c** palygorskite fibers and calcite grains (Aydemir 2001)

town (15,786.00 ha). Ekinci et al. (2004) studied two Vertisols formed on different physiographic units in Çanakkale (northwest Turkey). The first Vertisol was described in the Karamenderes flood plain (Troy) and the second one was formed on the Neogene old lake deposits in the Biga Plain (Çanakkale, northwest Turkey) (Table 9 and 10; Fig. 29 and 30). A cambic B horizon was determined in the second profile developed on the old Neogene deposits, with a medium-coarse texture and an angular block structure. The first Vertisol was classified as Aquic Haploxererts and the second as Typic Haploxererts.

Günal et al. (2004) studied Gleyic Vertisols (IUSS Working Group WRB 2015) fine, smectitic, active, mesic Ustic Epiaquerts) subjected to anthric saturation in a subhumid region of Turkey (Table 11; Fig. 31). The soils studied were deep, calcareous, clayey, and dark gray to grayish brown meeting the criteria for the Aquert suborder of Vertisols in Soil Taxonomy. Soils studied located on the western

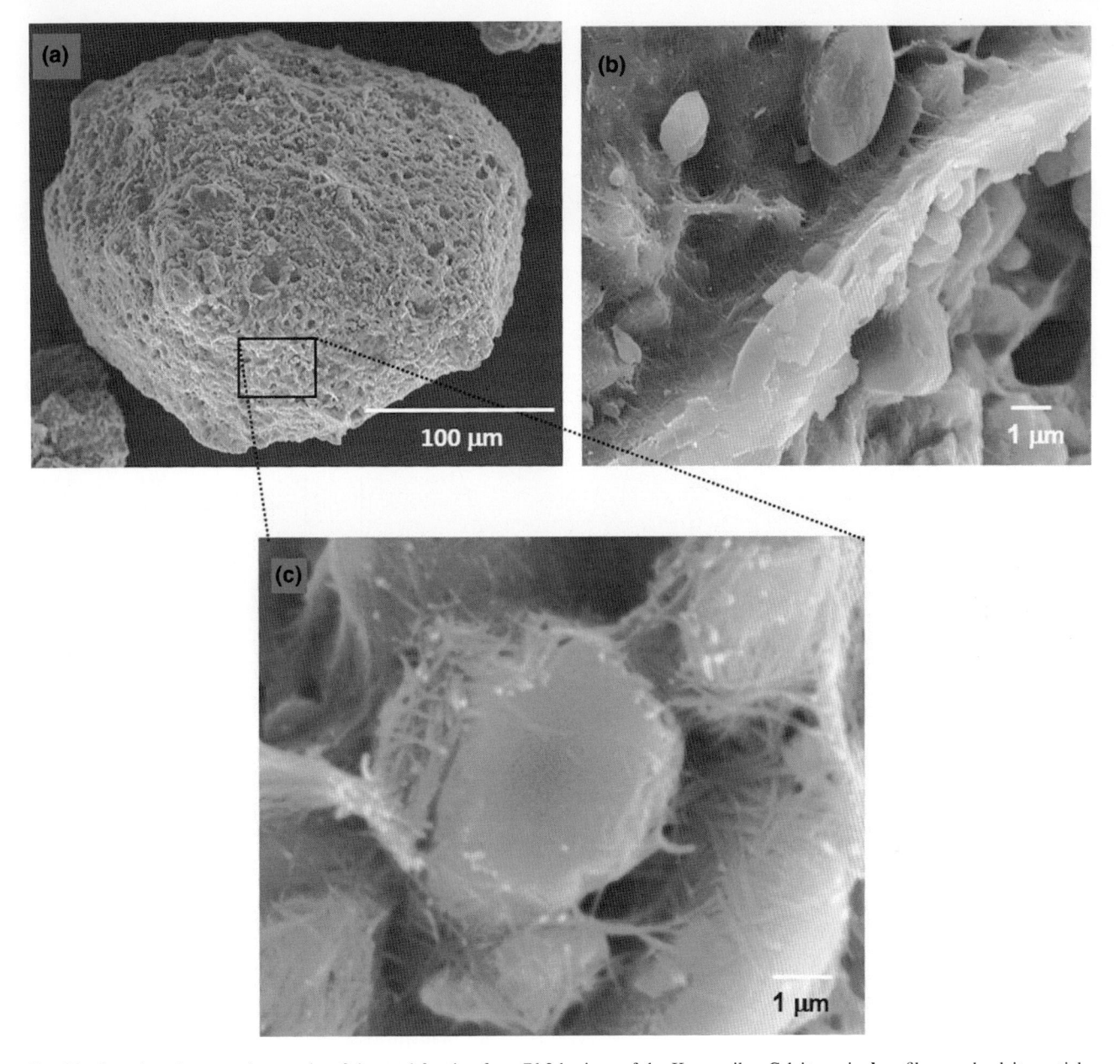

Fig. 22 Scanning electron micrographs of the sand fraction from Bk2 horizon of the Kısas soil. **a** Calcite grain, **b, c** fibers and calcite particles (Aydemir 2001)

edge of the Kızılırmak Delta, formed in a clayey Holocene alluvium underlain by ancient dune deposits consisting of approximately 75% of sand. Pedogenic carbonates were found at the contact zone of the two deposits in all the pedons studied. Depth to slickensides was from 20 to 30 cm and extended to the bottom of the clayey alluvial material. Soils studied had a coefficient of linear extensibility greater than 0.07 cm cm^{-1} in the solum. The cation exchange capacity values of these Vertisols were >20 cmolc kg^{-1} and smectite was the dominant clay mineral. Moderate to minor amounts of mica, chlorite, and kaolinite occurred throughout the profile. The very high clay contents (approximately 60%) of the upper sola and intensive agricultural practices (puddling and flooding) resulted in the saturation of the soils

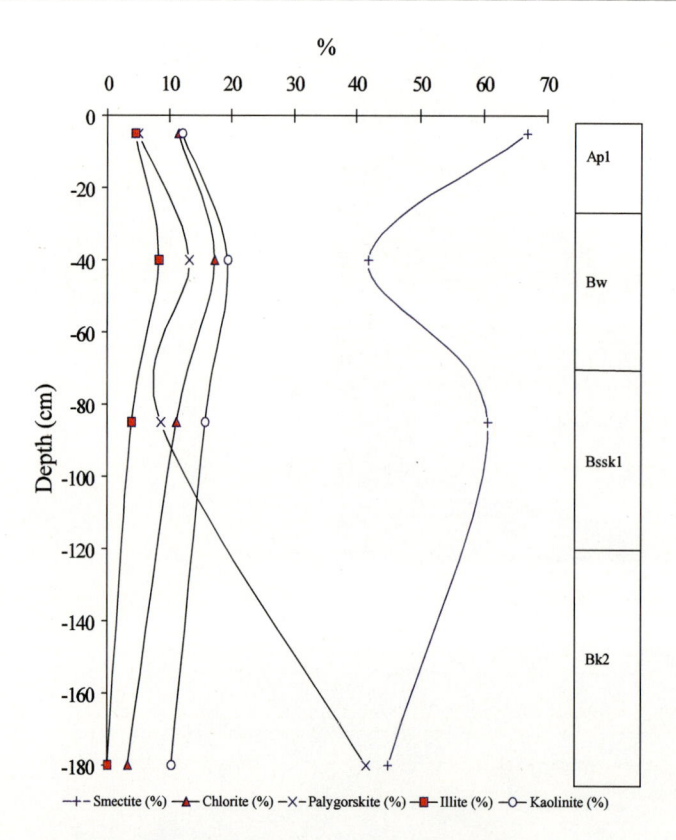

Fig. 23 Clay mineral distribution of the total clay fraction of the Kısas soil (Aydemir 2001)

approximately for five consecutive months which in turn resulted in reduction conditions to predominate the gray color to form in the 40 profiles. Seasonally irrigated flooding controls the redoximorphic features, saturation, and reduction of the rice-growing Vertisols in the study area.

4.5 Vertisols of Adana and Osmaniye, South Turkey

Vertisols of the Adana province in the Seyhan Plain (south Turkey) overlie the sediments of Seyhan and Ceyhan Rivers and are similar to the well-known Black Cotton Soils of India. These soils contain more than 70% clay-size particles where smectite is 80% predominant (Dinç et al. 1991) (Table 12).

The Vertisols of the Osmaniye province situated on the terraces of the Ceyhan River are composed of dark-colored fine soil materials (rich in opaque minerals) transported from the weathered basaltic soils located at the northern part of the Ceyhan plain as stated by Güzel and Kapur (1976) and Cangir (1982). The Vertisols of Osmaniye, the extensive fertile pastures of the past, consist of Ap/Ass1/Ass2/Ass3/C horizon sequences and are cultivated for irrigated crops such as cotton and vegetables today. They are high in total and free iron (Fe_2O_3) and ferrogenious smectite contents followed by illite and kaolinite as stated by Özkan and Ross (1979), and Cangir (1982). In contrast, Güzel and Wilson (1978) determined ferrogenious smectite even in some of the low-iron dark-colored Vertisols of the Ceyhan River terraces of different geologic ages.

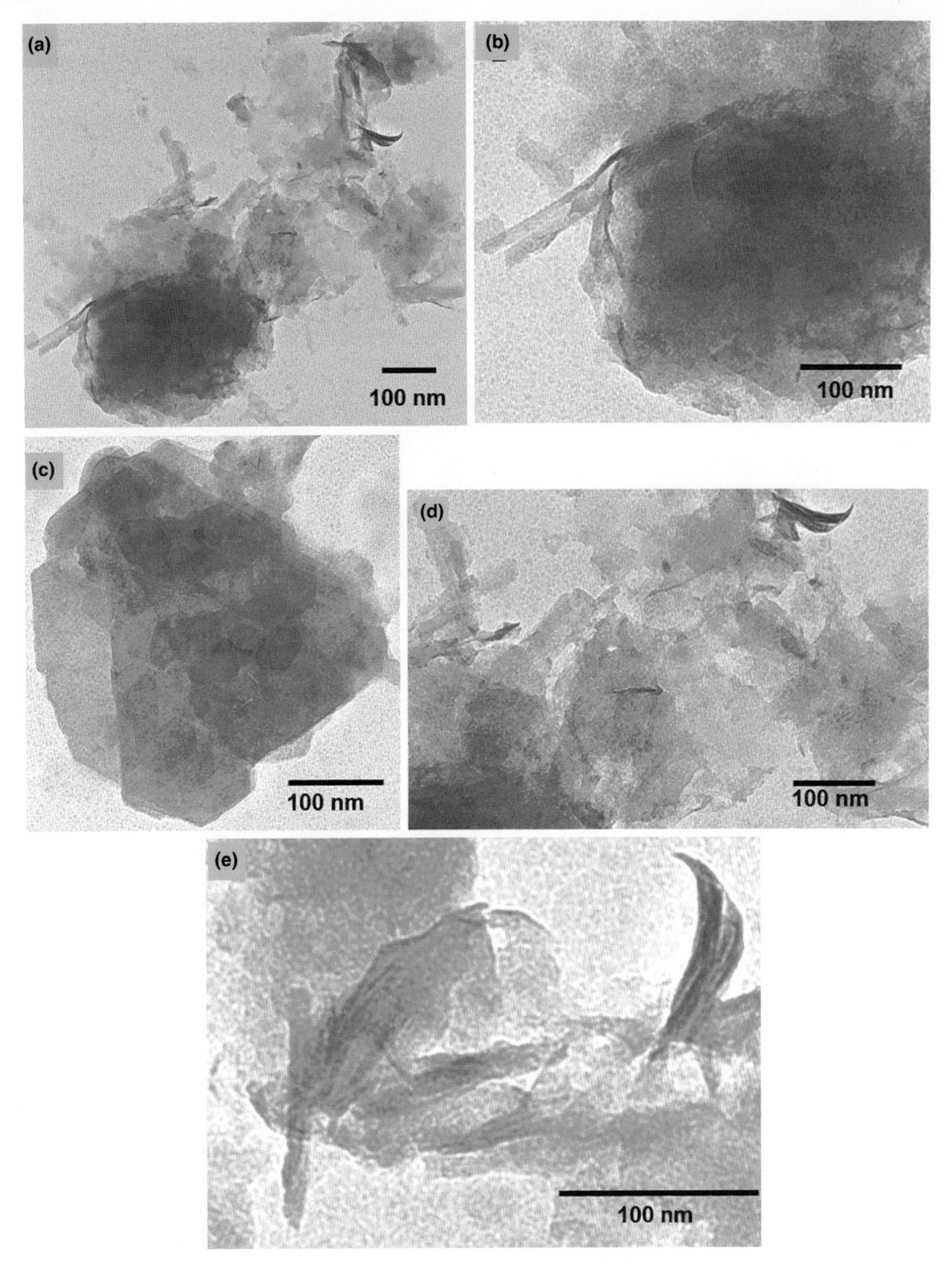

Fig. 24 Transmission electron micrographs of fine clay of the Ap1 horizon of the Kısas soil. **a**, **b**, **d**, **e** Deformed palygorskite fibers and reformed smectite in different magnifications, **c** hexagonal kaolinite particles (Aydemir 2001)

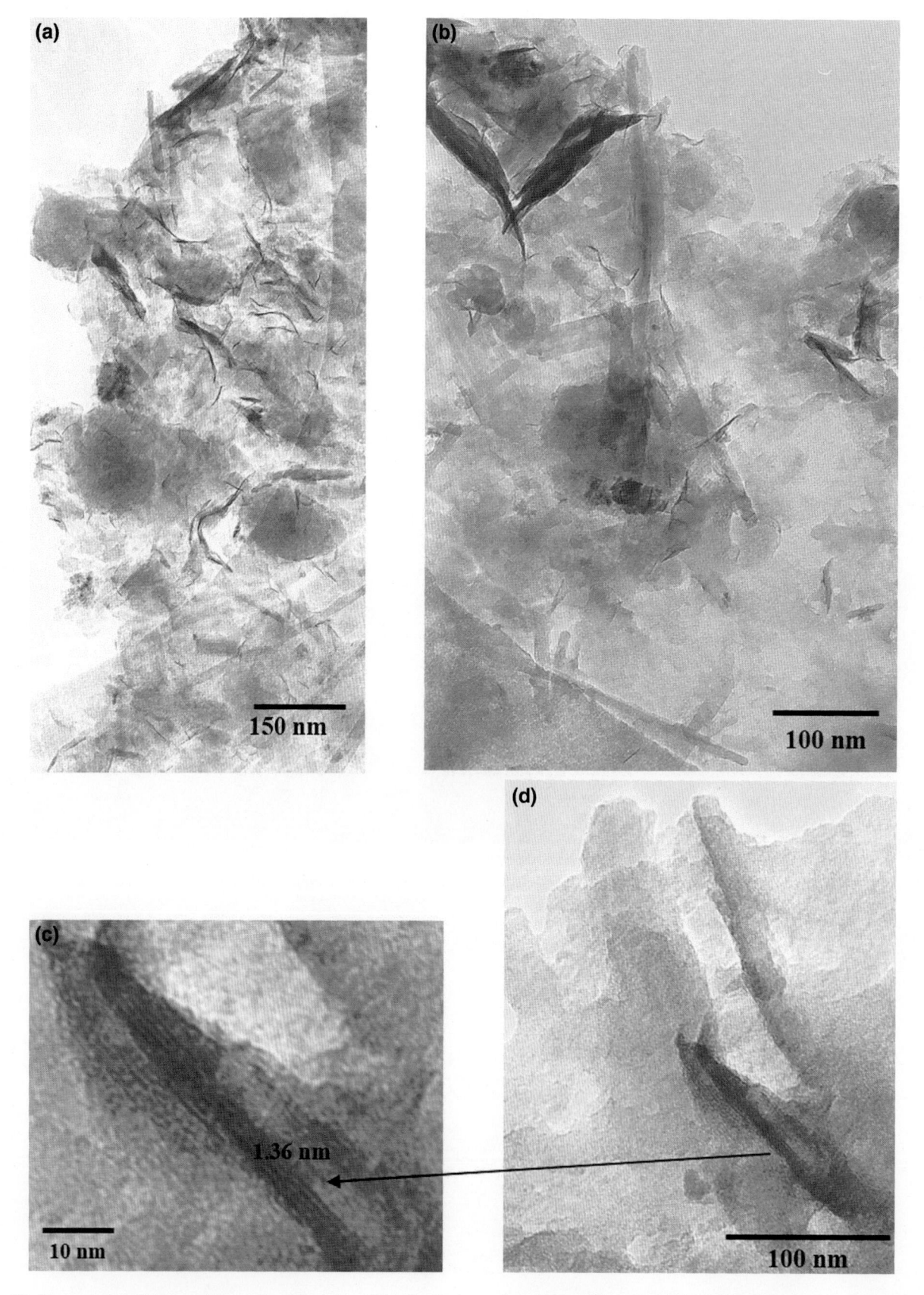

Fig. 25 Transmission electron micrographs of fine clay of the Bw horizon of the Kısas soil. **a**, **b**, **d** palygorskite fibers and smectite, **c** lattice fringes (Aydemir 2001)

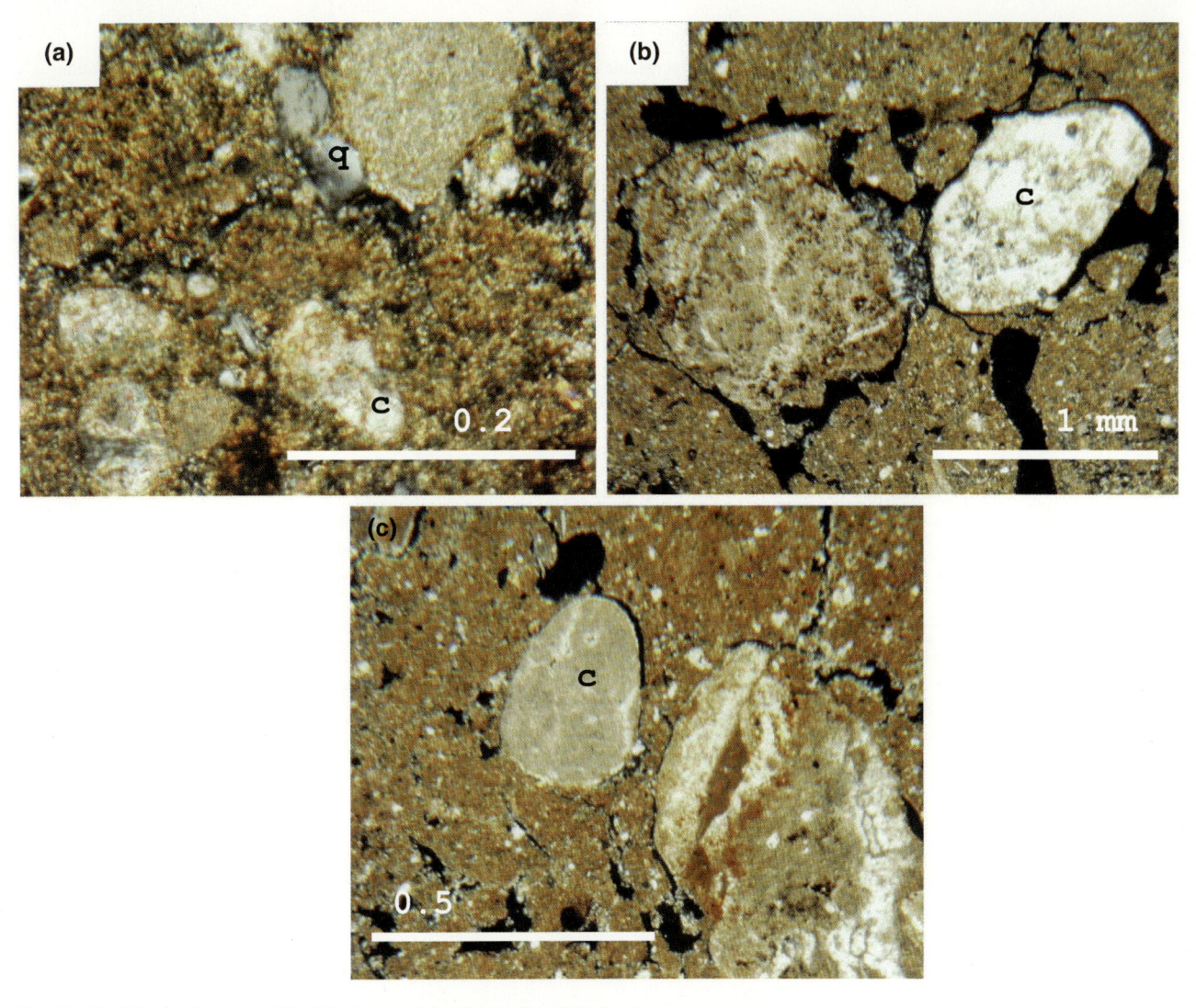

Fig. 26 Ovaliths in the Ap and Bw2 horizons of the Cepkenli soil (**a**, **b**, **c**)

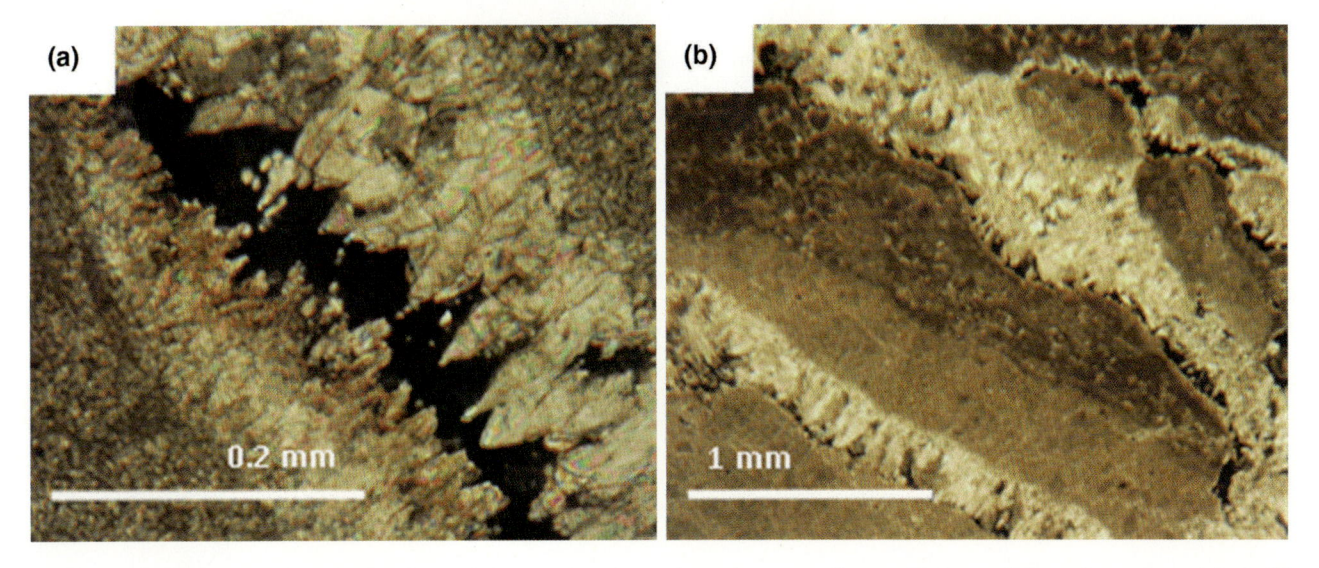

Fig. 27 Sparite formation, **a** in channels of fine micritic matrix and on calcrete fragment with crust (**b**) (*arrow* Calcrete frag and crust) in the Bk horizon of the Kısas soil (Aydemir 2001)

Table 7 Vertisol profile formed on Neogene clay-lime deposits in Bursa, northwest Turkey (Özsoy and Aksoy 2007)

Horizon	Depth (cm)	pH	EC (dS/m)	Exchangeable cations				CaCO$_3$	Particle size distribution		
				Na	K	Ca+Mg	CEC		Sand	Silt	Clay
		1:1		cmol$_c$ kg^{-1}				%			
Ap	0–26	6.65	0.045	0.27	0.97	50.12	51.36	0.39	27.37	15.72	56.91
Ass1	26–56	6.5	0.032	0.26	0.87	50.06	51.19	0.39	22.98	16.04	60.98
Ass2	56–80	7.29	0.041	0.33	0.95	60.86	62.14	3.63	22.75	15.73	61.52
AC	80–114	7.61	0.024	0.3	0.62	43.2	44.12	33.78	16.34	17.87	65.79
C	114+	7.84	0.017	0.29	0.28	23.22	23.79	63.56	31.11	22.87	46.02

Table 8 Some physical and chemical properties of a Vertisol from Çarşamba Plain, north Turkey (Yakupoğlu et al. 2010)

Horizon	Depth	pH	EC	Org. mat.	Exchangeable cations				CaCO$_3$	Particle size distribution		
					Na	K	Ca+Mg	CEC		Sand	Slit	Clay
	cm	1:1	dS/m	%	cmol$_c$ kg^{-1}				%			
Ap	0–20	7.96	0.58	68.7	68.7	0.03	0.02	67.3	15	6.5	38.5	55
Bss1	20–79	8.05	0.55	69.2	69.2	0.06	0.01	66.3	14	8.1	38.7	53.2
Bss2	79 +	7.95	1.49	67.3	67.3	0.17	0.02	65.7	7	11.2	32.2	56.6

Fig. 28 Vertisol in the Çarşamba plain (north Turkey) (Yakupoğlu et al. 2010)

Table 9 Some morphological, physical and chemical properties of Vertisols from Biga, Çanakkale, northwest Turkey (Ekinci et al. 2004)

Horizon	Depth (cm)	Color (moist)	Soil structure	COLE[a]	pH	EC (dSm^{-1})
Ap	0–16	10YR3/3	Fine granular	0.08	7.47	0.76
A1	16–38	10YR3/2	Coarse angular blocky	0.12	7.27	1.01
A2	38–70	10YR3/1	Coarse angular blocky	0.09	8.44	1.41
Bwss	70–106	10YR4/1	Strong coarse prismatic, angular blocky with slickensides	0.15	7.7	1.96
ACss	106–126	10YR3/3	Coarse angular blocky with slickensides	0.1	7.93	1.94
C1	126–145	10YR4/2	Massive	0.12	7.88	1.41
C2	145+	2.5Y3/2	Massive	0.13	774	1.76

Horizon	CEC (cmol$_c$ kg^{-1})	CaCO$_3$	Org. C. (%)	Clay (%)	Sand (%)	Silt (%)
Ap	28.5	0.05	1.35	42	20	38
A1	27.69	0.11	1.17	44	25	31
A2	30.13	0.04	0.81	44	29	27
Bwss	30.69	0.11	0.80	47	30	23
ACss	30.74	0.08	0.69	44	33	23
C$_1$	30.74	0.97	0.44	44	31	25
C$_2$	28.30	1.80	0.38	48	25	27

[a]Coefficient of linear extensibility

Table 10 Atterberg limits and some engineering properties of Vertisols from Biga, Çanakkale northwest Turkey (Ekinci et al. 2004)

Horizon	Clay (%)	COLE	Liquid limit	Plasticity limit	Activity	Cassagrande plasticity classification
Ap	67.0	0.09	72	33	0.58	High plastic inorganic clay
Ad	63.0	0.11	71	29	0.67	
Ass1	67.0	0.16	90	29	0.91	
Ass2	71.0	0.20	86	28	0.82	
C1	68.0	0.17	65	22	0.63	
C2	71.0	0.11	71	24	0.66	

5 Management

Sustainable management and appropriate time for cultivation practices are the critical detriments in the efficient use of Vertisols (Eswaran et al. 1988). The high water-holding capacity assists crops to survive and perhaps even to grow safer during prolonged dry seasons. The use of Vertisols in irrigated agricultural production needs management practices to be designed in order to overcome the physical problems (Eswaran et al. 1988). Vertisols are considered among the most fertile soils when appropriately managed, however, some of their physical properties such as the high clay contents, shrink–swell phenomena, deep cracks, and compactibility are undesirable especially for cultivation and engineering activities.

Although the first rain infiltrates quickly to considerable depths via large cracks (preferential flow), subsequent infiltration and permeability will be quite low due to the high

Fig. 29 Vertisol from Biga, Çanakkale, northwest Turkey (Ekinci et al. 2004)

Fig. 30 Sunflower cultivation on Vertisols in Biga, Çanakkale, northwest Turkey

expanding clay content and poor structure when wet. Drainage may be a problem and crops may become waterlogged. Poor trafficability when wet seriously interferes with the planting operations (Ahmad and Mermut 1996). If crops cannot be established during the rainy season in the Mediterranean or the Black Sea regions of the country, the high intensity of rain on unprotected soil may cause serious surface erosion by runoff even within field scale on slightly undulating Vertisols (Kanber et al. 2012). These soils are almost at saturation point during the wet winter seasons (November–May), and become very dry and desiccated in the summer (June–September) (Kanber et al. 2012). In this context, the Basaltic Vertisols of southeast Turkey require special management practices if irrigated (Fig. 32) for cotton. Consequently, irrigation/cropping of Vertisols requires special care to manage pedoturbation and deep cracks. Grazeland and rainfed cereal management would, therefore, be the best use for most Vertisols in the central and some parts of the southeast of the country due to tillage and irrigation con-

Table 11 Site profile description and some physical and chemical properties of a Vertisol from the Bafra Plain, Kızılırmak Delta, north Turkey (Günal et al. 2004)

Soil Survey Staff (2014)	Fine, smectitic, active, mesic Ustic Epiaquerts
IUSS Working Group WRB (2015)	Gleyic Vertisol
Soil series	Etyemezler
Location	Located between Yakakent and Alaçam towns, at Etyemezler Village
Coordinates	N41° 37′ 10.56″E 35° 34′ 12″
Elevation	20 m
Climate	Precipitation 727 mm, Mean air temperature 12.6 °C
Vegetation	Rice
Parent material	Alluvium
Geomorphic unit	Delta
Topography	Level to very gently undulating slopes, 0–2% slope
Water table	Deeper than 2 m
Drainage	Fair
Infiltration	Low
Stoniness	None
Soil Moisture Regime	Ustic
Soil Temperature Regime	Mesic

Horizon	Depth (cm)	Description
Ap	0–11	Dark grayish yellow (2.5Y 4/2) moist: clay, abrupt smooth Boundary, Massive, very friable, no effervescence: many fine roots, common iron concentrations as masses (10YR 4/6)
Agss	11–36	Grayish Brown (2.5Y 5/2) moist, clay, clear wavy boundary, Massive, very friable, none effervescence, common distinct slickensides, reduced matrix, many iron concentrations (7.5YR 4/4) as masses on the surfaces of cracks
Bgss1	36–66	Gray (5Y 4/1) moist, clay, gradual wavy boundary, wedge-shaped structure, friable, none effervescence, many prominent slickensides, reduced matrix, common iron concentrations as masses weak to moderate sizes
Bgss2	66–95	Gray (5Y 5/1) moist, Clay, gradual wavy boundary, wedge-shaped structure, friable, none effervescent, many prominent slickensides, reduced matrix, common iron concentrations as masses
BCgss	95–113	Grayish brown (5Y 5/2) moist, clay, clear wavy boundary, wedge-shaped structure, friable, slight effervescent, common distinct slickensides, reduced matrix, common iron concentrations as masses
2C	113–149	Light olive brown (2.5Y 5/3) moist, sandy clay loam, gradual wavy boundary, single grain, very friable, medium effervescent
2Ck1	149–180	Grayish brown (2.5Y 5/2) moist, sandy clay, gradual wavy boundary, single grain, very friable, strong effervescent
2Ck2	180–220+	Light yellowish brown (2.5Y 6/4) moist, sandy loam, single grain, loose, strong effervescent

Horizon	Depth (cm)	pH 1:1	EC (dS/m)	Org. C. (g/kg)	CEC*	Na	K	Ca	Mg	CaCO$_3$	Sand	Slit	Clay
					cmol$_c$ kg^{-1}					%			
Ap	0–11	7.8	0.97	0.78	22.38	2.23	1.18	16.2	2.77	11.8	18.9	18.5	62.6
Agss	11–36	7.7	1.28	0.73	39.05	1.85	0.96	31.65	2.85	12.5	18.4	19.1	62.5
Bgss1	36–66	7.7	0.83	0.52	20.24	1.9	0.98	30.96	3.07	11	18.5	16.6	64.9
Bgss2	66–95	7.8	1.34	0.44	25.08	1.69	0.69	29.76	2.65	6.5	18.4	14.1	67.5
BCgss	95–113	8	0.49	0.32	15.94	1.3	0.3	22.78	1.06	2	28.5	11.6	59.9
2C	113–149	8.1	0.43	0.18	6.09	0.76	0.22	23.3	0.11	7	43.5	11.6	44.9
2Ck1	149–180	8.2	0.41	0.18	6.74	0.98	0.29	17.85	0.07	18	68.5	4.1	27.4
2Ck2	180–220	8.2	0.5	0.15	7.38	0.73	0.14	25.55	0.11	21.5	51	11.6	37.4

* Extractable Ca may contain Ca from calcium carbonate or gypsum. CEC base saturation set to 100

Fig. 31 Land use of the Gleyic Vertisol in Alaçam, Samsun, north Turkey

Table 12 Some physical and chemical properties of the Gemisüre soil in the Çukurova Delta, Adana, south Turkey (Dinç et al. 1991)

IUSS Working Group WRB (2015)	Chromic Vertisols
Soil Survey Staff (2014)	Typic Chromoxerert
Soil series	Gemisüre
Location	2.5 km north of Gemisüre Village, 17 km southeast of Adana, south Turkey
Coordinates	36° 49′ 30″N 35° 27′ 22″E
Elevation	10 m
Climate	Semiarid, transitional Mediterranean
Vegetation	Cotton, maize
Parent material	Clay deposit
Geomorphic unit	Alluvial
Topography	Flat
Water table	180 cm
Drainage	Moderate
Infiltration	Moderate
Stoniness	No stones
Soil Moisture Regime	Xeric
Soil Temperature Regime	Mesic

Table 12 (continued)

Horizon	Depth (cm)	pH	EC	Org. mat. (%)	Exchangeable cations				CaCO$_3$	Particle size distribution		
					CEC	Na	K	Ca + Mg		Sand	Slit	Clay
		1:1	dS/m		cmol$_c$kg^{-1}				%			
Ap	0–21	7.6	0.12	1.53	36.4	2.5	2	31.9	23.5	2	26	72
Ad	21–36	7.8	0.093	1.47	37	2	0.6	34.4	21.8	3	24	73
Ass	36–78	7.6	0.178	1.34	37	3.9	0.6	32.5	23	3	22	75
C	78–120	7.6	0.33	1.07	39	8.3	0.7	30	22.4	4	19	77

Fig. 32 Cotton cultivation on Vertisols (Harran soils), Harran, Şanlıurfa, southeast Turkey

straints. Whereas, cotton cultivation for the Vertisols in the south and some Vertisols in the southeast of the country would be appropriate if coupled with suitable tillage and irrigation practices (Surge irrigation, Kanber et al. 2012).

Ultimately, as stated by Kapur et al. (2002), the sustainable management of the Vertisols as in the southeast Anatolian Irrigation Project area calls for a paradigm shift that would enable research outcomes to lead the implementations on land planning and allocation. The sustainable land management must be holistic and scientifically systems based, to include agronomic, crop and livestock based observations, but also monitoring of the linkages of these to the soil ecosystem (in this context the Vertisol, Calcisol, Cambisol and Luvisol ecosystems) and to the socioeconomic conditions of the area. Moreover, it should show change and specifically monitor how the resource base is maintained or enhanced (Eswaran et al. 2011).

Acknowledgements Prof. Salih Aydemir extends his sincere gratitude to Profs. JB Dixon (Texas A&M University, USA) and CT Hallmark (Texas A&M University, USA) for their invaluable efforts in the SEM laboratory and interpreting the images of Cepkenli and Kısas soils.

References

Akalan İ (1968) Comparison of the clay minerals and silt fractions of the Harran Reddish Brown Soil and the Mediterranean Grumusolic Lithosol. Ankara University Pub., Ankara University Press, Ankara, pp 4–22

Ahmad N, Mermut A (1996) Vertisols and technologies for their management. Dev Soil Sci 24:89–114 (Elsevier, Amsterdam)

Ahmad N (1983) Vertisols. In: Wilding LP, Smeck NE, Hall GF (eds) Pedogenesis and soil taxonomy (II. The Soil Orders). Elsevier, Netherlands

Arocena JM, Siddique T, Thring RW, Kapur S (2007) Investigation of lichens using molecular techniques and associated mineral accumulation on a basaltic flow in a Mediterranean environment. Catena 70:356–365

Atalay İ (1983) The geomorphology and soil geography of the muş plain and its surroundings. Ege University, Faculty of Letters, No: 24, İzmir (in Turkish)

Aydemir S (2001) Properties of palygorskite-influenced vertisols and vertic-like soils in the Harran Plain of Southeastern Turkey. Ph.D. Dissertation, Texas A&M University, College Station, TX, USA

Cangir C (1982) The morphology and genesis of Brown, Reddish Brown, Terra Rossa, Rendzina and Grumusolic soils formed on calcareous materials. Habilitation Thesis, the Faculty of Agriculture, University of Ankara, 135 p (in Turkish)

Çakmaklı M (2008). Genesis of the Harran Plain Soils (Relations on geology and pedology) Unpublished Ph.D. Thesis, Harran University Department of Soil Science and Plant Nutrition, Şanlıurfa (in Turkish)

Dengiz O, Sağlam M, Sarıoğlu FE, Saygın F, Atasoy Ç (2012) Morphological and physico-chemical characteristics and classification of Vertisol developed on Deltaic Plain. Open J Soil Sci 2:20–27. doi:10.4236/ojss.2012.21004. Published online March 2012 (http://www.SciRP.org/journal/ojss)

Dinç U, Senol S, Sayın M, Kapur S, Yılmaz K, Sarı M, Kara E (1991) Soils of the Harran Plain. TÜBİTAK Project No. 534, pp 1–10

Dinç U, Şenol S, Kapur S, Cangir C, Atalay İ (2001) Soils of Turkey, Çukurova University, Faculty of Agriculture, Pub. No: 51, Adana (in Turkish)

Driessen PM, Dudal R (1989) Geography formation, properties and use of the major soils of the world. Lecturer notes. Agric. University Wageningen and Catholic University Leuven, Leuven

Dudal R (1965) Dark clay soils of tropical and subtropical regions. Agric Dev Paper 83, FAO, Rome, Italy. 161 p

Ekinci H (1990) The applicability of soil taxonomy to the Turkish general soil map in Thrace, a case study. Çukurova University, Ph. D. Thesis, 218 P (in Turkish)

Ekinci H, Özcan H, Yiğini Y, Çavuşgil VY, Yüksel O, Kavdir Y (2004) Profile developments and some properties of vertisols formed on different physiographic units. International soil congress on natural resource management for sustainable development. Abstract book, June 7–10, Erzurum, Turkey, pp 71–76

Ergene A (1977) Physical, chemical and mineralogical properties of the major soil groups and their profile developments in Urfa, Gaziantep and Hatay provinces. Atatürk University Pub. No. 32, Atatürk University Press (in Turkish)

Eswaran H, Kimble J, Cook T (1988) In: Classification, management and use potential of shrink swell soils. Trans. of the Int. Workshop on the shrink swell soils. Oct 24–28, 1988, Nat. Bureau of Soil Survey and Land Use Planning, Nagpur, India, pp 1–22

Eswaran H, Berberoğlu S, Cangir C, Boyraz D, Zucca C, Özevren E, Yazıcı E, Zdruli P, Dingil M, Dönmez E, Akça E, Çelik İ, Watanabe T, Koca YK, Montanarella L, Cherlet M, Kapur S (2011) The anthroscape approach in sustainable land use. Sustainable land management, learning from the past for the future. Springer, Berlin, pp 1–50

Günal, H, Durak A, Akbaş F, Kılıç S (2004) Genesis and classification of vertisols as affected by rice cultivation. International soil congress. Natural resource management for sustainable development, June 7–10, Erzurum, Turkey

Güzel N, Kapur S (1976) Soil-geomorphology relationships on the terraces of the Ceyhan River. Annals of the Faculty of Agriculture, University of Çukurova, vol 2, pp 117–143

Güzel N, Wilson J (1978) Clay mineral studies of a Vertisol chronosequence in southern Turkey, p 405. Proceedings of the international clay conference, Oxford, England

Hocaoğlu ÖL (1973) Clay minerals of soils formed on basalts in Diyarbakır, Erzurum and Rize provinces. Atatürk University Pub. No. 33 (in Turkish)

İnce F, Kapur S (1985) Microstructure formation in vertic soils formed on basaltic parent rocks of south-eastern Turkey. Turk J Sci (Doğa) B-9(1):52–56

IUSS Working Group WRB (2015) World reference base for soil resources 2014, update 2015 international soil classification system for naming soils and creating legends for soil maps. World Soil Resources Reports No. 106. FAO, Rome, 203 p

Kanber R, Önder S, Ünlü M, Tekin S, Sezen SM, Diker K (2012) Different furrow management techniques for cotton production and water conservation in Harran Plain, Şanlıurfa. Turk J Agric For 36 (1):77–94

Kapur S (1981) Formation and classification of soils formed on the basaltic rocks of Şanlıurfa, Karacadağ, South East Turkey. Habilitation Thesis. University of Çukurova, Faculty of Agriculture, Adana, Turkey

Kapur S, Dinç U, Cangir C (1991) Inherited smectitic nodules in basaltic soils. Turk J Eng Environ Sci 15:259–264

Kapur S, Karaman C, Akça E, Aydın M, Dinç U, FitzPatrick EA, Pagliai M, Kalmar D, Mermut AR (1997) Similarities and differences of the spheroidal microstructure in Vertisols from Turkey and Israel. Catena 28(3–4):297–311

Kapur S, Eswaran H, Akça E, Dingil M (2002) Developing a sustainable land management research strategy for the southeast Anatolian Irrigation Project. In: P Zdruli, P Steduto, S Kapur (eds) 7th international meeting on soils with mediterranean type of climate (selected papers) Options Mediterraneennes, Seria A: Mediterranean Seminars No: 50, pp 326–334

Mermut AR, Acton DF, Tarnocai C (1990) A review of recent research on swelling clay soils in Canada. In: JM Kimble (ed) Proceeding of the sixth international soil correlation meeting. USDA-Soil Conservation Service, National Soil Survey Center, Lincoln, NE, pp 112–121

Mermut AR, Montanarella L, FitzPatrick EA, Eswaran H, Wilson M, Akça E, Serdem M, Kapur B, Öztürk A, Çullu MA, Kapur S (2004) 12th international meeting on soil micromorphology excursion book, 20–26 Sept 2004. EUR 21275 EN/1 (2004), Ispra Milano, 61 p

Mermut AR, Çullu, MA, Aydemir S, Karakaş S (2006) Excursion Book. 18th international soil meeting (ISM), soils sustaining life on earth (Managing soil and technology) 22–26 May 2006. ISBN: 975-96629-4-9

Munsuz N (1974) Clay minerals of some Great Soil Groups in Turkey determined by IR. Ankara University Pub. No. 523, Ankara University Press, 57 P (in Turkish)

Özkan I, Ross GJ (1979) Ferrogenious beidellites in Turkish soils. Soil Sci Am J 43(6):1242–1248

Özsoy G, Aksoy E (2007) Characterization, classification and agricultural usage of Vertisols developed on Neogene aged calcareous marl parent materials. J Biol Environ Sci 1(1):5–10

Soil Survey Staff (2014) Keys to soil taxonomy, 12th ed USDA-natural resources conservation service, Washington, DC. 372 p

SSSA (2008) Glossary of soil science terms. SSSA, WI

Stoops G (2003) Guidelines for analysis and description of soil and regolith thin sections. Soil Science Society of America Inc. Madison Wisconsin, 184 p

Tanju Ö (1981) A Research on the Bafra Province Grumusols. Ankara University, Faculty of Agriculture Pub. No: 773, 445, Ankara

Yakupoğlu T, Sarıoğlu FE, Dengiz O (2010) Morphology, physico-chemical characteristics and classification of two vertisol in Bafra and Çarşamba Delta Plains. Anadolu J Agric Sci 25 (S-1):67–73

Alisols-Acrisols

Hasan Özcan, Orhan Dengiz, and Sabit Erşahin

1 Introduction

The reference soil group of the Alisols encompasses strongly acid soils that have accumulation of high activity clays (less-weathered, high-effective surface area comprising smectite, vermiculite, mica/illite, chlorite) in the subsoil. They occur in humid and warm temperate regions, on parent materials that contain a substantial amount of unstable Al-bearing minerals. Ongoing hydrolysis of these minerals produces aluminum, which occupies more than half of the cation exchange sites. Hence, Alisols are unproductive soils under all but acid-tolerant crops. Internationally, Alisols refer to the earlier "Red-Yellow Podzolic Soils" (Oakes 1954) of Turkey and Indonesia, to Ultisols of Brazil with high-activity clays (Soil Survey Staff 2014), and to the "Ferialsols" and "sols fersiallitiques très lessivés" of France (IUSS Working Group WRB 2015).

The reference soil group of the Acrisols of the IUSS Working Group WRB (2015) holds soils that are characterized by accumulation of the low-activity clays in an "argic" subsurface horizon and a low-base saturation level. Acrisols also refer to the "Red-Yellow Podzolic soils" of Turkey (Oakes 1954) and Indonesia, "Podzolicos vermelho-amarello distroficos a argila de atividade baixa" of Brazil, the "Sols ferralitiques fortement ou moyennement désaturés" (France) and the Red and Yellow Earths" and several subgroups of Alfisols and Ultisols of the Soil Taxonomy (Soil Survey Staff 2014).

H. Özcan (✉)
Department of Soil Science and Plant Nutrition, Çanakkale Onsekiz Mart University, Çanakkale, Turkey
e-mail: hozcan@comu.edu.tr

O. Dengiz
Department of Soil Science and Plant Nutrition, Ondokuz Mayıs University, Samsun, Turkey

S. Erşahin
Department of Forestry Engineering, Karatekin University, Çankırı, Turkey

Alisols-Acrisols and Podzols are found at isolated localities as patches in the northeastern Black Sea region of Turkey. Conditions for podsolization are met in the far eastern Black Sea region of the country. However, the irregular jagged topography often prevents a complete podsolization due to water runoffs from the surface and/or to subsurface lateral flow along the slopes. Subsequently, water erosion and the lack of the plant cover on highly steeping slopes are the complementary reasons for incomplete podsolization. The widely distributed Yellowish Red Podzolic soils of the eastern Black Sea region designated by the 1938 classification system (Baldwin et al. 1938), meet the criteria of Alisols-Acrisols in the internationally accepted WRB soil classification system (IUSS Working Group WRB 2015).

2 Parent Materials and Environment

The northeastern Black Sea region, geographically and geologically defined as the "East Pontides" (Fig. 1) shelters the Alisols-Acrisols which are principally formed on magmatic rocks and especially sand stones (Özsayar et al. 1981; Eyüpoğlu et al. 2006).

The Pontide mountain ranges of the Black Sea region composed of sloping lands and abundant faulting and undulating topography has a humid temperate climate with high (2500–3000 mm annually, the highest rainfall in the country) and evenly distributed rainfall around the year. Snowfall is common between December and March and continues for a week or two. The Black Sea region is the most highly populated area with enhanced agricultural production related to the well-established and long-standing appropriate crop-soil ecosystems of the country (Table 1).

The eastern part of the Black Sea region consists of the highest mountain ranges encircling Turkey from the north. The mountain ranges are approximately 2000 m high in Samsun (middle Black Sea) and rise gradually towards the north and reach a maximum elevation at Kaçkar Mountains

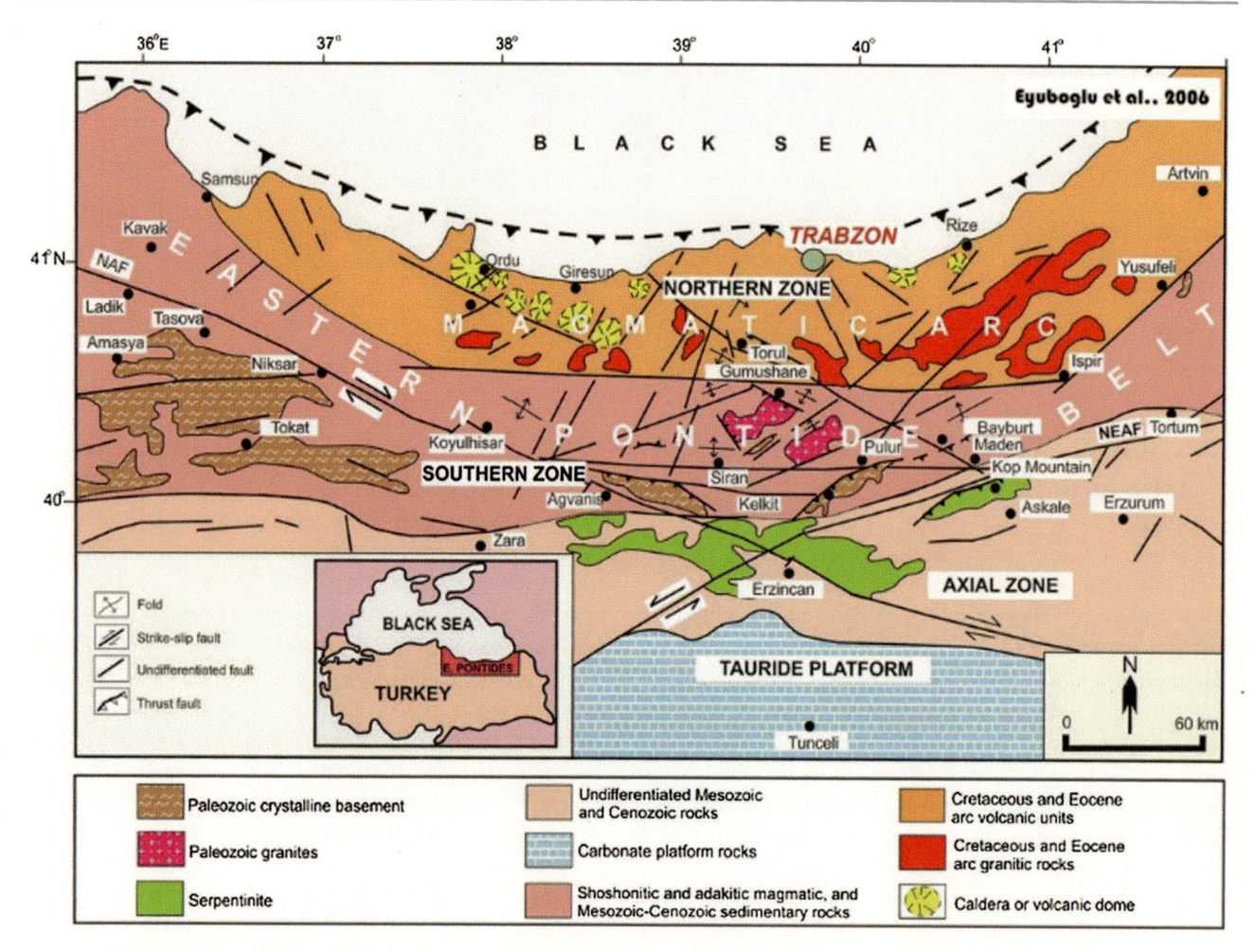

Fig. 1 The geology of the east Pontides (Eyüpoğlu et al. 2006)

Table 1 Some climatic data of the northeastern Black Sea region (1954–2013)

Province	Ordu	Giresun	Trabzon	Rize	Artvin
Mean annual temperature (°C)	14.27	14.52	14.77	14.23	12.15
Mean annual precipitation (mm)	1035.7	1263.5	798.8	2240.7	698.7

in the far northeastern part. The peaks of the northeastern Black Sea mountains (3000 m) can be reached within 30 km from the shore. Combined with the cascading streams from the mountains, the steep irregular slopes are responsible for the formation of a topography unique to this region. On these topographies, as already pointed out, the process of podsolization remains incomplete leading to the formation of Alisols-Acrisols in tea plantation areas and in most of the forested areas. However, Podzols can be found in some isolated highlands covered with spruce fir and where topography scarcely allows a complete podsolization in patches.

One-fourth of the Turkish forests are in the east and mid-Black Sea region. The tea production is the common agricultural activity at localities close to the shore and with slopes proper for tea plantation. Grasslands and forests are the common land use types at other locations, which are not suitable for tea cultivation. The forest understory is generally rich on the sea-faced (north-faced) slopes while the plant cover is sparser on the south-faced slopes.

3 Profile Development

Alisol-Acrisols in forested areas usually have O-A-E-Bt-C, and in the tea plantation areas A-E-Bt-C or A-Bt-C horizon sequences. Tillage in the tea-growing areas has caused the mixing of the O-A-E or A-E horizons and the formation of a thick A-horizon. The overall texture of the Alisol-Acrisol profiles is generally loamy due to their coarse-textured parent materials. There is evidence of fine material migration followed by deposition (illuviation) of clay-sized particles (Argic horizon) from the upper horizons to the lower due to the high precipitation in both Alisols and Acrisols (Figs. 2 and 3).

The rainwater, acidified by the acidic cations released from the decomposed organic matter, leaches the basic cations (Ca, Mg, Na, K) from the soil particle surfaces. Many studies (Müftüoğlu 1989; Akgül 1975; Aydın and Sezen 1990; Özyazıcı et al. 2013) revealed a base saturation of <50% in these soils due to the leaching of the basic cations following their replacement by acidic cations. Moreover, evidences of migration of even silt-sized materials/grains/aggregate alongside the clay to subsurface horizons were also determined. The primary criteria of designating these soils are the magnitude of their base saturations and CEC of their Bt horizons (Table 2).

Alisols and Acrisols are acidic soils with lower pH in their surface horizons compared to the deeper ones. However, in some cases, their pH varies in the profile depending on the organic matter content. The same is true for their CEC and base saturation. Many factors contribute to the

Fig. 3 Alisol-Acrisol in Hayrat-Trabzon, northeast Turkey

acidification in these soils. Extensive leaching under coniferous tree canopies has caused a complete podsolization in Karagöl-Şavşat, Artvin (Akgül 1975). Soils, formed on granite are/were subjected to a considerable leaching, due to their texture and high hydraulic conductivity and were ultimately acidified. A Spodic (albic-Soil Taxonomy) horizon is occasionally observed even in the areas with contemporary inadequate/partly adequate leaching.

The Argic Bt horizons of the Acrisols-Alisols are foreseen to occur with a base saturation of <50% where the CEC is >24 cmol$_c$/kg^{-1}. Akgül (1975) reported CECs ranging from 1.92 to 58.59 cmol$_c$/kg^{-1} and H+ ranging from 4.00 to 49.4 cmol$_c$/kg^{-1} in coniferous forest soils. Similarly, Müftüoğlu (1989) reported that the base saturation ranged from 16.2 to 46.1 in the tea-cultivated Acrisols-Alisols of the northeastern Black Sea region. Aydın and Sezen (1990) reported CECs ranging from 15.19 to 32.4 cmol$_c$/kg^{-1}, exchangeable Al + H from 2.2 to 13.2, organic matter from 1.92 to 6.67%, and pH from 3.6 to 5.1 (Fig. 4) in similar areas.

Çelik et al. (1999) stated that kaolinite and some illite (sericite) formation was responsible for the low pH of the

Fig. 2 Alisol at İnciköyü Village in Rize under tea (*Camellia siensis*) cultivation northeast Turkey

Table 2 Site description of an Acrisol in Hayrat, Trabzon, northeast Turkey

IUSS Working Group WRB (2015)	Umbric Acrisols Albic Clayic
Soil Survey Staff (2014)	Ochreptic Hapludults
Location	Hayrat, Trabzon, east Black Sea region (northeast Turkey)
Coordinates	37 T, 608149 E, 4 534 301 N
Altitude	25 m
Landforms	Hillside
Position	Slope
Aspect	–
Slope	16–20%
Groundwater	–
Effective soil depth	120+
Parent material	Sandstone
Köppen climate	Temperate rainy/humid temperate, no dry season
Land use	Tea cultivation
Vegetation	Tea (Camellia sinensis) and some forest trees

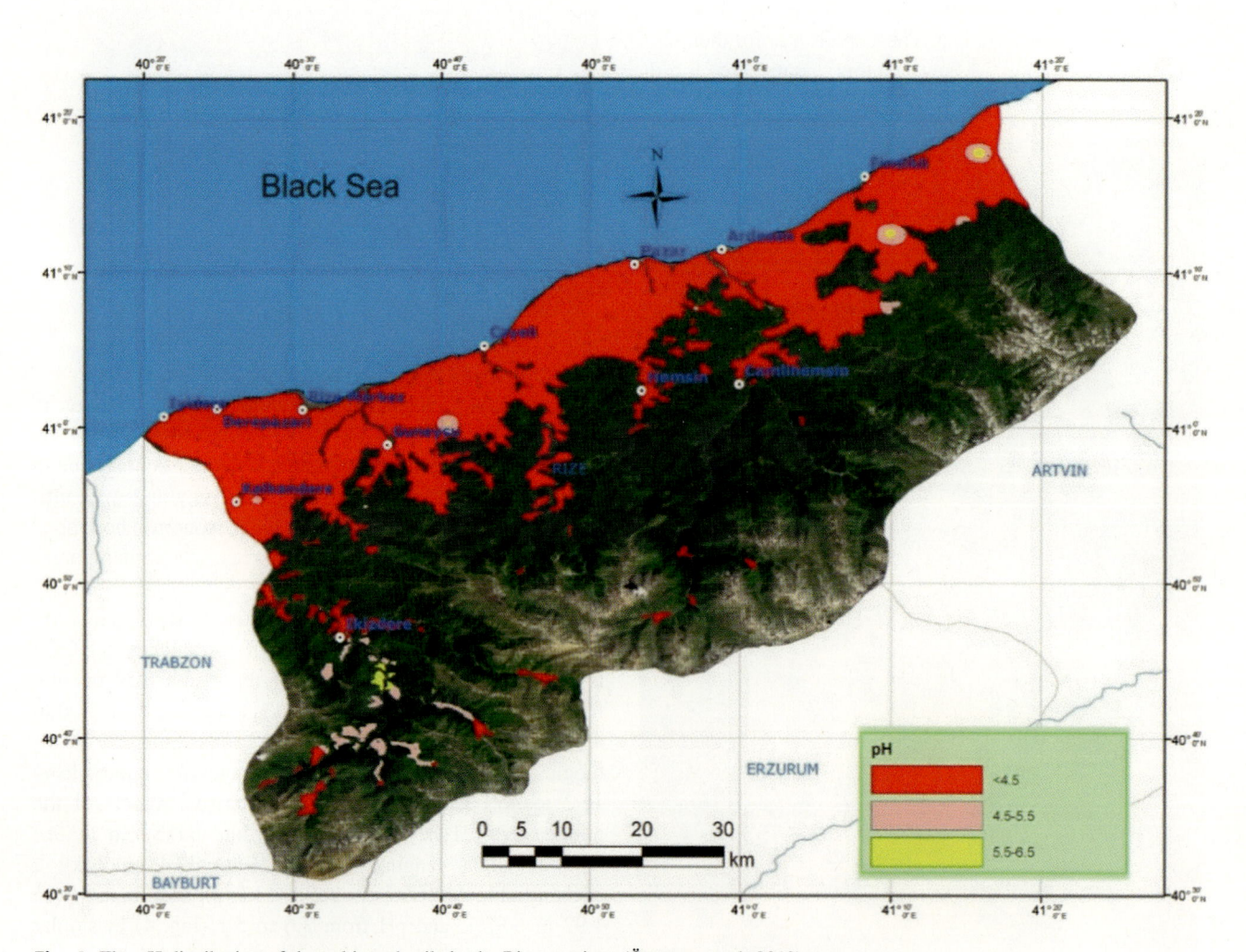

Fig. 4 The pH distribution of the cultivated soils in the Rize province (Özyazıcı et al. 2013)

soils of the region developed by hydrothermal alteration along the extensively distributed volcanic belt of the Pontides. Nonetheless, Mermut (1983) and Cangir et al. (1991) have determined smectite-rich soils with some kaolinite in moderately acidic Alisol-Acrisol (Ultisol-Krasnozem) profiles along the eastern Black Sea coast. This may indicate a contemporary long-enduring and ongoing weathering/soil development that may be insufficient in attaining a low pH and reaching the ultimate clay mineral (kaolinite) in the weathering sequence.

4 Distribution

Alisols-Acrisols together with patches of Podzols are located at mountain heights and north-facing slopes of the Black Sea region and cover a land area of 2.498.952.52 ha which is 3.25% of the whole agricultural land. These soils are derived from magmatic and sedimentary rocks, in the northeast and central parts of northern Turkey (Fig. 5). Acrisols are the major soils in Rize, Artvin, Trabzon, Giresun, and Ordu in decreasing order. Some representative profiles and their descriptions are given in Tables 3, 4 and Figs. 6, 7.

5 Management

Low pH, poor fertility, and toxic contents of Al in the root zone are the added constraints of the Acrisols-Alisols. However, fortunately, the cultivation of tea and hazelnuts are proven to be highly appropriate on Acrisols-Alisols since decades. The tea production is confined to a narrow coastal strip (180 km long and approximately 35 km wide) spanning from the Georgian border (Hopa) to the Araklı township of Trabzon (Müftüoğlu 1987; Özyazıcı et al. 2010). The tea-cultivated soils cover approximately 76,000 ha in the northeastern Black Sea region, of which 65% is in Rize, 21% in Trabzon, 11% in Artvin, and 3% in the Giresun and Ordu provinces (Müftüoğlu et al. 2010). However, according to Müftüoğlu (1987), the tea production is not economically feasible on the soils located west of Araklı (Trabzon). The soil pH is extremely low in the regions of Rize and Hopa due to the high precipitation, moist and rainy winters and warm, moist, and rainy summers, which stimulate the leaching of nutrients and basic cations from the soils. Moreover, studies conducted by Ülgen (1961) and (Kacar 2010) revealed that the extensively used ammonium sulphate fertilizers in the region decreased soil pH considerably in the tea soils. This

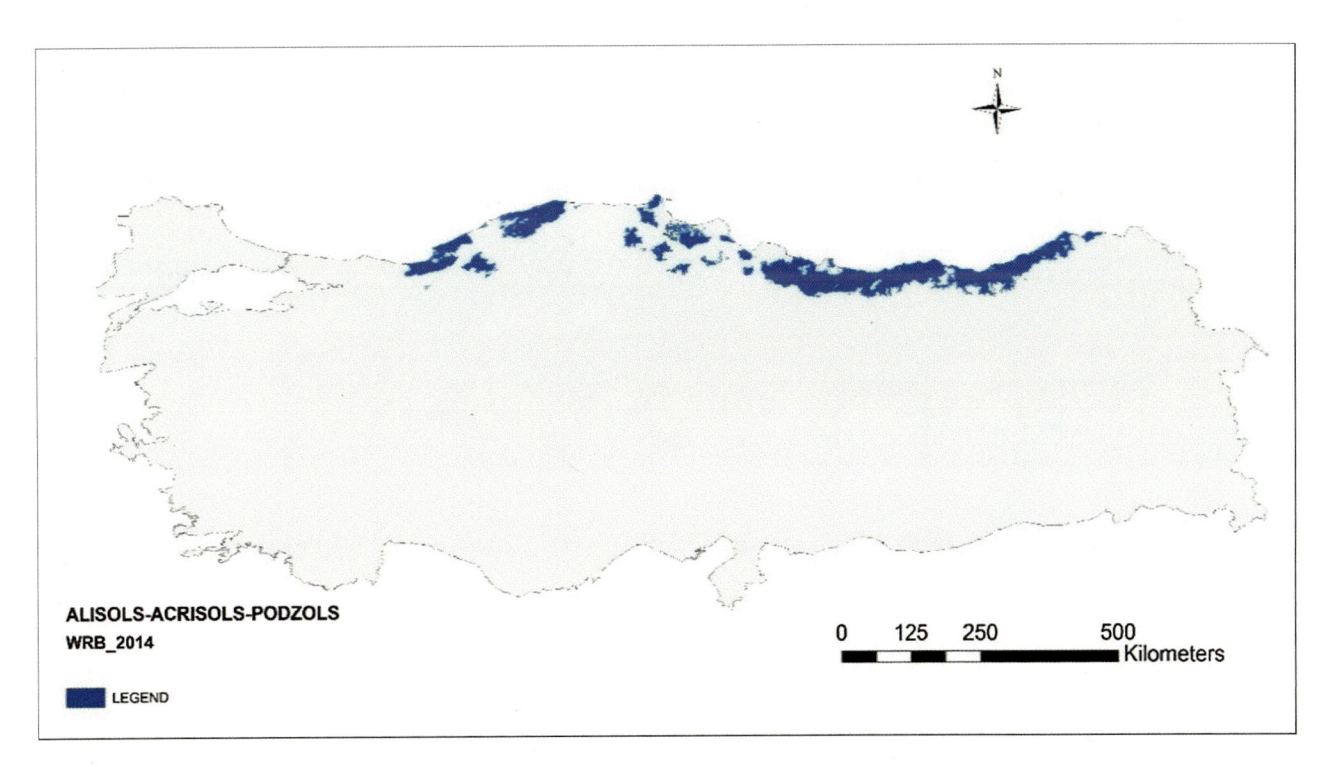

Fig. 5 Distribution of Alisols-Acrisols-Podzols in Turkey

Table 3 Site description and some physical and chemical properties of an Acrisols in the Of town, northeast Turkey (Müftüoğlu 1989)

IUSS Working Group WRB (2015)	Umbric Acrisols Albic Clayic
Soil Survey Staff (2014)	Ochreptic Hapludults
Location	Of province, Trabzon, east Black Sea region northeast Turkey
Coordinates	37 T, 608149E, 4 534 301N
Altitude	25 m
Landforms	Hillside
Position	Slightly undulating
Aspect	–
Slope	16–20%
Groundwater	–
Effective soil depth	120 cm+
Parent material	Sandstone
Köppen climate	Temperate rainy/humid temperate, no dry season
Land use and vegetation	Tea cultivation and some forest trees

Horizon	Depth cm	pH 1:1	Org mat (%)	Base saturation (%)	Na ($cmol_c kg^{-1}$)	K ($cmol_c kg^{-1}$)	Ca + Mg ($cmol_c kg^{-1}$)	CEC	Sand %	Slit	Clay
Ap	0–20	4.05	4.6	12.8	0.1	0.36	2.6	23.9	52.9	24.2	22.9
A2	20–54	4.12	3.74	6.08	0.11	0.21	2	38.0	60.9	16.2	22.9
Bt	54–74	4.36	1.73	6.35	0.07	0.08	2	33.7	38.9	24.2	36.9
Bw1	74–106	4.31	0.87	7.68	0.07	0.15	2.2	31.5	32.9	24.2	42.9
Bw2	106–136	4.64	0.67	27.03	0.11	0.26	10.8	41.3	30.9	26.2	42.9
C	136–+	4.71	0.25	46.12	0.2	0.36	21	21.6	42.9	26.2	30.9

Horizon	Color Dry	Moist	Mottles	CaCO₃ (%)	Roots	Texture
Ap	10YR4/3	10YR3/3	None	–	AF	SCL
A2	10YR4/4	10YR5/3	None	–	SF	SCL
Bt	7.5YR5/6	7.5YR4/4	None	–	TR	CL
Bw1	5YR 5/6	5YR4/4	Slightly	–	TR	C
Bw2	7.5YR6/4	7.5YR4/4	Slightly	–	–	C
C	Weathered sandstone					

[a] *AF* Abundant fine root, *S* Sparse fine roots, *TP* Tap roots

was also stated by the Provincial Agricultural directories (1986), revealing that 80% of the N fertilizers used in tea production was in the form of ammonium sulfate. The soil pH was observed as low as 2.92 (Özuygur et al. 1974) in some tea-growing soils and/or rather in the "Tea Soils" as named by the locals. Müftüoğlu and Sarımehmet (1993) reported that 68% of the tea soils in Rize had a pH < 5.4, whereas Özyazıcı et al. (2013) reported from their extensive study (by analysis conducted on 262 soil samples) that the pH of the tea soils in the Black Sea region ranged from 3.14 to 6.39. They also stated that soils with the lowest pH generally occurred in the Rize province. P deficiency is common in the tea soils, where approximately 85% of the tea soils in Rize were deficient in P (Müftüoğlu and Sarımehmet 1993) but generally rich in organic matter. Consequently, P fertilizers are widely used to meet the plant P supplies of these soils (Özyazıcı et al. 2010).

Tea grows well on quickly drained deep soils with pH varying from 4.5 to 5.5 and organic matter content of more than 2%. Shallow and compacted subsoils limit root growth for the tea plants and also make them vulnerable to draught during the dry periods and water-logged conditions in the

Table 4 Site-profile description and some physical and chemical properties of an Acrisol-Alisol in the Ordu province, northeast Turkey (Türkmen 2011)

IUSS Working Group WRB (2015)	Humic Acrisols-Alisols
Soil Survey Staff (2014)	Typic Dystroxerults
Location	Mount Abaz, Fatsa-Ordu, Black Sea region
Coordinates	37 T, 364641E, 4525376N
Altitude	770 m
Landforms	Medium gradient hill
Position	Lower slope
Aspect	–
Slope	12–20%
Drainage class	Good
Groundwater	–
Efficient soil depth	70 cm+
Parent material	Porphyric diorite, hydrothermally altered dacite
Stoniness	Increased stoniness with depth
Rockiness	Abundant unsorted rock fragments at 33 cm fromsurface
Land use and vegetation	Forest, natural habitat and tea (*Camellia sinensis*)

Horizon	Depth cm	Description
A	0–10	Brown (10YR4/3 dry), dark brown (10YR3/3 moist), loam, medium, fine, granular structure, slightly sticky, slightly plastic, crumble, no effervescence, many, medium, fine and few, coarse roots, clear, wavy boundary
Bt	10–33	Light yellowish brown (10YR6/4 dry), dark yellowish brown (10YR 4/4 moist), loam, weak, medium, sub-angular blocky structure, slightly sticky, slightly plastic, loose, no effervescence, fine gravely, common, medium, fibrous and few, coarse roots, clear broken boundary
C	33–76	Brownish yellow (10YR6/6 dry), yellowish brown (10YR 5/6 moist), loam, massive, slightly sticky, slightly plastic, slightly hard, no effervescence, few, medium, coarse roots

Horizon/ Depth	CaCO$_3$ (%)	Organic matter (%)	Sand (%)	Silt (%)	Clay (%)	Texture class
A 0–10	0.00	14.53	45.5	33.7	20.8	L
Bt 10–33	0.00	4.47	42.7	28.9	28.4	L
C 33–76	0.00	0.97	63.5	18.4	18.1	SL

Horizon/ Depth (cm)	pH	EC (dS/m)	Na	K	Ca (cmol$_c$kg^{-1})	Mg	CEC	Base saturation (%)
A 0–10	4.98	0.55	0.44	0.7	5.32	0.71	32.39	22
Bt 10–33	5.12	0.19	0.37	0.16	5.52	0.12	22.88	27
C 33–76	5.20	0.15	0.38	0.09	2.36	0.08	10.28	28

wet periods. Hard pans or concretions in the subsoil within 2 m depths should be absent, and the ground water table should not be closer to the soil surface for more than 90 cm in order to obtain good yields in tea. Despite all the knowledge developed on the regional Alisols-Acrisols, the area, as elsewhere in Turkey, is in need of a contextual sustainable catchment planning for improving the appropriate soil and water management practices coupled to economic management via powerful production unions as stated in the National Action Plan for Desertification of Turkey (NAP-D Turkey 2006).

The steep and irregular topography makes these soils extremely vulnerable to nutrient loss by deep percolation and soil erosion. Adequate annual rainfall and its appropriate distribution in the winter and spring, as it is in the eastern Black Sea region, is crucial for high tea quality and yield. Some of the precipitation falls as snow and the plants stay under snow for a sufficient period naturally controlling the pests. This makes the Turkish tea unique in the world in the sense that no pesticides are used contrary to the other tea-growing regions located within the equatorial circle. The temperature affects the tea yield due to its control on the rate of photosynthesis, growth, and dormancy. In general, the ambient temperature within 13 °C and 28–32 °C is adequate for tea production which is also enhanced by the higher relative humidity. All these variables are well-balanced for a successful tea cultivation in the region.

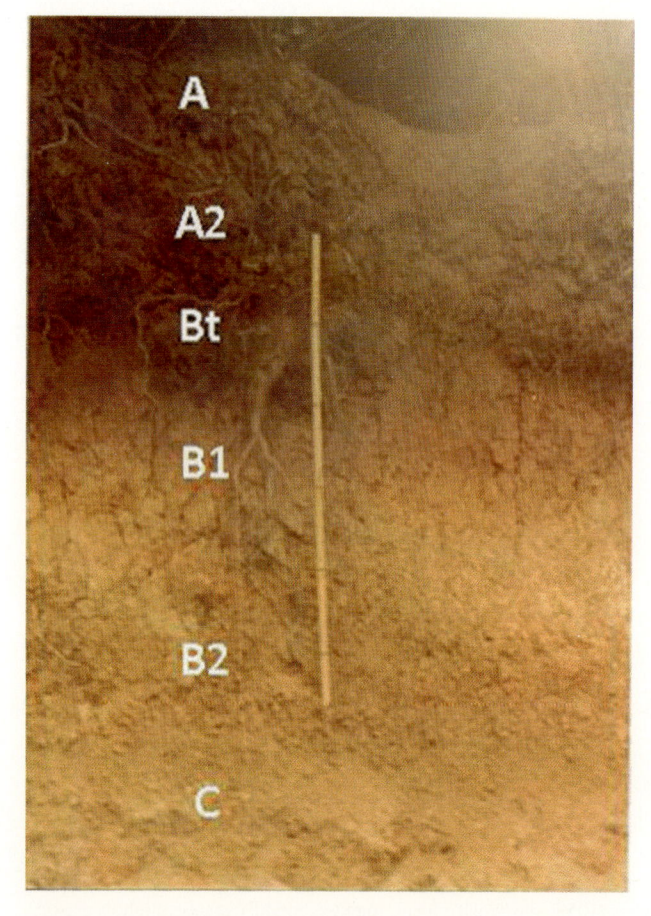

Fig. 6 An Alisol in the town of, Trabzon, northeast Turkey

Besides tea, hazelnuts are produced on the Alisols-Acrisols in the central Black Sea region and the region spanning from the Araklı township of Trabzon to the Terme township of Samsun. This area of the country produces approximately 65% of the world's total hazelnut production. The high-quality Turkish hazelnuts produced on the well-established and long-standing Alisol-Acrisol ecosystems are generally favored in the international markets. Hazelnut production is appropriate in this soil ecosystem due to the pH (around 6, slightly acid to neutral) and the fine texture (clay to clay loam). However, the soil pH is lower in some coastal zones, which is consequently unsuitable for hazelnut production. Çelik et al. (1999) reported soil pH values as high as 7.8 and as low as 4.9 for hazelnut soils. Lime ($CaCO_3$) is mainly applied to increase the pH of the hazelnut soils (as locally called), which are low in $CaCO_3$ and moderate in soil organic matter contents.

The hazelnut (filbert) tree is a natural bush species or a multi-trunked, shrubby tree and produces satisfactory crops only under mild climatic conditions. The tree cannot tolerate excessive dry summer heat, thus being limited in distribution to the central Black Sea and some isolated localities in the Western Black Sea regions. Most of the roots of the hazelnut tree are distributed in the top 60 cm of the soil. However, the active root system (the efficient soil depth) extends from 2 to 3 m of depth. In highly eroded, sloping, shallow areas the yield is often limited since root penetration is inhibited by hardpans, rock fragments, high water tables, or insufficient aeration.

Fortunately, contrary to many field and horticultural crops, hazelnut production is appropriate in most of the highly sloping soils with irregular topographies. Therefore, both topographical and climatic conditions are part of the suitable Alisol-Acrisol environment for the hazelnuts. High moisture contents may create problems in hazelnut production in some areas with especially poorly drained (poorly aerated) shallow or sandy soils. Thus, irrigation is not a proper practice in improving the suitability of these soils for

Fig. 7 An Alisol-Acrisol soil profile in the Ordu province, northeast Turkey (**a**), the Alisol-Acrisol environment of the Black Sea region (**b**) (Türkmen 2011)

higher hazelnut yields. In some flat areas with heavy textured soils, the tile drain irrigation system helps the crops to grow better. Subsoiling before planting generally increases plant adaptation in newly built orchards in the areas with compacted and dense subsoils, but in turn this practice may impede the structural development of the soil causing physical degradation, i.e., and loss of the physical quality of the soil.

References

Akgül E (1975) Comparison of the major properties of the soils of spruce ecologies (*Picea orientalis* and Carr.). Forest Res. Center Publications, Technical Bulletin No. 71, Ankara (in Turkish)

Aydın A, Sezen Y (1990) Effect of liming on some properties and the availability of some micro and macro nutrients of the soils of the East. Atatürk Univ Fac Agric J 21(1):94–95 (in Turkish)

Baldwin M, Kellogg CE, Thorp J (1938) Soil classification. In: soils and men: yearbook of agriculture. US Department of Agriculture, Washington DC. pp 979–1001

Cangir C, Kapur S, Başkaya HS (1991) Genesis of (Ultisols) of the eastern, Turkey. In: Yeşilsoy MŞ, Özbek H, Kapur S (eds) Mahmut Sayın Clay Mineral Symposium (Adana/Turkey) book of Proceedings, pp 59–63

Çelik M, Karakaya N, Temel A (1999) Clay minerals in hydrothermally altered volcanic rocks, eastern, Turkey. Clays Clay Miner 47(6):708–717

Eyüpoğlu Y, Bektaş O, Seren A, Nafiz M, Jacoby WR, Özer R (2006) Three-directional extensional deformation and formation of the Liassic rift basins in the eastern (NE Turkey). Geol Carpath 57(5):337–346

IUSS Working Group WRB (2015) World Reference Base for 2014, update 2015. In: International soil classification system for naming soils and creating legends for soil maps. World Soil Resources Reports No. 106. FAO, Rome. p 203

Kacar B (2010) Biochemistry and Fertilisation of Tea. Nobel Pub., Ankara, 356 P (in Turkish)

Mermut AR (1983) Micromorphogenesis of two Humults from northeastern Turkey. Soil Sci 136 (3)

Ministry of Environment and Forestry (2005) National Action Plan of Turkey (NAP-D of Turkey). In: Düzgün M, Kapur S, Cangir C, Akça E, Boyraz D, Gülşen N (eds) Ministry of Environment and Forestry, Pub. No:250. Ankara, p 110

Müftüoğlu NM (1987) Determination of acidification of tea soils by various periods and methods and evaluation of the data obtained. General Directorate of Tea Enterprises. Head of the Institute of Tea, Study Report of 1986, Rize, Turkey. pp 178–190 (in Turkish)

Müftüoğlu NM (1989) Research on microbiological activity of soils where tea is grown and factors affecting soil acidity. PhD Thesis. Aegean University Institute of Science, Soil Science Departments, Bornova İzmir, 118 P (in Turkish)

Müftüoğlu NM, Sarımehmet M (1993) Acidity status of tea grown soils in East. Aegean Univ J Fac Agric 30(3):41–48 (in Turkish)

Müftüoğlu NM, Yüce E, Turna T, Kabaoğlu A, Özer SP, Tanyel G (2010) The evaluation of some properties of tea soils in the Eastern. Aegean University Faculty of Agriculture. Joural. 5th Plant Nutrition and Fertilization Congress 3–7 June 2010. Proceedings 63–67

Oakes H (1954) The soils of Turkey. Turkey, Doğuş Press, Ankara, Turkey, FAO Ankara

Özsayar T, Gedikoğlu İ, Pelin S (1981) The paleontological data on the Artvin Province pillow lavas. K T Ü Earth Sci J 2(1–2): 2–37 (in Turkish)

Özuygur M, Ateşalp M, Börekçi M (1974) Methods to be used in determination of lime requirement in the East and a study on liming materials. TÜBİTAK Pub. No. 283, TOAG (in Turkish)

Özyazıcı G, Özyazıcı MA, Özdemir O, Sürücü A (2010) Some physical and chemical properties of tea grown soils in Rize and Artvin provinces. Anadolu J Agric Sci 25(2):94–99

Özyazıcı AM, Dengiz O, Aydoğan M (2013) Significance of soil reaction in tea soils and distribution of tea cultivated land in the. Journal 2(1):23–79 (in Turkish)

Soil Survey Staff (2014) Keys to 12th ed. USDA-Natural Resources Conservation Service, Washington DC. p 372

Türkmen F (2011) Genesis and classification of the Ordu province soils. Ankara University Graduate School of Natural Sciences, Ph. D. Thesis, Ankara (in Turkish)

Ülgen N (1961) The productivity and capability of the tea soils. Soil and Fertilizer Research Institute. Technical Pub. No 9, Ankara (in Turkish)

Podzols

Hüseyin Ekinci, Hasan Özcan, Orhan Yüksel, and Sabit Erşahin

1 Introduction

Podzols are soils with a spodic illuviation horizon under a subsurface horizon that has the appearance of ash and is covered by an organic layer; from Russian pod, underneath, and zola ash (IUSS Working Group WRB 2015). The Podzols in Turkey are found as patches under forest vegetation. The soils associated to Podzols are the Lixisols, Alisols, and some Acrisols. These soils have been classified as "Podzolic" by Akyürek (1987) and other earlier researchers that attempted to study the soils of the Black Sea region of the country with the highest annual precipitation. However, some true Podzols have formed at the east Black Sea Region, on the high slopes of Mounts Uludağ (Bursa Province, west Turkey) and Istranca (Yıldız-Mahya, the Thrace Region), as well as at some locations of the southern Amanos Mountains (Hatay Province, south Turkey). The high slopes, the non-coniferous vegetation cover, the dominant medium-fine textured parent materials, and the lack of an acidic soil environment prohibits the formation of widespread areas of Podzols in the country. These environmental conditions favor the formation of the rather patchy areas of Podzols in Turkey. The lack of the coniferous forests has been the major drawback in the formation of a thick organic matter layer, inturn being responsible for the strong leaching of the sesquioxides with little low organic matter.

H. Ekinci (✉) · H. Özcan · O. Yüksel
Department of Soil Science and Plant Nutrition, Onsekiz Mart University, Çanakkale, Turkey
e-mail: hekinci@comu.edu.tr

S. Erşahin
Department of Forestry Engineering, Çankırı Karatekin University, Çankırı, Turkey

2 Parent Materials and Environment

Podzols are present generally on quartzite (Fig. 1) and acidic igneous rocks and sandstones of the Istranca Mountains (Thrace northwest Turkey). Kantarcı (1989) has described Podzols formed on quartz-sericite schistic rocks in the same area. Dinç et al. (2001) described Podzols near the summit of Mount Uludağ overlying fine and coarse grained granodiorites. Podzols were also described on the granites and andesites of the east Black Sea Region (Akgül 1975; Atalay et al. 1985).

Podzols are also formed on other acidic igneous rocks such as the metagranites, granodiorites, and some schists located on hilly and undulated old topographic surfaces. They usually are formed above 700 m in Thrace and above 1700 m on Mount Uludağ (Irmak and Gülçür 1964). These Podzols are less leached when compared to the Podzols located on flat land. The annual mean temperatures of the Podzol areas in Thrace are between 12 and 13 °C and the rainfall is above 1000 mm. The prevailing climate here is similar to the climate encountered on the Black Sea shores, namely with cool summers and winters with rain throughout the year. The plant cover here is composed of various plant and tree species. The dominant plants are *Rhododendron ponticum, Hedera sp., Nephrolepsis sp., Fagus orientalis., Acer sp., Ephedral and Erica arborea* (Fig. 2), whereas the dominant tree species are Castanea sativa, *Qercus cerris,* Abies *bornmülleriana,* ve *Juniperus communis*. Podzols are also present under the *Picea orientalis, Pinus sylvestris* and *Abies nordmanniana* forests of the Central Black Sea Region with an annual rainfall of 1500 mm as well as in areas under *Picea orientalis covers* of the rainier and the north slopes of this region.

3 Profile Development

The Podzols in Turkey have mostly Oe-A-E-EB-Bs-CB/Cr profiles (Fig. 3). In general, organic horizons on the surface of the Podzols of the Thrace region are moderately

© Springer International Publishing AG 2018
S. Kapur et al. (eds.), *The Soils of Turkey*, World Soils Book Series,
https://doi.org/10.1007/978-3-319-64392-2_13

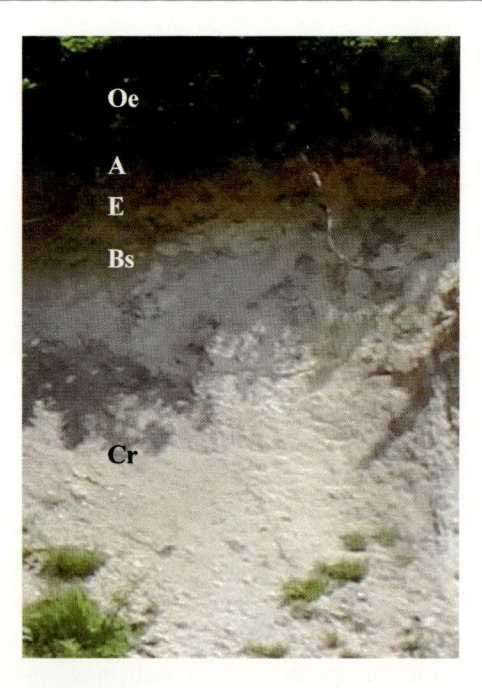

Fig. 1 A podzol profile on quartzites in the Yıldız Mountains, Thrace, northwest Turkey

decomposed. The organic horizons of the Podzols of Mount Uludağ are less decomposed due to the cooler climatic conditions. A spodic (Bs) illuvial horizon is present under the elluvial (E) at much warmer climatic conditions of the Thrace, west, and east Black Sea regions compared to the Podzols of the cooler Mount Uludağ conditions with a Bs horizon rich in humic substances. The slope positions and aspects are also significant in the development of the Podzols in Turkey as elsewhere. For example, leaching is more severe in the northeast slopes of the highlands forming much darker and reddish Bs horizons (Fig. 4) as also stated by Hunkler and Schaetzl (1997). Organic matter accumulation, slow decomposition, and the slower melting of the snow contribute on induce podzol formation depending on the exposition and local topography of the Mount Uludağ conditions. These, in turn, are responsible for the formation of the dark brown spodic horizon with low value and chroma (10YR 2/2–3/3) in Uludağ (Dinç et al. 2001).

The differences in the specific conditions (climate and topography) mentioned above are responsible for the reddish

Fig. 2 The environment of Podzols in the Yıldız Mountains with *Rhododendron ponticum, Hedera sp., Fagus orientalis*, northwest Turkey

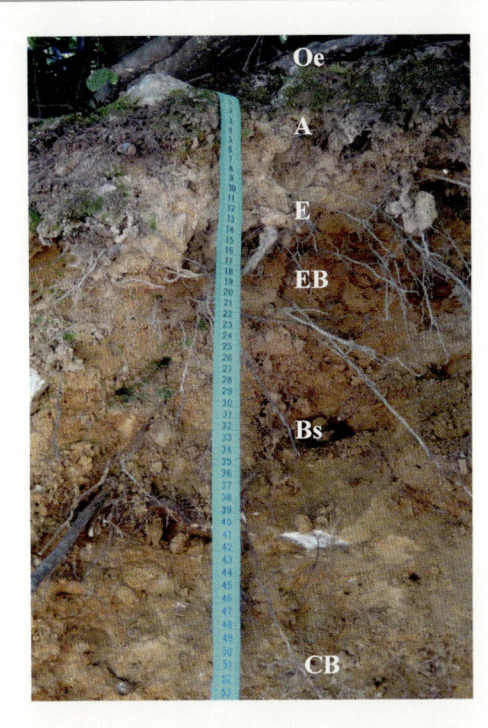

Fig. 3 The Podzol in Kadınkule, Kırklareli, Thrace, northwest Turkey

Fig. 4 A Podzol developed on the northeast slopes of the Yıldız Mountain, Thrace, northwest Turkey

Brown colors with high values and chroma (7.5YR 4/4–10YR 5/4) in the Thrace, and the light Brown colors of the spodic horizons in the west (Zonguldak province) (Fig. 5).

Fig. 5 A Podzol—like soil in the Zonguldak province, north Turkey

4 Characteristics

The pH values of the upper horizons in generally vary from 4.0–4.5 (1:2.5 in water) in the Podzols of Turkey (Table 1). The surface horizons are more acidic than the subsoil. The increased Fe_2O_3 contents are responsible for the strong red color in Thrace and Black Sea Podzols.

Podzols in the Thrace Region are, present generally on deposits of quartzites and their mineralogical properties mainly depend on the quartz contents. The clay contents of the elluvial horizons are below 15% but are above in the spodic horizons. The CEC of the Podzols is low as a consequence of their high degree of leaching which is also responsible for their low fertility.

5 Distribution

Podzols cover an estimated 485 million ha of land worldwide, mainly in the temperate and boreal regions of the North Hemisphere (IUSS Working Group WRB 2015). They are found in patches within the Alisols-Acrisols association in Turkey (see Fig. 5 Distribution of Alisols-Acrisols-Podzols in Turkey, Table 1, Distributions of WRB Groups in Turkey). They have also formed in the Yıldız Mountains area of Thrace (north west Turkey) especially in Kadınkule, Demirköy, Fatmakaya, and Kıyıköy sites.

Table 1 Site and profile description and some physical and chemical properties of a Haplic podzol in Kadınkule, Kırklareli, northwest Turkey

IUSS Working Group WRB (2015)	Haplic Podzols
Soil Survey Staff (2014)	Ultic Haplorthods
Location	Kadınkule-Kırklareli, northwest Turkey
Coordinates	Lat.: N 41° 44′ 26″ Lon.: E 27° 42′ 41″
Altitude	735 m
Landforms	Mountain, medium slopes
Position	Middle slope
Aspect	Northwest sloping hillside
Slope	20%
Drainage class	Well drained
Ground water	–
Eff. Soil depth	60 cm
Parent material	Quartzites
Köppen climate	Cool and rainy summers and winters
Land use	Forest
Vegetation	*Fagus sp., Rhododendron ponticum, Nephrolepsis sp., Erica sp., Acer sp., Hedera sp*

Horizon	Depth	Org. C (%)	pH H$_2$O	pH KCl	EC µS/cm	Ca	Mg	K	Na	H	Al	CEC (cmol$_c$ kg^{-1})
O	0–6	34.2	4.72	4.13	291							
A	6–11	6.8	4.52	3.41	124	8.75	4.62	0.79	0.22	13.2	0.39	47.8
E	11–19	0.82	4.15	3.53	84	0.86	0.69	0.61	0.7	4.92	2.78	13.3
EB	19–30	0.64	4.38	3.64	111	1.42	1.12	0.36	1.09	5.41	2.86	14.69
Bs	30–55	0.47	4.48	3.60	134	3.31	2.94	0.64	1.18	7.21	7.45	17.62
CB	55–86	0.09	4.11	3.75	96	4.15	3.14	0.12	0.9	3.76	7.13	8.83
Cr	86+											

Exchangeable cations (cmol kg^{-1}): Ca, Mg, K, Na, H, Al

Horizon	Depth	Sand (%)	Silt (%)	Clay (%)	Fe$_2$O$_3$ (%) Na–dithionite
O	0–6	–	–	–	
A	6–11	53.9	31.5	14.6	0.89
E	11–19	62.9	27.7	9.4	0.29
EB	19–30	57.1	30.6	12.3	1.17
Bs	30–55	64.7	19.1	16.2	3.03
CB	55–86	70.1	19.8	10.1	2.57
Cr	86+				

6 Management

Podzols are low in plant nutrients and may comprise a compacted solum at some sites. They are not suitable for cultivation due to their low pH and high slopes. On the other hand N, P, K, and micro nutrient fertilizations and lime additions enhance their agricultural productivity. They are generally covered by forest vegetation in Turkey composed of *Fagus orientalis, Carpinus betulus, Abies bornmülleriana, Juniperus comminis, Quercus cerris Pinus sylvestris*. Hazelnut (*Corylus avellane*) and tea (*Camellia sinensis*) are also cultivated on some local Podzols where topographical conditions are suitable (Fig. 6).

Fig. 6 Tea (*Camellia sinensis*) plantation, north Turkey

References

Akgül E (1975) Distribution of Picea orientalis Link. and Carr. and soil properties. Forestry Res Ins Pub Tech Bull No: 71. Şark Printers, Ankara, p 87 (in Turkish)

Akyürek İ (1987) General management planning of Turkey (soil Con. master plan) Min. of agriculture. Forestry and Rural Affairs, GD of Rural Services, Ankara, (in Turkish)

Atalay İ, Tetik M, Yılmaz Ö (1985) Ecosystems of North Eastern Anatolia. Aegean Geogr J 3:16–56

Dinç U, Şenol S, Kapur S, Cangir C, Atalay İ (2001) Soils of Turkey. Çukurova University, Faculty of Agriculture Publication No: 51, Adana (in Turkish)

Hunkler RV, Schaetzl RJ (1997) Spodosol development as affected by geomorphic aspect, Baraga country Michigan Soil Sci Soc Am J 61:1105–1115

Irmak A, Gülçür F (1964) Soils developed on granite parent materials in Uludağ. İstanbul Univ Fac Forestry J A 14(2):1–14

IUSS Working Group WRB (2015) World reference base for 2014, update 2015 international soil classification system for naming soils and creating legends for soil maps. World soil resources reports No. 106. FAO, Rome

Kantarcı MD (1989) Catenary changes in the genetic and environmental factors in a NS section of thrace. In: Proceedings of 10th scientific conference, publications No 5, Soil Science Society of Turkey, Kırklareli, (in Turkish)

Soil Survey Staff (2014) Keys to soil taxonomy, 12th edn. USDA-Natural Resources Conservation Service, Washington, DC, USA

Kastanozems

Ertuğrul Aksoy, Gökhan Özsoy, Ekin Ulaş Karaata, and Duygu Boyraz

1 Introduction

Kastanozems occur mostly in the short grass steppe belt environments with Chernozems. They have a chestnut-brown to brownish humus-rich surface horizon (thinner than the humus-rich horizons of the Chernozems) and they show prominent accumulation of secondary carbonates in their sub-surface horizons (IUSS Working Group WRB 2015). Kastanozems cover just over 3500 km^2 in the European Union which is less than one per thousands of its total soil resources. Kastanozems can be characterized by the qualifiers of Calcic, Haplic and Luvic. Calcic Kastanozems are the most widespread among them in Europe, covering nearly 95% of the total Kastanozems area (Toth et al. 2008).

Kastanozems were mapped by the extensive soil surveys conducted by the researchers of the Soil Science Department of Çukurova University at the leadership of Prof. Dr. Ural DİNÇ in the years from 1986 to 1995 and refined by Özden et al. (2001) aiming to establish the Soil Data of Turkey in collaboration with the Soil and Water Resources National Information Center (NIC) (Aksoy et al. 2010). Kastanozems were found to cover about 10.7% of the Turkish land area in association with the RendzicFluvisols/HaplicCambisols/Luvic Kastanozems (7.5%), Haplic Kastanozems/Haplic Cambisols (3.2%) and Dystric Leptosols/Haplic Kastanozems (0.036%).

2 Parent Materials

Kastanozems occur in/on a wide range of unconsolidated materials. A large part of all Kastanozems have developed in/on loess (IUSS Working Group 2015). In the south Marmara region Kastanozems have generally developed over calcareous materials, Neogene aged marls and clay-stones enriched by calcium carbonate (Fig. 1). Kastanozems have been also formed on Oligocene and Oligo-Miocene aged sandy clayey and Miocene aged calcareous deposits in the Tekirdağ province of the Thrace region (northwest Turkey) (Ekinci 1990) and on Pliocene gravels in the eastern Black Sea region (Aydınalp and FitzPatrick 2004).

3 Environment

Kastanozems are present generally on hilly and undulated rather young surfaces. They are usually found in the mid-altitude regions and on hilly and undulated topographies. However, Kastanozems that formed on flat lands are mostly characterized by profiles with a brown A-horizon of medium depth over a brown to reddish brown Cambic or Argic B-horizon and with lime accumulation in or below the B-horizon. Kastanozems usually occur in regions where the climate is relatively cool in the winter and hot in the summer. The annual mean temperature of the Marmara region where Kastanozems are widespread varies from 8 to 15 °C, and the rainfall is less than 800 mm. The mean annual precipitations of the other Kastanozem regions are 587 mm in Tekirdağ (Thrace) and 650 mm in Bursa (Marmara region). Many of the Kastanozems have developed under grass and forest vegetation and overlie highly calcareous parent materials and apparently are the pasture lands of the country (Fig. 2).

4 Profile Development

The Kastanozems in Turkey mostly have A B C horizon sequences (Fig. 3). Kastanozems comprise Cambic or Argic B horizons (20–35 cm) and lime/gypsum accumulations below their B horizons at depths of 35–55 cm. Diffuse

E. Aksoy (✉) · G. Özsoy · E.U. Karaata
Department of Soil Science and Plant Nutrition, Uludağ University, Bursa, Turkey
e-mail: aksoy@uludag.edu.tr

D. Boyraz
Department of Soil Science and Plant Nutrition, Namık Kemal University, Tekirdağ, Turkey

© Springer International Publishing AG 2018
S. Kapur et al. (eds.), *The Soils of Turkey*, World Soils Book Series,
https://doi.org/10.1007/978-3-319-64392-2_14

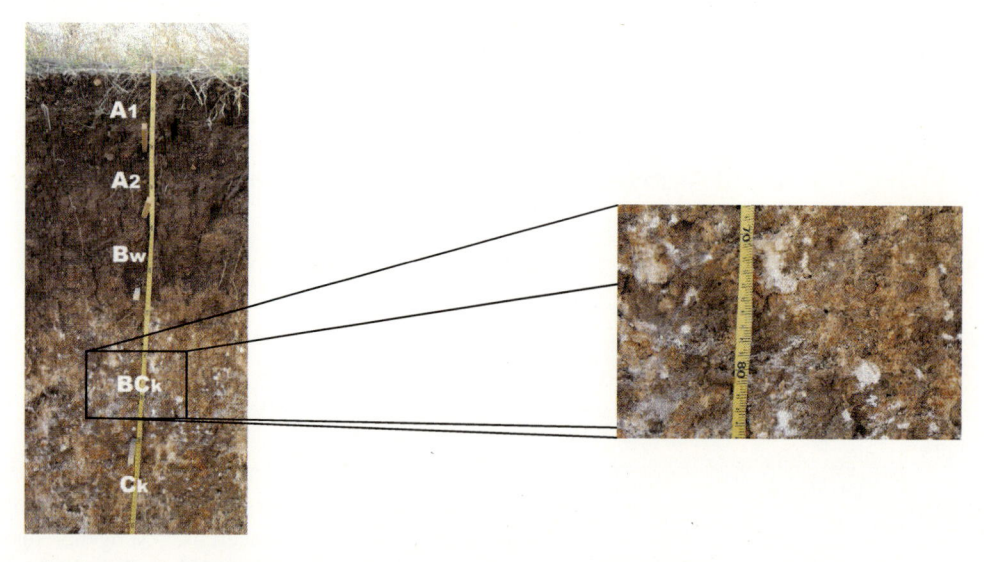

Fig. 1 Kastanozem developed on calcareous deposits in the Marmara Region, northwest Turkey

Fig. 2 Kastanozems in reforested grass and bushland (*Pinus pinea, Pinus brutia*-a, and Quercus Sp., *Palirius spina* and grasses-b), Bursa province, Marmara region, northwest Turkey

lime accumulations may occur at this depth due to weathering. Kastanozems are usually shallow but contain adequate organic matter levels. Their color ranges from brown to reddish brown (10YR 3/2–5YR 3/2) in the Bursa region (west Turkey) (Fig. 3). However, some of the Kastanozems in the Thrace region have dark brown colors due to their clay and organic matter contents. The texture of the Kastanozems in Turkey, depending on their parent materials is clayey, sandy clay loam or loam (Tables 1b, c, 2b, c and 3).

Most of the Kastanozems in Turkey are under agricultural use where a few decades ago the forests and bushes were cleared for agricultural use. Despite all the land/vegetation degrading activities conducted by the people, fortunately, some Kastanozems are still found under pasture and forest lands. The natural vegetation of the Kastanozem belt in western Turkey is dominated by Mediterranean short and poor grass vegetation which dries each summer. Percolated water enhanced by the spring rainfall leaches solutes from

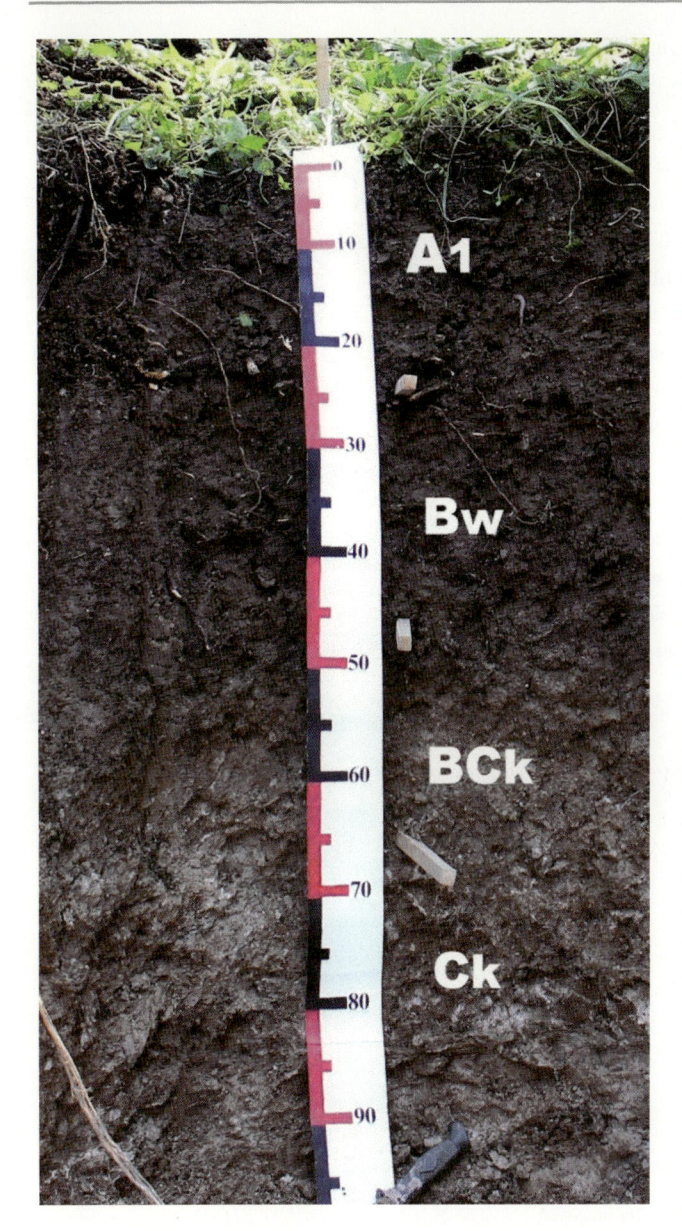

Fig. 3 Kastanozems in the Marmara Region, Bursa province, west Turkey

Kastanozems are usually developed on calcareous deposits in the Bursa province (northwest Turkey) but contain low amounts of lime in their surface soils which increases with depth. The organic matter content is generally measured in adequate levels but decreases with depth and varies from 2.0% to 4.0% in the surface soil (Özsoy and Aksoy 2012).

The moderate to high clay contents of the Kastanozem horizons vary due to the changing parent materials and increase at fluctuating modes in the subsoils compared to the surface. Their base saturation is thus 95% or more and the majority of the adsorbed cations are composed of Ca^{2+} and Mg^{2+}. The CEC of the Kastanozems is high as a consequence of their high clay and organic matter contents.

The darkest Kastanozem surface horizons in Turkey occur in the northwest of the country in the Thrace region. The structure of these soils varies from granular to fine—medium angular blocky.

The Kastanozems with the darkest surface horizons found in the Thrace region are mostly developed under forest cover with high organic matter contents (Fig. 2; Table 3).

6 Distribution

Kastanozems in Turkey can be observed in cool to warm temperate, semiarid to subhumid climatic regions, under grassland, bushland and cultivated lands which are recently gained by deforestation or from bushlands. They are located in the Thrace, Marmara, Aegean, Mediterranean and Eastern Turkey regions. Kastanozems are present in Korkuteli, Korkuteli-Bucak in Antalya, Gönen Isparta (Aydınalp and Aslan 2003a), Araç and Boyalı in Kastamonu (Aydınalp and Aslan 2003b),—Poyralı, Yenice in Kırklareli (Boyraz and Cangir 2009), Çölpan, Köşk, Erçek, Kaymaklı, Aktaş in Van (Boysan and Çimrin 2006), Akpınar Buzlukdağ in Kırşehir (Kibar et al. 2012) and in Çatalca, Silivri of İstanbul and Tekirdağ (Fig. 4). Kastanozems of Turkey cover a land area of about 2,219,766.04 ha which is 2.85% of the whole agricultural land (Fig. 4).

surface to the subsurface layers of the soil profile. Consequently, contemporary lime accumulates at a depth of approximately 50 cm.

5 Characteristics of Kastanozems

The pH of the surface horizons of the Kastanozems in Turkey varies from 6.5 to 7.5 (Tables 1b, c, 2b, c and 3) but may increase to around 8.0 at some depth. The reddish color of some of the Kastanozems of the Bursa province is due to their increased Fe_2O_3 contents (Özsoy 2001).

7 Management

Kastanozems are potentially fertile soils. They are rich in humus and comparatively rich in plant nutrient contents than the other soils of the country. This, together with their partly appropriate topographic properties, consequently leads to the clearing off of the natural vegetation of Kastanozems for cultivation. The periodic lack of soil moisture and the shallow depth of the Kastanozems is the main limiting factor for high yields. In addition, their high clay and $CaCO_3$

Table 1a Site-profile description of the Kastanozems in Bursa (Arız village), west Turkey

IUSS Working Group WRB (2015)	Calcic Kastanozems Chromic
Soil Survey Staff (2014)	Typic Calcixerolls
Location	Arız village, Bursa province, west Turkey
Coordinates	35 S, 603755 m E, 4455300 m N
Altitude	50 m
Landforms	Hilly
Position	Lower slope (foot slope)
Aspect	East to west
Slope	2%
Drainage class	Good
Groundwater	–
Eff. Soil depth	55 cm
Parent material	Clay-lime deposits
Köppen climate	Mediterranean
Land use	Pasture
Vegetation	Mediterranean type stumpy meadow plants

Table 1b Morphological properties of the Kastanozems in Bursa (Arız village), west Turkey

			Color				Consistency[c]						
Horizon	Depth	Boundary[a]	Dry	Moist	Structure[b]	Texture	Dry	Moist	Wet	Roots[d]	Rocks[e]	Effervescence[f]	Nodules[g]
A_1	0–20	g, w	5YR 3/3	5YR 3/2	mo, fn, ab	SCL	vh	fr	s, sp	c, fn	md, gr, a	nc	–
A_2	20–36	g, w	5YR 3/3	5YR 3/2	mo, fn, ab	SCL	vh	fr	s, sp	Fw, fn	md, gr, a	sc	–
Bw	36–55	c, w	5YR 4/6	5YR 3/6	mo, md, ab	CL	vh	fr	s, p	–	fw, gr, a	mc	–
BCk	55–90	c, w	5YR 4/6	5YR 4/6	mo, co, ab	L	h	fr	s, sp	–	fw, gr, a	hc	yes
Ck	90–115	c, w	5YR 4/6	5YR 4/6	mo, co, ab	L	h	fr	s, sp	–	fw, gr, a	hc	yes

[a]*c* clear; *g* gradual; *w* wavy
[b]*mo* moderate; *md* medium; *fn* fine; *co* coarse; *ab* angular blocky
[c]*vh* very hard; *vfr* very friable; *fr* friable; *fi* firm; *s* sticky; *p* plastic; *vp* very plastic
[d]*fw* few; *c* common; *m* many; *fn* fine; *md* medium; *co* coarse
[e]*md* medium (5–15%); *m* many (15–40%); *gr* gravel (0.2–6 cm); *r* round; *w* weathered
[f]*sc* slightly calcareous; *mc* moderately calcareous; *hc* highly calcareous
[g]diameter: 1–3 cm, common lime powder, decomposed by leaching and weathering processes

Table 1c Some physical and chemeical properties of the of the Kastanozems in Bursa (Arız village), west Turkey

Horizon	Depth (cm)	pH (H_2O)	EC (dS/m)	CEC[a] (cmol kg^{-1})	Exchangeable cation (cmol$_c$ kg^{-1})			CaCO$_3$ (%)	OM (%)	Particle size distribution (%)			Texture
					Na	K	Ca + Mg			Sand	Silt	Clay	
A_1	0–20	6.7	0.10	34.50	0.39	0.51	33.60	0.93	3.5	58.0	14.7	27.3	SCL
A_2	20–36	7.1	0.11	36.20	0.39	1.63	32.98	2.27	3.4	51.7	18.9	29.4	SCL
Bw	36–55	7.6	0.16	42.56	0.40	1.12	41.04	19.60	1.1	41.2	16.8	42.0	CL
BCk	55–90	7.5	0.12	28.07	0.39	0.35	27.33	24.12	0.7	37.0	37.8	25.2	L
Ck	90–115	7.9	0.10	24.5	0.51	0.51	23.51	33.90	0.6	49.6	33.6	16.8	L

[a]Cation exchange capacity

Table 2a Site-profile description of the Kastanozems in Bursa (Görükle Village), northwest Turkey

IUSS Working Group WRB (2015)	Calcic Kastanozems Chromic
Soil Survey Staff (2014)	Typic Calcixerolls
Location	Görükle village-Bursa (northwest Turkey)
Coordinates	35 S, 659945 m E, 4456515 m N
Altitude	110 m
Landforms	Hilly
Position	Summit
Aspect	North–west
Slope	4%
Drainage class	Good
Groundwater	–
Eff. soil depth	44 cm
Parent material	Calcareous claystone
Köppen climate	Mediterranean
Land use	Forested area
Vegetation	*Pinus brutia*

Table 2b Morphological properties of the Kastanozems in Bursa (Görükle Village), northwest Turkey

Horizon	Depth (cm)	Boundary[a]	Color Dry	Color Moist	Structure[b]	Texture	Consistency[c] Dry	Moist	Wet	Roots[d]	Rocks[e]	Effervescence[f]	Nodules[g]
A₁	0–22	g, w	10YR3/2	10YR3/2	st, md, ab	C	vh	fr	s, p	m, md, co	md, gr, r	sc	–
Bw	22–44	g, w	–	10YR3/2	st, md, ab	C	vh	fr	s, p	c, fn, md	md, gr, r	mc	–
BCk	44–67	c, w	–	10YR3/3	st, co, ab	C	–	vfr	s, p	fw, co	m, gr, r, w	hc	Yes
Ck	67+	–	–	10YR7/2	st, co, ab	C	–	fi	vs, vp	–	m, gr, r, w	hc	Yes

[a]*c* clear; *g* gradual; *w* wavy [b]*st* strong; *md* medium; *co* coarse; *ab* angular blocky
[c]*vh* very hard; *vfr* very friable; *fr* friable; *fi* firm; *s* sticky; *p* plastic; *vp* very plastic
[d]*fw* few; *c* common; *m* many; *fn* fine; *md* medium; *co* coarse
[e]*md* medium (5–15%); *m* many (15–40%); *gr* gravel (0.2–6 cm); *r* round; *w* weathered
[e]*md* medium (5–15%); *m* many (15–40%); *gr* gravel (0.2–6 cm); *r* round; *w* weathared
[f]*sc* slightly calcareous; *mc* medium calcareous; *hc* highly calcareous
[g]diameter:1–3 cm, common lime powder, decomposed by leaching and weathering processes

Table 2c Some physical and chemical properties of the Kastanozems in Bursa (Görükle Village), northwest Turkey

Horizon	Depth (cm)	pH (H₂O)	EC (dS/m)	CEC[a] cmol_c kg⁻¹	Exchangeable cation Na	K	Ca + Mg	CaCO₃ (%)	Organic matter (%)	Particle size distribution (%) Sand	Silt	Clay	Texture
A	0–22	7.4	0.4	48.08	0.2	0.77	42.44	3.75	2.1	33.2	17.9	48.9	C
Bw	22–44	7.5	0.3	46.41	0.21	0.58	42.36	7.9	1	30	16.8	53.2	C
BC_k	44–67	7.6	0.3	36.69	0.21	0.47	35.08	19.5	0.6	29.2	20	50.8	C
C_k	67+	7.8	0.2	28.32	0.16	0.17	28.29	44.49	0.2	32.6	25	42.4	C

[a]Cation exchange capacity

Table 3 Site description and some physical and chemical properties of the Kastanozems in Poyralı–Demirköy, Kırklareli, the Thrace region, northwest Turkey

IUSS Working Group WRB (2015)	Calcic Kastanozem
Soil Survey Staff (2014)	Udic Calciustoll
Location	Kırklareli–Poyralı–Demirköy
Coordinates	N 41°37' 99" E 27°36' 11"
Altitude	317 m
Landforms	Convex land
Position	Back slope
Aspect	–
Slope	6%
Drainage class	Well drained
Groundwater	+200 cm
Eff. soil depth	80 cm
Parent material	Calcareous clay loam sediments
Land use	Oak forest

Horizon	Depth (cm)	pH 1/2.5 H$_2$O	EC (dS/m)	OM (%)	CaCO$_3$ (%)	Na	K	Ca + Mg	CEC
A1	0–15	7.74	0.3	4.12	0.78	0.10	0.72	35.08	38.50
A2	15–33	7.78	0.2	2.95	1.17	0.09	0.62	34.50	37.48
AB	33–53	7.93	0.2	1.48	1.95	0.09	0.41	33.50	35.38
Bw	53–80	7.78	0.2	1.2	2.41	0.10	0.51	35.33	38.07
Ck	80–110	8.00	0.2	0.59	21.45	0.08	0.48	36.67	37.32

Exchangeable cations (cmol kg^{-1})

Horizon	Depth (cm)	Sand (%)	Silt (%)	Clay (%)	Texture	Color Dry	Color Moist
A1	0–15	32.6	33.9	33.5	CL	10YR 5/2	10YR 4/2
A2	15–33	27.9	23.6	48.5	C	7.5YR 4/3	7.5YR 4/4
AB	33–53	32.1	29.8	38.1	CL	7.5YR 4/4	7.5YR 4/6
Bw	53–80	27.9	27.7	44.4	C	7.5YR 5/3	7.5YR 4/3
Ck	80–110	34.1	27.7	38.2	CL	7.5YR 5/4	7.5YR 4/4

contents in the subsurface horizons may cause problems for some crops and for the tilling processes. Irrigation is nearly always necessary for high yields. However, utmost care must be taken to avoid the buildup of secondary salinization. Moreover, the high clay contents of the Kastanozems in Turkey demand appropriate management practices like minimum or reduced tillage, mulching, intercropping and terracing (for some areas) for sustainable cultivation (Özsoy and Aksoy 2007).

Wheat, barley, sunflower and irrigated food and vegetable crops are the principal crops grown on the Kastanozems of Turkey. Extensive grazing is another important asset of land but overgrazing is the serious snag for sustainability. Soil erosion by water is another serious problem where agriculture or overgrazing is the common land use type. The Mediterranean climate properties of the country, the parent materials of the soils and the slope factor create the suitable conditions for water erosion. Therefore, special priority should be given to the lands over 2% slopes for conservation.

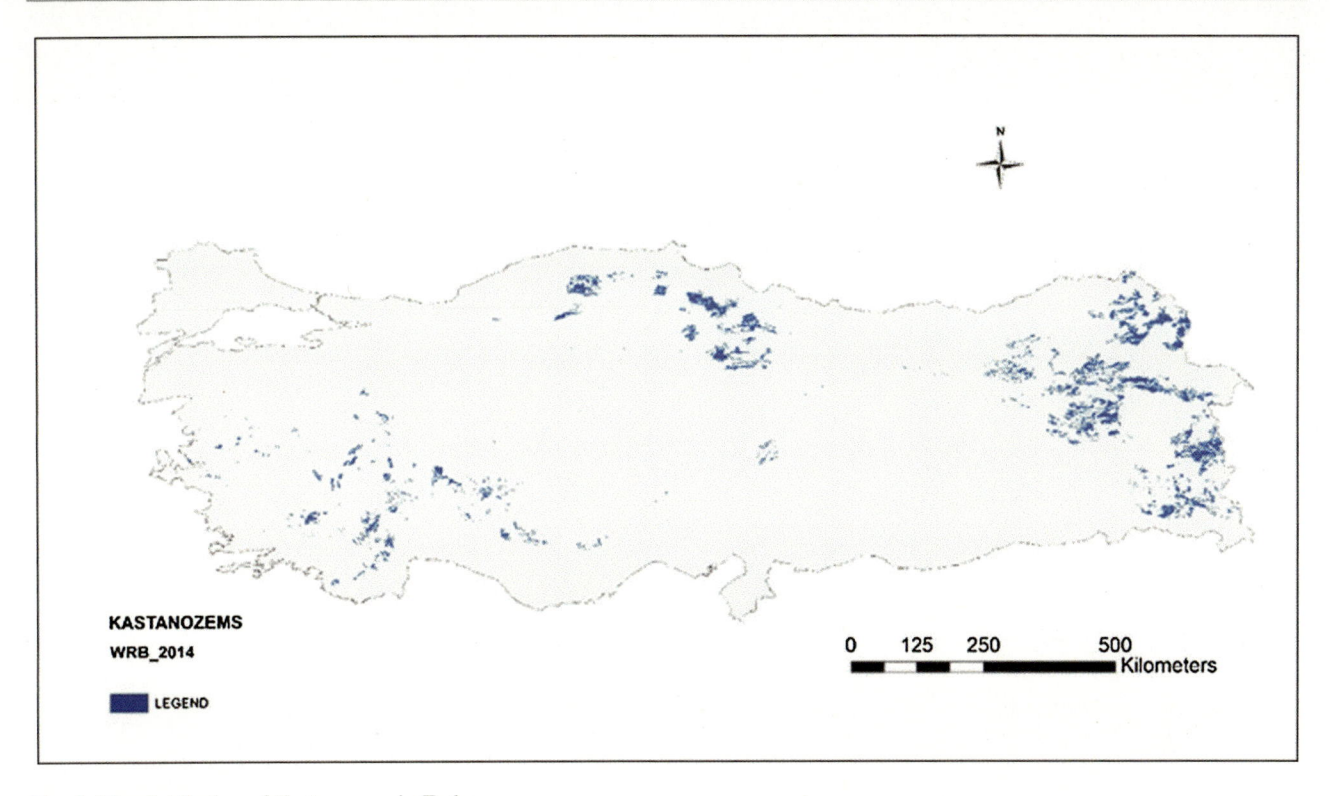

Fig. 4 The distribution of Kastanozems in Turkey

References

Aksoy E, Panagos P, Montanarella L, Jones A (2010) Integration of the Soli Database of Turkey into European Soil Database 1:1,000,000. Research Report. European Commission Joint Research Centre—Institute for Environment and Sustainability. Publications Office of European Union, Luxembourg, 45 p

Aydınalp C, Aslan Y (2003a) Classification of great soil groups in the Antalya basin, according to FAO/UNESCO (1990), FitzPatrick (1988) and USDA (1998) Systems. Anadolu J AARI 13(2):117–139 (in Turkish)

Aydınalp C, Aslan Y (2003b) Classification of great soil groups in the West Black Sea basin, FAO/UNESCO (1990), FitzPatrick (1988) and USDA (1998) Systems. Anadolu J AARI 13(1):188–200 (in Turkish)

Aydınalp C, FitzPatrick EA (2004) Classification of great soil groups in the east Black Sea basin according to international soil classification systems. J Cent Eur Agric 5(2):119–126

Boyraz D, Cangir C (2009) The management and classification of the typical soils of the Yıldız forest ecosystem. J Tekirdağ Agric Faculty, 6(1):65–77

Boysan S, Çimrin KM (2006) Determination of the fixation of the wheat growing soils in the Lake Van Basin. J Agron 5(2):196–200

Ekinci H (1990) The applicability of to the Turkish general soil map in thrace, a case study. Çukurova University, Ph.D. thesis, 218 p (in Turkish)

IUSS Working Group WRB (2015) World reference base for soil resources 2014, update 2015 international soil classification system for naming soils and creating legends for soil maps. World soil resources reports No. 106. FAO, Rome, 203 p

Kibar M, Kıymet D, Sarıoğlu FE (2012) The morphology, mineralogy, geochemistry and physical implications of foid bearing syenite and syenite-carbonate rocks contact zone soils. Eurasian J Soil Sci 2:69–74

Özden MO, Dinç U, Kapur S, Akça E, Şenol S, Dinç AO (2001) Soil geographical database of Turkey at a scale 1:1.000.000—4th approximation. Research Report. Agriculture Services Ministry National Information Center for Soil and Water Resources & Çukurova University Soil Science Department Collaboration, Ankara-Adana, Turkey

Özsoy G (2001) Soils of the Uludağ University campus area, their genesis and classification. Uludağ University, Institute of Applied Sciences. MSc thesis, Bursa

Özsoy G, Aksoy E (2007) Characterization, classification and agricultural use of developed on Neogen aged calcareous marl parent materials. J Biol Environ Sci 1(1):5–10

Özsoy G, Aksoy E (2012) Genesis and classification of some developed under forest vegetation in Bursa, Turkey. Int J Agric Biol 14:75–80

Soil Survey Staff (2014) Keys to soil taxonomy, 12th edn. USDA-Natural Resources Conservation Service, Washington, DC, USA, p 372

Tóth G, Montanarella L, Stolbovoy V, Máté F, Bódis K, Jones A, Panagos P, Van Liedekerke M (2008) Soils of the European Union. JRC Scientific and Technical Reports, Office for Official Publications of the European Communities, Luxembourg

Luvisols

Mustafa Sarı, Yusuf Kurucu, Erhan Akça, Muhsin Eren, Selahattin Kadir, Hikmet Günal, Claudio Zucca, İbrahim Atalay, Zülküf Kaya, Franco Previtali, Pandi Zdruli, Selim Kapur, and Ewart Adsil FitzPatrick

1 Introduction

Chromic and Rhodic Luvisols, otherwise known as the Red Mediterranean Soils or the Terra Rossa, show marked textural differences and genetic peculiarities within their profile. Controversial genetic characteristics observed in Luvisol profiles are related to the impacts of the successive climatic changes of the Quaternary and represent the scientific heritage of the Terra Rossa soils. Luvisols are the fragile patrimonies of the Mediterranean human and soil ecosystem enduring the millennia-long turmoil caused by land transformation and exploitation, warfare, and migrations which in turn have resulted in land degradation (Laouina et al. 2004). Luvisols are characterized with moderate weathering, subsequent leaching and illuviation of mainly 2:1 clays into B horizons and very fine-grained hematite (Fe_2O_3) induced reddening of the clays which coat the clays and the coarser particles, due to summer dehydration of free iron oxyhydroxides (Yaalon 1997).

These soils are best represented by the Luvisols in the WRB (IUSS Working Group WRB 2015) which is the soil type globally widespread. The surface horizon is depleted in clay while the subsurface 'argic/argillic' horizon is enriched with illuviated clay. A wide range of parent materials and environmental conditions lead to a great diversity in Luvisol properties. Other names used to describe Luvisols include Pseudo-podzolic soils (former Russian classification), Sols Lessivés (France), Parabraunerde (Germany), and Alfisols (Soil Taxonomy). Luvisols are often associated with the Leptosols in the highly eroded karstic land surfaces of Turkey. Often Leptosols, a different soil type that is also mentioned throughout this chapter, are the residues of the matured Luvisols, i.e., the exposed Bt horizons (after the topsoil has been totally removed by erosion) overlying

M. Sarı (✉)
Department of Soil Science and Plant Nutrition, Akdeniz University, Antalya, Turkey
e-mail: musari@akdeniz.edu.tr

Y. Kurucu
Department of Soil Science and Plant Nutrition, Ege University, İzmir, Turkey

E. Akça
School of Technical Sciences, University of Adıyaman, Adıyaman, Turkey

M. Eren
Department of Geological Engineering, Mersin University, Mersin, Turkey

S. Kadir
Department of Geological Engineering, Eskişehir Osmangazi University, Eskişehir, Turkey

C. Zucca
International Center for Agricultural Research in the Dry Areas, ICARDA,, Amman, Jordan
e-mail: C.Zucca@cgiar.org

İ. Atalay
Department of Forestry, Karabük University, Karabük, Turkey
e-mail: ibrahim.atalay@deu.edu.tr

Z. Kaya · S. Kapur
Department of Soil Science and Plant Nutrition, University of Çukurova, Adana, Turkey
e-mail: zkaya@cu.edu.tr

P. Zdruli
Mediterranean Agronomic Institute of Bari, CIHEAM, Bari, Italy

E.A. FitzPatrick
Department of Plant and Soil Science, University of Aberdeen, Aberdeen, UK

H. Günal
Department of Soil Science and Plant Nutrition, Gaziosmanpasa University, Tokat, Turkey
e-mail: hikmet.gunal@gop.edu.tr

F. Previtali
Department of Geosciences, University of Milan (Biocca), Milan, Italy

© Springer International Publishing AG 2018
S. Kapur et al. (eds.), *The Soils of Turkey*, World Soils Book Series, https://doi.org/10.1007/978-3-319-64392-2_15

deeply weathered karstic parent materials. These Bt horizons are defined as argic horizons and have been developed from the strongly decalcified and rubified soil materials/horizons.

2 Parent Material

Luvisols in Turkey occur on/in a wide variety of consolidated and unconsolidated materials including karstic and especially Miocene karstic/travertine limestones (Oakes 1957), marls and shales, and Quaternary calcretes (caliche) developed from mudflow/colluvial deposits. The calcrete/soil landforms that have ultimately reshaped the late Pleistocene-early Holocene terraces of the Mediterranean and Aegean regions of Turkey are also present throughout the Mediterranean Basin and the Middle East. They are more frequently classified as Calcisols and less frequently as Luvisols and/or Leptosols when overlying massive/hard calcretes (see Calcisols). Luvisols of Turkey have also developed on older surfaces such as the Permo-Triassic limestones of Ankara. The Luvisols of Mount Çal in Ankara central Turkey (the 'Ankara Clay') have been transported as mud-flows along a widespread glacis during the Pliocene–early Pleistocene and deposited in intermountain basins (ponds–lakes) and calichified (massive calcrete formation over calcite columns, Kaplan et al. 2014) during the mid to late Pleistocene. This glacis formation (Fig. 1) covers a large part of the south of the Ankara township (Hızalan 1953; Erol 1973; Mermut and Cangir 1978; Cangir and Kapur 1984; Yılmazer 1991).

3 Environment

Luvisols occur at warm temperate climatic zones with moderate rainfall and a distinct dry season. They are found at gently undulating or flat land overlying limestones, i.e., in karstic terrain (the karstic cracks/fissure systems) (Figs. 2, 3, and 4), or on some of the fine-textured old mudflow and/or colluvial deposits that evolved to calcretes (Erol 1984; Kapur et al. 1987; Atalay et al. 2014). Kapur et al. (1975, 1987, 1990) observed that Calcic Luvisols together with Luvic Calcisols frequently occur at Early (TH1) to Middle Pleistocene terrace (TH2) systems overlying or underlying massive weathered and/or weathering calcretes, whereas Luvic Calcisols occur on the Late Pleistocene to Early Holocene terraces (TH1-TH2, TL1-TL2) (Fig. 5).

4 Profile Development of Rhodic/Chromic Luvisols in Theory

4.1 The Residual Theory and Age

Several authors (Hızalan 1953; Kapur et al. 1975; Cangir and Kapur 1984) have mentioned the irrevocable significance of the residual theory in the formation of the Terra Rossa (Luvisols) of Turkey, i.e., paleo-formation from the weathering of the insoluble residue of the limestones from the Cretaceous to the Miocene. This theory was earlier stated by Montarlot De (1944), Reifenberg (1947), Kubiena

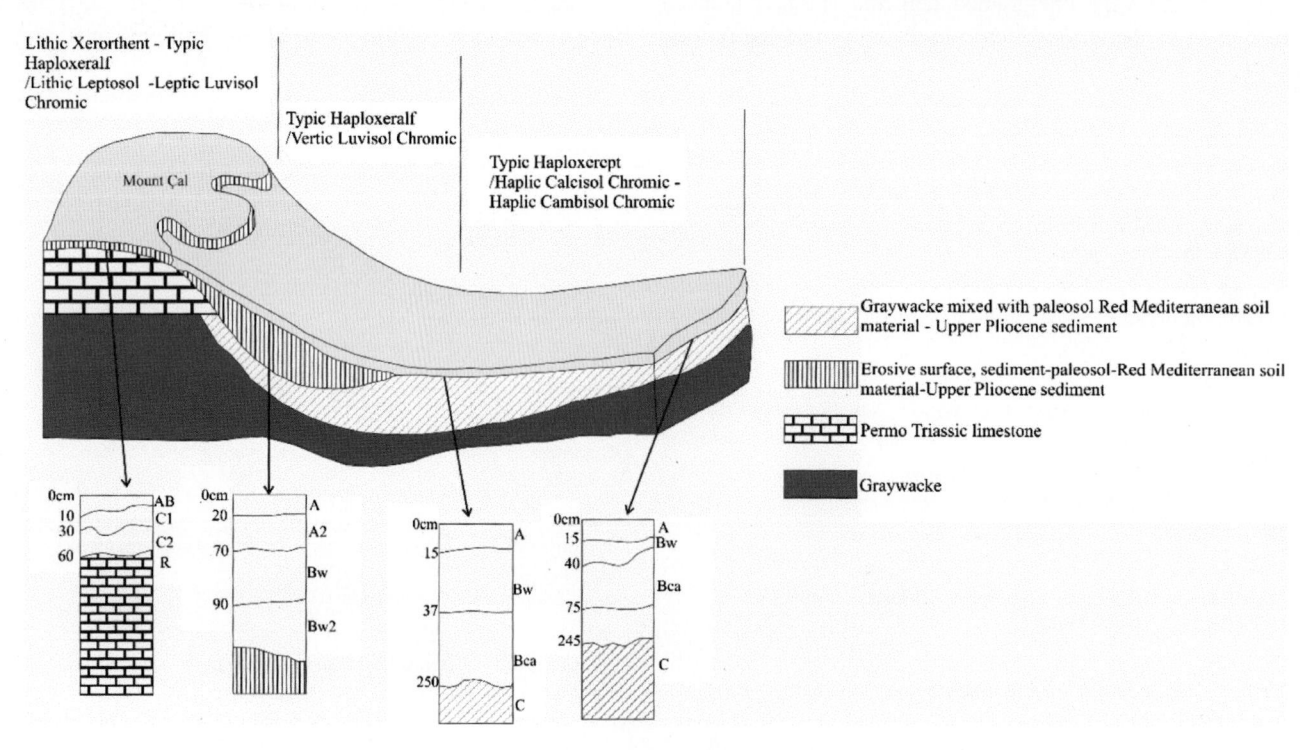

Fig. 1 The Leptosol—Luvisol—Calcisol—Cambisol toposequence of the 'Ankara Clay,' Ankara, central Turkey (Cangir and Kapur 1984)

Fig. 2 Formation of Luvisols in a karstic environment (Atalay 2011)

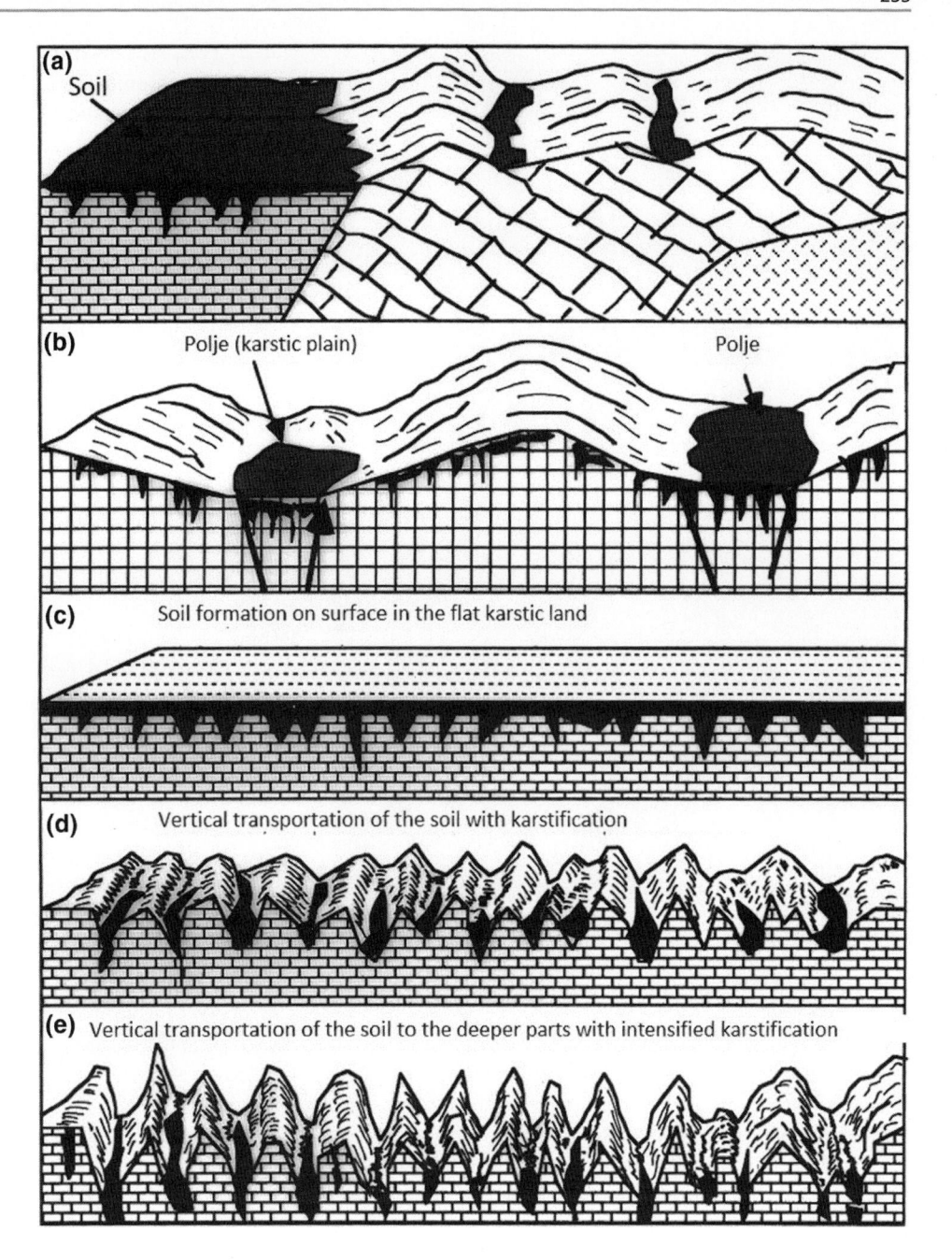

(1953), Durand (1959), Yaalon (1959), Smolikova (1963), Mancini (1966a, b), Lamouroux and Segalen (1969), Bronger (1978), and Bronger et al. (1983) for similar soils elsewhere. This theory is still disputed due to the probable ongoing contemporary formation of the Terra Rossa in the Mediterranean climatic zone. These authors proved that Mediterranean soils were mainly relict soils which have formed and developed under much moister tropical and subtropical climates of the late Tertiary and early Pleistocene periods. However, studies by Lamouroux (1971) and Verheye (1973) in Lebanon have illustrated that soil formation is still active today, and that they can at least be considered, as being formation-wise, in equilibrium with the present-day prevailing Mediterranean climate. Thomas and Southhard (2002) have also stated that the Xeralfs (Luvisols observed under xeric soil moisture conditions) on Pleistocene age landscapes in the central Valley of California are the weathering products of similar conditions to the current. Further, they state that these weathering processes may have been enhanced if cooler and wetter conditions prevailed during the Pleistocene. Ultimately, despite the numerous studies conducted on the subject, further research is needed on the formation of Luvisols/Red Mediterranean soils concerning the two different formation contexts related to age, i.e., the past and the contemporary. In this respect, Mancini (2002) stated that the formation of the Luvisols in the

Fig. 3 A Luvisol in Çeşme, İzmir (west Turkey) on a karstic terrain (Lm: Limestone)

mountainous areas and the Luvisols in the littoral flat plains would serve to understand the contemporary and paleo-climatic formation modes and their interrelations. The question on the age of Red Mediterranean Soils is still unsolved, and most probably both age-concerned theories hold a basis of truth. In this respect, Torrent (2004) refers to the intrinsic mosaic of Red Mediterranean soils observed in Spain and formed on both Pliocene *rañas* and Lower Pleistocene surfaces, represented by different pediments and river terraces (Verheye and de la Rosa 2006). Moreover, late Holocene evidence at sites of dwarf mammoth trampling related to illuviated clays in Sardinia (Italy) (2016, ongoing studies on Sardinian soils by F. Previtali—Milan-Bicocca University, C. Zucca—ICARDA, Amman, S. Madrau-Sassari University, Stefano Andreucci-Cagliari University, S. Kapur and E. Akça—Çukurova University) and well-accommodated macro-microstructure formation in the Luvisol dumps of Late-Roman periods in southern Turkey (Elaiussa Sebaste-Merdivenli Kuyu excavation site, University of Rome, La Sapienza) more or less document the contemporary formation capacity of the Luvisols (2016, ongoing studies by E. Akça-Adıyaman University and S. Kapur-Çukurova University).

4.2 The Aeolian Theory

However, studies conducted by Akalan (1957), Mermut et al. (1976), Kubilay et al. (1997), Kapur et al. (1991, 1993), and Fedoroff and Courty (2013) advocate the enrichment of the Luvisols by the addition of aeolian material from North Africa (Sahara Desert) and the Arabian Peninsula which is an ongoing process even today. The most striking episode among the numerous recorded is the one documenting a 50 g/m^2 dust fall to Adana on 16 March 1998. This process has been going on for millions of years after Sahara's formation (Yaalon 1997). On the other hand, Danin et al. (1983) calculated that it would take 2,000,000 years for a Terra Rossa soil in the Mediterranean basin to reach the depth of those near Jerusalem to accumulate by the dissolution and the weathering of 20 m of limestone. The validity of these earlier aeolian enrichment calculations were supported by P. Buring (1976, personal comm. Late Prof. of the Agricultural University of Wageningen, Netherlands), D.H. Yaalon (1993, personal comm. Late Prof. of the Hebrew University, Israel), N. Fedoroff (1993, 1996 and 2006, personal comm. Late Prof at Centre National de la Recherche Scientifique, Paris, France), A.R. Mermut (1990–2012, personal comm. Prof. Dr. in University of Harran, Şanlıurfa, Turkey), U. Dinç (1971–2000, personal comm. Late Prof. of the University of Çukurova, Adana, Turkey), F. Previtali (2000–2015, personal comm. Prof. of the University of Milan-Bicocca, Italy), P. Zdruli (2000–2012, personal comm. Prof. at MAI-B, Bari, Italy), and E.A. FitzPatrick (1973–2006, personal comm. Prof. of the University of Aberdeen, UK) in many occasions during field excursions and discussions in Turkey, Turkish Republic of Northern Cyprus, Syria, Tunisia, Morocco, and Italy.

Cangir (1982) also has stated that the calcic horizons of central Turkey formed from weathering of carbonate-rich materials transported by periodical aeolian action. This theory seems sound primarily due to the very low contents of acid insoluble residues in the west, south, and southeastern marine limestones of the Miocene in Turkey alike limestones elsewhere (Nihlen and Solyom 1986; Nihlen 1990; Pye 1992) that vary from 0.01 to 3% by weight (Kapur et al. 1976; Karaman 1995). Moreover, aeolian additions have been taking place in Turkey, Greece, and the Levant at least

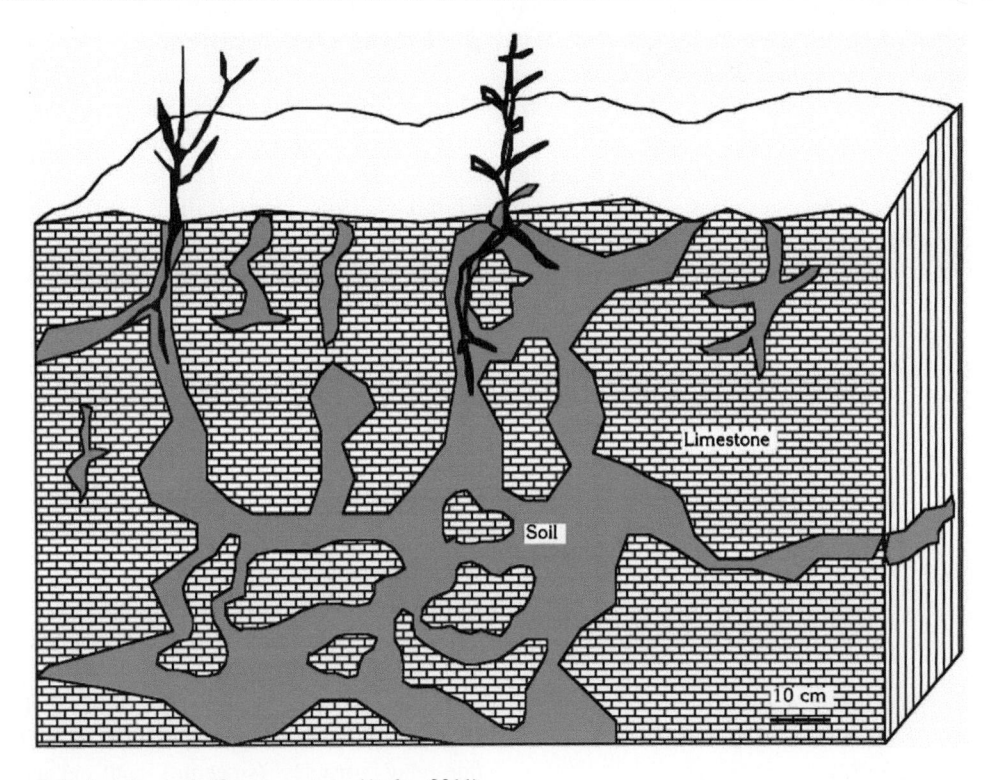

Fig. 4 Luvisols and the karstic cracks/fissure systems (Atalay 2011)

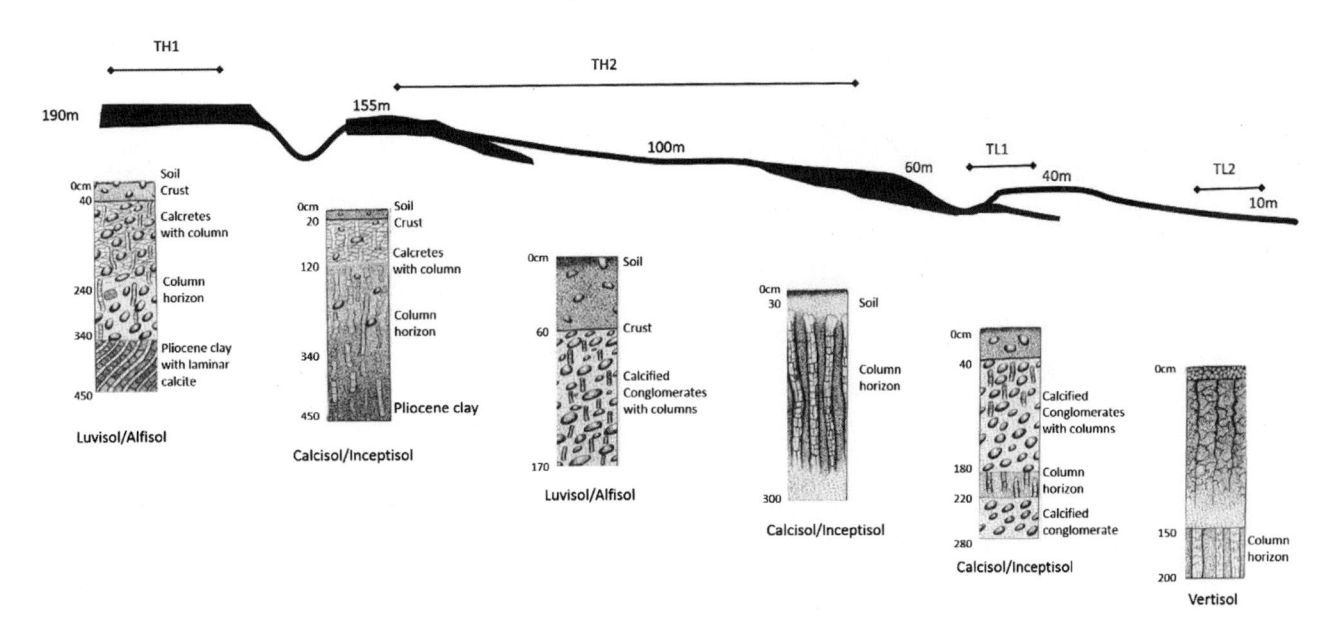

Fig. 5 The Luvisol—Calcisol—Vertisol toposequence as described by Kapur et al. (1975, 1987, 1990) in Adana, south Turkey

in the last 150,000 years (Yaalon and Lomas 1970; Yaalon and Ganor 1973; Macleod 1980; Yaalon 1997).

The contribution of the aeolian input from the Sahara (the east Erg in southern Tunisia) to the genetic properties, to the quality, and especially to the red color of the Red Mediterranean Soils (Rhodic/Chromic Luvisols) was first highlighted by Yaalon and Ganor (1973). Yaalon's views, based on a diligent study of meticulously analyzed dust samples collected for a period of 2-years from 30 well-selected stations, yielded contrasted outcomes to the dissolution theory of Reifenberg (1947). This highly significant pedologic contribution of D.H. Yaalon, as he had stated in 'The Yaalon Story' (Yaalon 2012), still needs further research concerning the particular climate that causes various strong and sudden winds which have an impact on the quantity and dispersion of the dust. The changes in the magnitude of the gusts and directions of the winds laden with Saharan dust have had significant impacts on the agricultural activities and the climatic stability of the Middle Eastern countries as well as the riparian countries of the Mediterranean Basin. Most regional soil scientists have concentrated on the subtle changes in the quality and fertility of the soils caused by the aeolian input and overlooked the evidences of the other detrimental effects of the Saharan dust. However, some of the recent studies conducted by the Turkish scientists on the effects of the Saharan dust on human health (activation of the Trigeminovascular System) and quality of precipitation (acidic and alkaline precipitation) deserve attention (Özsoy and Saydam 2000, 2001; Özsoy et al. 2000; Saydam and Şenyuva 2002; Sanın et al. 2005; Doğanay et al. 2009; Doğan et al. 2010; Bolay et al. 2010).

Fig. 6 A karstic (Miocene limestone/travertine) Luvisol in Güver, Antalya, Turkey

5 Profile Development

The Luvisols of Turkey with moderate eluviation/illuviation have A-Bt-C-R profiles (Altunbaş and Sarı 2009) (Figs. 6 and 7). The dominant Luvisols in the country are the Chromic/Rhodic Luvisols that have developed on karstic limestones (Table 1) with frequent Saharan dust input today and throughout the Quaternary period as stated by Yaalon and Ganor (1973) and Kubilay et al. (1997).

The aeolian materials of diverse origin contributing to the formation of the Luvisols and the Calcisols of the country are rich in quartz and calcite together with clay minerals (kaolinite, smectite, illite, and palygorskite). The aeolian materials were weathered by the organic acids released by the chemical and biological processes in the soils, most probably liberating the mobile elements of Ca and Na which were leached, while Mg, Fe (forming complexes with smectite), and K (fixed at interlayer sites by illite) showed moderate mobility and were likely enriched in the upper soil horizons (Özbek et al. 1979; Berner and Berner 1996; Kadir and Karakaş 2002; Kadir et al. 2008, 2013). The relative decrease of Ca in the upper horizons caused the dispersion of fine clay particles (smectite) and enhanced the illuviation/transportation (and in turn the rubefaction of the soil) of the Fe bearing smectitic clay particles to the lower horizons. This in turn has likely increased the Mg and K contents of the smectite + illite surfaces/interlayers in the upper horizons of the soils. Nevertheless, the relative increase of the organic matter and Fe either as a Fe (oxhydr-) oxide phase or Fe that substitutes some of the Mg in the octahedral site of clays, also increase the reddening/rubefaction at the upper soil horizons (Eren and Kadir 1999, 2013; Eren et al. 2015).

The dominant south westerly winds originating from the east Erg (part of the north Sahara Desert) in the south of Tunisia are laden with these sources of soil materials causing repeated calcification (recalcification) and fine grain/aggregate (of 10 μm average size) enrichment during spring and summer (Kubilay et al. 1997). The various shapes of quartz grains isolated from the Bw1 horizon of a Luvisol (Fig. 8; Table 2) formed on Miocene limestone in Reyhanlı (South Turkey) have proven the enrichment via aeolian material by Kapur et al. (1975), Kubilay et al. (1997, 2000), Koçak et al. (2012), and Kaplan et al. (2013). The grains vary from rounded ('feret' diameter ratio 1) to angular ('feret' diameter ratio 1.9) according to the classification of Pettijohn (1984).

The clay minerals determined in these soils (Tables 1 and 2) and other Luvisols developed on similar parent materials (highly crystalline Miocene limestones) are partly inherited from smectite, illite, and kaolinite present in the insoluble residues of the Miocene limestones in S Turkey (Kapur et al. 1975, 1987; Bal et al. 2003) and from palygorskite present in some crystalline limestones of the same area, primarily of

Fig. 7 Clay accumulation in Luvisol, Balcalı, Adana, South Turkey (plane-polarized light) (Kapur et al. 1991)

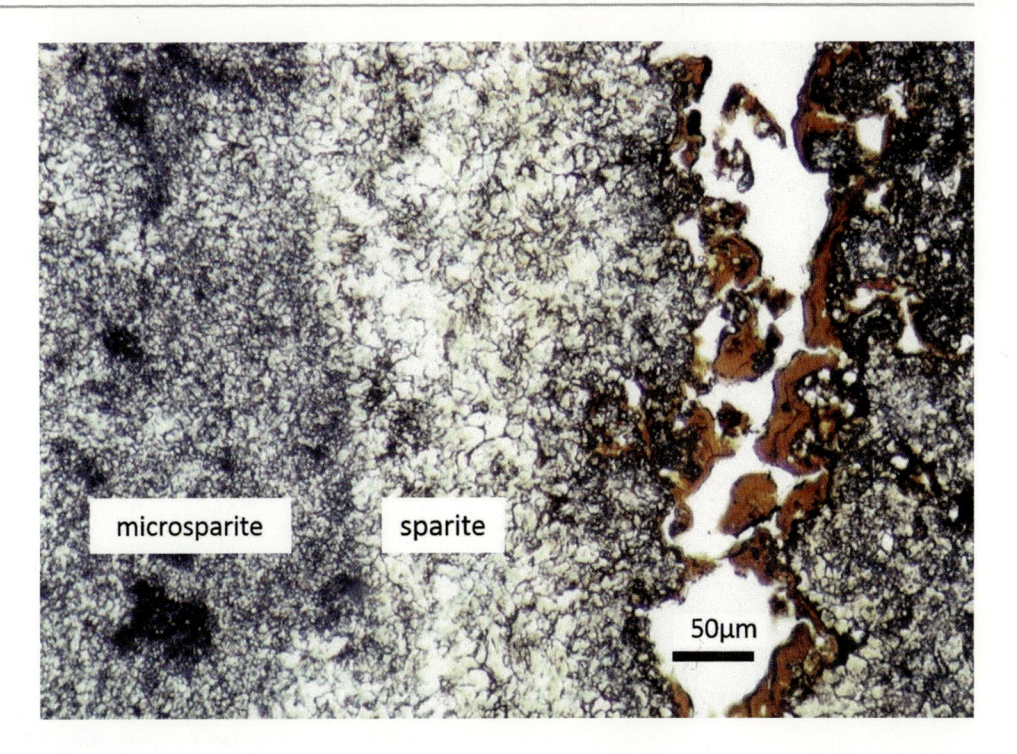

the Fatik limestone (Eocene—Oligocene) of Şanlıurfa, Southeast Turkey (Yılmaz 1999) (Fig. 9). The clay mineral suite of the dominant deep Luvisols formed on the Oligocene marine deposits of Tekirdağ (Thrace, northwest Turkey) is similar to the Luvisols developed on limestones in the other parts of the country (Cangir et al. 1987). The clay mineral suites of the Luvisols overlying the İslambeyli Formation and the Strandja (Istranca) Massif are rich in smectite and illite followed by kaolinite that originated from siliciclastic and pyroclastic materials.

The argic horizons of the Luvisols are generally reddish in color. This is due to the iron-rich smectitic (Ferrogenious beidellite, Özkan and Ross 1979) composition of the illuviated clay size particles deposited on mineral surfaces or in pore spaces as reddish clay coatings of a red hue (Fig. 10a, b).

Some reddish smectite-palygorskite clay coatings of the argic horizons in the Luvisols of the Adana province may most likely reflect neoformation and/or transformation of palygorskite from smectite (Fig. 11). These Luvisols are situated on the calichified conglomerates of the TH1 Villafrancian (late Pliocene to early Pleistocene) aged geomorphic calcrete/soil surfaces that commonly occur within the Luvisol-Calcisol-Cambisol toposequences (see Calcisols). This specific geomorphic continuum has been developed during the glacial–interglacial (earlier named as the pluvials–interpluvials) climatic fluctuations of the Pleistocene (Kapur et al. 1991).

Fedoroff and Courty (2013) have recently associated this phenomenon with the ongoing rubefaction and illuviation processes in soils where infiltration exceeds evapotranspiration which induces carbonate dissolution and leaching out of the profile. Consequently, goethite is altered to hematite, and the red color [outcome of present or earlier warm climates, IUSS Working Group WRB (2015)] is distributed throughout the profile by means of clay illuviation (Torrent 2004). Earlier studies conducted by Kapur et al. (1975) and Güzel et al. (1978) have stated that lepidocrocite was the fine sand size reddening mineral, whereas, maghemite (the readily dehydrated form of lepidocrocite) was the reddening material of the clay size fractions in the Luvisols (Terra Rossa) formed on calcretes (Adana, southern Turkey) and basalts (Kilis, southern Turkey). Tite and Mullins (1971) determined 35% of maghemite in the x-ray amorphous fraction of a Vertic Luvisol (Basaltic Red Soil) in the Mediterranean climatic region of Turkey formed on the Kilis basalts as the clay size reddening materials determined by magnetic susceptibility. Following iron mineral studies by Tite and Linington (1975) were conducted on the volcanic soils of different sites extending the knowledge on initial pedogenesis. Moreover, Eren and Kadir (1999, 2013), Eren et al. (2015) studied the iron minerals of paleosols and soil

Table 1 Site, profile description, and some physical and chemical properties of the Güver Soil (south Turkey)

IUSS Working Group WRB (2015)	Rhodic Luvisol
Soil Survey Staff (2014)	Typic/lithic Rhodoxeralf
Location	Skirts of northwest Güver cliff, Antalya, south Turkey
Coordinates	36°57′ 42″ N, 30°33′ 24″ E
Altitude	350 m
Landforms	Medium gradient hill, karstic
Position	Mid-slope
Aspect	–
Slope	6–12%
Drainage class	Well drained
Groundwater	None
Eff. Soil depth	90 cm
Parent material	Miocene limestone/travertine
Köppen climate	Mediterranean
Land use	Natural vegetation

Horizon	Depth (cm)	Description
A	0–15	Dark reddish brown (5YR 3/6 dry, 5YR 3/6 moist), clay, medium subangular blocky, sticky and plastic when wet, very slightly effervescent, dense stoniness within 10 cm, abundant fine roots, dense biological activity, 2–30 cm stones on surface, clear boundary
Bt1	15–33	Dark reddish brown (2.5YR 3/6 dry, 2.5YR 3/4 moist), clay, strong subangular blocky, hard when dry, friable when wet, sticky and plastic when wet, slightly effervescent, abundant fine roots, weak slickensides, clear smooth boundary
Bt2	33–52	Dark reddish brown (2.5YR 3/6 dry, 2.5YR 3/6 moist), clay, medium prismatic, very hard when dry, friable when wet, sticky and plastic when wet, sparse fine roots, slickensides, clear smooth boundary
Bt3	52–68	Dark reddish brown (2.5YR 3/4 dry), reddish brown (2.5YR 4/6 moist), clay, medium sub angular blocky, hard when dry, friable when wet, clear wavy boundary
C	60–90	Weathered crystalline limestone

Horizon	Depth (cm)	OM (%)	CaCO$_3$ (%)	Exchangeable bases (cmol$_c$ kg^{-1})				CEC*
				Ca	Mg	K	Na	
A	0–15	3.7	1.8	27.05	1.7	0.89	0.06	34.76
Bt1	15–33	3.2	1.6	27.20	1.61	0.46	0.10	35.96
Bt2	33–52	3.1	1.6	28.98	1.46	0.34	0.13	34.80
Bt3	52–68	2.8	3.8	35.49	1.37	0.31	0.12	32.39
C	68–90	2.2	35.7	35.89	1.05	0.34	0.17	26.06

Horizon	Color Dry	Color Moist	pH (H$_2$O—1:1)	Sand (%)	Silt (%)	Clay (%)
A	5YR 3/6	5YR 3/6	6.5	22.7	31.9	45.4
Bt1	2.5YR 3/6	2.5YR 3/4	6.2	19.7	28	52.3
Bt2	2.5YR 3/6	2.5YR 3/6	6.6	21.6	22.9	55.5
Bt3	2.5YR 3/4	2.5YR 4/6	7.1	21.3	19.3	59.4
C	–	–	7.4	35.3	23.3	41.4

* Extractable Ca may contain Ca from calcium carbonate or gypsum. CEC base saturation set to 100.

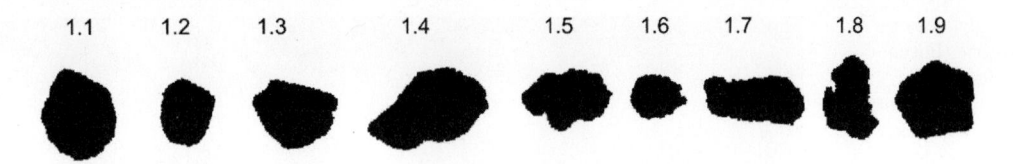

Fig. 8 Shapes of the quartz grains (50–150 μm) in the Bw horizon of a Luvisol from Reyhanlı (south Turkey)

Table 2 Site-profile description and some physical and chemical properties of the Reyhanlı Soil, Reyhanlı, Hatay, South Turkey (Kapur et al. 1975)

IUSS Working Group WRB (2015)	Rhodic Luvisol
Soil Survey Staff (2014)	Typic Rhodoxeralf
Location	2 km southwest of Reyhanlı town, Hatay, south Turkey
Coordinates	36°15′ 54″ N, 36°33′ 24″ E
Altitude	140 m
Landforms	Limestone plateau
Position	Mid-slope
Aspect	–
Slope	0–2%
Drainage class	Well drained
Groundwater	None
Eff. Soil depth	60 cm
Parent material	Limestone
Köppen climate	Mediterranean
Land use	Cultivated
Vegetation	Cereal

Horizon	Depth (cm)	Description
Ap	0–24	Dark reddish brown (2.5YR 2/4 moist, 2.5YR 3/4 dry), clayey, strongly granular to poorly developed angular blocky, hard when dry, sticky and plastic when wet, abundant fine roots, rare carbonate concretions, angular and slightly rounded, cherts and limestone fragments on the surface, clear and wavy boundary
Bt1	24–41	Dark reddish brown (2.5YR 2/4 moist), dark red (2.5YR 3/6 dry), clayey, well-developed medium angular, very hard when dry, sticky and plastic when wet, occasional fine roots, clear and smooth boundary
Bt2	41–60	Dark reddish brown (2.5YR 2/4 moist), dark red (2.5YR 3/6 dry), clayey, well-developed medium prismatic, very hard when dry, sticky and plastic when wet, rare fine roots, clear and irregular boundary
C1	60–78	Dark reddish brown (2.5YR 3/4 moist), clayey, moderately developed prismatic, very hard when dry, sticky and plastic when wet, occasional carbonate rock fragments and concretions, clear and wavy boundary
C2	78–91	Pink (5YR 7/4 moist, 5YR 8/4 dry), weathered limestone with red clay coatings on ped surfaces, and weathered rock fragments, clayey, well-developed medium granular, clear and wavy boundary

Horizon	Depth (cm)	OM (%)	CaCO$_3$ (%)	Exchangeable bases (cmol$_c$ kg^{-1})				
				Ca	Mg	K	Na	CEC
Ap	0–24	1.2	0.1	12.3	22	13.4	0.2	47.9
Bt1	24–41	1.1	0.6	22.5	21	5.3	2	50.8
Bt2	41–60	0.9	0.3	21.8	16.3	4.7	2.4	45.2
C1	60–78	1.0	3.9	28.9	5.2	4.8	2.6	41.5
C2	78–91	0.3	57.4	25.5	1.3	3.5	2.4	32.7

Horizon	pH (1:1)	Sand (%)	Silt (%)	Clay (%)
Ap	6.2	5.5	31.5	63
Bt1	7.0	4.3	19.2	76.5
Bt2	6.4	4.3	23.0	72.7
C1	7.3	16.5	16.8	66.7
C2	7.4	40.3	11.2	48.5

Fig. 9 Isolated palygorskite from Fatik limestone (Eocene-Oligocene), Şanlıurfa (SE Turkey) (cross polarized light) (archive of the late Prof. Dr. Mahmut Sayın, University of Çukurova, Adana, Turkey)

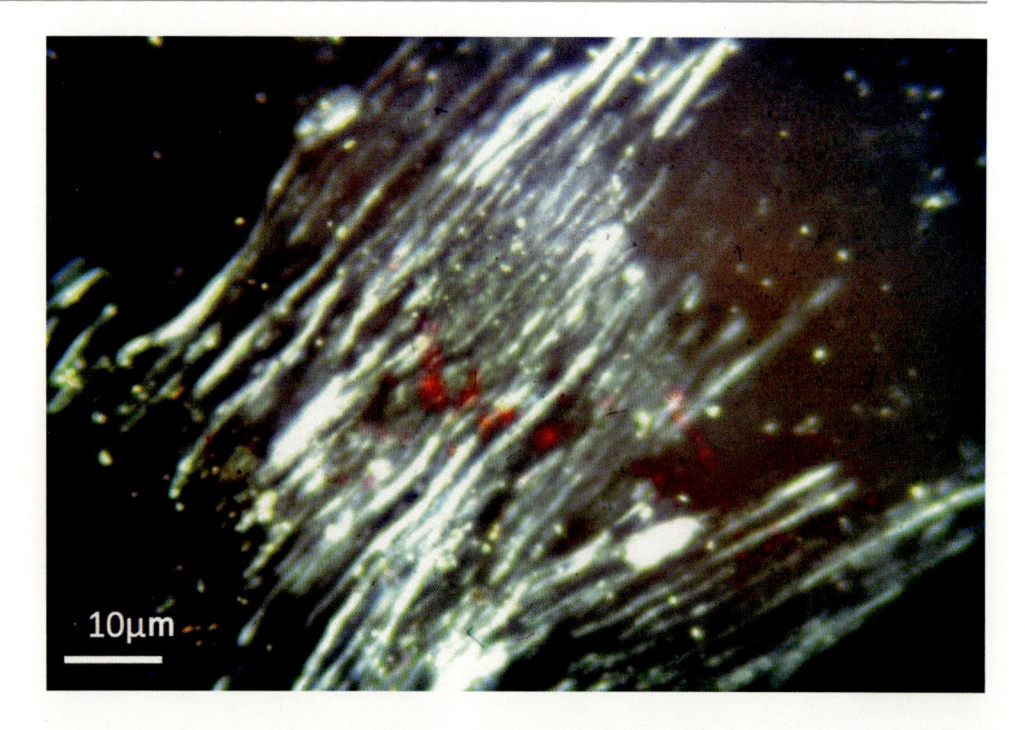

forming volcanic materials of southern Turkey revealing primary and initial pedogenesis.

6 Regional Distribution

Many detailed monographs on the distribution of the Red Mediterranean soils have been produced by Altunbaş and Sarı (1998), Atalay (1997), Bech et al. (1997), Darwish and Zurayk (1997), Yassoglou (1997), and Noulas et al. (2009). The Luvisols and the associated Leptosols generally form on sharp topographies as well as on flat or gently sloping landscapes under climatic regimes that range from semi-cold temperate to warm Mediterranean. Luvisols widely occur on the karstic terrains of the slopes of the Taurus mountains (the Taurides) throughout Turkey, and namely in the South (the Turkish Mediterranean coastal areas), Southeast (the Gaziantep-Şanlıurfa Plateaus and the Midyat Formation), Center, and West parts (the Aegean region) of the country, i.e., in the Mediterranean region of Turkey where Rhodic, Chromic, Calcic, and Vertic Luvisols are common on limestone or on the colluvial deposits of limestone and/or on weathering calcretes (Fig. 12). Luvisols of Turkey cover a land area of about 145,895,598 ha which is 2.05% of the whole agricultural land.

The widely distributed limestone/travertine terraces of Antalya (south Turkey) are among the best examples in the world for the karst/travertine surface features. The true Rhodic/Chromic Luvisols (Red Mediterranean Soils) of Turkey have developed within the depressions and/or crack systems of these topographic features. Soils have been reworked by the strong and deep lying root-system of the well-established Mediterranean maquis vegetation succeeding the earlier widespread Red Pine (*Pinus brutia*) forests that have been under high population pressure within millennia. This hardy vegetation regenerates with root suckers and deep roots that induce development of the Luvisols (Fig. 13) (Atalay 1997, 2008, 2011; Atalay and Efe 2008).

7 Management

The Luvisols of Turkey are mostly located under the Mediterranean climatic zone, and generally suitable for a wide range of agricultural uses. They have a high clay content and a well-developed soil structure suitable for orchards, greenhouses, vegetable and vineyard production, and/or grazing (Dinç et al. 2001). They are also suitable for legume rotations coupled with fallow for the preservation, improvement, and sustainability of the well- and

Fig. 10 Reddish clay coatings in the Reyhanlı soil (south Turkey), in plane-polarized light (**a**) and cross-polarized light (**b**)

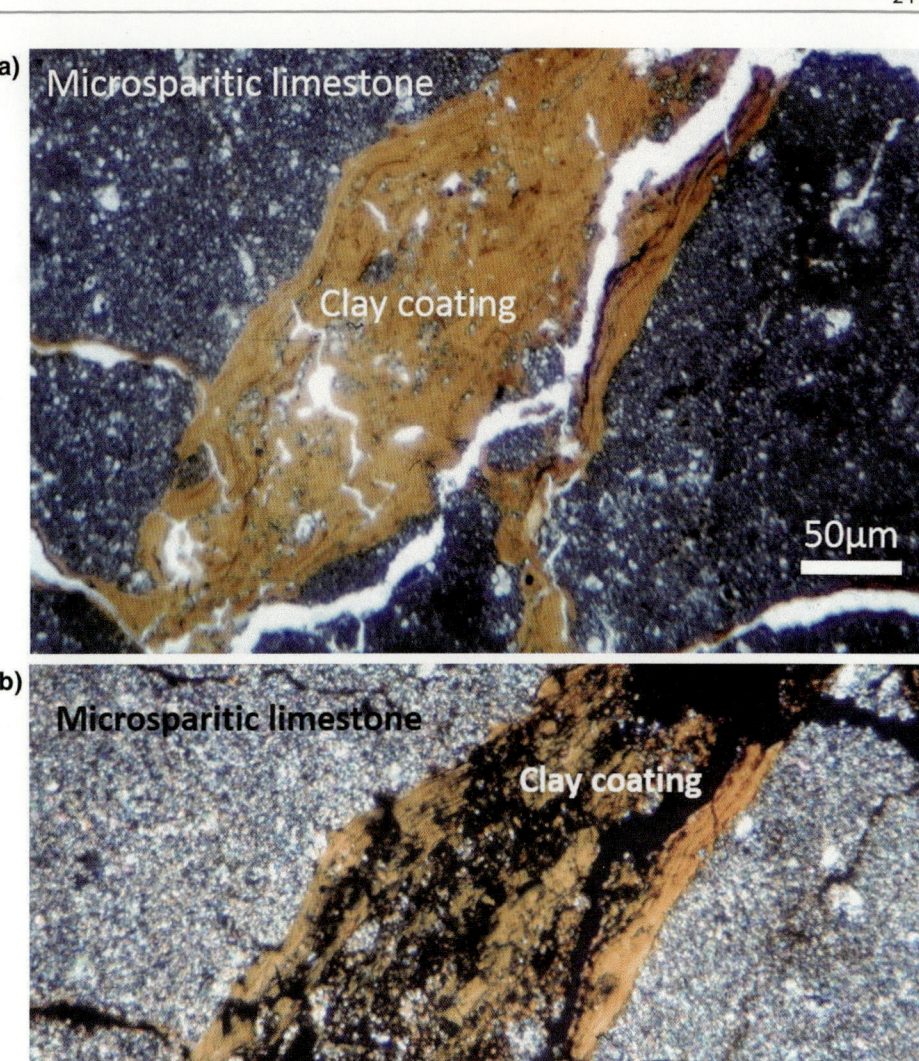

long-established stable microstructure, in the semiarid parts of Syria and Turkey (Kapur et al. 1975, 2007). Grazing is the most appropriate use of the extensive Leptosols/Luvisols located especially in the highlands of central and east Turkey, i.e., the mountainous Van (East Turkey) (see Calcisols) and Hakkari Luvisol/Leptosol ecosystems. Other moderately deep and shallow Luvisols and Leptosols are highly suitable for forest and maquis cover due to their clay mineral contents, soil structure, and porosity.

Luvisols are partly conserved at steep slopes standing against erosion and have developed in deep cracks along the main roots of the long-standing forest and horticultural trees

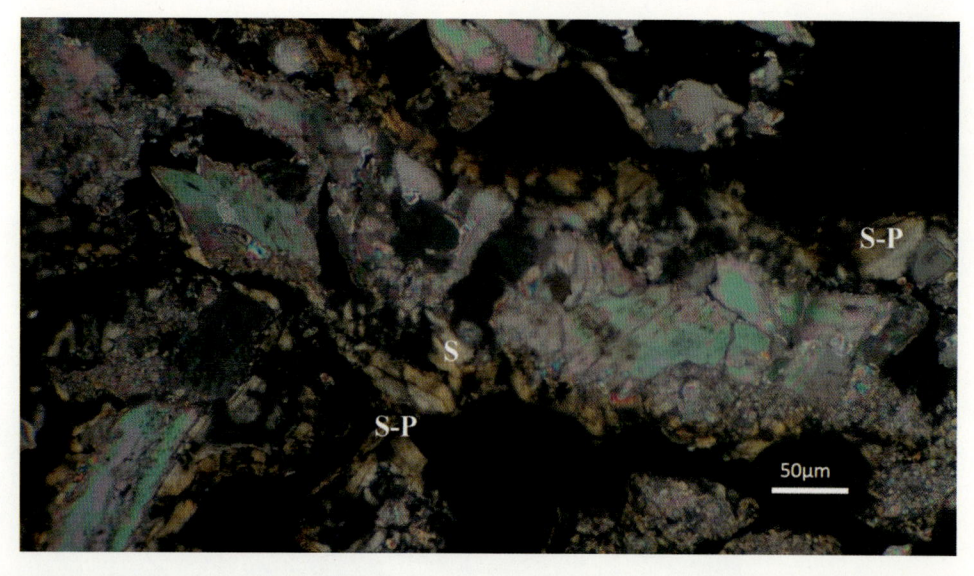

Fig. 11 Neoformed/transformed palygorskite from smectitic clay coating in argic horizon of the Luvisol of the TH1 Villafrancian surface in Adana, South Turkey (*S* smectite, *S-P* smectite-palygorskite) (cross-polarized light)

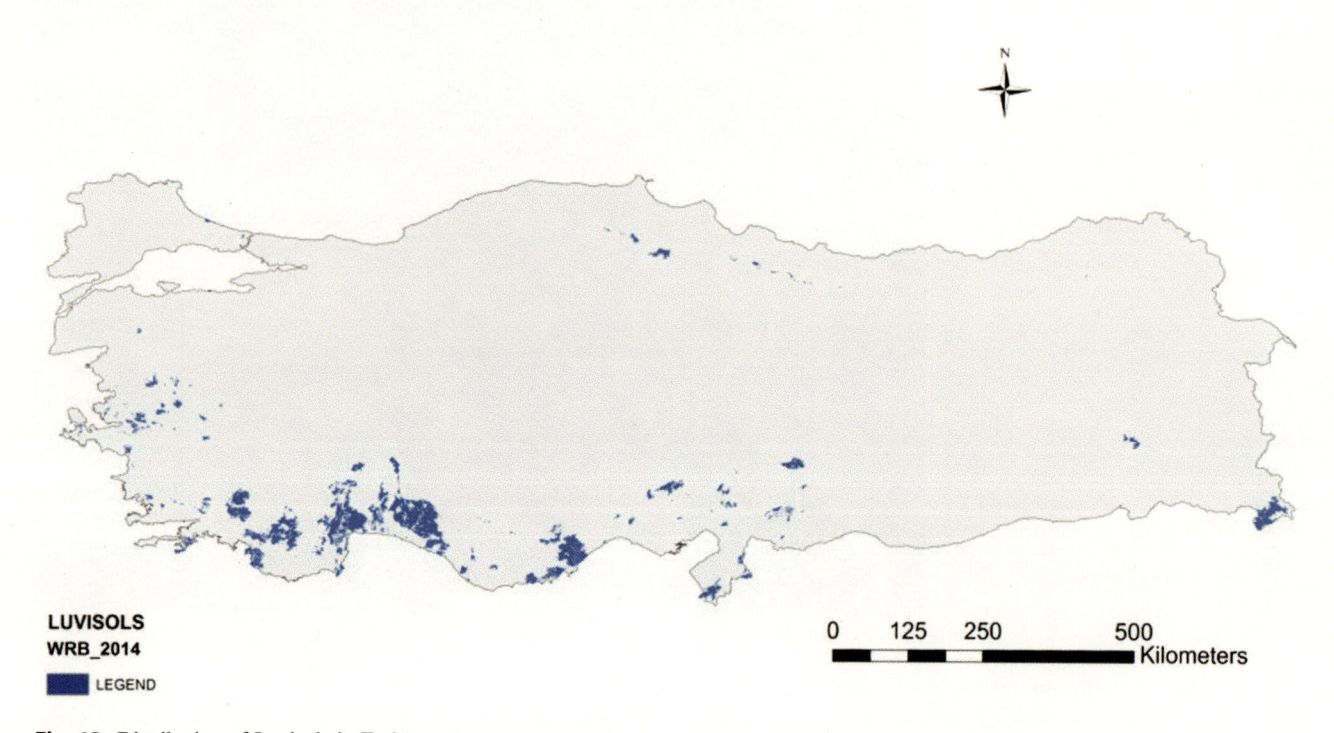

Fig. 12 Distribution of Luvisols in Turkey

(Fig. 14). The Luvisols overlying limestones and developing along the karstic cracks (dissolution chambers/channels/solution pipes) are responsible for the high water storage capacity of the Karst/Luvisol topography and are the naturally highly carbon sequestering sites of Turkey and of similar sites in the riparian countries of the Mediterranean Basin. Such soils have been designated as the 'tube soils' (a term coined by Prof. Dr. Mustafa Sarı, Akdeniz University, Antalya,

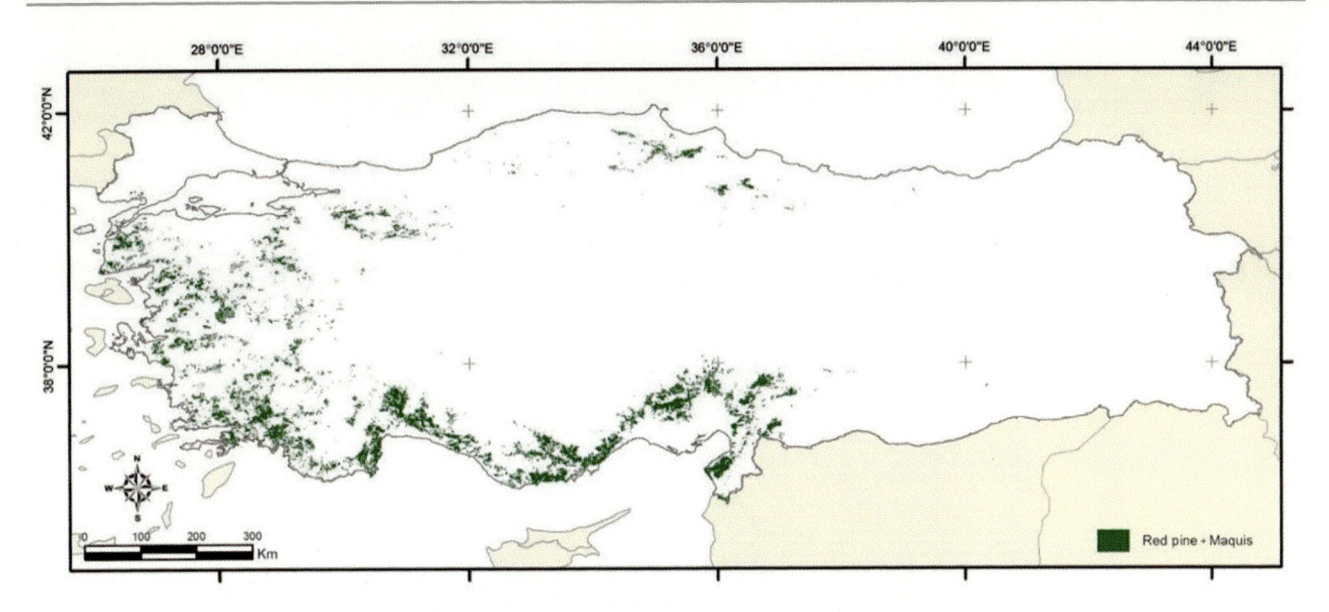

Fig. 13 The Red Pine (*Pinus brutia*) and maquis distribution in Turkey

Fig. 14 Luvisols (Tube soils), overlying and filling solution pipes in limestone (**a**) and developing along the karstic cracks (**b**) (south Turkey)

Turkey, Sarı et al. 1986) in Turkey especially for the traditional Luvisol/Leptosol ecosystems of the Aegean, Mediterranean, and southeast Turkey regions of the country (Fig. 14). These soils are significant because of their higher organic carbon contents compared to the shallow soils of the same regions overlying the karstic limestones of the Tauride slopes (Fig. 15).

Sequestration of soil organic carbon by the Chromic Cambisols of Mersin (south Turkey), under forest cover, at the elevations of 300–500 m, is enhanced by the activity of

Fig. 15 Olive vegetation on the Karst area at the skirts of the Taurides, showing deeply dissolved limestone and 'Tube soils,' Mersin (south Turkey)

the roots of the Red Pine (*Pinus brutia*), Black Pine (*Pinus nigra*), and the maquis vegetation via especially the increased fine pores and water retention capacity (Polat 2012). Whereas, the organic carbon sequestration in the 'Tube soils' (Luvisols) is enhanced by deeply penetrating pistachio and olive tree roots or canopies of the semiarid Mediterranean natural vegetation (Figs. 14 and 15).

Most of the karstic Luvisols are rich in kaolinite followed by smectite and illite. One of the primary agricultural advantages of the illite-rich Luvisols of the Mediterranean region of Turkey is their high K-retention ability and in turn their suitability in supporting the growth of the high carbohydrate producing crops such as the melon, watermelon, tomatoes, and cotton (Özbek et al. 1979). The high amounts of extractable and exchangeable K (depending on the weathering environment and cation concentrations of the clay surfaces) contents, related to the illite and montmorillonite contents in some of the Luvisols formed on calcretes and basalts, were also stated to be a significant merit in crop production by Güzel and Wilson (1981). Leptosols derived from the erosion of Luvisols are best used as forest land and well-managed grasslands or recreation areas. On the other hand, Leptosols overlying the hard and partly impermeable limestone eroded from calcretes/Calcisols are more suitable to olive orchards due to their clay mineral contents (smectite and palygorskite). However, Leptosols overlying weathered/weathering limestones, i.e., containing an appreciable solum—a sufficiently deep C horizon—may be allocated to rangeland management at appropriate sites (MEF 2006).

Many soil/crop-wise well-established Rhodic/Chromic Luvisol (Terra Rossa) areas/Luvisol ecosystems-Terroirs (parts of an Anthroscape), have been the major human development and settlement sites since the beginning of the Quaternary (Early Pleistocene) and particularly the classical periods and much earlier in the Neolithic Göbekli Tepe near Şanlıurfa, southeast Turkey (ca 12 K BP, Kromer and Schmidt 1998; Akça and Kapur 2014) (Fig. 16).

The ecosystems of Luvisols similar to the Fluvisol ecosystems were earlier or contemporary to the establishment of the first settlements of the Neolithic. These Luvisol ecosystems have especially been under the threat of soil/land degradation since the early human intervention. However, contemporary soil ecosystem disturbances have caused more intense degradation on the Luvisols. An appropriate example for this was the great job seeking migration of the 1960s, from the rural areas to the urban zones of Turkey that have caused intense and inappropriate urban development.

One of the other major examples of soil ecosystem disturbances and land degradation in the country was the unsuitable urban expansion progressing on the 'Ankara Clay.' This uniformly deposited widespread Luvisol is the renowned Ankara Clay rich in inherited kaolinite at the site of formation—Çal Mountain—and is followed by higher contents of smectite and illite in soils at farther locations of the glacis, i.e., farther to the source of the Ankara

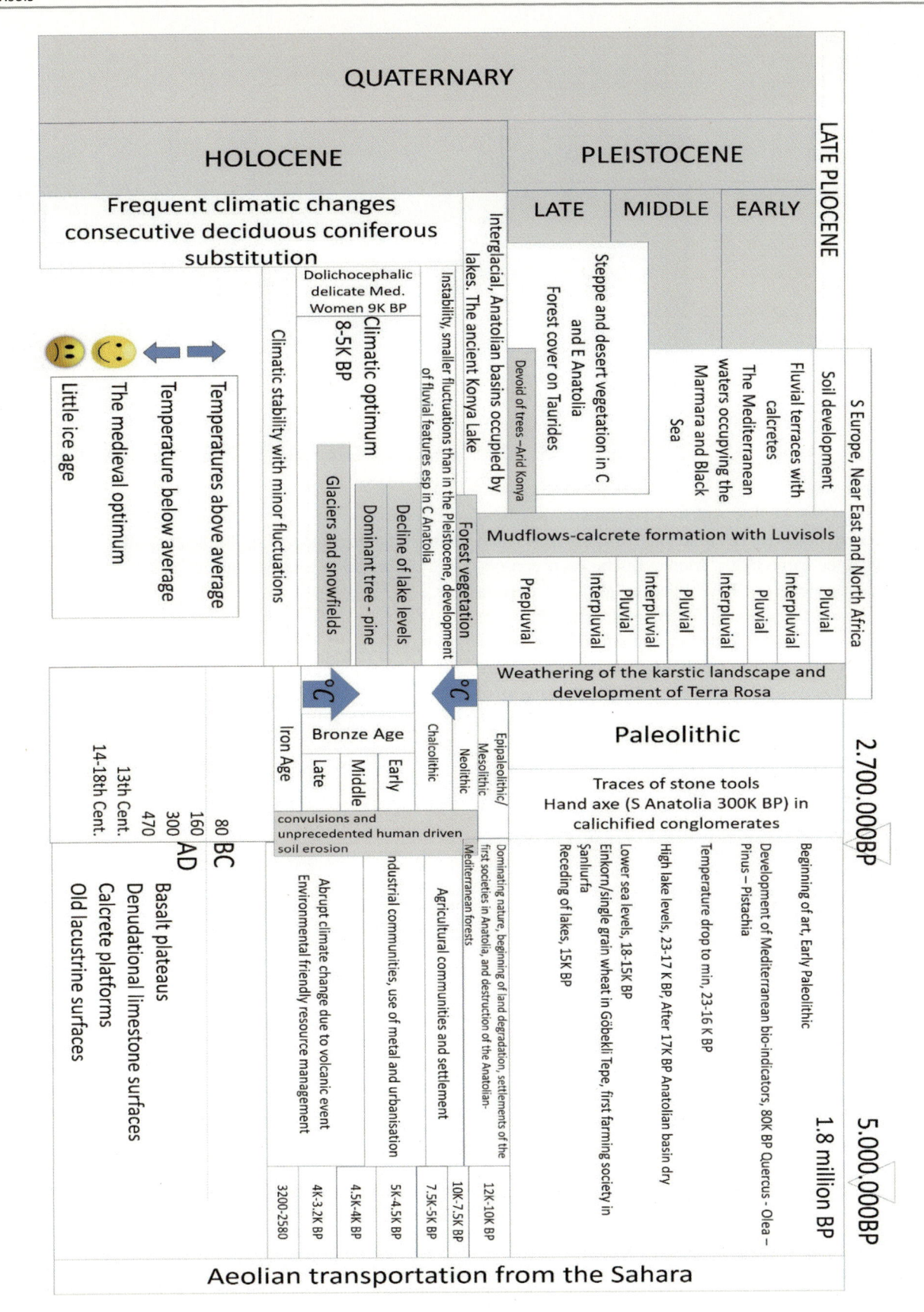

Fig. 16 The late Quaternary history of Turkey and the advent of the Rhodic/Chromic Luvisols (Terra Rosa) in the late Pliocene-early Pleistocene (modified from Kapur et al. 1999)

Fig. 17 Urban sprawl on the Ankara Clay-Luvisol ecosystem (archive of Prof. Dr. Koray Haktanır, University of Ankara, Turkey)

Clay/Luvisol, (Fig. 1). The Ankara Clay Luvisol has been inappropriately used as construction materials (tiles and bricks) since the early 1920s. Moreover, the deep consolidated calcretes along the S–W Pliocene glacis of Ankara were inappropriately allocated for urban development. The inappropriate urban development and misuse of these well-established soil/crop ecosystems—the Luvisols—were the primary examples of soil sealing in Turkey taking over the millennia-long traditional vine/grape, apricot, and almond Terroirs (Costantini et al. 2012) of Turkey.

The 'Ankara clay' Luvisols were also utilized as engineering materials at the foundations of the ever-developing constructions of Ankara. In this context Met et al. (2005) have stated that the 'Ankara clay' Luvisols with high cation exchange capacities (CEC) may be inappropriate for building infrastructures, whereas the lower CEC 'Ankara clays' may be appropriate for utilization as landfill liners. Kasapoğlu and Kiper (1985) have also mentioned the inappropriate properties of the Ankara Clay materials concerning tunnel constructions for underground transportation due to their high swelling capacity (high smectite contents) with water and breakdown (slaking due to kaolinite) after excess hardening (Cangir and Kapur 1984). Cangir and Kapur (1984) have mentioned the high probability of landslide occurrence along the layers of the Ankara Clay overlying massive calcretes and consequently the destruction of roads and buildings. Special high-cost technologies, which have been a great burden to the country's economy, are adopted in building skyscrapers (renovated settlements) and their reinforced foundations in order to avoid such landslides in the south of Ankara (Fig. 17).

Similar Luvisol ecosystems in Tekirdağ (northwest Turkey) earlier under forest cover were determined to be highly prone to landslides over a steeply sloping topography. This area alongside the Ankara Clay ecosystem highlights the susceptibility of the sloping Luvisols to infra and ultra-structures and the strong need to allocate them as nature conservation areas (Cangir et al. 1987).

References

Akalan İ (1957) Wind-borne soils (the 16 April 1957 dust storm episode). Ziraat Dergisi 158:8–12 (in Turkish)

Akça E, Kapur S (2014) The anatolian soil concept of the past and today. In: Churchman GJ, Landa ER (eds) The soil underfoot: infinite possibilities for a finite resource. CRC Press, pp 175–184

Altunbaş S, Sarı M (1998) Determination of relationships between the parent material and soil on Red Mediterranean soil. International symposium on Arid soils, İzmir, Turkey

Altunbaş S, Sarı M (2009) The Relationships of iron contents between red mediterranean soils and its parent material in Antalya province, Turkey. Mediterr Agric Sci 22(1):15–21

Atalay İ (1997) Red Mediterranean soils in some Karstic regions of Taurus Mountains Turkey. In: Mermut AR, Yaalon D, Kapur S (eds) Catena SI, Red Mediterranean soils, vol 28, Issue 3–4, pp 247–260

Atalay İ (2008) Relict and endemic plant species reflecting climatic changes in Anatolia. In: Ecology and environment from carpathians to Taurus Mountain. In: Atalay İ, Efe R, Ielenicz M, Balteanu D (eds) Proceedings of 5th Turkish-Romanian Academic Seminar

Atalay İ, Efe R (2008) Ecoregions of the Mediterranean area and the Lakes region of Turkey. In: Atalay İ, Efe R (eds) Proceedings of international symposium on geography

Atalay İ (2011) Soil formation, classification and geography, 4th edn. Meta Press, İzmir

Atalay İ, Efe R, Öztürk M (2014) Effects of topography and climate on the ecology of Taurus Mountains in the Mediterranean Region of Turkey. Procedia—social and behavioral sciences. In: 3rd international geography symposium, GEOMED 2013

Bal Y, Kelling G, Kapur S, Akça E, Çetin H, Erol O (2003) An improved method for determination of Holocene coastline changes around two ancient settlements in southern anatolia: a geoarchaeological approach to historical land degradation studies. Land Degrad Dev 14(4):363–376

Bech J, Rustullet J, Garrigo J, Tobias FJ, Martinez R (1997) The iron content of some red Mediterranean soils from northeast Spain and its pedogenic significance. In: Mermut R, Yaalon D, Kapur S (eds) Catena SI, Red Mediterranean soils, vol 28, pp 211–229

Berner EK, Berner RA (1996) Global environment: water, air, and geochemical cycles, 376 p

Bolay H, Doğanay H, Göktaş T, Çağlar K, Erbaş D, Saydam AC (2010) African Dust-Laden atmospheric conditions activate the trigeminovascular system. 62nd annual meeting of the American-Academy-of-Neurology. Neurology 74(9):A11–A12

Bronger A (1978) Climatic sequences of steppe soils from eastern Europe and the USA with emphasis on the genesis of the "argillic horizon". Catena 5(1):33–51

Bronger A, Ensling J, Gütlich P, Spiering H (1983) Rubefaction of terrae rossa in Slovakia: a Mösbauer effect study. Clays Clay Miner 31(4):269–276

Cangir C (1982) The morphology and genesis of Brown, Reddish Brown, Terra Rossa, Rendzina and Grumusolic soils formed on calcareous materials. Habilitation Thesis, the Faculty of Agriculture, University of Ankara, 135 p (in Turkish)

Cangir C, Kapur S (1984) Toposequential relationships between the Ankara-Dikmen paleosols (Ankara Clay) and the pedoliths. In: Proceedings of the 1st national clay symposium, 21–26 Feb 1983. University of Çukurova Publciation Adana, Turkey, pp 261–281 (in Turkish)

Cangir C, Kapur S, Ekinci H (1987) Landslide hazards and significance of the clay mineral contents of the Alfisols of Tekirdağ (north western Turkey) overlying marine deposits of the Oligocene. In: Proceedings of the 3rd national clay symposium, pp 177–188 (in Turkish)

Costantini EA, Bucelli P, Priori S (2012) Quaternary landscape history determines the soil functional characters of terroir. Quatern Int 265:63–73

Danin A, Gerson R, Garty J (1983) Weathering patterns on hard limestone and dolomite by endolithic lichens and cyanobacteria: supporting evidence for eolian contribution to Terra Rossa soil. Soil Sci 136(4):213–217

Darwish T, Zurayk RA (1997) Distribution and nature of red Mediterranean soils in Lebanon along an altitudinal sequence. In: Mermut R, Yaalon D, Kapur S (eds) Catena SI, Red Mediterranean Soils 28:191–202

Dinç U, Şenol S, Kapur S, Cangir C, Atalay İ (2001) Soils of Turkey. Çukurova University, Faculty of Agriculture Publication No: 51, Adana, 233 p (in Turkish)

Doğan TR, Saydam AC, Yeşilnacar M, Gencer O (2010) In-cloud alteration of desert-dust matrix and its possible impact on health: a test in southeastern Anatolia, Turkey. European Journal of Mineralogy 22(5):659–664

Doğanay H, Akçalı D, Göktaş T, Çağlar K, Erbaş D, Saydam AC, Bolay H (2009) African dust-laden atmospheric conditions activate the trigeminovascular system. Cephalalgia 29(10):1059–1068

Durand JH (1959) Les sols Rouges et les Croûtes en Algérie. Servlet Et. Science, Direct Hydraulics Equipment Rural, Birmandreis-Alger, Etude Générale No: 7, 187 p

Eren M, Kadir S (1999) Colour origin of upper Cretaceous pelagic red sediments within the Eastern Pontides, northeast Turkey. Geol Rundschau Int J Earth Sci 88:593–595

Eren M, Kadir S (2013) Color origin of red sandstone beds within the Hüdai Formation (Early Cambrian), Aydıncık (Mersin), Southern Turkey. Tur J Earth Sci 22:563–573

Eren M, Kadir S, Kapur S, Huggett J, Zucca C (2015) Color origin of Tortonian red mudstones within the Mersin area, southern Turkey. Sed Geol 318:10–19

Erol O (1973) The geomorphic units of Ankara. University of Ankara, Faculty of Letters, History and Geography Pub. No: 240. Ankara, pp 10–18 (in Turkish)

Erol O (1984) Neogene and quaternary continental formation and their significance for soil formation. In: Proceedings of the 1st national clay Symposium. University of Çukurova, Turkey, pp 24–28 (in Turkish)

Fedoroff N, Courty MA (2013) Revisiting the genesis of Red Mediterranean soils. Tur J Earth Sci 22:359–375

Güzel N, Kapur S, Özbek H (1978) A comparative pedological study of three Mediterranean Red Soils (terra-rossa) from southern Turkey, Vol 8, pp 236–253. Annals of the Faculty of Agriculture, University of Çukurova

Güzel N, Wilson MJ (1981) Clay-mineral studies of a soil chronosequence in southern Turkey. Geoderma 25(1):113–129

Hızalan E (1953) A comparative study on the Ankara-Dikmen and Mediterranean type of red soils. University of Ankara, Faculty of Agriculture Pub. No: 41, Ankara, 55 p (in Turkish)

IUSS Working Group WRB (2015) World reference base for soil resources 2014, update 2015. In: International soil classification system for naming soils and creating legends for soil maps. World Soil Resources Reports No. 106. FAO, Rome, 203 p

Kadir S, Karakaş Z (2002) Mineralogy, chemistry and origin of halloysite, kaolinite and smectite from Miocene ignimbrites, Konya, Turkey. Neues Jahrbuch für Mineralogie, Abhandlungen 177:113–132

Kadir S, Önen HP, Aydın SN, Yakıcıer C, Akarsu N, Tuncer M (2008) Environmental effect and genetic influence: a regional cancer predisposition survey in the Zonguldak region of Northwest Turkey. Environ Geol 54:391–409

Kadir S, Gürel A, Senem H, Külah T (2013) Geology of Late Miocene clayey sediments and distribution of paleosols clay minerals in the northeastern part of the Cappadocian Volcanic Province (Araplı-Erdemli), central Anatolia, Turkey. Tur J Earth Sci 22:427–443

Kaplan MY, Eren M, Kadir S, Kapur S (2013) Mineralogical, geochemical and isotopic characteristics of quaternary calcretes in the Adana region, southern Turkey: implications on their origin. Catena 101:164–177

Kaplan MY, Eren M, Kadir S, Kapur S, Huggett J (2014) A microscopic approach to the pedogenic palygorskite associated with Quaternary calcretes of the Adana area, southern Turkey. Tur J Earth Sci 23(5):559–574

Kapur S, FitzPatrick EA, Özbek H (1975) A pedological study of three soils from southern Turkey, vol 6. University of Çukurova, Annals of the Faculty of Agriculture, Adana, pp 73–98

Kapur S, Dinç U, Özbek H (1976) Mineralogical variations between two Miocene dolomitic limestones and the overlying weathered materials forming terra-rossas in Adana, Southern Turkey. University of Çukurova, Annals of the Faculty of Agriculture, Adana, pp 144–153

Kapur S, Çavuşgil VS, FitzPatrick EA (1987) Soil-calcrete (caliche) relationships on a Quaternary surface of the Çukurova region, Adana, Turkey. In: Fedoroff N, Bresson LM, Courty MA (eds) Soil micromorphology. L'Association Française pour l'Étude du Sol, pp 597–603

Kapur S, Çavuşgil VS, Şenol M, Gürel N, FitzPatrick EA (1990) Geomorphology and pedogenic evolution of quaternary calcretes in the Northern Adana Basin of Southern Turkey. Zeitschrift für Geomorphologie N F Bd Heft 1:49–59

Kapur S, Sayın, M, Gülüt, KY, Şahan S, Çavuşgil VS, Yılmaz K, Karaman C (1991) Mineralogical and micromorphological properties of Widely distributed soil series in the Harran Plain. TÜBİTAK TOAG-534, Ankara, pp 11–20

Kapur S, Yaman S, Gökçen SL, Yetiş C (1993) Soil stratigraphy and quaternary caliche in the Misis area of the Adana basin, southern Turkey. Catena 20:431–445

Kapur S, Ryan J, Akça E, Çelik İ, Pagliai M, Tülün Y (2007) Influence of Mediterranean cereal-based rotations on soil micromorphological characteristics. Geoderma 142(3–4):318–324

Kapur S, Atalay İ, Ernst F, Akça E, Yetiş C, İşler F, Öcal AD, Uzel İ, Şafak Ü (1999) A review of the late quaternary history of Anatolia. Special issue, Olba II, Mersin University Publication, pp 233–278 (in Turkish)

Karaman C (1995) The effect of the parent materials in the formation of the autochthonous terra rossa of the Mediterranean region. PhD Thesis. University of Çukurova, Graduate School of Natural and Applied Sciences. Adana, 140 p (in Turkish)

Kasapoğlu E, Kiper OP (1985) The Geo-technical properties of the Ankara clay. In: Gündoğdu N, Aksoy H (eds) Proceedings of the 2nd national clay symposium. Hacettepe University, pp 343–352 (in Turkish)

Koçak M, Theodosi C, Zampras P, Séguret MJM, Herut B, Kallos G, Nimmo M (2012) Influence of mineral dust transport on the chemical composition and physical properties of the Eastern Mediterranean aerosol. Atmos Environ 57:266–277

Kromer B, Schmidt K (1998) Two radiocarbon dates from Göbekli Tepe, South Eastern Turkey. Neo-Lithics 3(98):8–9

Kubiena WL (1953) The soils of Europe. The Murby and Cie, London, 317 p

Kubilay NN, Saydam C, Yemenicioglu S, Kelling G, Kapur S, Karaman C, Akça E (1997) Seasonal chemical and mineralogical variability of atmospheric particles in the coastal region of the Northeast Mediterranean. Catena 28:313–328

Kubilay N, Nickovic S, Moulin C, Dulac F (2000) An illustration of the transport and deposition of mineral dust onto the eastern Mediterranean. Atmos Environ 34:1293–1303

Lamouroux M, Segalen P (1969) Etude compare des produits ferrugineux dans les sols rouges et bruns mediterraneens du Liban. Sci Sol 1:38–63

Lamouroux M (1971) Etude de Sols formés sur Roches Carbonatées: Pédogenèse Fersiallitique au Liban. PhD Thesis, ORSTOM, Paris, 314 p

Laouina A, Watfeh A, Badraoui M (2004) The Red Mediterranean soils of Mamora: a fragile patrimony threatened by desertification. In: 8th international meeting on soils with Mediterranean type of climate. Extended abstracts, 71, Marrakech, Morocco

Macleod DA (1980) The origin of the red Mediterranean soils in Epirus, Greece. J Soil Sci 31:125–136

Mancini F (1966a) On the elimination of the term Mediterranean in soil science. In Transaction of the Conference on Mediterranean Soils, Society Espan. Ciencia de Suelo, Madrid, pp 413–416

Mancini F (1966b) Short commentary on the soil map of Italy. Soil Map Committee of Italy. Tipografia R. Coppini & C. Firenze, 80 p

Mancini F (2002) The study of Mediterranean soils: a difficult task. In: Zdruli P, Steduto P, Kapur S (eds) 7th international meeting on soils with Mediterranean type of climate (selected papers), vol 50. Option Mediterraneeanns, Series A: Mediterranean Semsom, pp 1–3

MEF (2006) National action program for combating desertification of Turkey (NAP-D TR). In: Düzgün M, Kapur S, Cangir C, Akça E, Boyraz D, Gülşen N (eds) Ministry of Environment and Forestry Publication No: 250 ISBN 975-7347-51-5

Mermut AR, Cangir C, Kapur S (1976) A study on the properties and provenance of the periodic windblown soil-dust material in Ankara. In: Publication of the Mineral Prospection and Research Institute of Turkey, Ankara, vol 40, pp 106–107 (in Turkish)

Mermut AR, Cangir C (1978) The formation of the Ankara Clay and its pedogenic properties. In: Proceedings of the symposium of the geological problems of Ankara. Proceedings of the Turkish Geological Society, Ankara (in Turkish)

Met I, Akgün H, Türkmenoğlu AG (2005) Environmental geological and geotechnical investigations related to the potential use of Ankara clay as a compacted landfill liner material, Turkey. Environ Geol 47(2):225–236

Montarlot De L (1944) Examen m'croscopiques d'un sol rouge de garrigues. Ann Agr 1

Nihlén T, Solyom Z (1986) Dust storms and eolian deposits in the Mediterranean area. GFF 108(3):235–242

Nihlén T (1990) Eolian processes in southern Scandinavia and the Mediterranean area. Lund University Press

Noulas C, Karyotis T, Charoulis A, Massas I (2009) Red Mediterranean soils: nature, properties and management of Rhodoxeralfs in Northern Greece. Commun Soil Sci Plant Anal 40:633–648

Oakes H (1957) The soils of Turkey. Doğuş Ltd. Press, Ankara, 180 p

Özkan İ, Ross GJ (1979) Ferrogenious beidellites in Turkish soils. Soil Sci Am J 43(6):1242–1248

Özbek H, Kaya Z, Derici RM, Kapur S (1979) Potassium availability of some soils of S. Turkey as related to clay mineralogy. Soils of Mediterranean type of climates and their yield potential. In: Proceedings of the 14th Colloquium International Potash Institute of Bern, pp 301–305

Özsoy T, Saydam AC (2000) Acidic and alkaline precipitation in the Cilician Basin, Northeastern Mediterranean Sea. Sci Total Environ 253(1–3):93–109

Özsoy T, Saydam AC (2001) Iron speciation in precipitation in the North-Eastern Mediterranean and its relationship with Saharan Dust. J Atmos Chem 40(1):41–76

Özsoy T, Saydam AC, Kubilay N, Salihoğlu İ (2000) Aerosolic nitrate and non-sea-salt sulfate over the Eastern Mediterranean. Glob Atmos Ocean Syst (GAOS) 7(3):185–228

Pettijohn FJ (1984) Sedimentary rocks. CBS Publishing, 628 p

Polat S (2012) Physical, chemical, mineralogical and micromorphological properties of cedar (Cedrus libani A. Rich.) and black pine (Pinus nigra Arnold.) soils in natural and planted karstic areas. Unpublished PhD thesis, University of Çukurova, Adana, 297 p (in Turkish)

Pye K (1992) Aeolian dust transport and deposition over Crete and adjacent parts of the Mediterranean Sea. Earth Surf Proc Land 17(3):271–288

Reifenberg A (1947) The soils of palestine. Thomas Murby & Co., London

Sanın S, Arısoy M, Tıpırdamaz R, Saydam AC (2005) Effect of Saharan dust on biodegradation of phenol by white rot fungi. Bull Environ Cont Tox 75(3):66–473

Sarı M, Dinç U, Şenol S, Kapur S (1986) Formation and classification of karstic soils, turkish soil science society. 9th Natl Meet Proc 4(6–1):6–9

Saydam AC, Şenyuva HZ (2002) Deserts, can they be the potential supplier of bioavailableiron. Geophys Res Lett 29(11). doi:10.1029/2001GL013562

Smolikova L (1963) Fossile gefleckte Böden in der Tschechoslowakei. Vibtnik üstfedniho üstavu geol 35:371–73

Soil Survey Staff (2014) Keys to soil Taxonomy, 12th edn. USDA-Natural Resources Conservation Service, Washington, p 372

Tite MS, Mullins C (1971) Enhancement of the magnetic susceptibility of soils on archaeological sites. Archaeometry 13(2):209–219

Tite MS, Linington WA (1975) Effect of climate on the magnetic susceptibility of soils. Nature 256(5518):565–566

Thomas J, Southard JR (2002) Water budgets in Xeric soil moisture regimes in California's Central Valley: can deep leaching occur under today's climate? In: Zdruli P, Steduto P, Kapur S (eds) 7th international meeting on soils with Mediterranean type of climate (selected papers), vol 50. Options Mediterraneennes, Serie A: Mediterranean Semsom, pp 96–108

Torrent J (2004) Mediterranean Soils. In: Hillel D (ed) Encyclopedia of soils in the environment. Elsevier, pp 418–427

Verheye W (1973). Formation, classification and land evaluation of soils in Mediterranean Areas, with Special Reference to Southern Lebanon. University Gent, 122 p

Verheye W, de la Rosa D (2006) Mediterranean soils. In: Verheye W (ed) UNESCO-EOLSS, encyclopedia of life support systems, Section 1.5.6. Dry lands and Desertification. Eolss Publisher, Oxford

Yaalon DH (1959) Classification and nomenclature of soils in Israel. Bull Res Coun Israel 8:91–118

Yaalon DH, Lomas J (1970) Factors controlling the supply and the chemical composition of aerosols in a near-shore and coastal environment. Agric Meteorol 7:443–454

Yaalon DH, Ganor E (1973) The influence of dust on soils in the quaternary. Soil Sci 11:146–155

Yaalon DH (1997) Soils in the Mediterranean region: what makes them different?. In: Mermut AR, Yaalon DH, Kapur S (eds) Red Mediterranean soils, vol 28. Special Issue, Catena, pp 157–170

Yaalon DH (2012) The Yaalon story, a passion for science and Zion. Maor Wallach Press, Jerusalem, 233 p

Yassoglou N, Kosmas C, Moustakas N (1997) The red soils, their origin, properties, use and management in Greece. In: Mermut R, Yaalon D, Kapur S (eds) Catena SI, Red Mediterranean soils, vol 28, pp 261–278

Yılmaz K (1999) The genesis of smectite and palygorskite in the harran plain soil series. Tur J Agr Forestry 23(3):635–642 (in Turkish)

Yılmazer İ (1991) The depositional environment of a pliocene aged outcrop unit in Ankara. Geol Eng 39:41–50 (in Turkish)

Regosols

Yusuf Kurucu, Mustafa Tolga Esetlili, Erhan Akça, and Mehmet Ali Çullu

1 Introduction

Regosols are developed on gravelly, sandy, silty, and clayey sediments rich in carbonate contents and transported by surface flow or by rivers and deposited to paleolakes and seas. The setting of sediments varies due to the transporting agents. Deposited materials exposed to the surface are easily eroded by tectonic movements. Based on their mode of formation, Regosols are found on hilly landscapes. The gradient of the slopes varies from slight (1–8%) to very steep (>45%) and have an undulating topography (Fig. 1). The horizon development of Regosols exposed to severe erosion is quite poor. The loose parent material and strongly sloping topography lead to severe erosion. Creepage and slumping commonly develop in Regosol areas with clayey parent materials during the rainy seasons. The Regosols of Turkey vary in soil properties due to their diverse parent materials of carbonate- rich sedimentary origin. This common carbonate-rich inherited property calls for the use of the Calcaric qualifier of the IUSS Working Group WRB (2015) for their designation as the Calcaric Regosols of Turkey.

2 Parent Materials

Parent materials of the Regosols are generally unconsolidated transported deposits. The parent materials of the Regosols and consequently the overlying soils are laminated, in variable depth and particle size and mineral contents due to the changing flow rates and flow periods of the transporting/depositing fluvial sources. The great fault lines and orogenic (mountain forming) belts have caused the development of the harsh topography in Turkey (Fig. 1) which in turn have created erosion problems and limited the formation of soils. This is the case of the Podzols in the Black Sea region and the Regosol ecosystems in the center, west, and east parts of the country.

3 Profile Development

The horizon development of Regosols is very limited where their surface horizons are in direct contact with the parent materials. The organic matter accumulation both on surface and throughout the profile is very low (<1%). The loose and granular particles dominant in the surface horizons of the Regosols give rise to low clay contents and consequently low aggregate stabilities. Regosols in Turkey have mainly A/C and A/AC/C horizon sequences revealing poor soil development (Fig. 2). The AC horizon sequences are common in areas with moderate slopes, where the boundaries of the horizons are slightly gradual and diffuse. The low water holding capacity of the Regosol surfaces is the major handicap in supporting vegetation and sequestering organic carbon. However, the Regosols in the Aydın (southwest Turkey) province under dense maquis vegetation might be conserved as the carbon sequestering soil areas also seeking advance in soil formation (Table 1).

Regosols are also defined in the Marmara Region of northwest Turkey. Regosols in the south of Marmara are mainly developed on calcium carbonate cemented Neogene parent materials (Özsoy and Aksoy 2011) and have shallow A-C horizons. They are located generally on smooth and wavy topographies with a slope range of 3–15%. The main texture varies from sandy clay loam to clay loam and clay with gravel contents of different origin and spherical shape throughout their profiles.

Y. Kurucu (✉) · M.T. Esetlili
Department of Soil Science and Plant Nutrition, Ege University, İzmir, Turkey
e-mail: yusuf.kurucu@ege.edu.tr

E. Akça
School of Technical Sciences, University of Adıyaman, Adıyaman, Turkey

M.A. Çullu
Department of Soil Science and Plant Nutrition, University of Harran, Şanlıurfa, Turkey

© Springer International Publishing AG 2018
S. Kapur et al. (eds.), *The Soils of Turkey*, World Soils Book Series,
https://doi.org/10.1007/978-3-319-64392-2_16

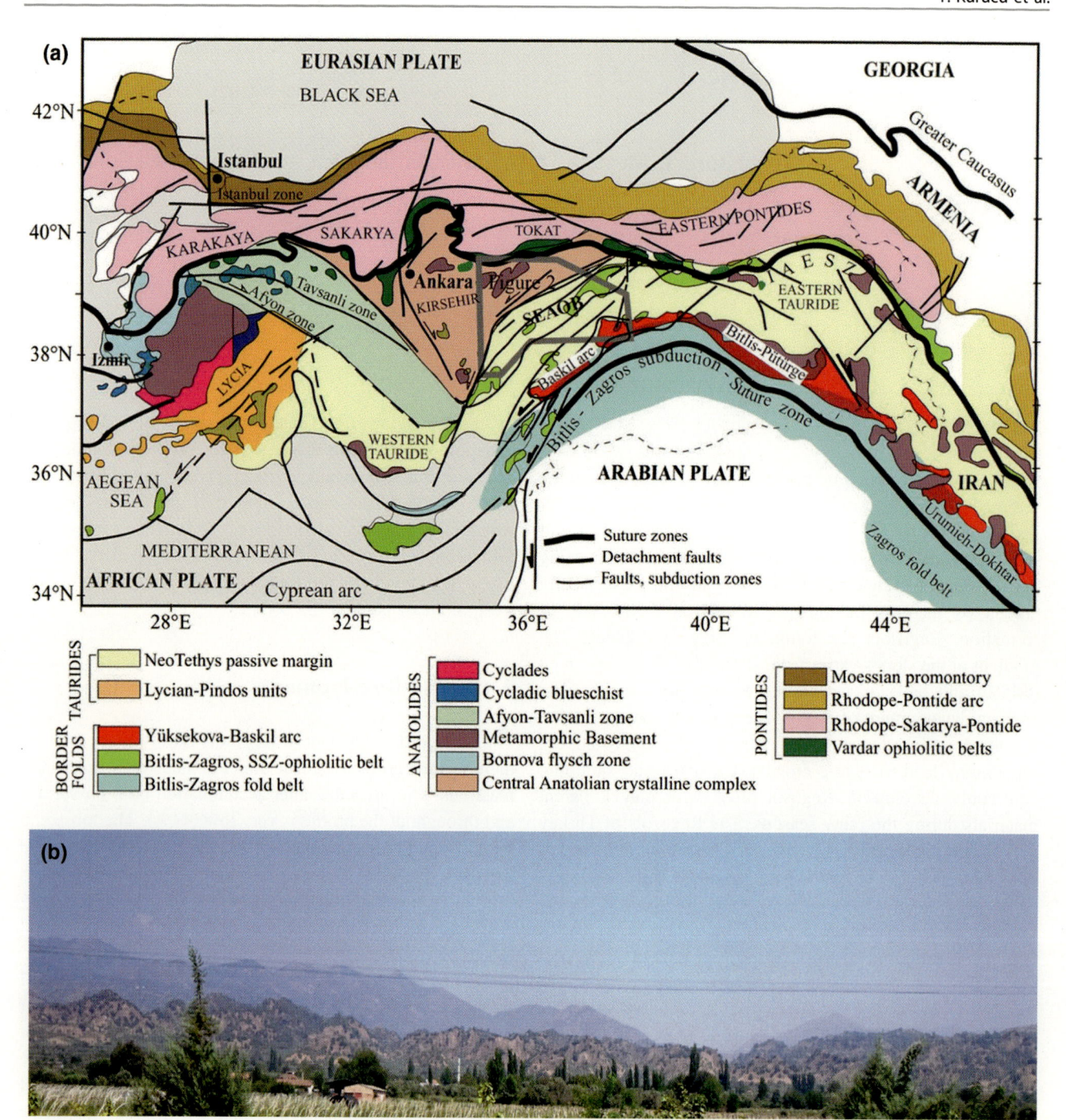

Fig. 1 The fault lines and orogenic belts of Turkey (Kuşcu et al. 2013) (**a**), regosol ecosystem located between the fault lines in Aydın, west Turkey (**b**)

The organic matter contents of the Marmara Regosols are low and vary from 0.15 to 1.99% decreasing with depth. Their cation exchange capacities increase especially in the clay rich horizons and are also affected by soil reaction, organic matter and clay contents fluctuating with depth from 4.83 to 55.19 cmolc kg^{-1} (Table 2).

4 Regional Distribution

Regosols cover an estimated 260 million ha worldwide, mainly in arid areas in Midwestern USA, northern Africa, the near east and Australia. The Regosols (Calcaric Regosols) of Turkey cover an area of approximately

Fig. 2 A Regosol profile in the Aydın district, west Turkey

989,711.55 ha (1.62% of the total land area of Turkey) and are found in central (63%), eastern (18.5%) and the Aegean (15%), Mediterranean (3%), and Marmara Regions (0.5%) (Fig. 3).

5 Environment

The majority of the Regosols in Turkey are found in the semi-arid climate zones of the country. The Regosols in the western and Mediterranean regions have been under the influence of the Mediterranean climate (Fig. 4). The precipitation in the Regosol ecosystem in the Aegean region is about 650 mm with an average annual temperature of 17–18 °C, whereas the precipitation of the Regosol ecosystems in central and eastern Turkey is about 400 mm annually with relatively lower temperatures of 10–12 °C. Due to their loose underlying parent materials Regosol areas are not preferred for settlement or industrial developments. Thus, they are mostly utilized for agricultural purposes or preferably abandoned to natural vegetation.

6 Management

Although Regosols are not well-developed soils, they are preferred for agriculture since their surface horizons and soft parent materials do not hinder root development. Regosols on sloping areas can be easily terraced for the cultivation of various crops (Fig. 5). Regosols in the Aegean and Marmara regions are mainly cultivated for olives (see Fig. 7 in Rendzic Leptosol Chapter), vine, and figs, whereas the dominant crop of the relatively drier central and eastern Turkish Regosol ecosystems is dominated by cereals accompanied by vineyards.

These soils are tilled intensively and inappropriately leading to the formation of plough pans. Inappropriate soil tilling by heavy equipment, especially in clay rich soils, cause poor root development and in the short run soil degradation (Özsoy and Aksoy 2011).

Agriculture and natural vegetation are found mixed on the sloping Regosols of the country where *Cistus sp., Quercus coccifera, Rosa canina, Pinus brutia, Thymus sp., Salvia officinalis, and Pistacia terebinthus* are the major natural plant species of the Regosol ecosystems (Atalay 2014, Bolca et al. 2012).

Özsoy and Aksoy (2011) reported that Regosols developed on steep slopes are subjected to erosion and have shallow profiles that usually inhibit cultivation. Regosols of the country of the drier and semi-arid ecologies also have moisture limitations generally requiring irrigation. These fragile soil ecosystems are highly recommended for forestry, pasture or recreation in order to prevent the decline in organic matter contents with a corresponding decrease in aggregate stability.

Yavuz and Bozdağ (2012) suggested that Regosols developed on freshly deposited alluvium or sands, i.e., unconsolidated materials of the Tersakan and Salt Lake regions (central Turkey), are not suitable for agriculture due their vulnerability to wind and water erosion. Regosols close to Ankara, central Turkey, with sparse vegetation are also not resistant to erosive processes (Tunç and Schroder 2010).

The soil loss on the shallow Regosols developed on moderate slopes in forest and rangelands poor in vegetation cover is well-above 10 t.ha^{-1} y^{-1} (Karaoğlu 2012). This reflects the need for a particular land use activity in landscapes where Regosols are dominant (Fig. 6).

Table 1 Site-profile description and some physical and chemical properties of Regosols in Germencik, Aydın, West Turkey

IUSS Working Group WRB (2015)	Haplic regosol
USDA Soil Taxonomy (2014)	Typic xerorthent
Location	Germencik, Aydın (West Turkey)
Altitude	86 m
Landforms	Undulating land
Position	Foot slope
Aspect	South
Slope	10%
Drainage class	Well drained
Groundwater	–
Eff. soil depth	31 cm
Parent material	Sediments
Köppen climate	Mediterranean
Land use and vegetation	Maquis

Horizon	Depth (cm)	Organic matter (%)	pH (H$_2$O)	CaCO$_3$ (%)	Exchangeable bases				CEC	BS (%)
					Ca + Mg	K	Na			
					cmol$_c$.kg^{-1}					
A	0–31	1.86	7.68	1.0	10.4	0.4	0.3		19	58
C1	31–82	0.46	7.86	1.0	7.9	0.3	0.2		44	61
C2	82+	0.36	7.99	2.3	9.8	0.3	0.3		16	65

Horizon	Color		EC (dS/m)	Structure	Stoniness (%)
	Dry	Moist			
A	2.5 YR4/8	2.5 YR3/4	0.02	Weak Granular	<1
C1	2.5 YR4/6	2.5 YR3/4	0.02	Moderate Blocky	<1
C2	2.5 YR4/6	2.5 YR3/4	0.03	Moderate Blocky	<1

Horizon	Sand (%)	Silt (%)	Clay (%)	Bulk density g.cm^{-3}
A	62	24	14	1.50
C1	64	22	14	1.51
C2	64	18	18	1.47

Table 2 Site-profile description and some physical and chemical properties of Regosols in Bursa, Northwest Turkey (Özsoy and Aksoy 2011)

IUSS Working Group WRB (2015)	Calcaric Regosol
USDA Soil Taxonomy (2014)	Typic Xerorthent
Location	Bursa, Marmara region, Northwest Turkey
Coordinates	40° 14′ 32″ N–28° 51′ 56″ E
Altitude	155 m
Landforms	Flat plain
Position	Foot slope
Aspect	Southeast
Slope	3%
Drainage class	Well drained
Groundwater	–
Eff. Soil depth	31 cm
Parent material	Calcareous claystone
Köppen climate	Mediterranean
Land use and Vegetation	Natural oak (*Quercus robur*)

Horizon	Depth (cm)	Organic matter (%)	pH (H$_2$O)	EC (dS/m)	CaCO$_3$ (%)	Stoniness (%)	Bulk density (g.cm^{-3})
A	0–30	1.59	6.44	0.31	0.64	<1	1.29
AC	30–45	0.62	7.39	0.34	21.70	<1	1.27
C	45+	0.38	7.42	0.22	35.55	<1	1.22

	Color		Structure	Root	Consistency
	Dry	Moist			
A	7.5 YR 3/4	7.5 YR 4/4	Strong, coarse, angular blocky	Common fine	Dry very hard, moist friable
AC	–	7.5 YR 3/4	Weak friable, subangular blocky	Common fine	Moist very friable
C	–	–	–	–	–

	Sand	Silt	Clay	Exchangeable Bases					
	(%)			Ca + Mg	K	Na	CEC	BS (%)
				cmol$_c$.kg^{-1}				
A	39.8	16.7	43.5	37.89	0.72	0.19	38.8	100
AC	35.6	17.7	46.7	37.23	0.38	0.22	37.8	100
C	24.9	19.5	55.6	27.66	0.33	0.22	28.2	100

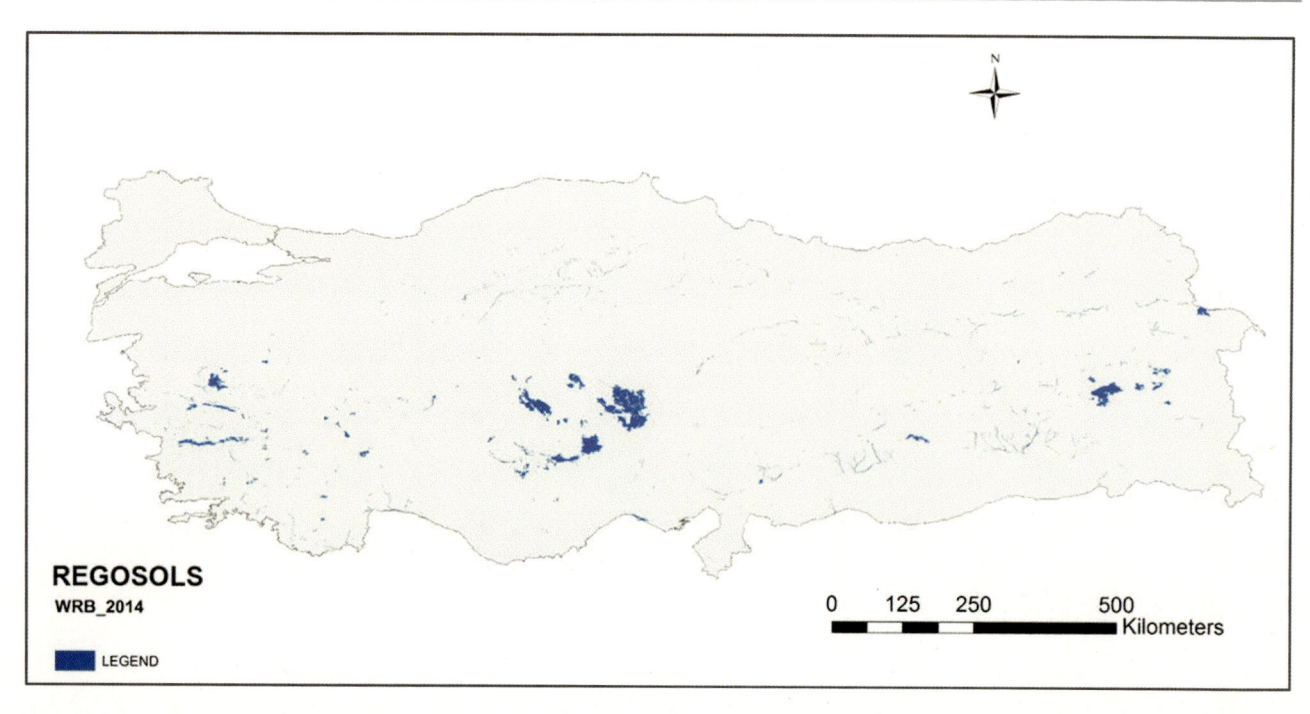

Fig. 3 Distribution of Regosols in Turkey

Fig. 4 A Regosol profile in the Taurides, south Turkey

Fig. 5 Terraced Regosol (Calcaric) for banana cultivation in Mersin, south Turkey

Fig. 6 Regosols under sparse forest cover, south Turkey

References

Atalay İ (2014) Ecoregions of Turkey. Ministry of Forestry Pub. No: 163. ISBN 975-8273-4-8, İzmir, Turkey, 96 p (in Turkish)

Bolca M, Altınbaş Ü, Kurucu Y (2012) A study on pedogenetical distribution of minerals in the pedon of Rhodoxeralf and Rendoll soils showing formation on different slope facets. Ege Univ J Fac Agric 49(3):229–238 (in Turkish)

IUSS Working Group WRB (2015) World reference base for soil resources 2014, update 2015 international soil classification system for naming soils and creating legends for soil maps. World Soil Resources Reports No. 106. FAO, Rome, 203 p

Karaoğlu M (2012) The evaluation of Iğdır soils for erosion. Iğdır Univ J Inst Sci Tech 2(1):23–30 (in Turkish)

Kuşcu İ, Tosdal RM, Gençalioğlu-Kuşcu G, Friedman R, Ullrich TD (2013) Late cretaceous to middle eocene magmatism and metallogeny of a portion of the Southeastern Anatolian orogenic belt, East-Central Turkey. Econ Geol 108(4):641–666

Özsoy G, Aksoy E (2011) Genesis and classification of Entisols in Mediterranean climate in Northwest of Turkey. J Food Agric Environ 9:998–1004

Soil Survey Staff (2014) Keys to soil taxonomy, 12th edn. USDA-Natural Resources Conservation Service, Washington, DC, p 372P

Tunç E, Schroder D (2010) Determination of the soil erosion level in agricultural lands in the western part of Ankara by USLE. Ekoloji 19:58–63 (in Turkish)

Yavuz F, Bozdağ A (2012) Evaluation of the obstacles against the economic development of Cihanbeyli rural area. In: ERSA conference papers (No. ersa14p1268). European Regional Science Association, Bratislava, Slovakia

Rendzic Leptosols

Yusuf Kurucu and Mustafa Tolga Esetlili

1 Introduction

Rendzic Leptosols are located only along the sloping and undulating topographies of the Aegean and Mediterranean parts of the Turkey. Erosion is the common threat for these soils especially when they are under cultivation. The average soil depth of the Rendzic Leptosols is around 30–35 cm. Their surface soil color is usually dark gray contrasting with the light color of their parent materials which are mainly soft carbonate rocks such as chalk and marls. The organic matter and lime contents are high at the surface and subsurface horizons. The other Leptosols of Turkey are the soils that have been eroded and subsequently formed/weathered following the removal of the top soils of the Calcisols, Cambisols, and Luvisols. They partly contain the residual soil materials of the B/C and/or C horizons of the Calcisols or Luvisols inherited from the calcretes and from the marine limestones (Miocene and older), respectively. In this respect, these Leptosols are separately discussed in the Calcisols and Leptosols chapters of this book.

2 Parent Materials

The Rendzic Leptosols of Turkey are formed on marl or chalk materials where they can directly overlie a paralithic contact (Fig. 1). Rendzic Leptosols also occur on low crystalline limestones such as the siliciclastic materials (clay and sand) containing limestone under forest cover. Moreover, the acidic root extracts of the coniferous trees were determined to have enhanced the decalcification of the parent materials giving rise to the formation of Rendzic Leptosols (Tunç and Schröder 2010) even in the low to moderate rainfall areas of Turkey. Bolca et al. (2012) determined the primary minerals in a Rendzic Leptosol in west Turkey under vegetation in which calcite and feldspars were the dominant minerals followed by lesser amounts of quartz and biotite (Table 1). The existence of iron oxide minerals (hematite) and hornblende indicated that the Rendzic Leptosols of west Turkey were good sources of iron as a plant nutrient.

3 Profile Development

Rendzic Leptosols of Turkey have well-developed but shallow A-horizons (Fig. 2). They are characterized by the accumulation of humified organic matter which is well mixed with lime and the inorganic fractions. The structure of the A horizon is crumby and strong granular. Their horizon sequences are A/C to A/AC/C and lack a B-horizon. They usually have high soil organic carbon (SOC) contents if covered by natural vegetation. Leptosols allocated for olive orchards have high SOC contents in the root-zones of the olive trees, but lower SOC contents and light colored A-horizons due to the SOC loss by erosion between the olive trees (Fig. 3) (Uğur et al. 2004). Most farmers use animal and/or green manures to increase fertility and combat erosion in Leptosols. Root distribution and biogenic activity are intense in all the Leptosols included in this chapter.

The color of the A-horizons of the Rendzic Leptosols is usually 10YR 3/3 or less when moist but the subsurface soils are of a lighter color due to the increasing carbonate content in the fine fractions. The high amounts of soluble carbonates in all the horizons are responsible for the high base saturations (more than 60%) throughout the pedons of the Rendzic Leptosols in Turkey. The high clay contents throughout the profiles of the Rendzic Leptosols increase in the subsurface horizons parallel to their carbonate contents (Table 2).

Y. Kurucu (✉) · M.T. Esetlili
Department of Soil Science and Plant Nutrition, Ege University, Bornova, İzmir, Turkey
e-mail: yusuf.kurucu@ege.edu.tr

© Springer International Publishing AG 2018
S. Kapur et al. (eds.), *The Soils of Turkey*, World Soils Book Series, https://doi.org/10.1007/978-3-319-64392-2_17

Fig. 1 Shallow Rendzic Leptosols on a paralithic contact of chalk (Çesme, İzmir, west Turkey)

Table 1 Primary minerals in a Rendzic Leptosol (west Turkey) (Bolca et al. 2012)

Minerals (%)	Horizon and depth (cm)		
	A (0–21)	AC (21–33)	C (33+)
Opaques	5.12	2.85	1.12
Opal	3.34	2.15	1.13
Hematite	1.25	0.9	0.85
Calcite	38.5	42.8	57.7
Illite	4.92	6.25	12.65
Feldspar	22.2	17.2	8.2
Quartz	4.93	3.66	3.6
Biotite	3.86	6.17	7.5
Muscovite	1.98	1.12	0.75
Hornblende	1.02	0.88	0.45
Hyperstene	0.85	0.5	0.2
Augite	1.38	1.52	0.95
Tourmaline	1.45	0.5	0.4
Other minerals	9.2	13.5	4.5

4 Regional Distribution of Rendzic Leptosols

Rendzic Leptosols are mostly present in the sub-humid and Mediterranean climate regions of Turkey (the west, northwest and south regions) with an average annual rainfall around 500–750 mm and cover an area of 839.379, 12 ha which is 1.05% of the whole agricultural land (Fig. 4).

5 Management

Rendzic Leptosols are mostly under shrub and forest vegetation in Turkey (Figs. 2, 5 and 6) and are appropriate carbon sinks with relatively higher SOC contents. Only 20% of the Rendzic Leptosols (135.000 ha) are available for cultivation due to their topographic conditions and shallow depth.

Fig. 2 Rendzic Leptosol in Manisa at reforested area, west Turkey

Fig. 3 Rendzic Leptosols in the Manisa District and the ancient olive groves (west Turkey)

Table 2 Site description and some physical and chemical properties of a Rendzic Leptosol, west Turkey

IUSS working group WRB (2015)	Rendzic Leptosol
Soil Survey Staff (2014)	Lithic Rendoll
Location	Soma, Manisa (west Turkey)
Altitude	174 m
Landforms	Undulating-sloping land
Position	Back slope
Aspect	South
Slope	8%
Drainage class	Well drained
Groundwater	None
Eff. Soil depth	27 cm
Parent material	Marl
Köppen climate	Mediterranean
Land use	Reforestation
Vegetation	Red and scots pines

Horizon	Depth (cm)	OM[a] (%)	Total N (%)	pH H$_2$O	EC dS/m	CaCO$_3$(%)	Sand (%)	Silt (%)	Clay (%)
A	0–17	9.34	0.03	7.49	0.07	8.5	54	22	24
AC	17–33	3.66	0.16	7.69	0.08	18.5	46	24	30

	Exchangeable bases and acidity				CEC	Base saturation
	Ca	Mg	K	Na		(%)
	cmol$_c$kg^{-1}					
A	20.4	8.4	2.5	0.4	35	91
AC	20.9	9.5	1.7	0.4	51	64

	Color		Consistency when dry	Structure
	Dry	Moist		
A	10YR3/2	10YR2/1	hard	medium to very coarse
AC	10YR4/2	10YR3/2	hard	coarse to very coarse

[a] Organic matter

The common Rendzic Leptosol vegetation is composed of *Quercus coccifera, Cistus sp, Pistesia sp, Rosa canina, Salvia sp, Capparis spinosa, Calicotome villosa, Ssarcopoterium spinosum* and *Pinus brutia*. Rendzic Leptosols are suitable only for a few crops that can match the vegetative and environmental properties of these natural plants. Thus, beyond doubt, the primary tree crop that has commonly coped with the soil and environmental characteristics of the Rendzic Leptosols together with the Regosols, Luvisols, Cambisols and Calcisols has been the olive tree for millennia in Turkey and in the Mediterranean basin (Figs. 3 and 7). Rendzic Leptosols of the Aegean region are also, to a lesser extent, cultivated for vineyards, tobacco and wheat production.

Rendzic Leptosols have also been utilized as meadow under canopies of the historical olive groves (age ca. 200 years) in the Aegean Region as a common practice (Fig. 8).

The allocation (planting distance/seeding density-optimally 10 m apart per tree, the suitable planting distance of the olive trees for Rendzic Leptosols as historically established) of the olive trees in these historical and/or comparatively contemporary olive groves is also appropriate for grazing of the small ruminants (especially goats due to the undulating topography) between the trees (Fig. 8).

However, this type of use may induce the risks of soil erodibility due to overgrazing, thus calling for a sustainable management programme concerning these particular meadows. Moreover, inappropriate ploughing between the trees and the removal of the natural grass might increase the risk of erosion and loss of the SOC of the Rendzic Leptosols. On the other hand, establishing contemporary olive groves, at

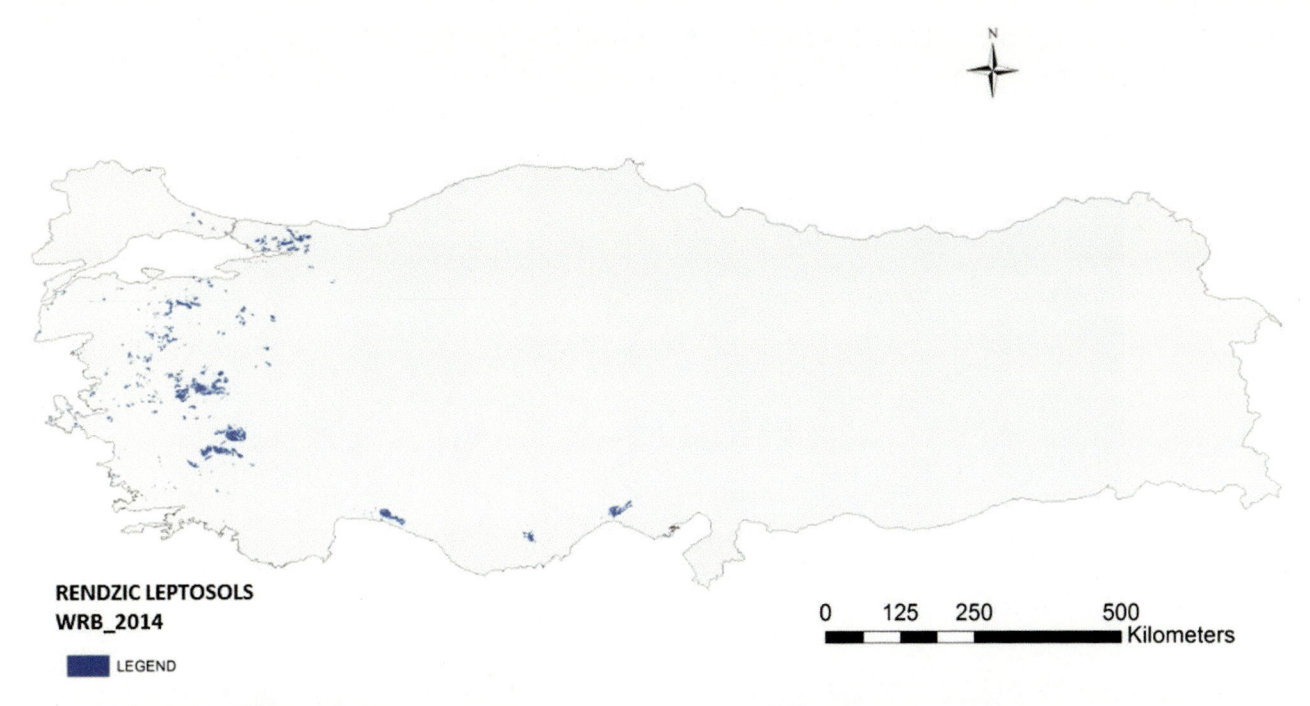

Fig. 4 Distribution of the Rendzic Leptosols in Turkey

Fig. 5 Natural vegetation (Medicinal plants, **a** *Thymus spp* and **b** *Salvia spp*), on Rendzic Leptosols, Çeşme, İzmir, western Turkey

illegally deforested sloping areas of Rendzic Leptosols, is the other prime threat in the Aegean and the Mediterranean areas of the country (Fig. 9). Unfortunately, these are the common practices implemented without appropriate land management programming ignoring the existing traditional sustainale land management (SLM) and C-management programmes and/or advices of the local authorities on appropriate land use. The improper use of the Rendzic Leptosols calls for the appropriate crop/forest allocations at a basin level (as stated in the National Action Plan of Turkey for combating Desertification empowered in 2006, recently revised according to the UNCCD 10-year Strategic Plan) which is encouraged by the recently increased state subsidies provided to the olive farmers.

Fig. 6 The distribution of oak (*Quercus coccifera*) roots in a dark brown Rendzic Leptosol in Bornova, İzmir (west Turkey)

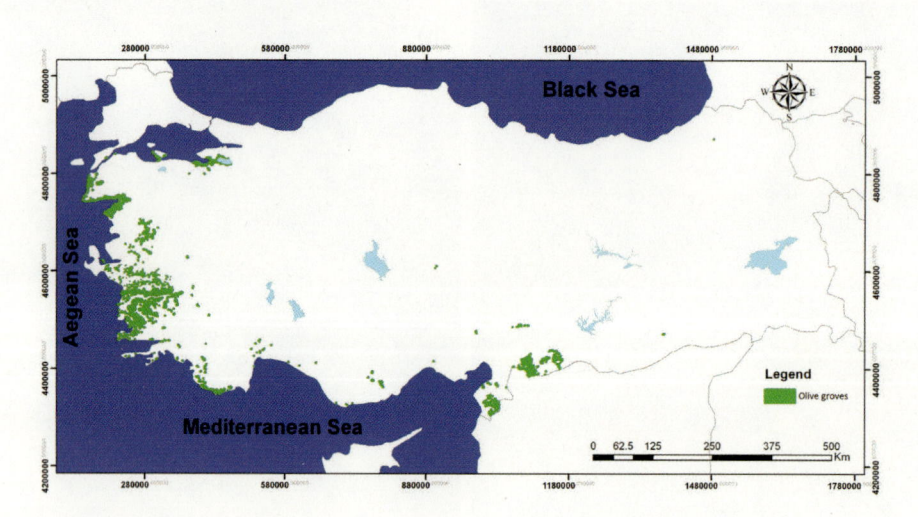

Fig. 7 Distribution of the olive groves in Turkey

In general, Rendzic Leptosols of the non-sloping areas (within or without natural meadows) have been cultivated mainly for olive trees and vineyards together with Regosols, Luvisols, Cambisols and Calcisols in western and southern Turkey (Fig. 7). Vine roots can penetrate the soft limestone materials of the Rendzic Leptosols and calcretes of the Calcisols and utilize the plant nutrients secured in the carbonate/clay mineral-rich matrices.

Fig. 8 A natural meadow under canopy in a historical olive grove (west Turkey)

Fig. 9 Deforested Rendzic Leptosols allocated to olive cultivation under erosion risk

References

Bolca M, Altınbaş Ü, Kurucu Y (2012) A study on pedogenetical distribution of minerals in the pedon of Rhodoxeralf and Rendoll soils showing formation on different slope facets. Ege University. J Fac Agric 49(3):229–238 (in Turkish)

IUSS Working Group WRB (2015) World Reference Base for Soil Resources 2014, update 2015 International soil classification system for naming soils and creating legends for soil maps. World Soil Resources Reports No. 106. FAO, Rome, 203 p

Soil Survey Staff (2014) Keys to Soil Taxonomy, 12th edn. USDA-Natural Resources Conservation Service, Washington, DC, 372 p

Tunç E, Schroder D (2010) Determination of the soil erosion level in agricultural lands in the western part of Ankara by USLE. Ekoloji 19:58–63 (in Turkish)

Uğur A, Saş MM, Yener G, Altınbaş Ü, Kurucu Y, Bolca M, Özden B (2004) Vertical distribution of the natural and artificial radionuclides in various soil profiles to investigate soil erosion. J Radioanal Nucl Chem 259(2):265–270

Solonchaks and Solonchak-Like Soils

Hasan Özcan, Mehmet Ali Çullu, Hikmet Günal, Hüseyin Ekinci, Mesut Budak, Ali Sungur, and Timuçin Everest

1 Introduction

Solonchaks are described as soils with a Salic horizon starting within ≤ 50 cm from the soil surface, and should not have a Thionic horizon starting within the ≤ 50 cm from the soil surface according to the IUSS Working Group WRB (2015). ISRIC (2014) defines Solonchaks as soils that occur in areas where saline groundwater rises near to the surface, or where through evapotranspiration salts from the soil water accumulate in the soil. The Soil Survey Staff (2014) defines saline soils with a Salic horizon of accumulated salt that is more soluble than gypsum in water. Thus, Alfisols (Natrustalfs) and Mollisols (Natrustolls) of the Soil Survey Staff (2014) can be the equivalents of the Solonchaks of the IUSS Working Group WRB (2015). The word Solonchak is derived from Russian "solončak" meaning salt marsh, from solonyĭ salty and from sol' salt. Eswaran et al. (2003) reviewed saline soils and reported that in the Russian soil classification the Halomorphic soils are recognized at order level, while in South Africa, no distinct provision has been made for saline soils. Fitzpatrick et al. (2002) developed a new classification for the Australian System which categorizes saline soils with their hydrological, soil water, and soil chemical properties. In the Australian system, the significant soil chemical features and the dominancy of sodium chloride are defined as the halitic property (Fitzpatrick et al. 2003). The French soil classification system defines saline soils as Salisols. The Solonchak soils and/or Solonchak-like soils have generally AC or A-B-C horizon sequences with frequent salt accumulation on the soil surface (Fig. 1).

Sönmez (2004) reported that the saline and alkaline areas were about 2% of the total area of the Turkish soils, 5.48% was arable land, and 17% were economically irrigable lands which now have increased to 5.27% following the irrigation projects initiated after the 1970s (Doğan 2014) (Table 1).

The first attempt on mapping the distribution and determining the characteristics of salt-affected soils of Turkey was accomplished by Çağlar (1949). Oakes' (1954) surveys were the next to determine the areal distribution of saline and sodic soils of the country. The TOPRAKSU of the Ministry of Agriculture and Rural Affairs completed the soil map of Turkey at a scale of 1:200,000 from 1966 to 1971. This map consisted information on the salt-affected lands and salinity levels of Turkey. These changes in salinity levels were revised from 1982 to 1984 by the Ministry of Agriculture, General Directorate of Rural Affairs. According to this national survey the salt-affected soils were indicated to cover an area of 1.517.695 ha where the alkaline soils were determined to cover 0.5% of the total salt-affected lands (Sönmez 2011).

The salt-affected soils of Turkey are widespread in the Konya (Driessen 1970; De Meester 1970), the Aegean (Saatçı and Tuncay 1971; Atatanır et al. 2010; DSİ 1966), Lower Meriç (Cangir ve Boyraz 2000), Seyhan (Dinç et al. 1990), Iğdır (Oruç 1970), Bafra (Cemek et al. 2006), Harran (Çullu et al. 2010), Amik (Kılıç et al. 2008; Ağca et al. 2000), Emen (Budak, 2012) basins as well as in the Malya (Munsuz 1969; Çullu et al. 1995), Erzurum (Akgül and Şimşek 1996) Erzincan (Bahtiyar 1971), Reyhanlı (Atasoy and Çeçen 2014), and Acıpayam and Salihli (Bayramin et al. 2004; Sönmez 2011) areas.

Fig. 1 Saline soils in Şanlıurfa, the Harran plain (southeast Turkey)

Table 1 Salt-affected soils of Turkey in 1971 (Sönmez 2004)

Salinity classes	Area (ha)	%
Slightly Saline	614,617	41.0
Saline	504,603	33.0
Alkaline	8641	0.5
Slightly saline and alkaline	125,863	8.0
Severe saline and alkaline	264,958	17.5
Total	1,518,722	100.0

2 Parent Materials

The geologic salt deposits (evaporites etc.), as the outcrops of the ancient lake deposits in the Konya Basin such as the Lake Tuz, are the remnants of the Tethys Sea (Fig. 2). The major geologic groups representing the sources of the salt are those from marine sediments, volcanic deposits, and weathered rocks. Old marine sediments occur everywhere around and under the Konya Basin. They are Upper Eocene, Oligocene, or Miocene in age and contain highly soluble salts, or gypsum, or both (Driessen 1970). According to Şengör and Yılmaz (1981) some sedimentary rocks in the vicinity of Sivas and Seyitgazi–Eskişehir are the parent materials of the saline soils/solonchaks and the lands bearing the risk of secondary salinity by irrigation. These are recognized as the soils with primary salinity enriched by secondary salinity, i.e., the outcome of the excess and repeated irrigation by the use of the fossil saline waters related to the geologic salt deposits. Salinity is mainly a secondary event in Turkey building up from the accumulation of the salts in soils following the excess use of irrigation water or groundwater in central and southeast Turkey and the Mediterranean region. These lands are the "Gene Management Zones", determined by the Ministry of Agriculture and Forestry for the protection of the traditional crops of Turkey and primarily to sustain the halophytes of the central, east and southeast parts of the country (Kapur et al. 2003) (Fig. 3). These traditional crop zones were mapped onto the pedo-geologically developed paleo-saline and potential saline areas determined during field and laboratory studies, in order to verify the irrevocable protection of the natural central and southeast habitats of Turkey. These are the Solonchak and Solonetz-like Solonchak ecosystems of the country, i.e., the saline halophyte and cereals ecosystems (Dinç et al. 1991; Kapur et al. 2003) (Fig. 4).

Fig. 2 The periodically drying and cultivated seasonally dried parts of Lake Tuz (Salt Lake), the remnant of the Tethys Sea in central Turkey

3 Environment

Solonchaks in Turkey generally occur in the slightly arid to semiarid areas of the country where the rainfall is insufficient to leach soluble salts from the soil. Salinity problems can also be observed in irrigated land, particularly when the irrigation water quality is marginal or worse. The sources of salinity in Turkey are the locally weathered parent materials, the surface and subsurface waters, and the human activities. Because of the topographical structure and dry climate conditions, wide lands at the Great Konya Basin are salt affected.

The climate of the Great Konya Basin is semi-arid, with a dry hot summer and a cool relatively wet winter. The annual precipitation barely exceeds 300 mm, since both the prevailing northerly winds and the common southerly are dry, the first comes from a dry area, the latter loses its moisture in the Taurus Mountains so that in the Basin its relative humidity is below 50% (Driessen 1970). The salinity of the Great Konya Basin increases downward from the fringes toward the center. The degree and type of salinity depend on the hydrology of the area. There is a flow of groundwater through the previous limestones at the fringe and through the permeable sand and gravel bodies of the alluvial fans. Seasonal fluctuations in the water table are therefore high and consequent salinization in these zones is severe. In the flat center the water table and salinity do not fluctuate much, but vary with slight differences in surface topography. Salinity is internal in most soils but areas of seepage occur and large depressions are the external Solonchaks (De Meester 1970). Salinity in the Great Konya Basin occurs mainly in three areas: the west central part of the Basin, a depression north of the Town of Karapınar, and the central east part of the Basin. The north and south are not salt affected (Driessen 1970), (Fig. 4).

Generally, salinity increases with decreasing altitude, except for the central west part, where this trend is disturbed by irrigation. The low parts also have the highest water table, with very high electrical conductivity. There salinity is caused mainly by supply of salts from subsurface flow from the higher soils toward the central depression. This distribution of salts has consequences for agriculture, but as salinity and alkalinity are both correlated with topography and the water status, these consequences are not always distinguishable. The topography of the Basin favors horizontal drainage from the higher areas along the fringes toward the center, particularly near irrigated areas (Driessen 1970). Besides the Great Konya Basin, salinity has also been extensively tackled in the Adana Province (the Çukurova plain, south Turkey), where Kapur et al. (1995) revealed the existence of salt deposits of the Messinien period. The alkaline pH and Mg-rich Miocene–Pliocene shallow marine formations of the Handere sediments in the Çukurova Plain soils have large amounts of salts (Figs. 5 and 6) (Kaplan et al. 2014).

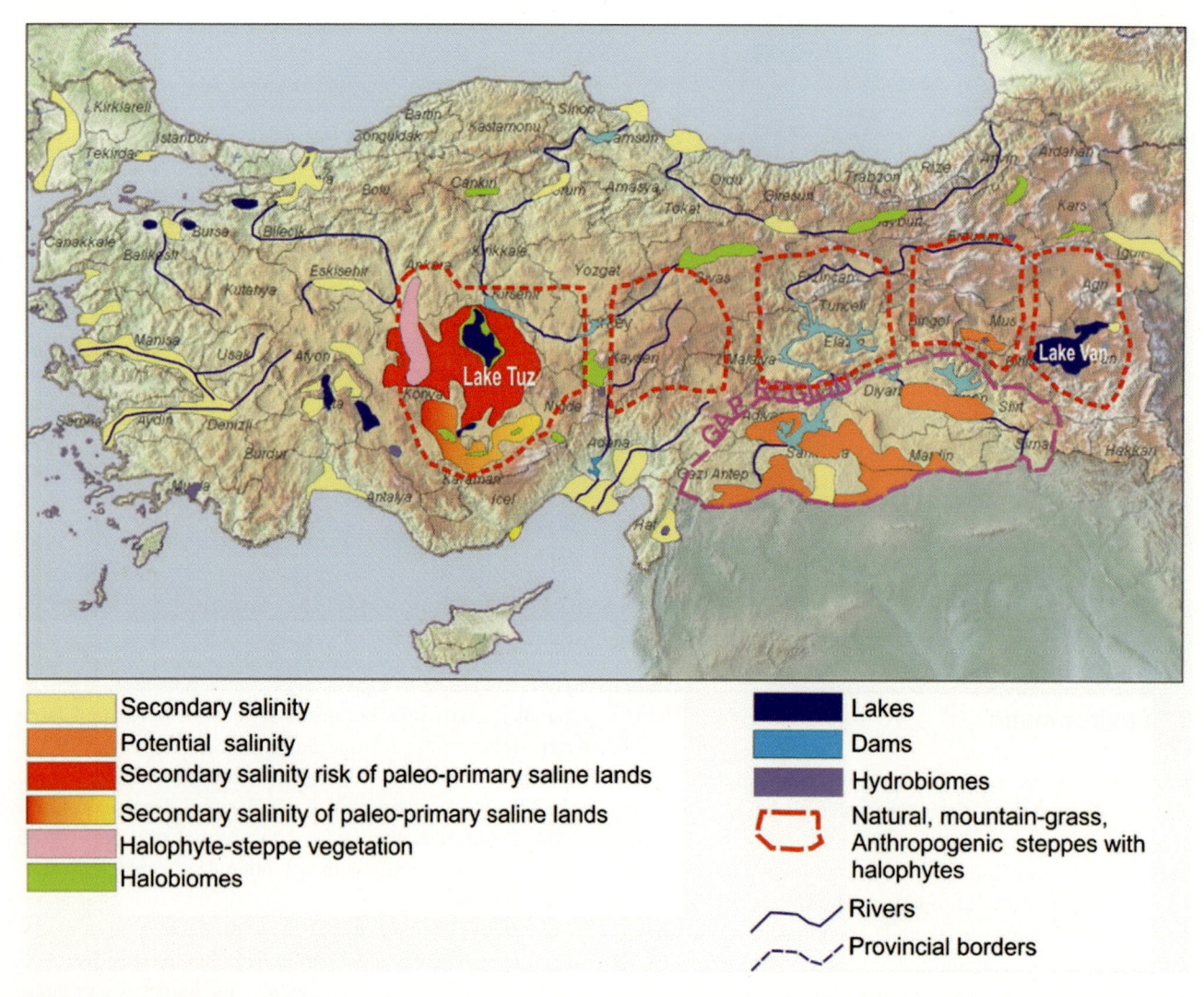

Fig. 3 Gene management zones/traditional crop management zones in Solonchak, Solonetz-like Solonchak ecosystems (Kapur et al. 2003)

4 Profile Development

4.1 Solonchaks (Solonchak-Like Soils) in the Konya Basin, Central Turkey

The widely distributed saline areas of central Turkey developed in the Great Konya Basin are the prime examples of the soils closest to be classified as Solonchaks. In almost all salt-affected soils of the Great Konya Basin, both compounds of sodium chloride and sodium sulfate occur together with magnesium compounds. In the west of the basin, where sulfates predominate, a $Na_2SO_4–H_2O$ system develops in puffed solonchaks and in parts develops in the flooded solonchaks. In the puffed solonchaks, sodium sulfate continues to accumulate in the wet and cold winter relative to other salts and is not completely leached during the wet period.

The salt may accumulate at the soil surface by evaporation of the water to form external Solonchaks in temporary ponds or along the shores of the permanent lakes and consequently a nearly flat depression may remain from the ponds covered with a white crust. External Solonchaks may also form in this environment where groundwater reaches the surface. They form from the evaporation accompanied by accumulation of the salts, on or in the uppermost soil layer mainly containing sodium sulfate and sodium chloride (Buringh 1960).

The Solonchak-like soils of central Turkey consist of profiles with Az-Bw-Ckm-C horizon sequences (Table 2). Since the lands of this environment are not naturally drained, water ponds developed on the soil surfaces for a long period of time. However, these ponds are either evaporated in the drier seasons or water may be slowly leached along the solum. Nevertheless, in both cases, salt deposited by the surface water is left on the surface or somewhere in the soil profile. Soils similar to the soils in Bor, Niğde (Figs. 6, 7 and 8) are the flooded Solonchaks that formed by evaporation of the

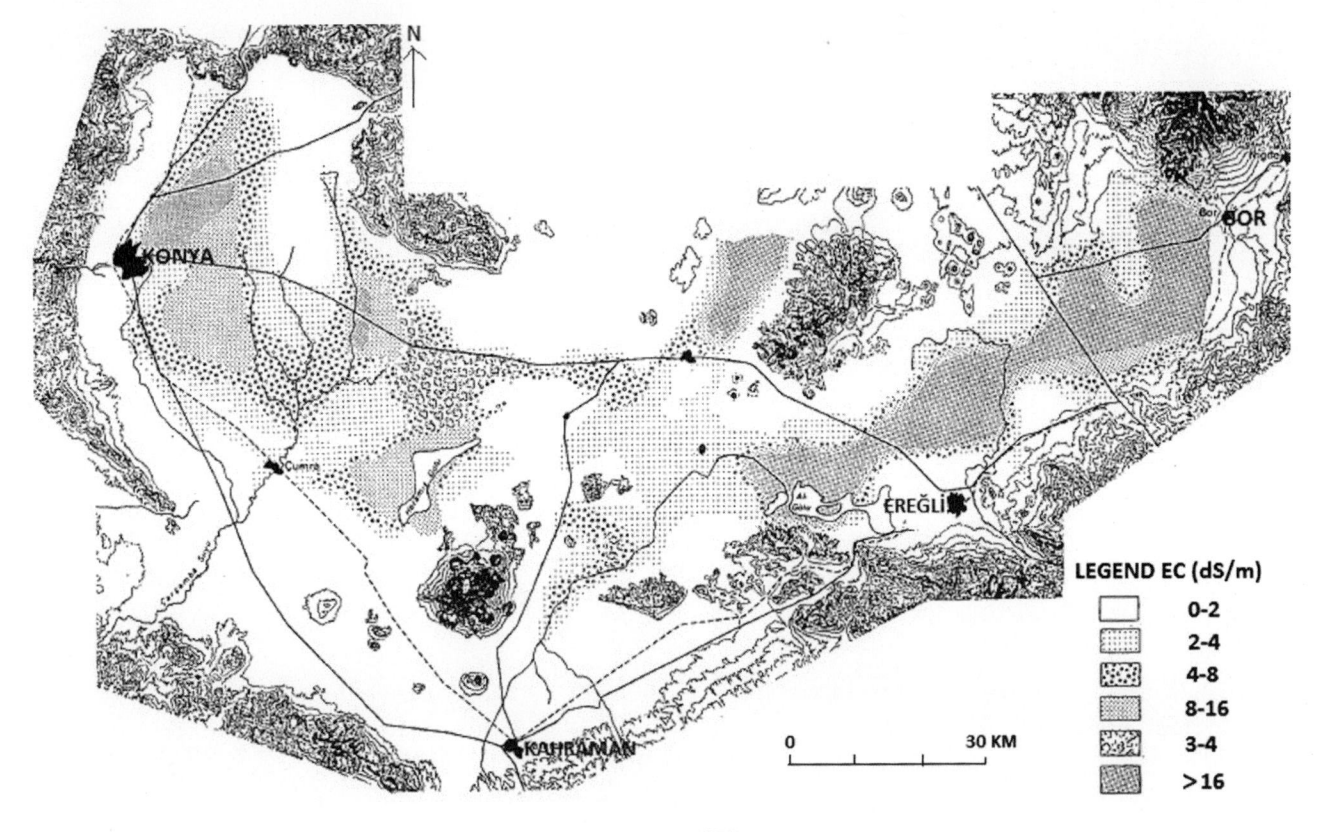

Fig. 4 Distribution of soil salinity at the Great Konya Basin (De Meester 1970)

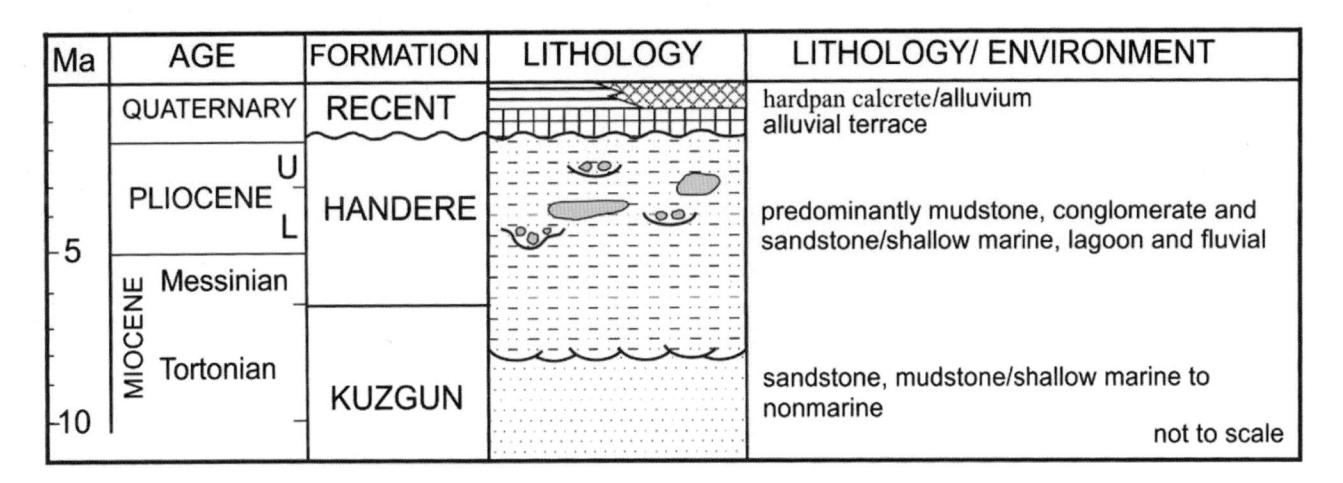

Fig. 5 Simplified stratigraphic column of the Çukurova plain (Kaplan et al. 2014)

saline water standing in depressions. This saline water is derived from seepage and run-off, and fills the depressions in the spring period forming the strongly salt-affected areas that are very common in the Emen Plain of central Turkey. In temporary ponds, the salt accumulates at the soil surface by evaporation of the water to form the external solonchaks where only a nearly flat depression may remain covered with a white crust (Fig. 6) (Budak 2012). De Meester (1970) indicated that salts in the Konya Basin are from marine sediments, volcanic deposits, and weathering of rocks. He stated that the lack of discharge from the basin means that water can only be removed by evaporation which induces salinization.

Halophytes (*Halimione verrucifera, Plantago crassifolia, Taraxacum farinosum, Puccinellia distans (Jacq.) Parl. Subsp. Distans*) are the indicators of high salinity in the Leben soils. The surface crust of the Leben soils become puffy following extensive grazing by small ruminants which

Fig. 6 Soil surface of Leben soils located in Bor town of the Niğde Province, central Turkey

makes the soils vulnerable to wind erosion. Although the EC of the surface horizon in the Leben soils does not meet the salic horizon requirement of the IUSS Working Group WRB (2015) (≥ 8 dS m^{-1} if the pH water of the saturation extract is ≥ 8.5), Budak (2012) reported that the EC of the Leben soils varies from 2.08 to 26.7 dS/m with a mean value of 9.58 dS/m. This points out the need to consider the spatial variability of the EC values when classifying these soils. In this context, Budak (2012) determined a high coefficient variability (77.24%) for the EC values (for 20 soil samples) within the Leben soils as accepted by Cambardella et al. (1994). The spatial variability of these EC values caused the development of uneven patterns in the distribution of salinity within even the same soil map. Consequently, the spatial variability of the soil properties particularly in saline environments, as in the Leben soils (Calcaric Sodic Solonchaks), should be taken into consideration to increase the efficiency of the management practices. At this point, it can be stated that the traditional soil surveys consider the uniformity of the soils within mapping units and overlook their intrinsic spatial variability. Soil surveys do not also have appropriate sampling designs to present quantitative estimates regarding spatial variability within and across map units (Lin et al. 2005). This shortcoming causes confusions in classifying soils as in the Leben soils and emphasizes the importance of spatial variability. Soil classification systems (the WRB and Soil Taxonomy) usually consider the characteristic properties of the typical soil profiles and overlook the spatial variability information of particular soils with varying micro-topographic conditions of the saline environments.

4.2 Solonchaks (Solonchak-Like Soils) in the Kavak Delta, Northwest Turkey

As given in Table 3, the Solonchaks of the Kavak delta (northwest Turkey) have horizon sequences of Az-AC-C1-Cg (Fig. 9; Table 3). They contain secondary salts accumulated via the crystallization of salt minerals following repeated excess irrigation with salt-laden fossil waters and repeated solution crystallization of the already existing primary salt crystals being present in the particular environments of the west.

4.3 Solonchaks (Solonchak-Like Soils) of the Çukurova Plain, South Turkey

The reclamation of the saline soils following the irrigation drainage activities (the cause of the salinity buildup at some locations of the region) in the Çukurova region (Adana, south Turkey) was studied by LANDSAT imagery for over a

Table 2 Site profile description and some physical and chemical properties of the Solonchak-like Leben soil in Bor town of the Niğde Province, central Turkey (Budak 2012)

IUSS Working Group WRB (2015)	Calcaric Sodic Solonchaks
Soil Survey Staff (2014)	Fine silty, super active, Mesic, Petrocalcic Calcixerepts
Location	16 km northwest of Bor town of Niğde Province, central Turkey
Coordinates	37° 47′ 42″N–34° 20′ 30″E
Elevation	1162 m
Climate	Arid/semi arid
Vegetation	Major plant cover is composed of Halimione verrucifera, Plantago crassifolia, Taraxacum farinosum, Puccinellia distans (Jacq.) Parl. Subsp. Distans (Fig. 8)
Parent material	Shallow marine deposits
Geomorphic unit	Old lake bed
Topography	Flat
Erosion	Wind erosion
Water table	1.80 m
Drainage	Very good
Infiltration	High
Stoniness	No stones
Soil moisture regime	Xeric

Horizon	Depth (cm)	Description
A	0–22	2.5Y7/3 (dry) light yellow, 2.5Y5/4 (moist) yellowish brown, moderate medium subangular blocky structure and very fine granular structure, clay loam, strongly effervescent, rare hairy roots and common tap roots, surface crust with salts, very weak aggregate structure, abrupt smooth boundary
Bw1	22–39	10YR6/4 (dry) dull yellow orange, moderate weak subangular blocky structure, clay loam, salt crystals on ped surfaces, strongly effervescent, moderately common tap roots, abrupt smooth boundary
Bw2	39–64	10YR6/3 (moist) dull yellow orange, thick weak platy structure, sandy loam, very hard when moist, plastic and sticky when wet, strongly effervescent, rare hairy roots, abrupt smooth boundary
Bw3	64–89	10YR 4/2 (moist) grayish yellow brown, very thick strong platy structure, sandy loam, very hard when dry and moist, strongly effervescent, moderately common tap roots, moderately common vesicular pores, rare calcium carbonate accumulations, abrupt smooth boundary
Ckm	89–107	10YR8/2 (dry) light gray, thick platy and coarse strong blocky structure, loamy sand, very hard when moist, strongly effervescent, rare hairy and tap roots, broken, platy cemented calcium carbonate, abrupt smooth boundary
C2	107–146	10YR8/2 (moist) light gray, single grain, clay loam, very hard when moist, Strongly effervescent, rare hairy roots, common vesicular pores, abrupt smooth boundary
C3	146–216+	5Y7/2 (moist) light gray, single grain, fragile when moist, not plastic and sticky when wet, clay loam, strongly effervescent, rare hairy roots, abrupt smooth boundary

Horizon	Depth (cm)	Clay %	Silt	Sand	Texture Class
A	0–22	34.4	23.9	41.8	CL
Bw1	22–39	33.1	22.2	44.8	CL
Bw2	39–64	18.6	23.4	58.0	SL
Bw3	64–89	13.4	18.7	68.0	SL
Ckm	89–107	13.1	6.4	80.5	SL
C2	107–146	19.4	19.7	61.0	CL
C3	146–216	17.4	16.7	66.0	CL

Horizon	Depth (cm)	pH Saturation Paste	EC (dS/m)	CaCO$_3$ (%)	Organic Mat. (%)	CEC (cmol$_c$ kg^{-1})	Boron (mg/kg)	P$_2$O$_5$ (kg/ha)
Az	0–22	8.44	6.24	19.24	1.86	25.6	10.98	56.8
Bw1	22–39	8.83	10.55	18.88	0.92	19.7	4.26	1.6
Bw2	39–64	8.19	10.91	13.07	0.66	27.6	1.81	32.5

(continued)

Table 2 (continued)

Horizon	Depth (cm)	pH Saturation Paste	EC (dS/m)	CaCO$_3$ (%)	Organic Mat. (%)	CEC (cmol$_c$ kg^{-1})	Boron (mg/kg)	P$_2$O$_5$ (kg/ha)
Bw3	64–89	7.86	7.12	14.89	0.62	26.0	1.90	10.0
Ckm	89–107	8.23	1.33	36.31	0.43	5.8	1.58	37.2
C2	107–146	8.36	0.5	34.13	0.37	10.0	0.94	82.9
C3	146–216	8.20	0.24	19.61	0.04	7.3	0.13	37.2

Fig. 7 Halopyhtes grown in Leben soils **a** *Halimione verrucifera*, **b** *Plantago crassifolia* **c** *Taraxacum farinosum* **d** *Puccinellia distans (Jacq.) Parl. Subsp. Distans*

period of 30 years at two 3-year sampling shifts, namely from 1954 to 1956 and from 1981 to 1984 by Dinç et al. (1991) (Table 3).

Consequently, the reclaimed soils of the region were grouped into three classes: (i) almost totally reclaimed soils found in the associations of bajadas, river terraces, and bottom lands; (ii) partially reclaimed soils located at the lower end (the southernmost area) of the region comprising saline marshy soils of the tidal marshes and soils with increased salinity and/or salinity alkalinity of the Sand Dune associations and the Saline Marshy Soils in the southeast most tip of the region. Şatır et al. (2010) in an updating–monitoring study conducted in the same area determined that the salinity of the upper northern part of the Çukurova region that developed at the above-mentioned periods (Table 4) was almost completely reclaimed, whereas the soils of the middle parts of the Çukurova Delta revealed partial reclamation following the completion of the drainage infrastructure. The southern most part faced increased salinity around the lagoonal lakes (lowest parts of the delta) which were spread to wider areas in the last decade, due to the accumulation of the drainage waters derived from the upper

Fig. 8 Soil profile of the Leben soils (**a**) and closer look to some horizons (**b**) in Bor town, Niğde Province, central Turkey

delta. The contemporary crop pattern of the delta changed from cotton to maize compared to the crop pattern of the area earlier studied by Dinç et al. (1991) in the upper and lower parts and to peanut cultivation in the lowermost parts of the delta.

4.4 Solonchaks (Solonchak-Like Soils) of the Harran Plain, Southeast Turkey

The Akçakale soils (Solonchak-like soils) are formed at the lower parts of the Harran plain under shallow and saline water with a Az-Bk-Cy-horizon sequence (Table 5).

The surface horizon of the Akçakale soil does not meet the criteria of a salic horizon according to the Soil Taxonomy (Soil Survey Staff 2014) in which a salic horizon (at least 15 cm) should have an electrical conductivity (EC) equal to or greater than 30 dS/m in the water extracted from a saturated paste. To this extent, the Akçakale soils can only be classified as Inceptisols (Sodic Calcixerepts) in Soil Taxonomy, whereas they can be classified as Calcic Solonchaks (Vertic) by the WRB (IUSS Working Group of WRB 2015) due to the difference in the requirement of the salic horizon in the latter (the salic horizon requirement of the WRB is stated as "an electrical conductivity of the saturation extract (ECe) at 25 °C of is ≥ 15 dS m^{-1} if the pH water of the saturation extract is ≥ 8.5"). Çullu et al. (2002) monitored the change in soil salinity from 1987 to 2000 in the Harran plain and determined significant increases in salt accumulation during the 13-year monitoring period. Especially after 1995 following furrow irrigation, the soil salinity buildup was about 30% higher than the salinity increase before 1995. The soil salinity change was from none to slight, slight to moderate, and moderate to severe. Strongly saline soils were particularly located at the northern part of the Harran Plain in 1987, but soils with similar salinity increase were determined in the middle and south parts in 1997. Almost all strongly saline soils were located in the south part close to Akçakale town in 2000 (Çullu et al. 2000), whereas they were spread over a wider area in the center and northwards of Akçakale town in the salinity monitoring periods of 2004 and 2009 due to the increase of

Table 3 Site description and some physical and chemical properties of a Solonchak in northwest Turkey

IUSS Working Group WRB (2015)	Fluvic Solonchaks Hypersalic
Soil Survey Staff (2014)	Oxyaquic Ustifluvents
Location	Kavak Delta, Saros Bay, Çanakkale, northwest Turkey
Coordinates	35 T, 486,236 E, 4,496,449 N
Altitude	3 m
Landforms	Flood plain-delta
Position	Light Pit-Flat
Aspect	–
Slope	0–1%
Drainage class	Very poor
Groundwater	90 cm
Eff. soil depth	75 cm
Parent material	Alluvium
Climate	Mediterranean
Land use	Natural habitat
Vegetation	Halophytes (*Arthrocnemum fruticosum, Halimione portulacoides, Halocnemum strobilaceum, Salicornia patula, Salsola kali* and *Salsola ruthenica*) (Fig. 10)

Horizon	Depth (cm)	OM (%)	pH	EC (dS/m)	CaCO$_3$(%)	CEC (cmol$_c$ kg^{-1})
Az	0–16	2.77	7.5	65.8	7.34	15.06
AC	16–25	0.79	7.94	10.25	7.88	10.14
C1	25–40	0.63	7.67	7.89	6.65	4.46
Cg	40–60	0.29	7.86	5.42	6.88	3.42
2C	60+	0.21	7.84	7.10	7.73	2.17

Horizon	Color Dry	Color Moist	Mottles	Roots	Structure	Texture (%) Clay	Silt	Sand
Az	2.5Y4/2	2.5Y3/2	None	M	MS-Platy	12.2	40	47.8
AC	–	2.5Y4/2	Few	M	Single-grained	12.2	4.0	83.8
C1	–	2.5Y4/6	Moderate	S	Single-grained	4.2	8.0	87.8
Cg	–	2.5Y5/3	Few	–	Single-grained	8.2	6.0	85.8
2C	–	2.5Y5/4	Few	–	Single-grained	6.16	4.0	89.8

M Medium, *S* Sparse, *MS-platy* Medium–small

irrigated lands and excessive use of irrigation water (Table 2; Fig. 11) (Çullu et al. 2002, 2010).

Fluctuations in the water table within 1–2 m and capillary movement of saline water to the soil surface have contributed to salt accumulation at various soil depths (Çullu et al. 2010). Irrigation initiated at 1995 triggered soil salinity at different levels and salts moved to the soil surface (Fig. 12a). The high summer evaporations caused the rapid formation of the salt minerals in the surface horizons in turn causing yield and economical losses. However, the Agrarian Reform of the Ministry of Food, Agriculture and Livestock initiated a project to combat salinization and accordingly sought to reclaim a 10,000 ha strongly salt-affected land (Bilgili et al. 2013) of the southeast of the country. Within this project, drainage channels were constructed to leach secondary salinity from the Akçakale soils, which most likely will reverse the process of Solonchak formation in the future.

5 Management and Regional Distribution

5.1 The Çukurova Delta

Irrigation water often contains salinity that builds up as water moves across the landscape, or the salts may come from human-induced sources such as the municipal runoffs and/or industrial discharges (Çukurova plain, some plains in the Aegean Region, Nilüfer and Ergene River basins). Özcan and Çetin (1996) stated that saline areas increased fivefold

Fig. 9 Soil profile and surface terrain covered by Halocnemum Strobilaceum in the Kavak Delta (Saros Bay northwest Turkey)

due to shallow groundwater and poor-quality drainage water used in some parts of the project area (Lower Seyhan Irrigation Project) with no irrigation infrastructures during the period from 1960 to 1980. A detailed soil survey carried out by Dinç et al. (1995) showed that shallow and saline groundwater, originated from upstream-irrigated areas, was the major cause of soil salinity in the Çukurova Delta. Here, the transportation of the soluble salts to the Mediterranean Sea via drainage outlets is mostly insufficient due to the lack of on-farm development practices, irregular canal maintenance, topography (low slope angle), and siltation by increased erodibility in the drainage canals (Özcan and Çetin 1996). Özcan and Çetin (1998) compared the groundwater depth, groundwater salinity, and soil texture both in irrigated and nonirrigated areas of the east Mediterranean coastal region of Turkey and found that soil salinity was affected when groundwater salinity was over 4 dS/m. The nonirrigated area was affected by 65–100% more than the irrigated area within the same groundwater salinity levels ignoring groundwater depths. Moreover, they found a linear relationship between the groundwater salinity and soil salinity, depending on the groundwater depth. The soil salinity in the clayey soils was twice as much as it was in the nonirrigated and over 55–60% in the irrigated area in comparison to the medium- and light-textured soils, provided the groundwater depth was 0–150 cm and its salinity >15 dS/m.

Despite the construction of the contemporary drainage systems, recent research has revealed the need to manage halophytes in the lowermost part of the delta in order to protect the vast delta wetland habitat (Çakan et al. 2011). The upper parts are cultivated for cotton, maize (sought to decrease in expense of cotton via state subsidies), and partly for peanuts in the lowermost delta requiring much less chemicals and suitable for saline Arenosols (leveled sand dunes).

5.2 The Konya Basin

The drainage and well waters in the Konya basin (central Turkey) used for irrigation are partly responsible for the salinity of the soils allocated for wheat and sugar beet. The parent materials of the soils in the Konya basin are mainly ancient lake deposits with high contents of $CaCO_3$ and

Fig. 10 Salt crust on surface (**a**) in the Kavak Delta, Saros northwest Turkey covered by *Halocnemum strobilaceum* (**b, c**)

Table 4 Distribution of salt-affected areas in the Çukurova Plain (south Turkey) (Dinç et al. 1991)

Salinity changes	Slightly saline (<4 dS/m)		Moderately saline (4–8 dS/m)		Strongly saline (>15 dS/m)		Total area
	ha	%	ha	%	ha	%	ha
1954–1956	19,982	5.9	29,053	8.6	56,602	16.8	105,638.9
1981–1984	35,941	10.6	17,759	5.2	7197	2.1	60,898

gypsum. Consequently, well and drainage waters are not suitable for irrigated agriculture in the major parts of the plain. The drainage conditions are also very poor, precipitation varies from 250 to 350 mm/year, and evaporation is over 1500 mm. Thus, high salinity buildup is inevitable in the Konya Plain due to the limited leaching of the waters coupled with a dry climate and high evaporation. Earlier soil survey reports (GDRS 1978) document a saline land of about 510 000 ha in this area. The Konya plain (average altitude of the plain is 1020 m and the water surface of Lake Tuz is 905 m above sea level) is a worldwide example for soils suffering primary salinity that is documented by the numerous lakes and the Great Salt Lake (Lake Tuz). These are the remnants and the evidence of the pre-Quaternary/Quaternary central Turkey inner sea that dried out throughout the Holocene period. Moreover, Konya is a

Table 5 Site description and some physical and chemical properties (Çullu et al. 2000) of a Solonchak-like (Akçakale soils) in southeast Turkey

IUSS Working Group WRB (2015)	Calcic Solonchaks (Vertic)
Soil Survey Staff (2014)	Sodic Calcixerepts
Location	Akçakale, Şanlıurfa, southeast Turkey
Coordinates	37 T, 498,520 E, 4,063,829 N
Altitude	345 m
Landforms	Ancient lacustrine
Position	Flat
Aspect	–
Slope	0–1%
Drainage class	Poor
Groundwater	95 cm
Eff. Soil depth	70 cm
Parent material	Lacustrine
Climate	Semi arid
Land use	Natural habitat
Vegetation	Halophytes

Horizon	Depth (cm)	OM (%)	pH Sat. Ext.	EC (dS/m)	$CaCO_3$ (%)	CEC ($cmol_c$/kg)	ESP (%)	Exc. Na (me/100 g)
Az	0–22	0.94	7.53	28.25	18.25	34.4	28.88	9.92
A2	22–45	0.90	7.74	12.40	18.15	37.4	24.38	9.12
Bk	45–67	0.92	7.74	12.40	23.15	36.6	30.16	11.04
Cy1	67–80	0.94	7.54	9.90	20.85	35.4	29.71	10.52
Cy2	80–100	0.91	7.66	6.42	20.00	32.9	23.06	7.59

Horizon	Depth (cm)	Clay (%)	Silt (%)	Sand (%)
Az	0–22	63.85	25.2	10.95
A_2	22–45	64.9	25.2	9.90
Bk	45–67	63.85	27.3	8.85
Cy1	67–80	64.9	27.3	87.80
Cy2	80–100	59.65	32.5	7.85

closed basin due to the topography of this ancient geomorphic system and is surrounded by all directions by highlands and all the transported water from the highlands to the basin adds up for more than 40% of the total groundwater of Turkey.

The drainage activities of the salt-bearing waters from the upper parts of the Konya Basin and the Çukurova Deltas have raised and are still raising the groundwater level to the soil surface on the lower lands and ditches. The poor drainage conditions, i.e., the high groundwater and the upward movement of saline groundwater, followed by evaporation at the surface cause the formation of the saline soils in the Konya Basin. External Solonchaks of the Konya Basin can be reclaimed by leaching the salts using low saline irrigation water as performed in the Leben soils of Bor town in the Niğde Province (Fig. 13). The low organic matter contents of the reclaimed soils need the application of best management practices to prevent further degradation. Thus, in order to obtain the desired outcomes from the reclamation of saline soils, the spatial variability of the EC values should be taken into consideration. Further, Lin et al. (2005) indicated that the spatial variability also strongly influences the reliability of the results of logical, empirical, and physical models of soil and landscape processes.

5.3 The Harran Plain

Salinization related to excess irrigation and inappropriate irrigation systems is a significant problem in the Harran

Fig. 11 Salinity changes from 2004 to 2009 in the Harran Plain, southeast Turkey (Çullu et al. 2010)

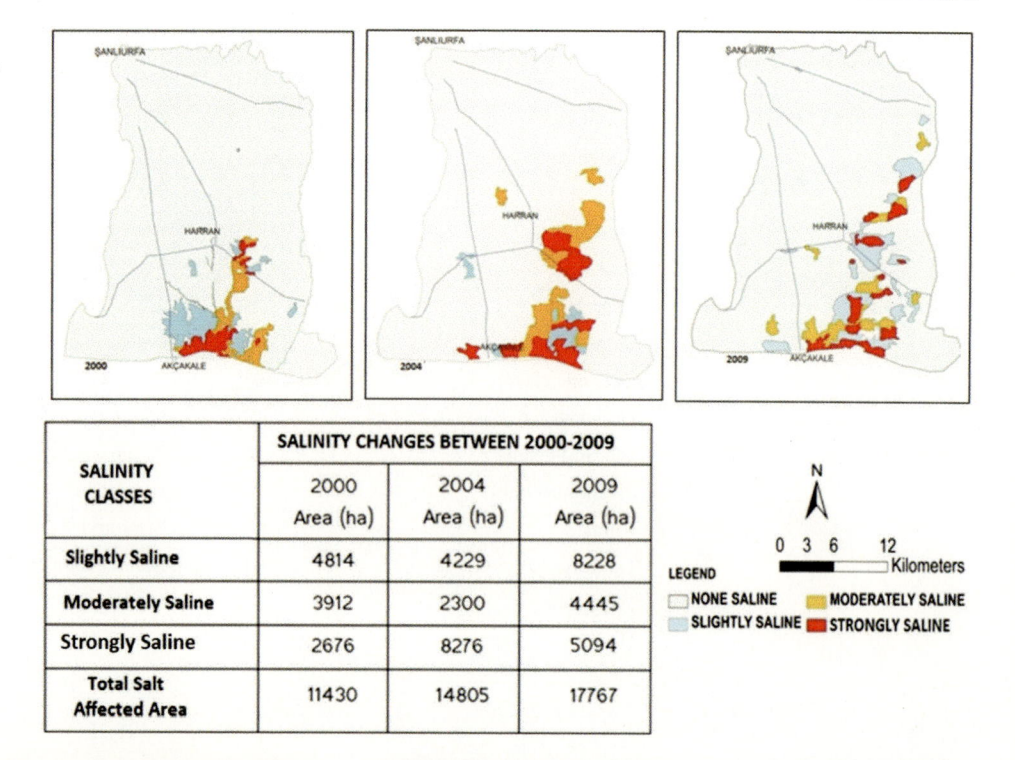

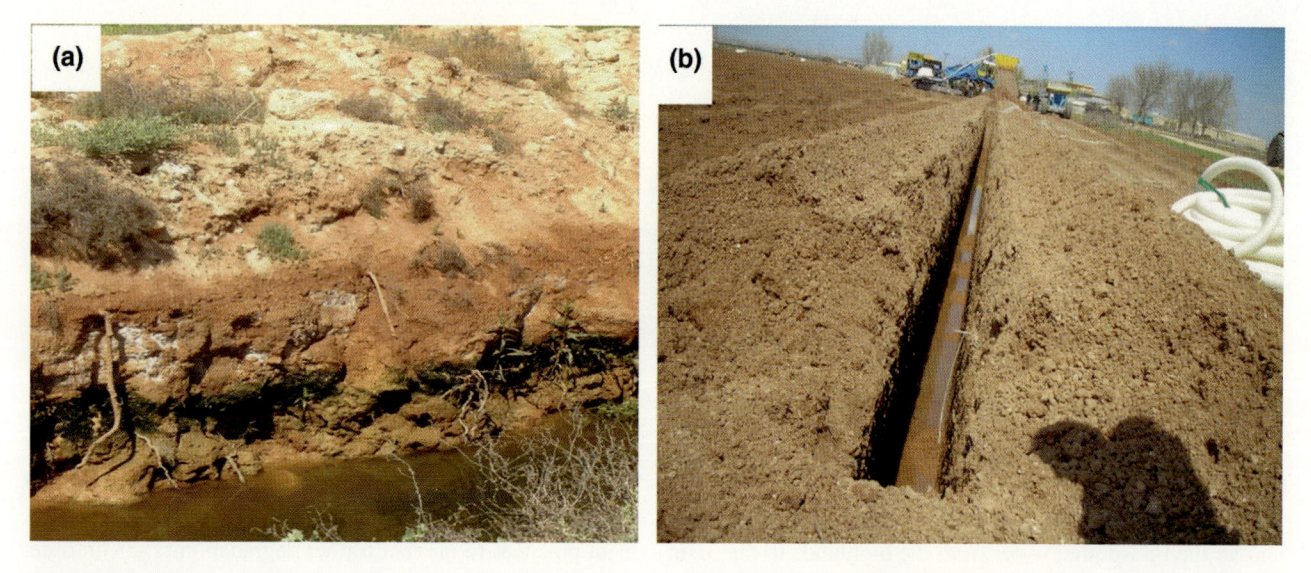

SALINITY CLASSES	SALINITY CHANGES BETWEEN 2000-2009		
	2000 Area (ha)	2004 Area (ha)	2009 Area (ha)
Slightly Saline	4814	4229	8228
Moderately Saline	3912	2300	4445
Strongly Saline	2676	8276	5094
Total Salt Affected Area	11430	14805	17767

Fig. 12 Salt-affected soil with high saline water table (**a**) and drainage infrastructure (**b**) leaching salts from the Akçakale soils in the Harran Plain, southeast Turkey

Plain (Fig. 14). More than a 40,000 ha area is under the threat of salinity today in this area, whereas the salinity of the waters of the Euphrates River is less than 0.3 dS/m and the annual average evaporation is about 2200 mm and the precipitation varies from 300 to 500 mm. The Agrarian Reform of the Ministry of Food, Agriculture and Livestock has launched a project seeking to reclaim the Harran plain saline areas via drainage infrastructure in the near future (Anonymous 2015) as implemented in the first part of the area (Fig. 12).

Fig. 13 Reclamation of the Leben soils, Bor town, Niğde Province, central Turkey

Fig. 14 Salinity buildup and degradation of the surface soil and vegetation in the irrigated Harran plain, Şanlıurfa Province, south Turkey

References

Ağca N, Doğan K, Akgöl A (2000) Distribution of soil salinity and alkalinity in Amik Plain. Univ Mustafa Kemal Univ, Ann Fac Agric 5:29–40 (in Turkish)

Anonymous (2015) Strategic plan, 2013–2017. Publication of the Ministry of Food, Agriculture and Livestock, Turkey, 132 p

Akgül M, Şimşek G (1996) Basic soil surveys of Daphan plain III. Detailed soil map and report. Atatürk Univ J Agri Fac 27(1):74–88 (in Turkish)

Atasoy A, Çeçen R (2014) Salinity problem in Reyhanlı town. J Turk Geogr 62:21–28 (in Turkish)

Atatanır L, Aydın G, Yorulmaz A (2010) The determination of salt affected soils using satellite data and GIS in Söke plain, International conference on soil fertility and soil productivity, Berlin

Bahtiyar M (1971) A research on soil genesis, characteristics and reclamation of sodic soils of Erzincan Ada. Atatürk University, Faculty of Agriculture, Soil Science Department, pp 95–98 (in Turkish)

Bayramin İ, Yalçın OZ, Tunçay T, Samray HN (2004) Remediation of the salt affected soils and their economic value, an example from Ayrancı-Karaman, international soil congress on natural resource management sustainable development, 7–10 June 2004. Erzurum, Turkey

Bilgili AV, Çullu MA, Aydemir S, Aydemir A, Almaca A (2013) Probability mapping of saline and sodic soils in the Harran plain using a non-linear kriging technique. Eurasian J Soil Sci 2:76–81

Budak M (2012) Genesis and classification of saline alkaline soils and mapping with both classical and geostatistical techniques. (PhD Dissertation) No: 322692. Gaziosmanpaşa University, Institute of Science. Tokat, Turkey, 254 p (in Turkish)

Buringh P (1960) Soils and soil conditions in Iraq. Baghdad, Republic of Iraq, Ministry of Agriculture, 322 p

Cambardella CA, Moorman TB, Parkin TB, Karlen DL, Novak JM, Turco RF, Konopka AE (1994) Field-scale variability of soil properties in central Iowa soils. Soil Sci Soc Am J 58(5):1501–1511

Cemek B, Güler M, Arslan H (2006) Determination of salinity distribution in Bafra plain using geographic information system (GIS), Atatürk University. J Agri Fac 37(1):63–72 (in Turkish)

Çakan H, Yılmaz KT, Alphan H, Ünlükaplan Y (2011) The classification and assessment of vegetation for monitoring coastal sand dune succession: the case of Tuzla in Adana, Turkey. Turk J Bot 35:697–711

Çağlar KÖ (1949) Soil science. Ankara University, Agricultural Faculty Publication, Ankara (in Turkish)

Cangir C, Boyraz D (2000) Status of the saline and sodic soils of the lower Meriç valley in İpsala flood plain. Proceeding of international symposium on desertification, June 13–17, Konya Turkey

Çullu MA, Dinç U, Şenol S, Öztürk N, Çelik İ, Günal H (1995) Mapping the saline and alkaline soils using remote sensing data. İlhan Akalan Soil and Environment Symposium. 1:163–172 (in Turkish)

Çullu MA, Çelik İ, Almaca A (2000) Degradation of the harran plain soils due to irrigation. Proceedings of international symposium on desertification, 13–17 June 2000, Konya, Turkey, pp 193–197

Çullu MA, Almaca A, Şahin Y, Aydemir S (2002) Application of GIS for monitoring soil salinization in the Harran Plain, Turkey. Turkish soil science society, international conference on sustainable land use and management, Çanakkale, Turkey 326–332

Çullu MA, Aydemir S, Almaca A, Öztürkmen AR, Sönmez O, Binici T, Bilgili AV, Yılmaz G, Dikilitaş M, Karakaş S, Sakin E, Şahin Y, Aydoğdu M, Aydemir A, Çeliker M (2010) Mapping the salinity maps of Harran plain and estimation the effects of salinity on crop yield losses. Turkish Republic Priemership SEP Regional Development Administration. Project Report, Şanlıurfa, Turkey (in Turkish)

De Meester T (1970) Soils of the Great Konya Basin, Turkey. Agricultural research reports 740. Centre for Agricultural Publishing and Documentation, Wageningen, 302 p

Driessen PM (1970) Soil salinity and alkalinity in the Great Konya Basin, Turkey. Department of Tropical Soil Science, Agricultural University, Wageningen. ISBN 9022003086, 257 p

Dinç U, Şenol S, Kapur S, Sarı M, Derici MR, Sayın M (1991) Formation, distribution and chemical properties of saline and alkaline soils of the Çukurova Region, southern Turkey. Catena 18 (2):173–183

Dinç U, Sarı M, Şenol S, Kapur S, Sayın M, Derici MR, Çavuşgil V, Gök M, Aydın M, Ekinci H, Ağca N, Schlicting E (1990). Çukurova Region Soils. Çukurova University. Agricultural Faculty. Yardımcı Ders Kitabı No: 26:171 p. Adana (in Turkish)

Dinç U, Sarı M, Şenol S, Kapur S, Sayın M, Derici MR, Çavuşgil V, Gök M, Aydın M, Ekinci H, Ağca N, Schlichting E (1995) Soils of the Çukurova Region, University of Çukurova, Faculty of Agriculture, No. 26, Adana, Turkey, 210 p (in Turkish)

Doğan O (2014) Turkish soils productivity project. Ministry of forestry and water affairs. General Directorate of Combating Desertification and Erosion. Ankara, 62 p (in Turkish)

DSİ (1966) II. Region DSİ General Directorate. Çine-menderes project, planning drainage report for Aydın-Söke Plains (in Turkish)

Eswaran H, Rice TJ, Ahrens R, Stewart BA (2003) (Eds.) Soil classification: a global desk reference. CRC Press, Boca Raton, FL. 280 p

Fitzpatrick RW, Cox JW, Munday B, Bourne J (2002) Development of soil-landscape and vegetation indicators for managing waterlogged and saline catchments. Aust J Exp, Agric

Fitzpatrick RW, Powell B, McKenzie NJ, Maschmedt DJ, Schoknecht N, Jacquier DW (2003) Demands on soil classification in Australia. In: H Eswaran, T Rice, R Ahrens, BA Stewart (eds) Soil classification: a global desk reference. CRC Press LLC Boco Raton, pp 77–100

GDRS (1978) Land resources of Turkey. Ministry of rural affairs and cooperatives, General directorate of rural services (GDRS), Soil Survey Department, Ankara, 55 p

ISRIC (2014) World soil distribution (accessed on 30 12 2014) http://www.isric.org/aboutsoils/world-soil-distribution/solonchaks

IUSS Working Group WRB (2015) World reference base for soil resources 2014, update 2015 International soil classification system for naming soils and creating legends for soil maps. World Soil Resources Reports No. 106. FAO, Rome, 203 p

Kaplan YM, Eren M, Kadir S, Kapur S, Huggett J (2014) A microscopic approach to the pedogenic formation of palygorskite associated with quaternary calcretes of the Adana area, southern Turkey. Turkish J Earth Sci 23:559–574

Kapur S, Şenol M, Karaman C, Akça E, Güvercin E (1995) Stratigraphy and clay mineralogy of Messinien deposits of the Çukurova region. İlhan Akalan soil and agriculture symposium, vol I. S. A367-A78, Ankara (in Turkish)

Kapur S, Akça E, Özden DM, Sakarya N, Çimrin KM, Alagöz U, Ulusoy R, Darıcı C, Kaya Z, Düzenli S, Gülcan H (2003) Land degradation in Turkey (RJA Jones, L Montanarella Eds.) The JRC enlargement action land degradation. Contributions to the international workshop—land degradation, 5–6 Dec 2002, Ispra, Italy, pp 303–318

Kılıç Ş, Ağca N, Karanlık S, Şenol S, Aydın M, Yalçın M, Çelik İ, Evrendilek F, Uygur V, Doğan K, Aslan S, Çullu MA (2008) Detailed soil surveys of the Amik plain, fertility studies and land use planning, project of state planning organization (DPT) 2002 K 120480

Lin H, Wheeler D, Bell J, Wilding L (2005) Assessment of soil spatial variability at multiple scales. Ecol Model 182(3):271–290

Munsuz N (1969) Factors affecting the formation of saline and sodic soils of Malya state farm and Reclamation strategies. Ankara University. Publications of Agricultural Faculty. Ankara. No. 336. 110 p (in Turkish)

Oakes H (1954) The soils of Turkey. Republic of Turkey. Ministry of agriculture. Soil conservation and farm irrigation div. Public. Ankara, No: 1, 180 p

Oruç N (1970) Some of physical and chemical characteristics of saline and sodic soils of Iğdır plain. Publication Center of Ataturk University. No: 80. Publications of agricultural faculty No: 27 research Pub. No 8. Soil Science, pp 48–49 (in Turkish)

Özcan H, Cetin M (1996) Land degradation induced by high groundwater level and soil salinity in the Fourth Stage Project Area of Yüreğir Plain: a case study. In: S Kapur, E Akça, H Eswaran, G Kelling, C Vita-Finzi, AR Mermut, AD Öcal (eds) Proceedings of the 1st international conference on land degradation. International working group on land degradation and desertification (IWGLDD-ISSS), Çukurova University Press, Adana, Turkey, pp 93–99

Özcan H, Çetin M (1998) The relationship between groundwater and soil salinity in the Eastern Mediterranean Coastal Region, Turkey. M. Şefik Yeşilsoy International Symposium on Arid Region Soil. Share our experiences to conserve the land. International Agro-hydrology Research and Training Centre, 21–24 Sept, İzmir, Turkey, pp 370–375

Saatçı F, Tuncay H (1971) Researches on saline and alkaline soils of Aegean Region. Aegean University. Publication of Agricultural Faculty No: 173. Bornova, İzmir-Turkey (in Turkish)

Şatir O, Berberoğlu S, Kapur S, Erdoğan A, Dönmez C, Şatır NY, Nagano T, Akça E, Tanaka K (2010) Soil Salinity Mapping Using CHRIS-PROBA Hyperspectral Data, Hyperspectral Workshop 2010 from CHRIS-Proba to PRISMA & EnMAP and Beyond, ESA-ESRI, 17–19 Mar 2010. Frascati, Italy

Şengör C, Yılmaz Y (1981) Tethyan evolution of Turkey: a plate tectonic approach. Tectonophysics 75:181–241

Soil Survey Staff (2014) Keys to soil taxonomy, 12th ed. USDA-Natural Resources Conservation Service, Washington, DC, 372 p

Sönmez B (2004) Salinity amelioration studies in Turkey and saline soil management. Proceedings of Salinity Management, 20–21 May 2004 Ankara, pp 157–162 (in Turkish)

Sönmez B (2011) Reclamation and management of saline and sodic soils. Train Scope Sci Wisdom 134:52–56 (in Turkish)

Lixisols

Orhan Dengiz, Hasan Özcan, and Sabit Erşahin

1 Introduction

The name of Lixisol (IUSS Working Group 2015) is derived from the Latin *"lixivia"* meaning washed out substances and strongly weathered soils, in which clays are leached/migrated to an *argic* B horizon with moderate to high base saturation. These soils are generally observed in tropical, subtropical, or warm temperate climates with a pronounced dry season, notably on old eroded or deposited surfaces. Lixisols were formerly included in the 'Red-Yellow Podzolic soils' (e.g., Turkey and Indonesia), 'Podzolicos vermelho-amarello eutroficos a argila de atividade baixa' (Brazil), 'Sols ferralitiques faiblement désaturés appauvris' and 'Sols ferrugineux tropicaux lessivés' (France), 'Red and Yellow Earths,' 'Latosols,' or classified as oxic subgroups of Alfisols (Soil Survey Staff 2014).

2 Parent Materials

The soils are mainly formed on unconsolidated, strongly weathered, and strongly leached, finely textured materials. In addition, they are located on andesitic and clay stone material in Turkey.

3 Profile Development and Characteristics

The horizons order of Lixisols is mostly A-E-Bt-C or A-Bt-C in Turkey. On slopes and on other surfaces subject to erosion, the argic accumulation horizon may be exposed or remain at shallow depth. Many Lixisols have ochric surface horizons over a brown or reddish brown argic Bt-horizon that often lacks clear evidence of clay illuviation other than a sharp increase in clay content over a short vertical distance. The overlying eluvial (E) horizon, when still present, is commonly massive and very hard when dry (hard setting). Stone lines are common in the subsoil. They have low aggregate stabilities (no pseudo-sand structures like Ferralsols because of higher pH), higher base saturation, and accordingly somewhat stronger structure than normally found in Acrisols but slaking and caking of the surface soil are still serious problems. The moisture holding properties of Lixisols are slightly better than that of Ferralsols or Acrisols at similar clay and organic matter contents.

The physical, chemical, and morphological characteristics of the Lixisols commonly found in the Black Sea region are given in Tables 1 and 2. The descriptions and analyses of the soils were conducted by Dengiz et al. (2008) to evaluate the genesis and classification of the soils of the hazelnut cultivation areas in the Ünye-Tekkiraz District in the Ordu province (the central Black Sea region) (Fig. 1) and the protected intra-forestal meadows in Gölköy (Ordu, the central Black Sea region) (Fig. 2).

4 Environment and Distribution

Lixisols are commonly observed in colluvial landscapes, gently sloping plains, flat and gently sloping uplands merging into small narrow valleys. And also in the areas where colluvial materials have creeped to the lower small stream settings of the mountain systems at the central and western parts of the Black Sea region. Lixisols of Turkey, mainly Chromic, Calcic, Clayic cover a land area of about

Table 1 Site description and some physical, chemical, and morphological characteristics of the Yaycı Lixisols described in Ünye-Tekkiraz (North Turkey)

IUSS Working Group WRB (2015)	Clayic Lixisol
Soil Survey Staff (2014)	Vertic Haplustalf
Location	Ünye-Tekkiraz (Ordu)
Coordinate	4538861 N-342632 E (37-TM m)
Elevation	341 m
Parent Material	Clay stone
Precipitation	1162,4 mm
Temperature	14.2 °C
Land Use/land cover	Cultivated, hazelnut
Slope	5–15%
Drainage	Poor

Horizon	Depth (cm)	pH[a]	EC[a] (dSm^{-1})	CaCO$_3$ (%)	Org. mat. (%)	CEC (cmol$_c$kg)	Exchangeable Cations cmol$_c$kg^{-1}			
							Na$^+$	K$^+$	Ca^{++}	Mg^{++}
A1	0–10	7.15	0.266	2.88	7.52	47.09	0.16	1.01	41.19	2.56
Bt1	10–45	7.21	0.16	1.32	1.6	47.2	0.38	0.99	42.31	2.21
Bt2	45–77	7.46	0.149	1.28	1.45	47.12	0.54	1.37	42.28	2.25
Ck	77+	7.65	0.138	39.6	0.79	32.59	0.35	0.36	30.71	0.26

Horizon	Color Dry/Wet	Particle size (%)			Texture Class	Coarse fragments (>2 mm %)	Field capacity (%)	Wilting point (%)	Available water (%)
		Clay	Silt	Sand					
A1	2,5 Y 5/4 / 2,5 Y 5/3	42.7	24	33.3	C	0.88	41.5	29.1	12.4
Bt1	2,5 Y 6/4 / 2,5 Y 5/4	63	17.4	19.6	C	0.02	45.5	29.2	16.3
Bt2	2,5 Y 6/3 / 2,5 Y 5/4	61.4	20.4	18.2	C	0.04	44.3	27.2	17.1
Ck	2,5 Y 8/4 / 2,5 Y 7/4	27.6	46.8	25.5	C	2.12	36.5	22.2	14.4

[a]pH and EC are measured in saturation paste

385,311.14 ha which is 0.50% of the whole agricultural land (Fig. 3). Lixisols have mainly formed on the unstable geomorphic surfaces prone to erosion with low organic matter contents. Their organic matter content is generally low on the steeper slopes where erosion exposes the lighter color. Topography is an important mediator of soil formation as it redistributes infiltrating water and in so doing it affects the development of the soil.

5 Management of Lixisols

Both soils described in this chapter (the Tekkiraz and Gölköy soils) are the major management areas for the Lixisol ecosystem and the main income generating sites for hazelnut and tea (Figs. 1, 2, 4, and 5). The hazelnut production sites of this region also comprise the intra-forestal pastures. Notwithstanding, these are the hazelnut-pasture mixed ecologies of northern Turkey. Intra-forestal pastures are a common land use type in the country due to the historical forest settlements in the provinces especially in the north and other parts of the country. These rural populations of the north in contrast to many other parts of Turkey have been the major factor in protecting this mixed human reshaped and sustained the natural ecology, more specifically the Lixisol ecosystem highly appropriate for hazelnut production. The higher organic matter contents of the surface horizons of the two Lixisols is a strong indicator of the protected and/or the well-managed soil ecosystem of the Lixisols (Tables 1 and 2).

However, despite the unsuitable (undulating) topographic conditions of the region, a wide range of commercial farming is also profitable in the area, due to the climatic conditions together with the appropriate properties of the

Table 2 Site description and some physical, chemical, and morphological characteristics of Eğribel-Gölköy Soil

IUSS Working Group (2015)	Haplic Lixisol
Soil Survey Staff (2014)	Typic Haplustalf
Location	Eğribel-Gölköy (Ordu)
Coordinate	383812 E. 4512362 N (37-UTM m)
Series name	Eğribel
Elevation	1134
Parent material	Andesitic
Precipitation	1041 mm
Temperature	14 °C
Land Use/land cover	Pasture
Slope	6–12%
Drainage	Good

Horizon	Depth (cm)	pH[a]	EC[a] (dSm^{-1})	CaCO$_3$ (%)	OM (%)	CEC (cmol$_c$kg^{-1})	Na$^+$	K$^+$	Ca^{++}	Mg^{++}
A	0–12	7.9	0.54	0.37	3.86	33.15	0.32	1.17	29.87	1.79
Bt1	12–44	7.0	0.16	0.51	1.57	27.59	0.37	0.65	22.07	1.44
Bt2	44–67	7.3	0.38	0.37	1.31	29.81	0.34	0.46	20.61	1.39
Bt3	67–100	6.8	0.06	0.44	0.84	28.15	0.35	49	17.67	1.42
IIC	100–150	6.9	0.09	0.44	0.37	56.75	0.35	0.58	36.88	1.7

Horizon	Color Dry/Wet	Clay	Silt	Sand	Texture class	Coarse fragments (>2 mm %)	Field capacity (%)	Hydraulic conductivity (cm h^{-1})
A	7,5 YR 4/3 / 7,5 YR 3/3	25	31.5	43.5	L	44.4	40.8	2.8
Bt1	7,5 YR 5/4 / 7,5 YR 4/4	41.1	35.1	23.8	C	18.1	37.7	0.6
Bt2	7,5 YR 5/4 / 7,5 YR 4/4	37.3	31,3	31.4	CL	34.2	35	0.7
Bt3	7,5 YR 5/4 / 7,5 YR 3/4	37.4	29.2	33.4	CL	44.4	35.1	0.54
IIC	7,5 YR 5/4 / 7,5 YR 4/4	33.3	27.3	39,4	CL	59.3	33.5	0.59

[a]pH and EC are measured in saturation paste

Lixisols (Tables 1 and 2). The mountains are moderately high and steep, with elevations rarely exceeding 1500 meters. Cereals (mainly wheat), vegetables, and variable fruits are produced on the foothills of the colluvial landscapes, piedmonts, piedmont slopes, delta plains, alluvial fans, basin floor remnants, backshores, plateaus, various terraces, valley sides, valley floors, and stream terraces. As precipitation is most common in the spring and fall, most of the cultivated soils are irrigated with the waters diverged from the nearby streams and rivers. Some of these streams become ephemeral during the drought years. Some Lixisols, on the mountain floors, plains, and similar landscapes are allocated to urban settlements, railroads, highways, and other infrastructures. Lixisols on the highlands and plateaus are mainly allocated for forestry and grazing.

Cultivated Lixisols on the sloping topographies are highly prone to water erosion. Conventional tillage applied down the slope together with the decreased organic matter has been causing the degradation of the soil structure, increased soil compaction, and decreased soil fertility. Overgrazing and poorly managed grazing are the main causes of decreased grassland efficiency and grass quality. However, the tendency to soil degradation is greatest under intensive cropping, depleting soil fertility and yields of corn and rice. Some Lixisols have a poorly developed soil structure that makes these soils highly prone to crusting,

Fig. 1 The Yaycı Soils in a hazelnut orchard in Ünye-Tekkiraz (Ordu)

Fig. 2 The Eğribel-Gölköy Soil in a protected grassland in the Gölköy (Ordu) intra-forestal area

accelerated erosion, compaction, and other degradative processes.

A community-based approach via well-established farmers unions to rehabilitating the land and improving crop production based on soil ecological/terroir and/or the Anthroscape contextual principles can improve fertility and sustain the Lixisols. In the Samsun region, a significant portion of Lixisols is used for maize cultivation. In general, no-till and other types of conservation tillage practices in corn may result in increased organic matter and improved structure of these soils (Lal 1995). The use of conservation tillage will help reduce soil erosion in other field and horticultural crops grown on the gently to moderately sloping topographies. Improved soil structure and increased soil organic matter through conservation tillage can result in increased infiltration rates and soil fertility, which are necessary for rehabilitating the physical, chemical, and biological qualities of Lixisols. However, some disadvantages of no-till, such as the lower response to N fertilizers due to microbial immobilization of some fertilizers in the surface soils, should be considered before its application (Lal 1995). P fertilizers are equally effective in no-till and plow applications. Incorporation of crop residues is highly recommended to maintain and/or improve the soil physical and chemical conditions for sustainable corn and rice

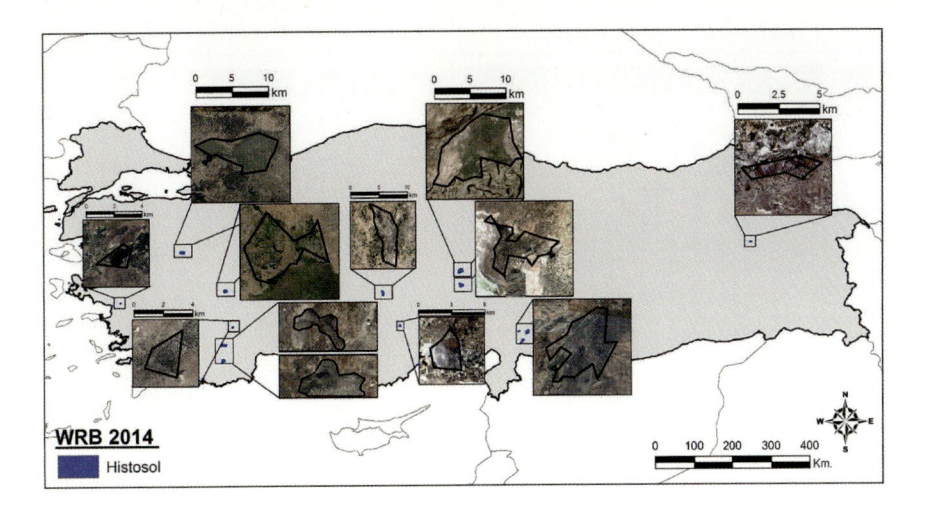

Fig. 3 Lixisol distribution in Turkey

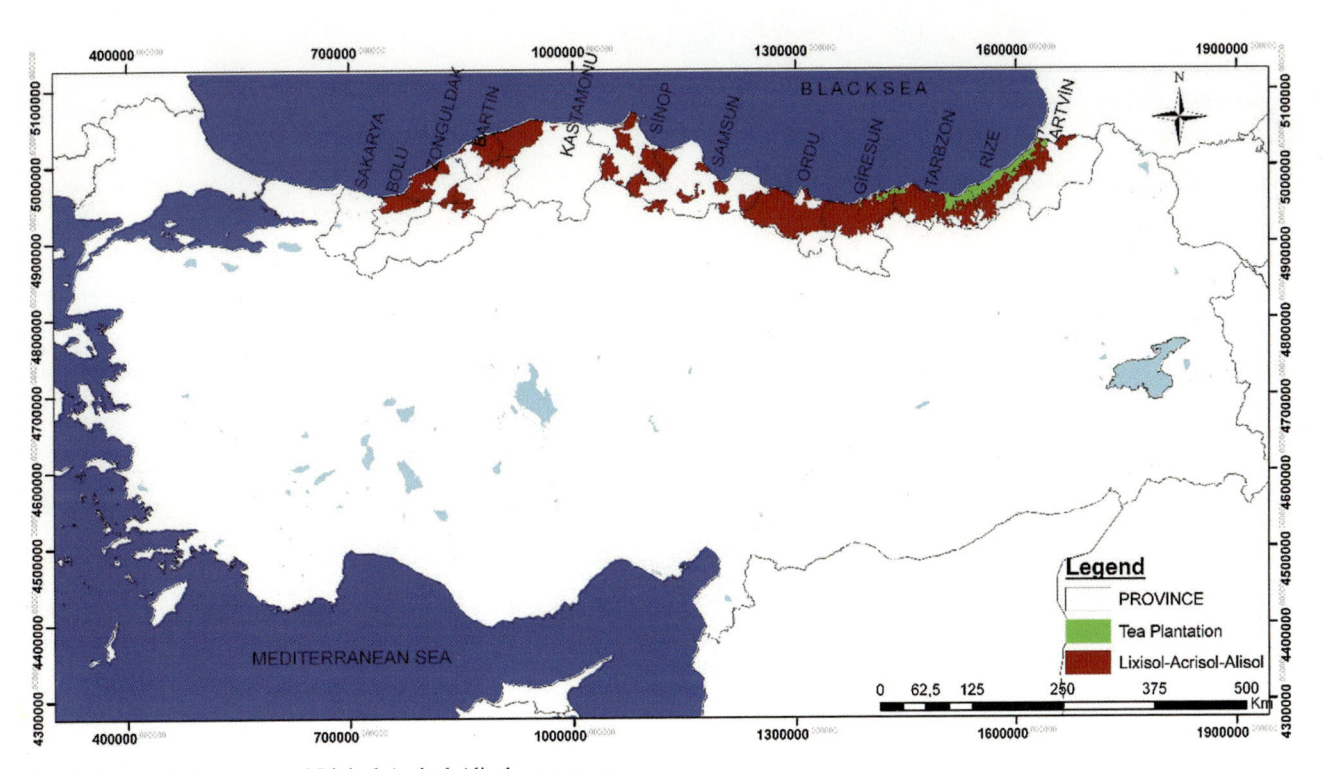

Fig. 4 Tea cultivated areas and Lixisol-Acrisol-Alisol ecosystems

production. Crop rotation can be adopted in the soil fertility depleted areas to revert the trend of declining yield due to continuous cropping.

Turan et al. (2009) reported the increase of plant available water capacity of the soil following the construction of terracing and the application of high amount organic fertilizers in the initial years of the tea plantation in Lixisols of the Black Sea Region. Turan et al. (2009) also observed increases in the water resistant aggregate stability in Lixisols with the use of organic amendments and organic matter. However, the authors outlined reduction in the organic matter and loss of soil nutrients within the forest ecosystem due to the harvesting of herbaceous plants growing beneath the alders for fodder or animal bedding on Lixisols.

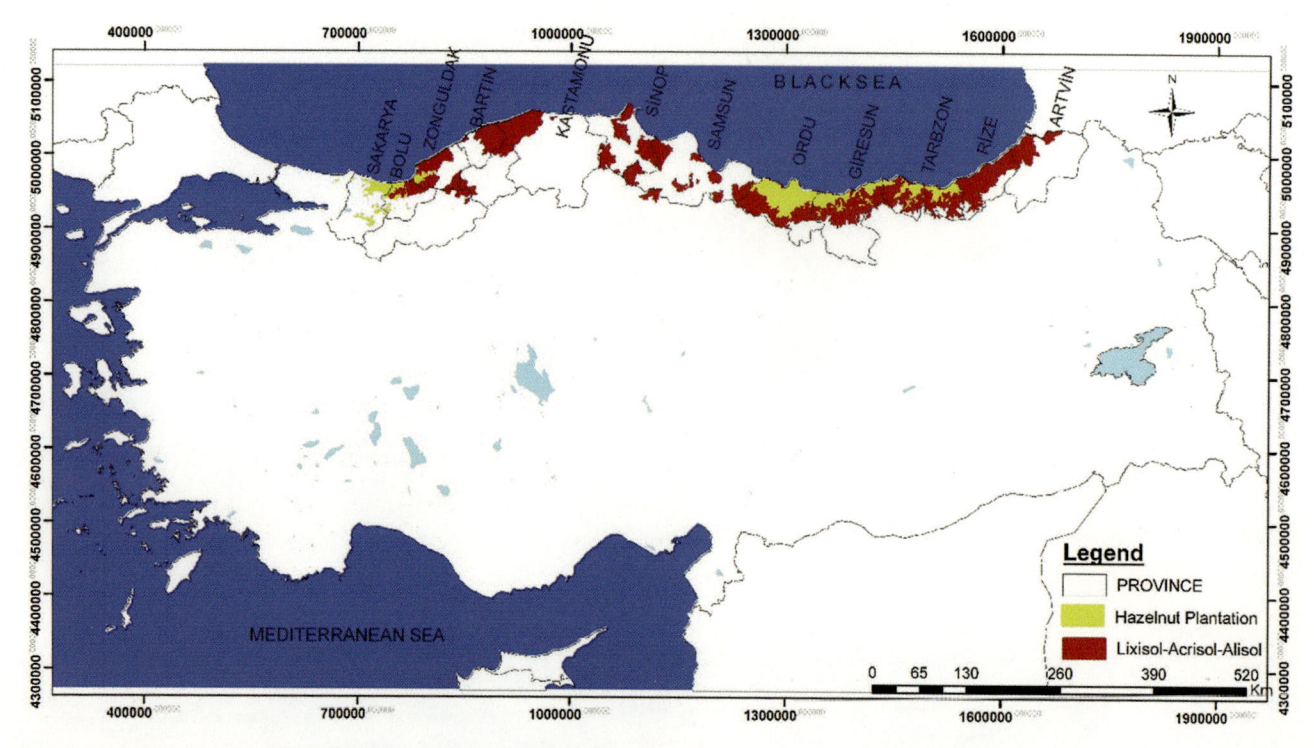

Fig. 5 Hazelnut cultivated areas and Lixisol-Acrisol-Alisol ecosystems

The water holding capacity of the Lixisols decreases and increased runoff is determined in Lixisols due to the removal of organic litter from forests for different purposes (Yüksek 2001).

Moreover, Lixisols on the mountain slopes, backshores, and alluvial plains are under the threat of sealing, which is one of the primary drawbacks for the whole of Turkey. Unfortunately, a considerable part of Lixisol land has been sealed under urban sprawl and highway constructions. This and many similar malpractices and inappropriate land/soil use is proceeding despite the Soil Conservation and Land Use Law (Law No: 5403) and the National Action Plan of Desertification of Turkey as well as the activities strongly requiring soil protection and sustainable land management by the recently completed UNCCD 10-year strategy plan (UNCCD 2014).

References

Dengiz O, Özdemir N, Tuğrul Y, Öztürk E (2008) Basic soil properties and soil classification of Hazelnut cultivation area in the Eastern Black sea region. Case Study, Ünye-Tekkiraz District. In: International meeting on soil fertility land management and agroclimatology, pp 357–366

IUSS Working Group WRB (2015) World reference base for soil resources 2014, update 2015 International soil classification system for naming soils and creating legends for soil maps. World Soil Resources Reports No. 106. FAO, Rome. 203 p

Lal R (1995) Tillage systems in the tropics. Management options and sustainable implications. FAO, Rome, Italy. 70 p

Soil Survey Staff (2014) Keys to soil taxonomy, 12th edn. USDA-Natural Resources Conservation Service, Washington, DC, p 372 p

Turan YK, Ceyhun G, Filiz YK, Esin EYK (2009) The effects of land-use changes on soil properties: the conversion of alder coppice to tea plantations in the humid northern Black Sea region. Afr J Agric Res 4(7):665–674

UNCCD (2014) Land degradation neutrality. Resilience at local, national and regional levels. www.unccd.int/Lists/SiteDocumentLibrary/Publications/Land_Degrad_Neutrality_E_Web.pdf

Yüksek T (2001) Investigations on soil erodibility and some properties of the soils under different land use types in the Pazar Creek watershed, near Rize-Turkey. Karadeniz Technical University. Sci. Inst. Ph.D. Thesis. Trabzon, Turkey (in Turkish)

Arenosols

Hasan Özcan, Hüseyin Ekinci, Erhan Akça, Osman Polat, Muhsin Eren, Selahattin Kadir, Ali Sungur, Timuçin Everest, Franco Previtali, and Selim Kapur

1 Introduction

The Reference Soil Group of the Arenosols consists of sandy soils. These are the soils developed from residual sands, in situ formed after weathering of old usually quartz-rich soil material or rock, and soils developed in recently deposited sands as they occur in deserts and beach lands. Arenosols in Soil Taxonomy include Psamments (Soil Survey Staff 2014). Other international soil names for Arenosols are the 'siliceous, earthy and calcareous sands' and various 'podzolic soils' (Australia), 'red and yellow sands' (Brazil), and the Arenosols of the FAO Soil Map of the World (Anonymous 2015a).

In the study of Arenosols, one should primarily distinguish desert dunes from coastal dunes. Dunes can be mobile or fixed where fixed dunes are older (Herrmann 2007). Dunes are land formations of sand of heights ranging from 1 to 500 m which have been shaped by the wind. These topographical features are found where large masses of sand have accumulated in the desert or long beach environments. Dunes can originate where the wind power is sufficiently strong to transport unconsolidated sediments of weathering residues and especially sand. Desert dunes can form where arid conditions prevail and consequently plant cover is sparse or even absent (Klijn 1990). Wind does play a role in the abrading and rounding of sand grains, where inland dune fields have more rounded grains than coastal dunes due to the increased distance of movement and collision by wind to the surrounding heights (Ahlbrandt 1979).

2 Parent Materials and Environment

In central Turkey and especially in the Konya basin, during the Pleistocene and Holocene, several periods of sand mobilization have been recognized in the present day arid and semiarid areas linked to aridification trends (Rognon 1994). In the present semi-arid climate of the high plateaus of central Turkey, some fossil sand dunes related to the lakes occupying the central parts of the numerous closed depressions can be used to study the chronology of the upper Pleistocene and Holocene climate changes in the Konya basin (Erinç 1962; Erol 1991; Kuzucuoğlu 1993). Several sand flats located on the northern shores of the late Pleistocene paleolake of the Konya plain (central Turkey) are related to changes in lake levels (Kuzucuoğlu et al. 1998). Sand sources in the Konya basin are the alluvial fans developed on the southern borders of the plain formed from west to east by the rivers Çarşamba, Karaman, Deliçay, İbrala, and Serpak. During the upper Pleistocene, sand erosion and accumulation occurred for at least two periods, resulting in two parabolic dune systems. These systems cover flat areas corresponding to two different levels. These levels are the older 'north-İsmil dune system', at 1030–1050 m in altitude, which covers an erosional surface of the Neogene limestone plateaus and the younger 'south-Karapınar dune system', at 995–1010 m in altitude,

H. Özcan (✉) · H. Ekinci · A. Sungur · T. Everest
Department of Soil Science and Plant Nutrition, Çanakkale Onsekiz Mart University, Çanakkale, Turkey
e-mail: hozcan@comu.edu.tr

E. Akça
School of Technical Sciences, University of Adıyaman, Adıyaman, Turkey

O. Polat
Ministry of Forestry and Water Affairs, Eastern Mediterranean Forestry Research Institute, Tarsus, Turkey

M. Eren
Department of Geological Engineering, Mersin University, Mersin, Turkey

S. Kadir
Department of Geological Engineering, Eskişehir Osmangazi University, Eskişehir, Turkey

F. Previtali
Department of Geosciences, University of Milan (Bicocca), Milan, Italy

S. Kapur
Department of Soil Science and Plant Nutrition, Çukurova University, Adana, Turkey

© Springer International Publishing AG 2018
S. Kapur et al. (eds.), *The Soils of Turkey*, World Soils Book Series, https://doi.org/10.1007/978-3-319-64392-2_20

covering the marly bottom surface of the dried upper Pleistocene lake. The flat surfaces buried by the dune systems correspond to two different levels of the bottom of the Konya paleolake. Both dune systems were formed during the dry periods which were contemporaneous to or post-dated to the shrinking lake (Kuzucuoğlu et al. 1998). The dune systems of the northern shores of the Konya plain are the result of climatic changes during the upper Pleistocene and the Holocene. The aeolian landforms are also related to the Konya basin and subbasin lacustrine systems. In the 1960s, the moving sand dune system which was threatening the town of Karapınar, located on the northern border of the Konya plain (Doğan 1992; Doğan and Kuzucuoğlu 1993) in central Turkey, was stabilized by planting and by controlled grazing. This program, started in 1962, has proved a success.

Coastal dunes accrete at a significant extent in prograding deltaic environments and sheltered bays on Turkey's coastline (Özcan et al. 2010). The total area of coastal dunes in the country is 46,583 ha, covering 10.1% of the coastal areas. Coastal dunes in Turkey are found in 110 different localities, occupying ca. 845 km out of the total 8333 km of Turkey's coastline (Uslu 1989). Furthermore, much of the 110 coastal dune areas have been severely degraded since the 1980s, corresponding to the reduction of 5000 ha of land (Uslu 1989). The existing national literature on coastal dunes of Turkey contains considerable knowledge on several aspects of dunes, such as dune vegetation interactions and soil formation on dune sands (Serteser 2002, 2004), stabilization of dune areas (Tekinel and Çevik 1972), dune degradation by sand extraction for various purposes (Atik and Altan 2004; Akbulak et al. 2008), mobilization of dunes due to shoreline changes (Mater and Turoğlu 2002), morphological studies which involve the delta sand dune formation areas of the main rivers (Ozaner et al. 1992), heavy mineral accumulations in the coastal dunes (Erkal 2005) and geomorphological evolution, physicochemical characteristics, and classification of the sand dune fields (Dinç et al. 2001; Özcan et al. 2010).

3 Regional Distribution

There are two types of sand dune formations according to their sources. These are: (i) coastal sand dunes generated from marine sources and (ii) aeolian or inland dunes forming from the weathering of the rocks (Atay 1964; Acatay 1959).

According to the 1/100,000 scaled soil map of Turkey based on the Great Soil Groups, in 1975, there were 110 coastal sand dune areas in the country. Distribution of these sand dunes by regions are, the Marmara region with 46 sand dunes and 11,274 ha area, the Mediterranean region with 36 sand dunes and 19,525 ha area, the Black Sea region with 12 sand dunes and 4136 ha area, and the Aegean region with 11 sand dunes and 1023 ha area (Anonymous 2015b). The sand dune areas of Turkey under the effect of the coastal winds, waves, rivers that especially discharge to the sea, are the locations of Yumurtalık, Akyatan (Adana), Orta Kumluk (Turan-Emeksiz Forest plantation area) Tarsus, Silifke, Kazanlı (Mersin), Samandağ (Antakya), Side-Sorgun, Serik and Finike, Demre (Antalya), Kalkan (Ovagelmiş)-Fethiye (Kumluova) (Muğla), Terkos-Ağaçlı-Kilyos and Şile-Ağva (İstanbul), Karasu-Acarlar (Sakarya), Sarıkumköyü (Sinop), Kavak-Bozcaada (Çanakkale), Kasatura (Tekirdağ), Çarşamba, and the Bafra plain (Samsun). The major aeolian/inland sand dune areas in Turkey are the Konya-Karapınar, Manisa-Beyoba/Sazoba and Iğdır/Aralık.

The size of the areas under the influence of wind erosion in Turkey is around 506,309 ha. The successful attempt to stop wind erosion at a 13,000 ha area in Konya-Karapınar was initiated in 1962. Sand dune stabilization structures were initiated in a total area of 11,000 ha in the country by the Ministry of Agriculture (the State Erosion Control Centre, Karapınar, Konya, central Turkey). The wind erosion problem prevails from slight to very strong at about 465,913 ha inland sand dune areas. About 70% (322,475 ha) of these areas are in the Konya province and 103,000 ha area in the Karapınar town of Konya (central Turkey). The other widely distributed erosion threatened areas by erosion are in Niğde (122,741 ha—26.34%), Kayseri (12,894 ha—2.77%) Iğdır (2910 ha—0.62%), Mersin (2552 ha—0.55%), and Sakarya (2342 ha—0.50%) (Kayalık 2007).

4 Profile Development

Arenosol profile descriptions are given below from the widespread Kavak delta (Saros Bay, northwest Turkey) (Figs. 1, 2 and 3; Tables 1, 2 and 3), the Kapıköy in Seyhan delta (Adana, south Turkey) (Figs. 4, 5, 6, 7 and 8; Tables 4 and 5), and the Karapınar dune area (central Turkey) (Figs. 9 and 10; Tables 6 and 7).

Fig. 1 Coastal sand dune (fore dune) partially covered with vegetation (*arrow*) (*Ammophila arenaria* L, *Pancratium maritimum* L.) in the Kavak delta (Saros bay, northwest Turkey)

4.1 Arenosols in the Kavak Delta Saros Bay in Northern Turkey

The coastal sand dunes of the Kavak Delta are partially covered with Ammophila arenaria L. and Pancratium maritimum L., whereas the recently stabilized dunes are under Ophrys sphegodes. The stable dunes consist of dense heaps of Juncus maritimus Lam. and sparse Bromus hordeaceus L. and Bromus tectorum L. along with Juncus maritimus Lam. and Elymus elongatus communities (Figs. 1, 2 and 3; Tables 1, 2 and 3).

4.2 Arenosols in the Seyhan Delta Kapıköy in Southern Turkey

The Arenosols defined below represent the majority of the coastal Arenosols on the stable dunes of the country extending from Adana to Antalya along the Mediterranean coast of Turkey (Fig. 4).

The increase in Mn, Fe, Zn, Cu contents (already available in the ophiolitic minerals of the sand dunes) in the A/O horizon of the Calcaric Arenosol is associated to the increase of organic matter in the stone pine rhizosphere soils since 1973. The Kapıköy-Karataş (Akyatan wetland) area of sand dunes was declared as a natural park in 1970 based on the Ramsar Wetlands Agreement (3 February 1971). The site is located along the southern coasts of Adana (south Turkey) (Figs. 4 and 5) covering a total area of 7000 ha.

Studies conducted on the fundamental properties and formation of this sand dune area revealed high organic matter contents where the highest were determined in the silt fractions collected from all the plots. The studied plots were partly conserved by the conserved and enhanced natural sand dune vegetation and partly allocated to stone pine plantations at the different periods of 1973, 1980, 1989 and 1997. The highest contents of C-sequestration determined in the Arenosols of the 1980 stone pine plantation plot were followed by the 1973 stone pine, natural vegetation, 1989 and 1997 stone pine plots, respectively. Increased C contents were also determined to be associated with lower levels of pH, higher levels of available P contents and available Mn, Fe, and Zn inherited from commonly distributed ophiolitic/volcanic mineral contents (Fig. 6).

Silt size fractions increased parallel to the age of the stone pine plantation most probably due to the dissolution/weathering effects of the organic acids (exudated at the rhizosphere) on the abundant feldspars of sand size (Yaktı et al. 2004). Humic compounds were determined to be highest at the 1980 stone pine plantation site followed by the conserved/natural vegetation plots and 1973, 1989, 1997 stone pine plots, respectively (Table 5). The high ratios in the HC/OM (36.56%) of the 1997 plot are most probably due to the better management practices performed by the project team or the slightly better climatic conditions favoring humification in the site of the plot (Fig. 4; Table 5). However, the highest humification level (42.65%) determined in the natural vegetation site reflects the optimal conditions for humic compound development in an undisturbed and long-enduring rhizosphere (Table 5).

The C/N ratios were unusually high due to the high rate transportation/mixing of the N-bearing plant residues—the decomposed needles of the stone pines within the Arenosol pedon (Table 5).

In situ humification was determined in the two selected—1973 and 1980—plots with the highest C/N contents (Table 5) together with plant residues at variable decomposition levels in the Karataş Arenosols. Fungal hyphae (mycelia) (Fig. 7) were observed to surround and stabilize sand particles as well as large spheroidal humic aggregates (app. 250 μm) (Fig. 8) and humic compounds surrounding and binding/bridging mineral grains, revealing initial soil development via sequestered C at different forms of decomposition and mineralization. SEM analyses aided in determining the nearly pure high C and O contents of the

Fig. 2 Recently stabilized sand dune profile (**a**) and vegetation (**b, c**) (*Ophrys sphegodes*) in the Kavak delta (Saros bay, northwest Turkey)

humic aggregates with trace amounts of probable Na, Mg, Ca, and Fe precipitations on surfaces (Fig. 8). Although high and almost similar amounts of C were determined to be sequestrated on the sand dunes covered with natural vegetation, the 1980 and 1973 plots and the natural vegetation plots proved to be a successful story for sand dune management and stabilization of the Arenosols. Additionally, the high income-generating stone pine crop seems to be preferable, if allocated to selected plots for sand dune/wetland management (Yaktı et al. 2004).

4.3 Arenosols in the Konya Basin, Karapınar, and Central Turkey

Central Turkey has been under severe wind erosion threats for millennia due to its steppe vegetation and overgrazing of small ruminants. In the course of overgrazing, the lagoonal sands of the ancient Konya lake were shifted by wind action to form sand dunes at certain localities such as the sand dune area of Karapınar. The land reclamation initiated in the 1960s led to the stabilization of the sand dunes inducing the development of the Arenosols (Akça et al. 2010).

Groneman (1968) has described two profiles on active dunes in 1965 with available analytical data belonging to the Meke sand dune (now the Meke soil). He stated that there was no pedogenic evidence in the Meke sand dune since there were no plantation activities on the sandy material to initiate pedogenic development. However, by a more recent monitoring study concerning soil formation conducted in this area by Akça et al. (2010) at the same profile site, pedogenesis was determined within the sand dune converting it to a soil—the Meke soil (Figs. 9 and 10). The changing pedogenic parameters within the 45-year period of Groneman's (1968)

Fig. 3 Stable (old) sand dune profile (**a**) dense heaps of *Juncus maritimus Lam.* and sparse mixtures of *Bromus hordeaceus* L. and *Bromus tectorum* L. (**b**) mixed communities of *Juncus maritimus Lam.* and *Elymus elongatus* (**c**) in the Kavak delta (Saros bay, northwest Turkey)

(Table 6) sampling time of 1965 and Akça et al. (2010) study in 2010 were: (i) the slight change in color (2.5Y 5/2 in Groneman 1968; and 2.5Y 5/3 in Akça et al. 2010), (ii) the increase in organic matter and phosphorous contents together with the cation exchange capacities of the horizons. On the contrary, textures, pH, and $CaCO_3$ contents remained unchanged (Tables 6 and 7). The magnitudes of the humic acids determined in Akça (2001) (Table 7) were also the proofs of pedogenesis as stated by Trevisan et al. (2010).

The Meke soil is deep and excessively drained on flat to gently sloping physiography with calcareous aeolian non-saline sand as parent material. The plant cover on the Meke soils varies from pine to acacia and natural vegetation (Fig. 9).

Table 1 Site description and some physical and chemical properties of the coastal dune Arenosol in the Kavak delta, Çanakkale (Özcan et al. 2010)

IUSS Working Group WRB (2015)	Brunic Arenosols
Soil Survey Staff (2014)	Typic Xeropsamments
Location	Kavak delta, Saros Bay-Çanakkale (northwest Turkey)
Coordinates	35°48′ 62.50″ E, 44°96′ 24.8″ N
Altitude	1–4 m above sea level
Landforms	Delta (coastal sand dune)
Position	From ridge to flat and depression
Aspect	–
Slope	Variable, in fault areas: 0–1%, in ridge areas: 10–15%
Drainage class	Good
Groundwater	140 cm deep from the surface
Effective Soil depth	140 cm in the stable sand dune
Parent material	Fluvial
Köppen climate	Mediterranean
Land use	Natural habitat
Vegetation	*Ammophila arenaria* (L.) Link, *Pancratium maritimum* L. (in the fore dune area) and, *Leucojum aestivum* L., *Lagurus avatus* L., *Catapodium rigidum* (L.), *Juncus maritimus* lam. and sparse mixtures of *Bromus hordeaceus* L. and *Bromus tectorum* L. (in the stable dune area)

Horizon	Depth (cm)	pH (1:1)	EC (dS/m)	CaCO$_3$ (%)	Organic matter (%)
A	0–5	7.2	0.1	7.8	0.15
C	5–30	7.0	0.08	4.3	0.17
2C	30–70	7.3	0.08	5.2	0.02
3C	70–80	7.7	0.3	3.2	0.01
4C1	80–125	7.4	0.31	6.0	0.14
4C2	125+	7.5	0.3	4.6	0.01

Horizon	Depth (cm)	2–1 mm	Grain size distribution (%)			Texture (%)		
			1–0.5 mm	0.5–0.2 mm	<0.2 mm	Clay	Sand	Silt
A	0–5	0.0	1.6	71.16	27.24	0.4	95.6	4.0
C	5–30	0.04	2.26	67.88	29.82	4.0	94.8	1.3
2C	30–70	0.0	6.42	72.12	21.46	0.0	98	2.0
3C	70–80	0.1	5.46	71.48	22.96	2.2	92.6	5.2
4C1	80–125	0.74	14.28	65.02	19.96	0.0	98	2.0
4C2	125+	0.22	8.52	76.22	15.04	6.2	92.6	1.3

The pH of the Meke soil varies from 7.8 to 8.2 (Table 7) and the low CEC along the solum is four fold higher in horizon A than horizon C (Table 7). The organic matter content of the A horizon is moderately high, but null in horizon C. Root distribution under natural vegetation is significantly dense eliminating wind erosion effects on the Meke soil surface, but they are prone to water erosion because of the absence of structural development at the surface horizons.

The significant change in CEC determined after the erosion control activities can be attributed to the accumulation of organic matter (Tables 6 and 7). The difference of CaCO$_3$ contents in Groneman's (1968) Meke sand dune and Akça's

Table 2 Some physical and chemical properties of the recently stabilized sand dune Arenosol in the Kavak Delta (Özcan et al. 2010)

Horizon	Depth (cm)	pH (1:1)	EC (dS/m)	$CaCO_3$ (%)	Organic Matter (%)
A	0–5	7.4	0.14	1.7	1.01
AC	5–20	7.6	0.14	3.5	0.16
C	20–45	7.5	0.08	4.0	0.10
2C	45–70	7.5	0.87	3.2	0.10
3C	70+	7.8	0.90	4.4	0.07

Horizon	Depth (cm)	2–1 mm	Grain size distribution (%)			Texture (%)		
			1–0.5 mm	0.5–0.2 mm	<0.2 mm	Clay	Sand	Silt
A	0–5	0.28	1.78	50.12	47.8	2.4	91.6	6.0
AC	5–20	0.00	1.46	44.80	53.7	4.0	93.3	2.7
C	20–45	0.00	1.56	55.08	43.4	2.0	98.0	0.0
2C	45–70	0.04	0.76	48.24	51.0	3.9	94.5	1.6
3C	70+	0.08	1.66	54.40	43.9	2.0	94.0	4.0

Table 3 Some physical and chemical properties of the stable Arenosol in the Kavak Delta (northwest Turkey) (Özcan et al. 2010)

Horizon	Depth (cm)	pH (1:1)	EC (dS/m)	$CaCO_3$ (%)	Organic matter (%)
A	0–10	7.4	0.15	2.7	1.26
C	10–16	7.4	0.10	2.5	0.19

Horizon	Depth (cm)	2–1 mm	Grain size distribution (%)			Texture (%)		
			1–0.5 mm	0.5–0.2 mm	<0.2 mm	Clay	Sand	Silt
A	0–10	0.16	1.56	51.58	46.70	6.36	88.36	5.28
C	10–16	0.04	0.40	46.96	52.60	6.24	91.76	2.0

(2001) Meke soil may most likely be due to the dissolution of CaCO3 based on the increased plant cover and in turn due to the organic acids/exudates excreted by the plant roots in the rhizosphere.

5 Management

Arenosols occur at different topographical locations in Turkey and have a multipurpose use. They are partly irrigated in the semi-arid parts of the country, namely in southern (Adana, Antakya, Antalya and Mersin), western (Manisa), central and eastern (Konya and Iğdır) Turkey. During the last 76 years, about 46% of the sand dune areas of the Seyhan Delta (Adana) have been shifted to agricultural use (Kapur et al. 1999; Bal et al. 2003) (Fig. 11), whereas, they are managed as extensive grazelands in Konya and Iğdır and as forest areas in northern Adana and Muğla. Their use for eucalyptus-cypress-stone pine forests at the inland dunes of the Adana (the Kapıköy Forest) and Tarsus (the Karabucak Forest) coast has proven to be successful soil carbon sequestration areas (carbon sinks) (Yaktı et al. 2004; Polat and Kapur 2010). Moreover, the ongoing initial microstructural development in the root zones of the Arenosols, bound to the increased C-sequestration, points out to the indispensable need of protection of the overground natural vegetation cover enhancing microbial activity laden with mycorrhizal populations (Figs. 7, 8 and 12). Consequently, despite the rightful and undisputable need to preserve the stable sand dunes as plant/soil ecosystems (the Arenosol ecosystems, which act as high carbon sequestration areas), public demands for income generation should also be regarded for cash crop cultivation of primarily stone pine and low pesticide demanding peanuts on unlevelled/leveled sand dunes/Arenosols as part of a sustainable soil management program. These forest and field crops would also mitigate climate change by their rapid and high SOC sequestration capacity in their root zones (Mollenhauer et al. 2002).

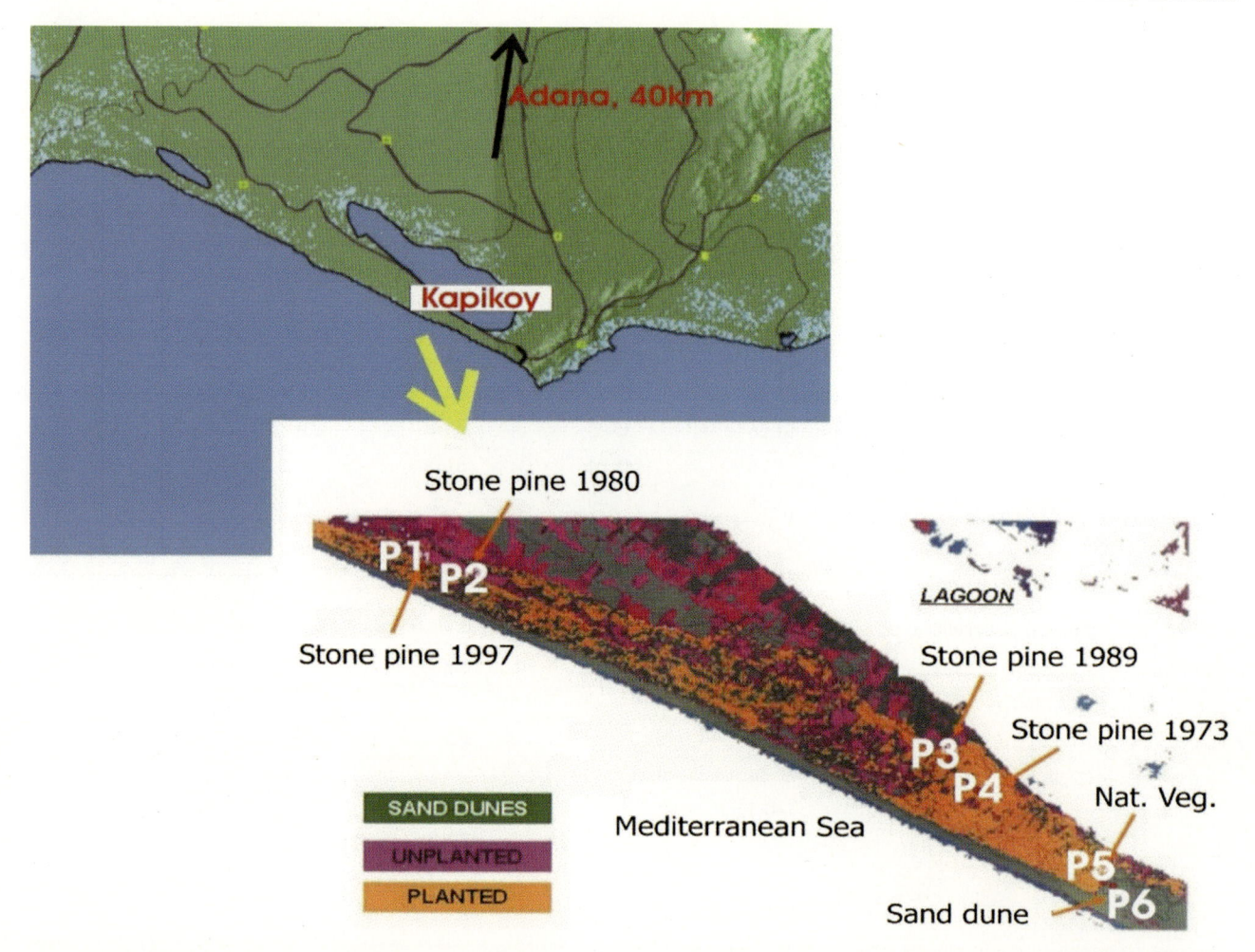

Fig. 4 Location of the Kapıköy-Karataş sand dune area (south Turkey) (Stone pine plantations in 1997 (P1), 1980 (P2), 1989 (P3), 1973 (P4), natural vegetation (P5), sand dune (P6) areas

Moreover, sand dune stabilizing natural vegetation is being observed to most likely permanently sequester SOC around old and deeper decomposed/humified roots which were coated by a carbonate capsule (calcified fine material—calcite—surrounding the decomposed root) in the sand dunes of Alghero, NW Sardinia in Italy, a climatically and topographically similar Arenosol ecosystem to Adana-Karataş (Profs. S. Madrau, University of Sassari, Italy and ICARDA, Amman, Private Comm.).

The SOC contents of such leveled Arenosols were increased several folds in 20 years under governmental forest management projects (Fig. 11). Monitoring of SOC in coastal and inland Arenosols has revealed eightfold increases within 40–50 years (Yaktı et al. 2004; Akça et al. 2009; Polat and Kapur 2010). Polat and Kapur (2010) have determined high fulvic acid contents in Arenosols (leveled stable sand dunes) in the State Tarsus Forestry Research Center experimental plots under red pine canopies (established in 1961, 1971 and 1981) (Table 8) most likely caused by the high decomposition rates of red pine litter. This is stated to consequently lead to the development of fulvic acid-bound (preserved against decomposition) polysaccharides/carbohydrates which in turn preserve fulvic acids against decomposition. On the other hand, the humic acid contents of these Arenosols have most likely enhanced the development of the stable aggregates (microstructural units-MSUs) laden with micropores which increase water and plant nutrient retention. The spheroidal to oval humin compounds (10 µm) held in and/or apart from the

Fig. 5 The stabilized sand dunes of the Akyatan lagoon in Kapıköy, Adana (south Turkey)

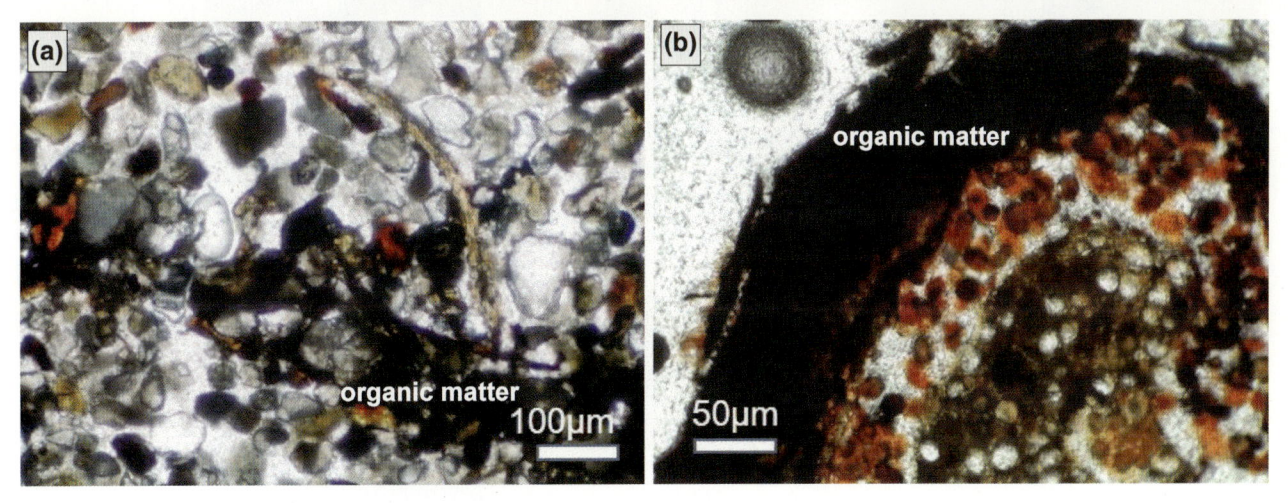

Fig. 6 Organic matter accumulation in the sand consisting ophiolitic/volcanic mineral grains (**a**) close-up view of decomposing organic residues with arthropod activity (**b**) (Yaktı et al. 2004)

stable aggregates act as plant nutrient retaining sites as well. Moreover, the fine mycorrhizal (arbuscular) hyphae (20–80 AM/100 g) embracing or surrounding the abundant porous aggregates formed from finer sand and silt fractions in the surface layers of the Arenosols are also enhancing the formation of these stable aggregates/MSUs (Fig. 12a, b).

Unfortunately, many such coastal sand dune areas have been degraded by sand extraction for use in constructions to build secondary houses and summer resorts in the Mediterranean and Thracian coastlines of Turkey (Antalya, Muğla, Manisa, Mersin, Tekirdağ, Edirne, and İstanbul). Stable sand dunes and/or Arenosol ecosystems have also been studied for historical progressive coastal erosion (retrogradation) and build-up/sedimentation (progradation) by Bal et al. (2003) and Akça et al. (2009) along the Mediterranean coasts of Turkey and especially in Antakya (Samandağ, southern Turkey) (Fig. 13).

The fluctuating trends of progradation/deposition and retrogradation/erosion are the indicators of the magnitude of the historical erosion (delta development) based on

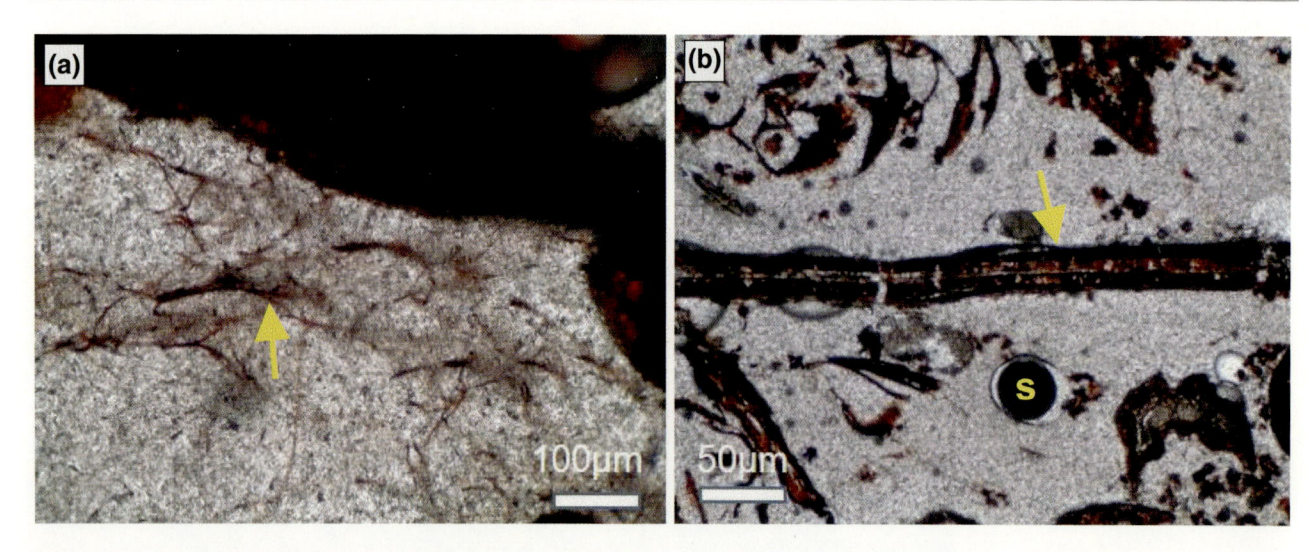

Fig. 7 Fungal hyphae (mycelia, *arrow*) developing in a Calcaric Arenosol in the A/O horizon (**a**) decomposing plant residues (pine needles, *arrow*) with spheroidal humic aggregate(s) in the A/O horizon (**b**) (Yaktı et al. 2004)

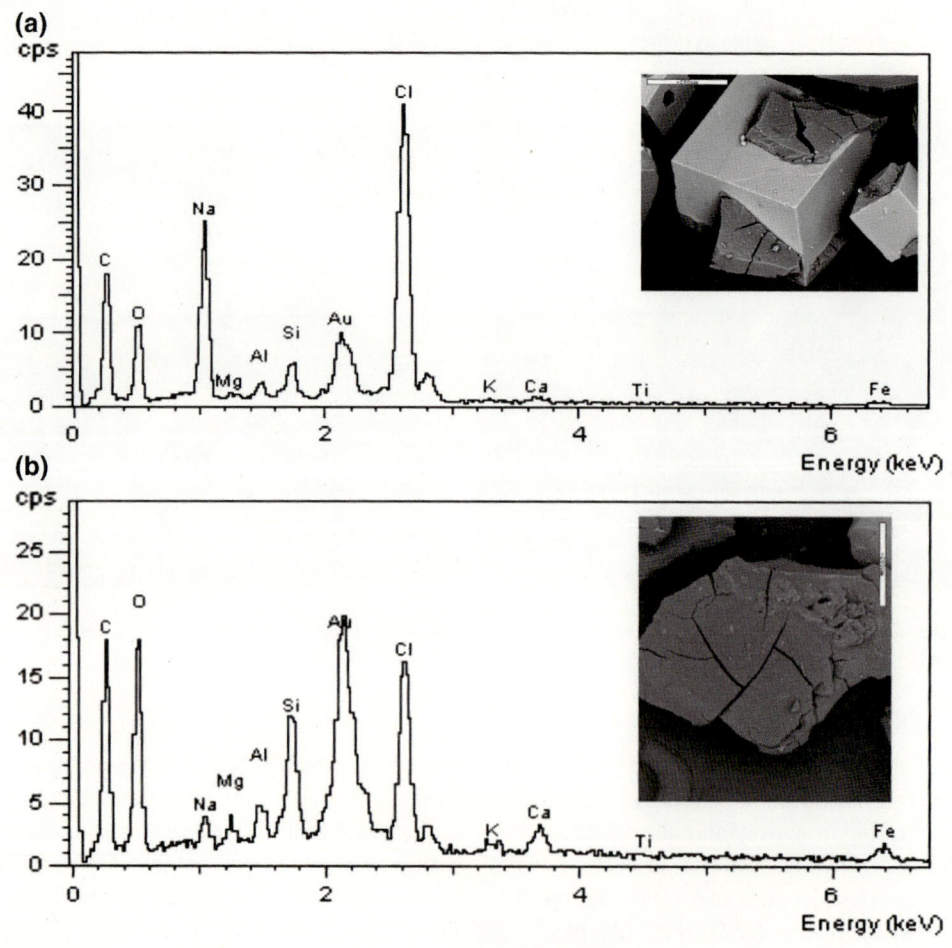

Fig. 8 SEM images—microprobe analyses of humic aggregates on sodium chloride crystal (**a**), and humic aggregate with cracking surface (**b**) (Yaktı et al. 2004)

Table 4 Site description and some chemical and physical properties of the Kapıköy soil, Adana (south Turkey), developed on the Akyatan sand dune site stabilized in 1973 (Yaktı et al. 2004)

IUSS Working Group WRB (2015)	Calcaric Arenosols
Soil Survey Staff (2014)	Typic Xeropsamments
Location	Akyatan lagoon, Kapıköy, Adana (south Turkey)
Coordinates	36°37′ 35.60″ N, 35°11′ 51.36″ E
Altitude	2–4 m above sea level
Landforms	Delta (coastal sand dune)
Position	From ridges to the flat land and depressions
Aspect	–
Slope	Slightly north to south
Drainage class	Good
Groundwater	In the stable sand dune it is 140 cm deep from the surface
Effective Soil depth	+200 cm
Parent material	Sand dune
Köppen climate	Mediterranean
Land use	Afforested area
Vegetation	Stone pine (*Pinus pinea*)

Horizon	Depth (cm)	pH (1:1)	EC (dS/m)	CaCO$_3$ (%)	CEC (cmol$_c$ kg^{-1})	Mn (ppm)	Fe	Zn	Cu	OM (%)	Particle size distribution		
											Clay	Sand	Silt
A/O	0–4	7.28	0.8	21.19	8.87	8.22	2.84	0.69	1.17	3.26	2.4	95.3	2.3
C	4+	7.72	1.6	23.04	5.92	4.34	1.84	0.15	1.06	1.53	0.7	98.2	1.1

Table 5 The change of organic matter and humin compounds in the Kapıköy, Adana (south Turkey) Arenosol *Pinus pinea* afforested area in 1973, 1980, 1989, 1997 (Yaktı et al. 2004)

Plantation (date)	Horizon	Depth (cm)	HC	HA (% (g/g))	FA	OM	HC/OM (%)	HA/FA (% (g/g))	C (%)	N (%)	C/N
1973	A/O	0–4	1.06	0.41	0.65	3.26	32.50	0.63	1.9	0.04	46.2
	C	4+	0.51	0.10	0.41	1.53	33.33	0.25	0.9	0.01	98.8
1980	A/O	0–6	1.30	0.61	0.69	4.34	29.95	0.89	2.5	0.04	61.5
	C	6+	0.61	0.15	0.46	1.99	30.65	0.32	1.2	0.01	96.4
1989	A/O	0–3	0.75	0.22	0.54	2.57	29.33	0.41	1.5	0.02	93.4
	C	3+	0.54	0.14	0.40	1.96	27.24	0.35	1.1	0.01	126.6
1997	A/O	0–2	0.93	0.32	0.61	2.53	36.56	0.52	1.5	0.02	63.9
	C	2+	0.60	0.17	0.43	1.66	36.14	0.39	0.9	0.01	87.7
Nat. Veg.	Surface		1.19	0.40	0.80	2.79	42.65	0.51	1.6	0.04	45.1
Sand Dune	Surface		0.47	0.17	0.31	1.47	32.24	0.54	0.9	0.02	53.4

HC humic compounds, *HA* humic acid, *FA* fulvic acid

Fig. 9 The Meke soil landscape in Karapınar, Konya, central Turkey

deforestation for timber during societal conflicts and to the peace established among the Greco-Roman City-State Colonies and the settlers of the aftermath. Retrogradation is also a contemporary action induced by the obstruction of the natural sedimentation process in deltaic areas by the construction of upstream dams. In this context, a coastline erosion of about 60 meters inland was measured for a period by comparing earlier maps to the contemporary in the Samandağ coast of Antakya (S. Turkey) (unpublished data, Profs. F. İşler and S. Kapur) (Fig. 14). The Arenosol areas, ecosystems/inland sand dune ecosystems in central Turkey (Konya-Karapınar area), suffer land subsidence and recent doline (sink hole) formation and degradation of the landscape as land patches due to the overuse of the shallow groundwater (Akça et al. 2012) (Fig. 15). The most prominent evidence concerning the overuse of the groundwater for crop irrigation can be observed via the fall of the groundwater levels of the area measured by DSİ (State Hydraulic Works) which is about 3.7 meters/year as stated by Yılmaz (2010). The water level of the Meke Lake, which is about 20–30 cm at present (2015), decreased from the 1 m level of 2004, is the other evidence of water shortage in the area coupled with salt accumulation on the lake surface (Fig. 16).

More developed precursor MSUs/aggregates than the ones determined by Polat and Kapur (2010) in the leveled and planted sand dunes/Arenosols of the Tarsus State Forestry Research Center experimental plots under red pine canopies were determined in the root zones of the ameliorated terrestrial (beach remnants of the ancient central Anatolian Quaternary lake) dune soils Arenosols (Typic Torripsamments) of Karapınar (State Erosion Control Center, Karapınar, Konya), Konya, central Turkey (Okur 2010, 2010). The development of the precursor microstructural features (Figs. 17, 18 and 19) was enhanced most likely by the organic exudates of the under canopies of the 50-year-old almond plantation root zones laden with natural plants such as Festuca ovina, Tragopogon longirostris, Cirsium arvense, Polygonum cognatum, Lamium amplexicaule, Descurainia sophia, Minuartia hamata, Onobrychis armena, Noaea mucronata, Scleranthus annuus, Alyssum strigosum, Alhagi pseudalhagi, and Anchusa azurea.

Fig. 10 The Meke soil in Karapınar, Konya, central Turkey

Table 6 Some physical and chemical properties of Groneman's (1968) Meke sand dune, Karapınar, Konya, central Turkey

Horizon		C1	C2	C3
Depth (cm)		0–0.5	0.5–70	70–170
pH (1:1)		7.8	7.6	7.8
EC (dS/m)		0.45	0.50	0.51
CEC (m/100 g)		2.0	3.0	3.0
CaCO$_3$ (%)		76.5	77.7	78.3
Texture (%)	Sand	91.5	90.7	72.6
	Silt	2.8	2.5	7.2
	Clay	5.7	6.8	20.8

Table 7 Site description and some physical and chemical properties of the Meke soil, Karapınar (central Turkey) (Akça 2001)

IUSS Working Group WRB (2015)	Calcaric Arenosol		
Soil Survey Staff (2014)	Xeric Torripsamment		
Soil series	Meke		
Location	Karapınar research farm, ministry of rural affairs at the end of the NS main track, 5 km southward of Kindam on a barchan dune at the windward side. Profile perpendicular on the crest line of the barchan		
Elevation	1009 m above sea level		
Climate	Arid, transitional to the Mediterranean semi-arid		
Vegetation	Dense natural vegetation, mainly herbaceous		
Parent material	Aeolian sand		
Geomorphic unit	Aeolian plain		
Topography	Moderately steep, short slope, flat to gently sloping		
Erosion	Wind erosion, strong		
Water table	Deeper than 2 m		
Drainage	Excessive		
Infiltration	Rapid		
Stoniness	No stones		
Soil moisture regime	Aridic marginal to Xeric		
Soil temperature regime	Mesic		
Epipedon	Ochric 0–5 cm		
Field classification	Fine sandy, carbonatic, mesic, Xeric Torripsamment		
Horizon		A	C
Depth (cm)		0–4	4–100
pH		7.8	8.2
EC (dS/m)		0.060	0.050
CEC (cmolc kg^{-1})		8.7	2.1
CaCO$_3$ (%)		65.4	66.3
P2O5 (kg/ha)		8	2
C (%)		1.0	0
Humic acid (%)		0.04	0
Particle size distribution (%)	Sand	92.3	94.7
	Silt	2.9	3.7
	Clay	4.8	2.1
Texture class		Sand	Sand

Arenosols

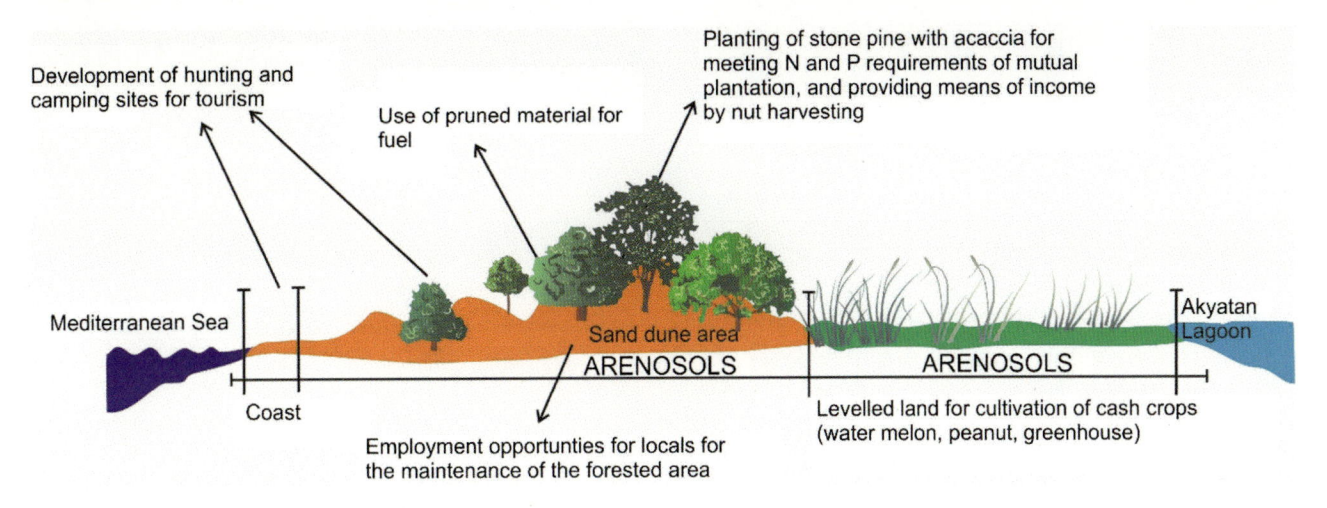

Fig. 11 The ameliorated sand dune area management in the southern Mediterranean coast of Turkey (Adana) (Yaktı et al. 2004)

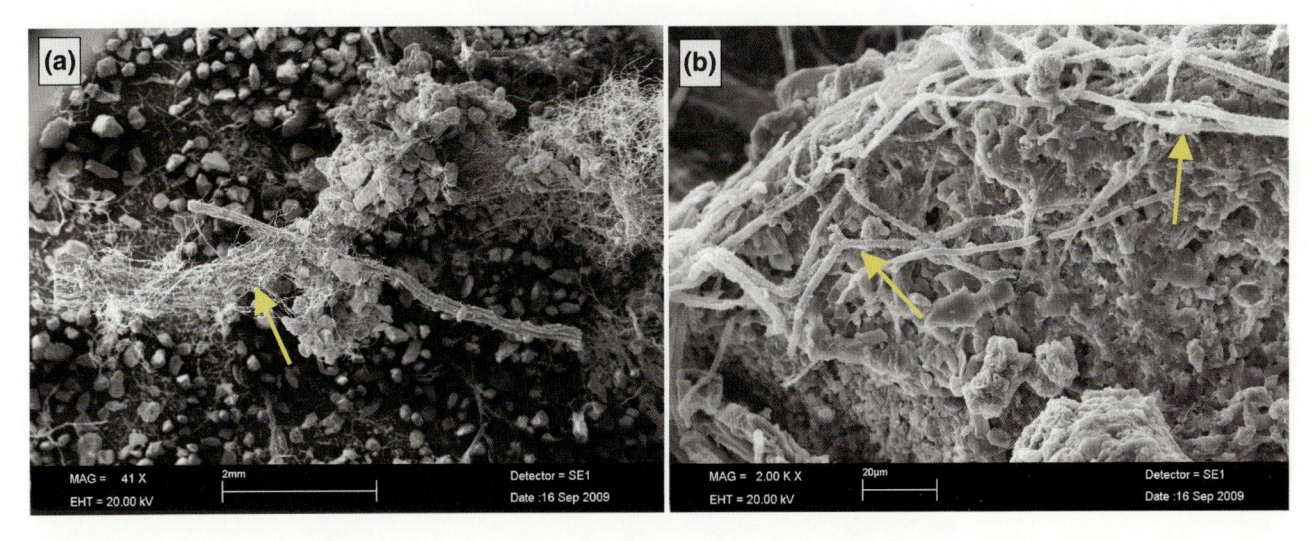

Fig. 12 SEM images of: intermatted fine roots (*arrow*) and AM hyphae initiating MSU/aggregate formation in the red pine Arenosol site established in 1974 (**a**), close-up view of mycorrhiza hyphae/fine roots (*arrows*) surrounding an aggregate under the red pine canopy established in 1981 (**b**) (Polat and Kapur 2010)

Table 8 Humin compounds and organic matter contents of an Arenosol in the Tarsus State Forestry Research Center experimental plots under red pine canopies established in 1961, 1971, and 1981 (Polat and Kapur 2010)

Year of plantation	Horizon	Humic matter (HM) (HA + FA)	Humic acid (HA) (g/100 g soil)	Fulvic acid (FA) (g/100 g soil)	Organic matter (%)	HM/OM (%)	HA/HM	FA/HM
1961	Ah	2.96	0.056	2.904	5.085	58.2	1.90	98.10
1974	Ah	2.68	0.064	2.616	3.705	72.3	2.40	97.60
1981	Ah	3.00	0.037	2.963	4.892	61.3	1.23	98.77

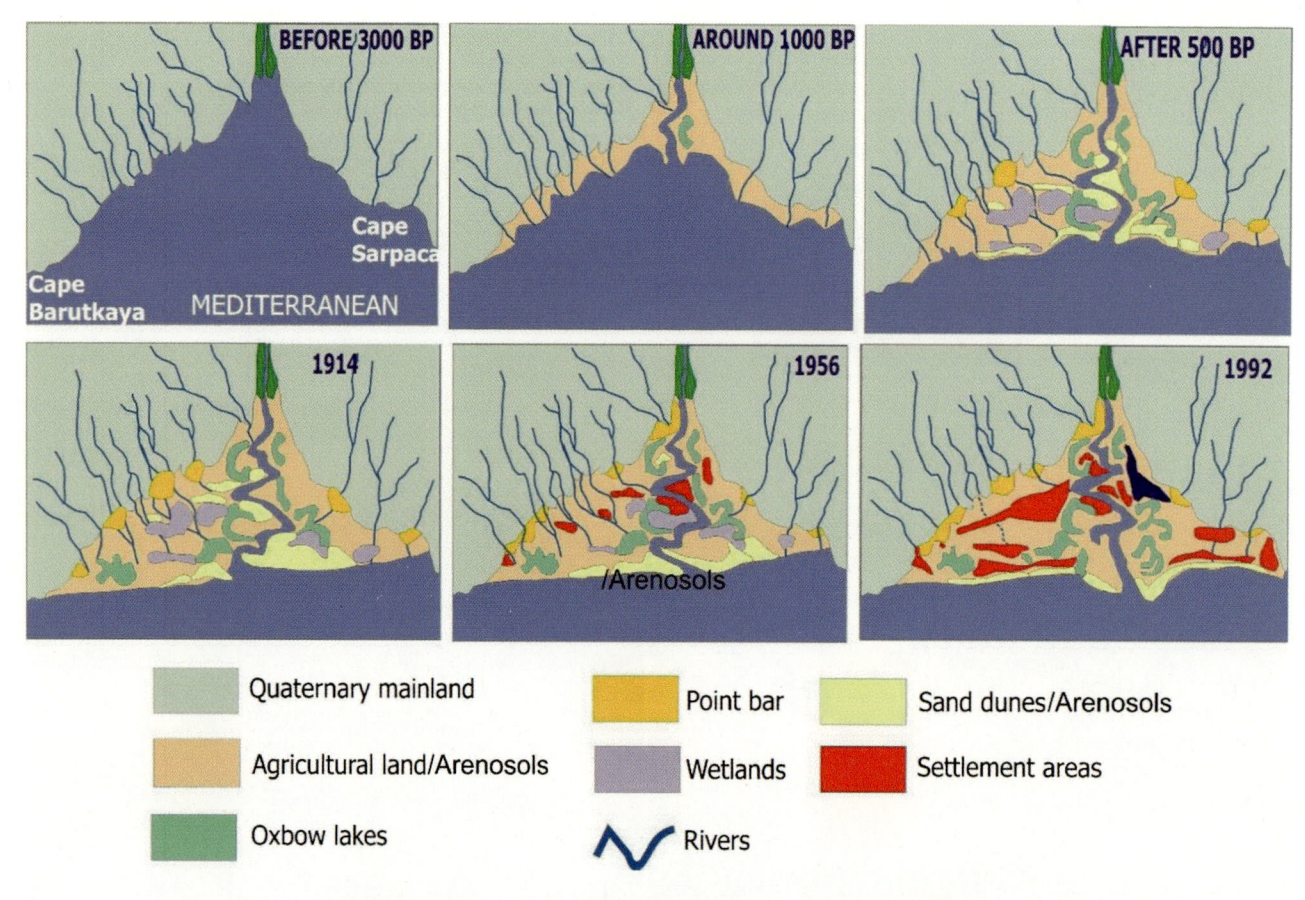

Fig. 13 Progradation/sedimentation leading to Arenosol formation in southern Turkey (Akça et al. 2009)

Fig. 14 Sand dune movement to town center in Antakya, Samandağ, southern Turkey

Fig. 15 The İnobası sinkhole in Karapınar, Konya Basin, central Turkey developed in 2007

Fig. 16 The water level (*arrow*) of the Meke lake in 2004 (**a**) and the lowered water level in the lake associated with salt deposition in 2015 (**b**)

The enhanced soil quality indicators in this context were the organic matter contents of the litter accumulated on the surface and in the A horizon of the almond-Arenosol ecosystem within the 50-year amelioration period. The organic matter was 38.6% (organic litter) and 1.8% (from 0.1%) respectively together with an increase of 82 kg/ha of phosphorus (Okur 2010).

Natural plant covers (*Acantholimon* spp., *Scabiosa* spp., *Astragalus microcephalus*) enhanced by fencing in earlier grazed Arenosols of the Karapınar (State Erosion Control Center, Karapınar, central Turkey) were also equally successful in enhancing soil quality in the A horizons of the same Arenosol ecosystem by increasing the organic matter contents from 0.1 to 1.24% and from 0.1 to 1.21%, respectively. The abundant precursor MSUs/aggregates of the A—horizons/rhizospheres of the Arenosols ameliorated by forage crops were most probably developed and stabilized by the organic exudates of the dense forage root system (Figs. 20 and 21) (Okur 2010).

Fig. 17 The organic matter accumulation and structure development in Arenosols under acacia (**a**) and almond (**b**) canopy (Okur 2010) in Karapınar, Konya Basin, central Turkey

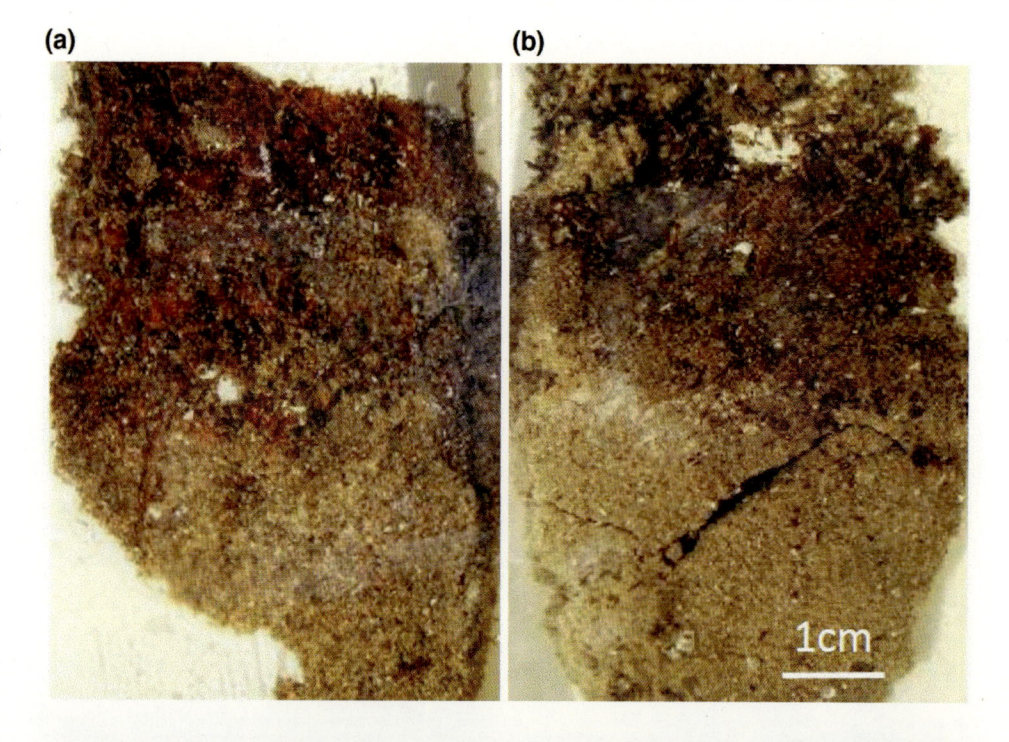

Fig. 18 MSUs/aggregate development under acacia canopy around fine roots (SEM image) in Arenosol rhizosphere, in (Okur 2010)

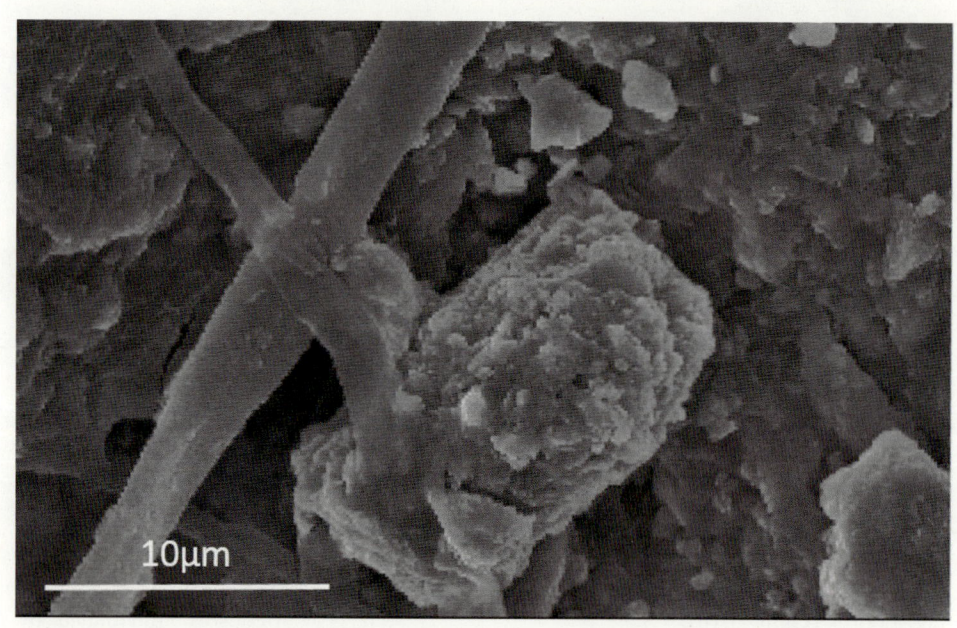

Fig. 19 The frequent faunal excrements developing under the almond canopy of the Arenosol rhizosphere (Okur 2010)

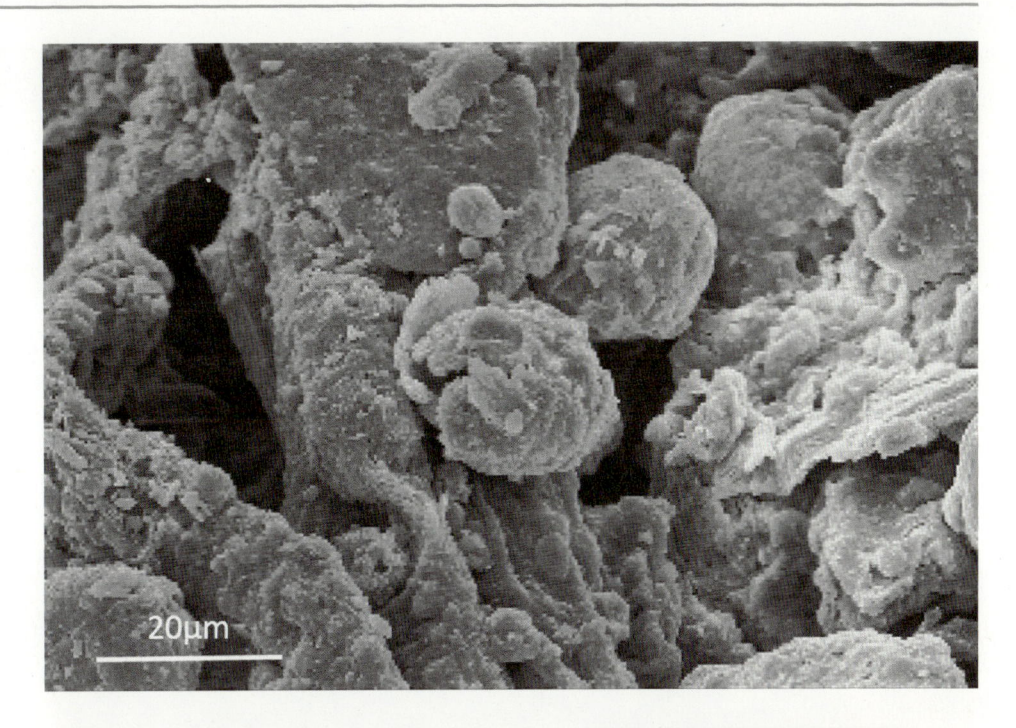

Fig. 20 The dense root network of Astragalus microcephalus in the understories of the Karapınar Arenosol, Konya Basin, central Turkey (Okur 2010)

Fig. 21 MSUs/aggregates development in Astragalus microcephalus root zone (SEM image) of the Karapınar Arenosol (Okur 2010)

References

Acatay A (1959) Forest conservation, Publication of the İstanbul University, Faculty of Forestry, No: 62. İstanbul. 313–318

Ahlbrandt TS (1979) Textural parameters in aeolian deposits. In: McKee E (ed) A study of global sand seas. Washington, US, pp 21–52. Geological Survey Paper 1052

Akbulak C, Erginal AE, Gönüz A, Öztürk B, Çavuş CZ (2008) Investigation of land use and coastline changes on the Kepez Delta (NW Turkey) using remote sensing. J Black Sea Medit Environ 14 (2):85–96

Akça E (2001) Determination of the soil development in Karapınar erosion control station following rehabilitation. PhD Thesis. University of Çukurova, Institute of Natural and Applied Sciences, Adana, Turkey

Akça E, Bal Y, Kapur S, Eswaran H (2009) Human-induced late holocene degradation of the Asi (Orontes) Delta, Samandağ, Southern Turkey. Catena Spec Issue Adv Geoecol 18–26. ISBN 3-923381-54(8)

Akça E, Kapur S, Tanaka Y, Kaya Z, Bedestenci HÇ, Yaktı S (2010) Afforestation effect on soil quality of sand dunes. Pol J Environ Stud 19(6):1109–1116

Akça E, Katısöz Ö, Takashi K, Akihiro K, Nagano T, Koca YK, Kapur S (2012) A decade of change in land use and development of sinkholes in Karapınar, C. Anatolia. In: 8th international soil science congress on land degradation and challenges in sustainable soil management, 15–17 May 2012, İzmir Turkey, pp 222–226

Anonymous (2015a) ISRIC. World Soil Information. Arenosols. http://isric.org/isric/webdocs/docs/major_soils_of_the_world/set3/ar/arenosol.pdf. Last Accessed in 6 Dec 2015

Anonymous (2015b) Sand dune and coastal dunes. http://www.agaclar.net/forum/doga-cevre-ekoloji-gida-hukuk-ve-politikalari/14482.htm (In Turkish). Last Accessed in 6 Dec 2015

Atay İ (1964) A research on the determination and reforestation of the coastal sand dunes of Turkey. OGM Publication 39/385 Ankara, 112 p

Atik M, Altan T (2004) Ecologically important biotopes in South Antalya Region and their comparison with the European Union Natura 2000 Habitats. Akdeniz Univ J Agr Fac 17(2):225–236 (in Turkish)

Bal Y, Kelling G, Kapur S, Akça E, Çetin H, Erol O (2003) An Improved method for determination of holocene coastline changes around two ancient settlements in Southern Anatolia: a geoarchaeological approach to historical land degradation studies. Land Degrad Dev 14(4):363–376

Dinç U, Şenol S, Kapur S, Cangir C, Atalay İ (2001) Soils of Turkey. Çukurova University, Faculty of Agriculture, Publication No: 51, Adana, 185 p

Doğan O (1992) Regional development through alleviating wind eroded soil in Karapınar, Turkey. OECD Report. Working Group on Soil and Land Management, Paris, 17 p

Doğan O, Kuzucuoğlu C (1993) Wind erosion in Anatolia. Fighting measures and results obtained in the Karapınar (Konya) region. Communication presented at the Jan de Ploey Memorial Symposium, Leuwen

Erkal T (2005) Search of titan in sand dunes: the Karasu (Sakarya) example. Quaternary Symposium of Turkey TURQUA-V. ITU The Eurasian Earth Sciences Institute, 2–5 June 66-70

Erinç S (1962) On the relief features of blown sand at the Karapınar surroundings in central Anatolia. Rev Geogr Inst İstanbul 8:113–130 (in Turkish)

Erol O (1991) The relationship between the phases of the development of the Konya-Karapınar Obruk and the Pleistocene Salt Lake and Konya pluvial lakes, Turkey. Annals of the Institute of Marine Science and Geography 7:5–49 (in Turkish)

Groneman AF (1968) The soils of the Wind Erosion Control Camp Area Karapınar, Turkey. Agricultural University Wageningen, The Netherlands, 161 p

Herrmann HJ (2007) Aeolian transport and dune formation. Lect Notes Phys 705:363–386

IUSS Working Group WRB (2015) World reference base for soil resources 2014, update 2015. In: International soil classification system for naming soils and creating legends for soil maps. World Soil Resources Reports No. 106. FAO, Rome. 203 p

Kapur S, Eswaran H, Akça E, Dinç O, Kaya Z, Ulusoy R, Bal Y, Yılmaz T, Çelik İ, Özcan H (1999) Agroecological management of degrading coastal dunes: a major land resource area in Southern

Anatolia. In: Proceedings of the 4th international conference on the Mediterranean coastal environment, vol 1, 9–13 Nov 1999, Antalya, pp 8–101

Kayalık P (2007) Wind erosion event in Turkey, to examine the studies on Karapınar (Konya) sample area and suggestions. Ege University Graduate School of Natural Sciences, MSc Thesis, İzmir, 224 p (in Turkish)

Klijn JA (1990) Dune forming factors in geographical context. Dunes of The European Coasts. In: Bakker THW, Jungerius PD, Klijn JA (eds) Geomorphology-hydrology-soils, vol 18. Catena Supplement, pp 1–14

Kuzucuoğlu C (1993) The climatic significance of the upper pleistocene and holocene aeolian sand flats around Karapinar-Konya. Preliminary results. İTÜ Quaternary Workshop, İstanbul, pp 12–16

Kuzucuoğlu C, Parish R, Karabiyikoglu M (1998) The dune systems of the Konya Plain (Turkey): their relation to environmental changes in Central Anatolia during the Late Pleistocene and Holocene. Geomorphology 23(2–4):257–271

Mater B, Turoğlu H (2002) The geomorphological change in Göksu Delta, Reasons and Results. In: Proceedings of the 4th national conference of coast and marine zones of Turkey 5–8 Nov 2002, vol 2, pp 1249–1259. Dokuz Eylül University İzmir

Möllenhauer K, Taysun A, Frede HG (2002) C-factors of the universal soil loss equation for olive plantations in Western Anatolia (Central Aegean Region). J Plant Nutr Soil Sci 165(3):313–319

Okur M (2010) Effects of grazing land crops on the soil quality under texture using the land in historical center Anatolia. MSc Thesis (unpublished), University of Çukurova, Graduate School of Applied and Natural Sciences, Adana, Turkey, 140 p (in Turkish)

Okur O (2010) Changes of soil quality of historic Karapınar (Konya) desertification soils under long-term Almond and Acacia. MSc Thesis (unpublished), University of Çukurova, Graduate School of Applied and Natural Sciences, Adana, Turkey, 106 p (in Turkish)

Ozaner S, Gates MH, Özgen I (1992) Dating the coastal Dunes of Karabasamak District (İskenderun Bay) by geomorphological and archaeological methods. Ministry of Culture, General Directorate of Monuments and Museums VIII. Archaeometry Outcomes Meeting Ankara, pp 357–367 (in Turkish)

Özcan H, Erginal AE, Akbulat C, Sungur A, Bozcu M (2010) Physico-chemical characteristics of coastal Dunes on the Saros Gulf, Turkey. J Coastal Res 26(1):132–142

Polat O, Kapur S (2010) Soil quality parameters in Arenosols under Stone Pine (*Pinus Pinea* L.). Plantations in the Turan Emeksiz sand dune Area. Çukurova Univ Inst Sci J Sci Eng 3(2):167–174. Adana (in Turkish)

Rognon P (1994) Biographie d'un Désert: Le Sahara, 2ème éd., L'Harmattan Publication, Paris, 347 p

Serteser A (2002) Sakarya coastal dune vegetation cover—soil relations. In: Proceedings of the 4th national conference of coast and marine zones of Turkey, 5–8 Nov 2002, vol 1, pp 57–65, Dokuz Eylül University, İzmir (in Turkish)

Serteser A (2004) The evaluation of Ceyhan Delta (Adana) Coastal Dunes for Plant Cover-Soil relation. In: Proceedings of the 5th national conference of coast and marine zones of Turkey, 4–7 May 2004, vol 1, pp 17–24. Çukurova University, Adana (in Turkish)

Soil Survey Staff (2014) Keys to Soil Taxonomy, 12th ed. USDA-Natural Resources Conservation Service, Washington, DC. 372 P

Tekinel O, Çevik B (1972) A research on stabilization of Mediterranean coastal dunes and their Management. Univ Çukurova Ann Agr Fac Adana 3(1–2):119–120 (in Turkish)

Trevisan S, Francioso O, Quaggiotti S, Nardi S (2010) Humic substances biological activity at the plant-soil interface: from environmental aspects to molecular factors. Plant Signal Behav 5 (6):635–643

Uslu T (1989) Geographical informations on Turkish coastal dunes—European Union for Dune Conservation and Coastal Management pub: Leiden. 60 P

Yaktı S, Akça E, Kapur S. (2004) Management of coastal sand dunes by an agroecosystem approach. In: Zdruli P, Liuzzi GT, Hikmet EF (eds) Workshop proceedings of determining an income-product generating approach for soil conservation management. Marrakech, Morocco, 12–16 Feb 2004. CIHEAM, pp 227–234

Yılmaz M (2010) Environmental problems caused by ground water level changes around Karapınar. Ankara Univ Environ Sci J 2 (2):145–163

Gleysols

Hasan Özcan, Hüseyin Ekinci, Ali Sungur, and Timuçin Everest

1 Introduction

A Gleysol according to the IUSS Working Group WRB (2015) is a wetland soil that, unless drained, is saturated with groundwater for sufficient periods to develop a characteristic gleyic color pattern. This pattern is essentially made up of reddish, brownish, or yellowish colors on the surfaces of the soil particles (*peds*) and/or in the upper soil horizons mixed with grayish/bluish colors inside the peds and/or deeper in the soil. Gleysols are also known as *Gleyzems* and *meadow soils* (Russia), *Aqu*-suborders of Entisols, Inceptisols and Mollisols (Soil Survey Staff 2014), or as *groundwater soils* and *hydromorphic soils*.

2 Parent Materials and Profile Development

Gleysols occur on a wide range of unconsolidated materials, mainly fluvial, basic to acidic marine to lacustrine sediments of Pleistocene or Holocene age, nearly in all climatic regions. Similar to organic soils they are found in depression areas and low landscape positions with shallow groundwater in almost all geological parent materials (Greenlee 1981). According to the soil map of Turkey prepared by the Toprak Su (GDRS 1987), 2,775,110 ha of cultivated land has drainage problems in the country. This is 3.6% of the total land area (77,797,140 ha) of the country (Table 1).

2.1 Gleysols in the Kavak Delta (Northwest Turkey)

Gleysols of the Kavak delta are found adjacent to the tidal shorelines with clay accumulation features on their surfaces (Fig. 1). The Typic Fluvaquents in northwestern Turkey have gleyic horizons as shallow as 18 cm from the surface (Figs. 2 and 3; Table 2), even their A-horizons with 2.5Y Munsell color reveal a table water limiting effect on crop production. Thus, the majority of the Gleysols unless drained are covered by natural vegetation (Fig. 4).

2.2 Gleysols in Adana, South Turkey

Gleysols are poorly drained soils with profiles reflecting the influence of waterlogging for significant periods. The outcomes of these conditions are the gleyed horizons having dull gray to olive, greenish or bluish-gray moist colors, frequently accompanied by prominent usually rust-colored mottles of localized oxidation and reduction of hydrated iron oxides-ferrihydrites in the Gleysols of the Karataş delta, Adana, south Turkey (Fig. 5). The Gleysols and Fluvisols of the Karataş delta developed as part of the bottom sediments of the paleo-lagoon (after the regression of the Mediterranean Sea) located along Seyhan River (the contemporary Fluvisol–Gleysol environment filled by sediments in the late Holocene) which was used most likely for transportation during the late Hittite period (ca. 750 BC). Hittite statues made from hard crystalline basalt and limestone materials were discovered intact after being buried under wind-blown and fluvial soils/sediments (Fluvents/Gleysols) in the contemporary Karataş delta area. In comparison to statues of similar age, found in the same area, that was left on the surface, the fact that these artifacts were preserved in the Gleysol environment meant that they were protected from the effects of long period weathering (Jones et al. 2005). Moreover, the late Hittite statues of known age have been excellent indicators of the beginning of soil accumulation/formation and development of the late Holocene Karataş delta (Unpublished data, Akça and Kapur, University of Çukurova, Departments of Soil Science and Plant Nutrition and Archaeometry, Adana, Turkey) which was about

H. Özcan (✉) · H. Ekinci · A. Sungur · T. Everest
Department of Soil Science and Plant Nutrition, Çanakkale
Onsekiz Mart University, Çanakkale, Turkey
e-mail: hozcan@comu.edu.tr

© Springer International Publishing AG 2018
S. Kapur et al. (eds.), *The Soils of Turkey*, World Soils Book Series,
https://doi.org/10.1007/978-3-319-64392-2_21

Table 1 Areas with drainage problems in Turkey (GDRS 1987)

Drainage class	Area (ha)	Area (%)
Moderately well drained	1,689,358	61.0
Somewhat poorly drained	776,312	28.0
Poorly drained	283,381	10.0
Very poorly drained	26,064	1.0
Total	2,775,115	100.0

Fig. 1 Clay accumulation on the surfaces of the tidal shoreline (Kavak Delta, northwest Turkey)

Fig. 2 Gleysol in the Kavak Delta, Saroz Bay (northwest Turkey)

Fig. 3 Gleysol land cover (Hordeum marinum huds) in the Kavak Delta (Saros Bay, northwest Turkey)

3000 years BP as stated by Evans (1971), Erol (1997) and Gürbüz (1999).

2.3 Gleysols in the Gönen Delta, the Marmara Region, Northwest Turkey

The Marmara region with a relatively higher precipitation level is rich in river networks and delta development around the Marmara Sea. The Gönen River Delta soils are acidic to neutral Gleysols (Özşahin 2013) and rich in organic matter (Table 3). High organic matter and phosphorus contents of the Gleysols make them preferable arable soils after being drained.

3 Management

Based on the approach of the 1950s, studies were initiated concerning the prevention of malaria and gaining cultivated land by drying the major wetlands of Turkey (Fig. 6). After this initiative, the total amount of wetlands lost in the country was 236,538 ha, in turn leading to the decline of the Gleysol environment. Figure 7 illustrates the distribution of the dried major wetlands in the country.

Thus, the main reasons of the drainage problems in Turkey concerning the Gleysols are the inappropriate irrigation practices utilized against the rules of nature, which means that all irrigation activities on Gleysols would be problematic. Insufficient or poor drainage systems or the lack of the drainage infrastructure together with the wild flooding irrigation systems are the cause for the increasing groundwater levels and the occurrence of the poor drainage following the so-called reclamation of the widely distributed wetlands of the country. There are two major reasons for Gleysol formation in Turkey as elsewhere, namely the topographic conditions and the inappropriate irrigation systems creating drainage problems in some irrigation project areas that were wetlands or ancient lakes in the past. Topographically, parts of lowlands and depression areas (widespread in Turkey due to the graben-horst systems) are

Table 2 Site-profile description and some physical and chemical properties of a Gleysol in the Kavak Delta (northwest Turkey)

IUSS Working Group WRB (2015)	Fluvic Gleysols
Soil Survey Staff (2014)	Typic Fluvaquents
Location	Kavak Delta, Saroz Bay-Çanakkale, northwest Turkey
Coordinates	35 T, 486 434 E, 4 492 719 N
Altitude	3–4 m
Landforms	Delta
Position	Bottom
Aspect	–
Slope	0–1%
Drainage class	Very poor
Groundwater	55 cm
Eff. Soil depth	80 cm
Parent material	Fluvial
Köppen climate	Mediterranean
Land use	Natural habitat
Vegetation	*Mentha pulegium* L., *Hordeum marinum* Huds. var. *Marinum*, *Aeluropus littoralis* (Gouan.) Parl. *Juncus maritimus* Lam

Horizon	Color		Mottles	Roots	Structure	Texture (%)		
	Moist	Wet				Clay	Silt	Sand
A	2.5Y2/1	–	None	Intensive, thin	Granular	55	24	21
ACg	2.5Y4/1	10YR 5/3	Intensive	Sparse	Massive	58	24	18
Cg1	–	G25/10BG	Groundwater	Sparse	Massive	58	26	16
Cg2	–	G27/5BG	–	–	Massive	55	28	17

Horizon	Depth (cm)	Organic matter (%)	pH (H$_2$O)	EC (dS/m)	CaCO3 (%)	CEC (cmol$_c$kg^{-1})
A	0–18	3.84	7.64	0.87	16.38	21.05
ACg	18–39	2.26	7.74	0.49	15.61	19.75
Cg1	39–62	1.10	7.79	0.41	15.92	10.58
Cg2	62+	0.96	7.84	0.36	19.78	9.13

Fig. 4 The natural vegetation of Gleysols in the Kavak delta, northwest Turkey

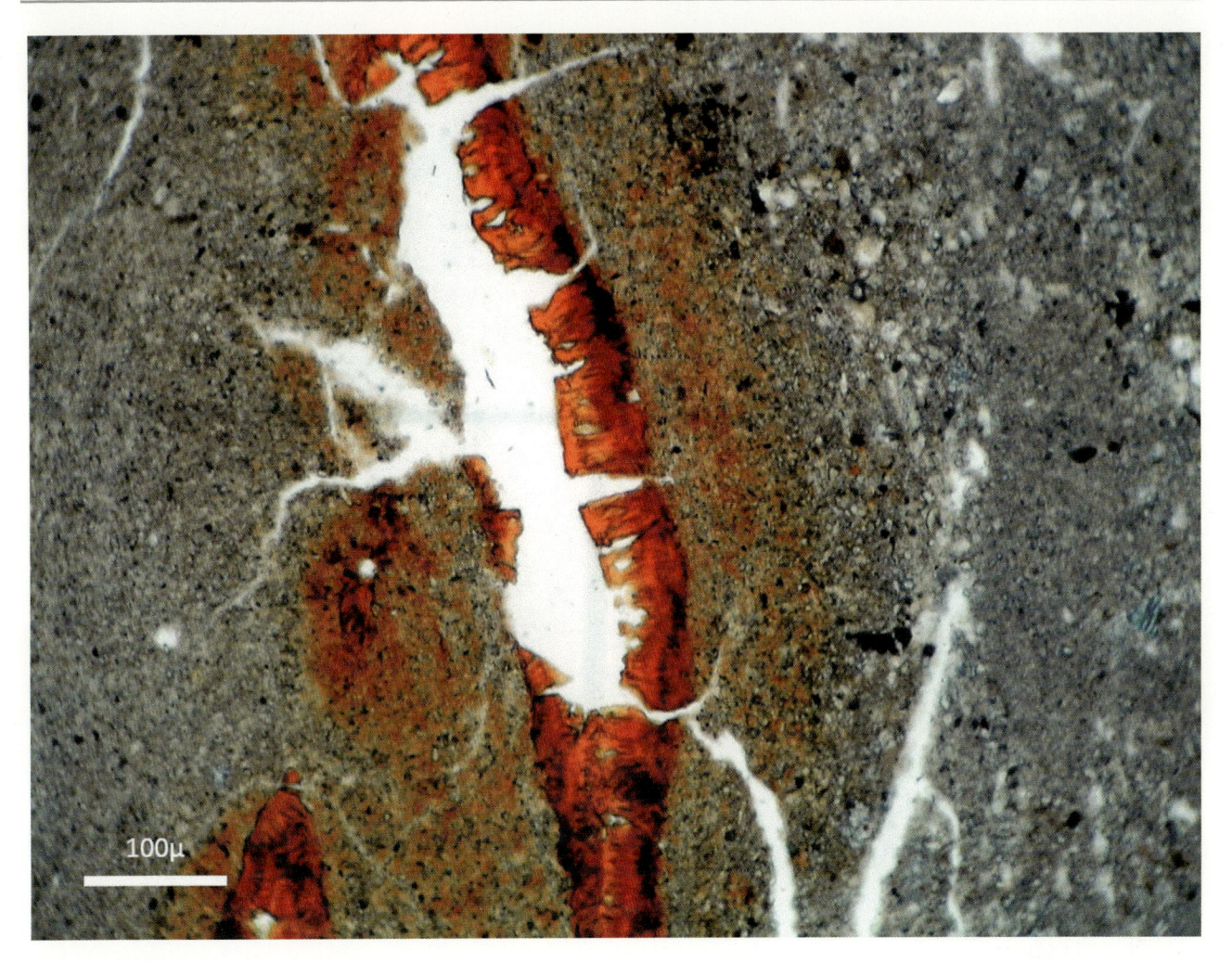

Fig. 5 Iron oxide accumulation/crystallization (ferrihydrite) in the subsurface of a Gleysol in the Karataş Delta (south Turkey)

rich in water which is discharged from the highlands and the vicinity ultimately creating the appropriate natural environments for Gleysol formation.

A detailed land survey carried out by Dinç et al. (1995) showed that shallow groundwater, excess irrigation water use and poor drainage systems were the reason for the development of the Gleysols in the Çukurova Delta (southern Turkey). Irregular maintenance, topography (low slope angle) and siltation by increased erodibility in the drainage canals were stated to be the other major reasons for the formation of the gleyic soils in the Çukurova delta by Özcan and Çetin (1996).

Gleysols have formed since decades and are commonly distributed in the poorly drained lands of the Konya Basin and Çukurova, Göksu, Gediz, Menemen, Bafra, Çarşamba, Kavak, Meriç Deltas as well as in the Gönen-Bursa, Aydın-Söke, Denizli-Acıpayam, Burdur-Yarıköy Plains. Despite the inappropriate topographic and soil conditions these areas were irrigated for higher income generation, ultimately creating salinity problems and causing the loss of soil quality. The expected fall of the yields was encountered after the threshold of the maximum amounts of harvest was reached in almost a decade after irrigation started. A prominent example for this is the SE Anatolian irrigation project area (seeking to irrigate 1.7 million ha of land via 13 large dams) which has been highly prone to Gleysol formation and inappropriate for irrigation due to its deep and highly clayey (smectitic) cracking soils that were traditionally appropriate for rainfed horticultural crops (pistachio and olives) and rangeland management of small ruminants.

The total area under irrigation in Turkey is about 5.1 million ha. All the irrigated land requires surface or subsurface drainage of different magnitudes. Although regular maintenance and cleaning of silt and weeds from open (main) drainage canals are periodically carried out by the DSİ, stabilization of canal embankments, siltation and weed problems are still not successfully accomplished in many parts of the country. Siltation and weeds occupying the

Table 3 Site description and some average physical and chemical properties of the horizons of a Gleysol profile, the Gönen Delta (northwest Turkey) (Özşahin 2013)

IUSS Working Group WRB (2015)		Fluvic Gleysols
Soil Survey Staff (2014)		Typic Fluvaquents
Location		Gönen River Delta, northwest Turkey
Coordinates		40.320582°N, 27.126136°E
Altitude		5 m
Landforms		Delta
Position		Bottom
Aspect		–
Slope		0–1%
Drainage class		Very poor
Groundwater		65 cm
Eff. Soil depth		110 cm
Parent material		Neogene—Quaternary fluvial
Köppen climate		Mediterranean
Land use		Natural habitat
Vegetation		*Cultivated land (rice, maize, wheat)*
Horizon		
A	Water saturation	51
	Texture	Clayey loam
	CaCO$_3$ (%)	–
	Organic matter (%)	4.19
	pH	6.39
	Available Phosphorus (kg/ha)	77
	Available Potassium (kg/ha)	455

Fig. 6 A dried wetland for gaining arable land in Isparta, southwest Turkey

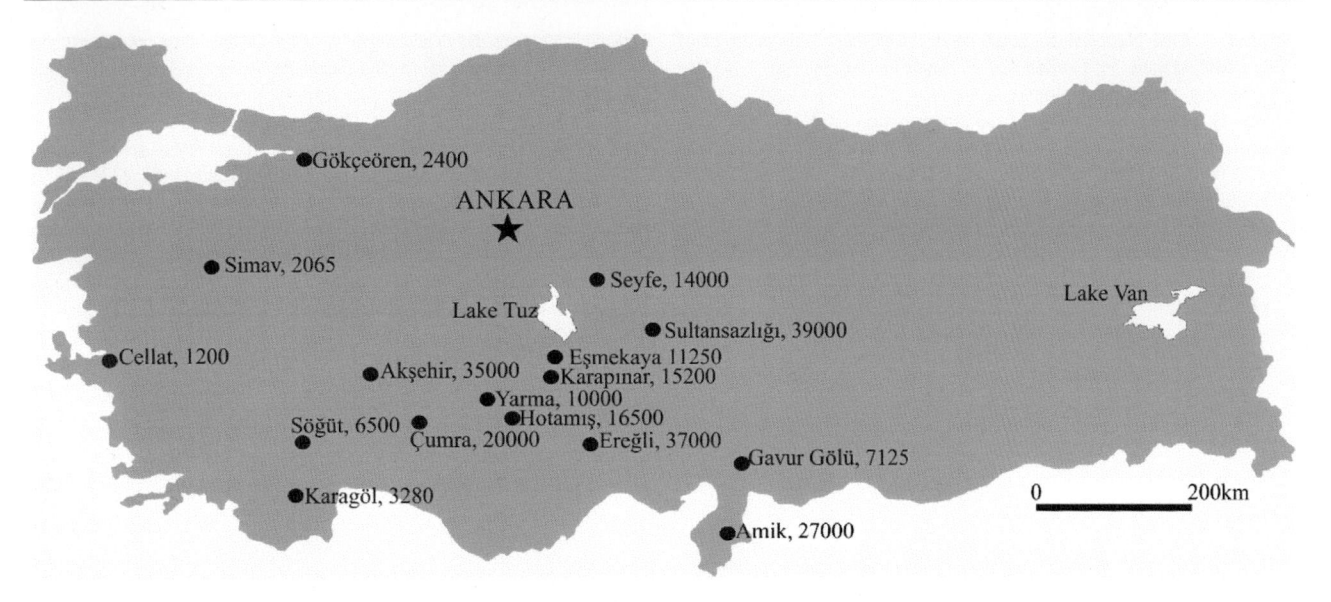

Fig. 7 Some dried lakes/wetlands and lakes of Turkey (ha) since the 1960s

drainage canals are the cause of the ponding of the water in the canals which is fed back to the land increasing the groundwater levels.

Based on the reconnaissance soil survey of Turkey accomplished by the TOPRAKSU from 1966 to 1971, the country had drainage problems at a land of about 2,775,115 ha. The drainage problems were determined to occur at about 1,968,814 ha arable land and at about 803,161 ha land in non-arable areas. These areas were stated to be 623,446 ha in Konya, 83,331 ha in Samsun, 74,177 ha in Sakarya, 62,528 ha in Antalya and 51,599 ha in Bursa. The drainage problem areas were smaller in Adana, Burdur, Kütahya, Eskişehir and Van, which were totally about 30,000 ha based on the report of the GDRS.

Based on the soil survey conducted in the Şanlıurfa--Harran plain in 1978, the area with poor drainage was 2474 ha with a groundwater depth varying from 0 to 2 m before irrigation. After irrigation this increased to about 28,000 ha together with a potential problem in a land of 8000 ha (Özkaldı et al. 2004). Nowadays, there exists a problem area of more than 40,000 ha with a high groundwater level in the Harran plain (Kanber and Ünlü 2008).

Wetness is the main limitation of Gleysols. Gleysols in deltas are mainly covered with natural swamp vegetation (Fig. 3) and are used for grazing and at some places for rice cultivation such as in the Meriç, Kavak, Çarşamba and Bafra deltas. Artificially drained Gleysols are used for arable cropping, dairy farming and poplar cultivation in Turkey. However, environmental scientists and land managers still remain without negotiation with the agricultural or forestry scientists/experts concerning the appropriate use of Gleysols, where the former have consistently sought a management system based on the abandonment of these soils to their natural succession and to land conservation rather than employing income-generating well managed grazing and appropriate cropping. In this respect halophytes cultivation, as part of the wetlands management, together with rice production would be an appropriate crop pattern for Gleysol ecosystems.

References

Dinç U, Sarı M, Şenol S, Kapur S, Sayın M, Derici MR, Çavuşgil VS, Gök M, Aydın M, Ekinci H, Ağca N, Schlichting E (1995) Soils of Çukurova Region, University of Çukurova, Faculty of Agriculture, No. 26, Adana, Turkey (in Turkish)

Erol O (1997) The noetectonic and geomorphological evolution of the Çukurova region. Geosound-Earth Sci. J University of Çukurova Special Issue. 127–134

Evans G (1971) The recent sedimentation of Turkey and the adjacent Mediterranean and Black Seas: a review. In: Campbell AS (ed) Geology and history of Turkey: Tripoli Petroleum Explorer Soc. Libya, pp 385–406

Greenlee GM (1981) Guidebook for use soil survey reports of Alberta Provincial Parks and Recreation Areas. Earth Sciences Report 81-1. Alberta Research Council, Edmonton, Alberta. 66 p

Gürbüz K (1999) An example for river course changes on a delta plain: Seyhan Delta (Çukurova plain, southern Turkey). In: Bozkurt E, Rowbotham G (eds) Aspects of Geology of Turkey II. Geol J Special Issue. 34(1–2): 211–222

IUSS Working Group WRB (2015) World Reference Base for soil resources 2014, update 2015 International soil classification system for naming soils and creating legends for soil maps. World Soil Resources Reports No. 106. FAO, Rome. 203 p

Jones A, Montanarella L, Jones R (2005) Soil Atlas of Europe, European Soil Bureau Network. European Commission, 128, Office of the Official Publications of the European Communities, L-2995 Luxembourg

Kanber R, Ünlü M (2008) Drainage and irrigation problems in Turkey: a general review. In: 5th world water forum regional readiness process, DSİ National Regional Water Meetings, Proceedings of Irrigation-Drainage Conference Preparation Adana, pp 1–45 (in Turkish), 10–11 Apr 2008

Özcan H, Çetin M (1996) Land degradation induced by high groundwater level and soil salinity in the fourth stage project area of Yüreğir plain: a case study. In: Kapur S, Akça E, Eswaran H, Kelling G, Vita-Finzi C, Mermut AR, Öcal AD (eds) Proceedings of the 1st international conference on land degradation. International Working Group on Land Degradation and Desertification (IWGLDD-ISSS), Çukurova University Press, Adana, Turkey, pp 93–99

Özkaldı A, Boz B, Yazıcıoğlu V (2004) Drainage problems and solution suggestions in Southeastern Anatolia Irrigation Project area. In: Proceedings of the salinity problems in irrigated area symposium, Ankara, pp 97–106 (in Turkish), 20–21 May 2004

Özşahin E (2013) Discussion of geographical survey in respect of soil features of Gönen river delta. EKEV Akad J 57:233–245 (in Turkish)

Soil Survey Staff (2014) Keys to soil Taxonomy. In: USDA-Natural Resources Conservation Service, 12th edn, Washington, DC, 372 p

TOPRAKSU (1987) The planning of Turkey for soil management, The general directorate of soil and water, Ankara, p 26 (in Turkish)

Histosols

Orhan Dengiz, Hasan Özcan, and Zülküf Kaya

1 Introduction

Common names for histosols are *peat soils, muck soils, bog soils, and organic soils* (FAO/ISRIC 2006). Many histosols belong to : *Moore, Felshumusböden* and *Skeletthumusböden* (Germany), *Organosols* (Australia), *Organossolos* (Brazil), *Organic order* (Canada), and *Histosols* and *Histels* (United States of America). The name of histosol is derived from Greek "histos" meaning tissue. Thus, histosols are recognized by the slow transformation of plant remains through biochemical disintegration and formation of humic substances creating a surface layer of the mold with or without prolonged water saturation. Histosols are generally defined as soils which have accumulated partially decomposed of plant and animal residues under anaerobic conditions (Fitz-Patrick 1971). According to the Keys of Soil Taxonomy (2014), histosols are slightly decomposed to well-decomposed organic materials that are saturated with or artificially drained, containing organic carbon more than 12% by weight excluding living roots. The Reference Soil Group of the Histosols comprises soils formed in organic material. These vary from soils developed in predominantly moss peat in boreal, arctic, and subarctic regions, via moss peat, reeds/sedge peat (fen) and forest peat in temperate regions to mangrove peat and swamp forest peat in the humid tropics.

2 Environment

Histosols are found at all altitudes, but the vast majority occurs in lowlands. Conservation of the wetlands in arid and semiarid regions is very important for the natural life and environment. Draining of peat swamps for agricultural use or the extreme decline of the water table levels in these areas will result in the annihilation of organic soils that serve as natural resources. Peatland ecosystems are highly susceptible to changes in the water balance and retaining the natural balance, which is the key to long-term conservation of the individual peatland sites. There is a strong interrelationship between plants, water, and peat where one component is affected, eventually all components change in the peatland ecosystem. Modification of the peatland's hydrology directly affects peatland's the flora and fauna. On the longer run, changes in the hydrology influence the hydraulic soil properties. However, the most drastic and negative change performed on the histosols are the activities of the local people in Turkey that initiated the burning of the peat as fuel besides exploiting it as manure and the occurrence of the subsidence processes in the study area enhanced by drainage activated oxidation (Dengiz et al. 2009).

3 Profile Development

Factors that cause peat to accumulate are similar all over the world, but different types of peatlands develop because of differences in climate, topography, soil type, and plant species. In upland areas where temperatures are low and rainfall/fog is evenly spread over the year, rain-dependent "ombrogenous peat" (high moor peat) may form where microbial activity is depressed by severe acidity, oligotrophy, and organic toxins. Peat formations in Turkey are generally known as basin peats (Dinç 1974). Here, the peat accumulates in a basin where the rate of plant production exceeds the rate of its decomposition. And, thus, complete plant decomposition is prevented in areas where

O. Dengiz (✉)
Department of Soil Science and Plant Nutrition, Ondokuz Mayıs University, Samsun, Turkey
e-mail: odengiz@omu.edu.tr

H. Özcan
Department of Soil Science and Plant Nutrition, Çanakkale Onsekiz Mart University, Çanakkale, Turkey

Z. Kaya
Department of Soil Science and Plant Nutrition, University of Çukurova, Adana, Turkey

waterlogging occurs (Fig. 1). Basin peat formations are fertile compared to peat formed under cool and wet climatic conditions, which have high acidity, low nutrient status, and low microbiological activity (Dengiz et al. 2009).

There are many classification systems for peat, each geared to the objectives of the disciplines responsible for their development. Generally, physical, chemical, and physicochemical properties of peat such as texture, organic and minerals contents, pH, color, water content, and degree of decomposition could serve as a basis for its classification (Bujang et al. 2011). Other classifications systems may be based on the origin of peat and the field conditions during deposition. Most of the time, peats are classified on the basis of the constituent plants rather based on their texture and composition (Davis 1947). The degree of decomposition is usually assessed by using the von Post (1922) scale. The von Post scale is a classification system which is based on a number of factors such as botanical composition, degree of humification, and the color of peat water after squeezing. In this classification system, there are ten degrees of decomposition ranging from H1 (very fibrous) to H10 (very few fibers), which represent the state of decomposition/decay of the organic plant remains. The higher the number in the von Post scale, the higher is the degree of decomposition. The degree of decomposition is determined to be ten degrees based on the appearance of soil water coming out upon

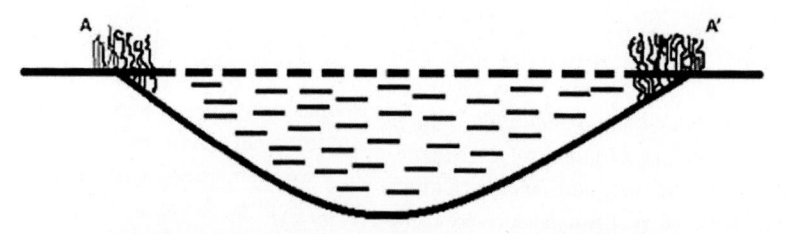

(a) Plant (phragmites com.) thrive around lake edges

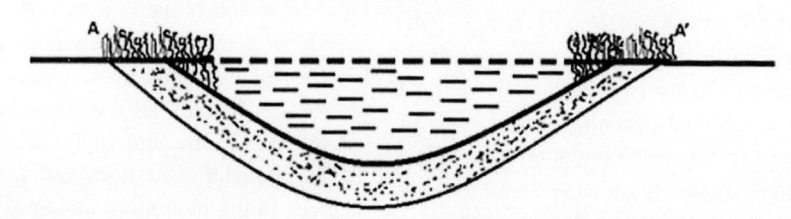

(b) Dead plant residues sink to bottom in anaerobic environment (gyhttja formation)

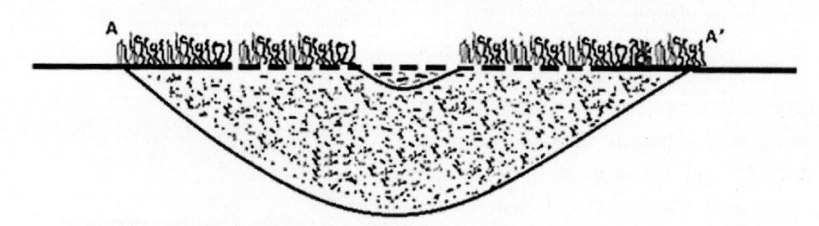

(c) A thick mat-covering of organic material develops as the lake continues to fill in and shrink

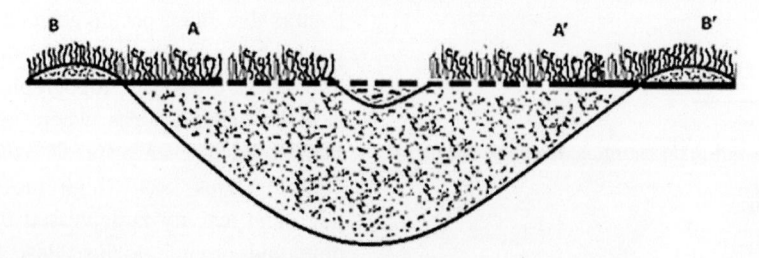

(d) Lake completely fills. The buried organic material continues transforming into peat and formation of new peat lands (B-B')

Fig. 1 Stages for geogenetic accumulation of organic soil parent material in Kahramanmaraş-Türkoğlu Lake (Dinç 1974)

squeezing the soil in the hand. Baran and Çaycı (1996) determined the decomposition degree of peats collected from 14 different areas, namely from Ağrı, Muş, K.Maraş, Mersin, Antalya, Burdur, Bolu, Afyon, Niğde, Konya, Kayseri, and Trabzon using the von Post scale. The Trabzon (Sürmene), Bolu and Muş peats were at the lowest decomposition degrees, whereas, the Antalya, Afyon, and Kayseri were at the highest decomposition level.

In the American Society for Testing and Materials (ASTM 1990) the classification of organic soils is based on their ash and organic matter contents, where a soil is classified as peat if the organic content is more than 75% and ash content is less than 25%. The fiber and organic matter contents varied from 12.5 to 91.5% and from 4.3 to 91.5%, respectively, in the Bolu-Yeniçağa peatland (Dengiz et al. 2009) whereas the Akköy-Sakarya peatland was reported fiber content of the peats were averagely 16.5% (Cayci et al. 2011). As compared at Bolu-Yeniçağa peats, the fiber contents were very low due to the high rate organic matter decomposition. Kılıç and Saray (2006) determined the fiber contents of the peats located in Kayseri-Karasaz describing two peat soil profiles classified as sapric medihemists and hydric medihemists with fiber contents in the hemic range (46.1–62.3%).

The amount of the water content depends on the origin, degree of decomposition, and the chemical composition of the peat. The high natural water holding capacity is due to the soil structure characterized by organic coarse particles (fibres), which can hold a considerable amount of water since the soil fibers are very loose and hollow. The high water content is also due to the low bulk density of the peat leading to high buoyancy and high pore volumes. Moreover, the chemical properties of peat are affected by the chemical composition of peat's components, the environment in which they were deposited and the extent of decomposition. Generally, peats are in an acidic condition and the pH value of the soils often lies between 4 and 8 in the Yeniçağa peat area (Table 1; Fig. 2) (Dengiz et al. 2009). According to Table 1, the hand-rubbed colors of each horizon of the profiles were generally brownish black (5YR 2/4, 2/2. 7.5YR 3/2, 2/2), indicating a high degree of chemical decomposition in the soil surface. The pH and ECe are 5.38–7.92 and 0.50–3.80 dSm^{-d}, respectively. Both pH and EC values had irregular distribution in all pedons and the pH ranged from slightly acid to slightly alkaline. A relationship was determined between the CEC, organic matter, and the degree of decomposition. The decomposition degree was found to decrease, where the organic matter and consequently the CEC increased in the studied soils. The CEC ranged from 37 to 222 cmol/kg. The high CEC of the fibric horizons was due to the high organic matter contents of the soils. The bulk density ranged from 0.09 to 0.78 g cm^{-3} and was related to the organic matter contents and the mineralization rates. The highest bulk density (0.78 g cm^{-3}) was found in the sapric (Oa) horizon of profile 1. The bulk density values of surface horizons in all profiles were higher than the subsurface horizons on account of mineralization. Water holding capacities were as low as 140–330% in the sapric horizons of the four pedons, where the fibric horizons had higher water holding capacities (730–1200%).

4 Regional Distribution

The distribution of peat deposits is extensive and can be found in many countries throughout the world. The peatlands consist of nearly 5–8% of the earth's land surface and nearly 60% of the wetland of the world is peat. The total extent of histosols in the world is estimated at some 325–375 million ha, the majority located in the boreal, subarctic, and low arctic regions of the northern Hemisphere. According to IUCN UK Peat Land Programme (2011), peatlands are found in at least 175 countries and cover around 4 million km^2 or 3% of the world's land area. According to the world soil map (van Engelen et al. 2005), histosols do not exist in Turkey, where an area of 45.069 km^2 is designated as gley soils. Almost all peatlands have been drained in the country to prevent the spread of malaria and for gaining land for cultivation According to Byfield and Özhatay (1997) the original peatland area in Turkey was approximately 240 km^2, of which a current area of 30 km^2 has remained (http://www.ipcc.ie/wpturkey.htlm). The main distribution areas of Histosols in Turkey (5,845.06 ha, 0,02%) are located in the provinces of Bolu, Sakarya, Trabzon, Denizli, Kayseri, Kahramanmaraş, Konya, Burdur, Antalya, Muş, Adıyaman, Antakya, Hakkari, Ağrı, and Kars (Fig. 3).

5 Management

Wetlands throughout the world are very important for natural life and the environment. Draining of peat swamps for agricultural use or the extreme decline of the water table level in these areas will result in the annihilation of organic soils that serve as natural resources. Peatland ecosystems are highly susceptible to changes in the water balance and retaining the natural balance is the key to long-term conservation of individual peatland sites. There is a strong interrelationship between plants, water, and peat. When one component is affected, eventually all components change in the peatland ecosystem. Modification of the hydrology of the peatland's directly affects the peatland flora and fauna. On the long run, the changes in hydrology affect soil hydraulic properties. The general trend and will to intensely drain the peatlands of Turkey that in turn induced pedogenesis and the degradation of the peat, severely affected the peatland

Table 1 Site description and some physical and chemical properties of the organic soils in Bolu-Yenicağ peatlands (North Turkey)

IUSS Working Group WRB (2015)				Profile 1: Hemic Fibric Histosols (Eutric)					
				Profile 2: Fibric Histosols (Eutric)					
				Profile 3–4: Subaquatic Fibric Histosols (Eutric)					
Soil Survey Staff (2014)				Profile 1: Hemic Medifibrist					
				Profile 2: Typic Medifibrist					
				Profile 3–4: Hydric Medifibrist					
Location				Yeniçağ, Bolu (north Turkey)					
Coordinates				36 T, 32°, 2′, 8.9808″ E, 40°, 46′, 20.6688″ N					
Altitude				990 m					
Landforms				Lake (wetland)					
Position				Depression					
Aspect				–					
Slope				0–1%					
Drainage class				Very poor					
Groundwater				–					
Parent material				Organic + fluvial					
Köppen climate				Mediterranean					
Land use				Natural habitat (peatland, wetland)					
Vegetation				*Carex, phragmites,* and brush wood					
Pedon	Horizon	Depth (cm)	Org. mat. (%)	pH	EC (dS/m)	CEC (cmol/kg)	Color (Rubbed)	Fiber (%)	B. D (g/cm^3)	MWC (%)
1	Oa	0–11	36.0	6.35	1.20	104	7.5YR2/2	<10	0.78	140
	Oi	11–42	88.5	6.26	0.80	212	7.5YR2/2	87	0.20	1143
	Oe	42–67	81.0	5.96	1.08	197	5YR2/1	43	0.17	840
	Oi$_1$	67–94	71.5	5.38	2.24	201	7.5YR3/3	75	0.19	780
	C	94–108	12.5	6.86	2.80	37	5Y 5/1	–	–	–
	Oi$_2$	108–136	57.0	6.90	3.80	169	7.5YR3/3	48	0.13	745
	C	+136	34.5	7.18	1.70	100	5Y 4/1	–	–	–
2	Oa	0–20	42.5	7.19	1.42	97	7.5YR3/3	<10	0.54	152
	Oi$_1$	20–31	74.0	7.49	1.00	192	5YR 3/4	77.0	0.16	889
	Oe	31–51	80.5	7.59	0.80	208	7.5YR3/2	45.0	0.17	770
	Oi$_1$	51–64	75.0	7.64	0.62	201	5YR 2/2	58.0	0.12	1250
	Oi$_2$	64–91	57.0	7.10	0.80	147	7.5YR3/2	75.0	0.12	790
	Oi$_3$	91–130	56.5	6.82	0.50	141	7.5YR3/4	70.0	0.13	730
3	Oa	0–15	32.0	7.92	0.80	84	10YR 3/3	<10	0.62	140
	Oe	15–31	78.0	7.04	1.30	210	7.5YR3/2	39.4	0.20	580
	Oi$_1$	31–63	91.0	7.02	2.40	222	7.5YR2/3	83.8	0.12	1170
	Oi$_2$	+63	76.5	6.88	0.60	188	10YR 4/4	87.5	0.09	1200
4	Oa	0–16	57.5	7.32	1.20	161	7.5YR2/3	<10	0.32	330
	Oe	16–28	55.1	6.96	1.00	144	5YR 2/2	21.3	0.22	645
	Oi	28–71	85.0	6.50	1.30	195	5Y 5/2	88.1	0.13	943
	Cg	71–85	22.0	6.37	2.20	44	10YR 3/3	–	–	–
	Oi	85–108	91.5	6.05	3.20	220	7.5YR2/2	91.5	0.15	825
	Cg	108–160	29.0	6.97	2.20	61	5Y 5/3	–	–	–

MWC Maximum water content, *CEC* cation exchange capacity, *BD* dry bulk density

Fig. 2 Bolu-Yenicağ lake (a), profile (2) of a Typic Medifibrist (fibric histosol (Eutric) (b) (Dengiz et al. 2009)

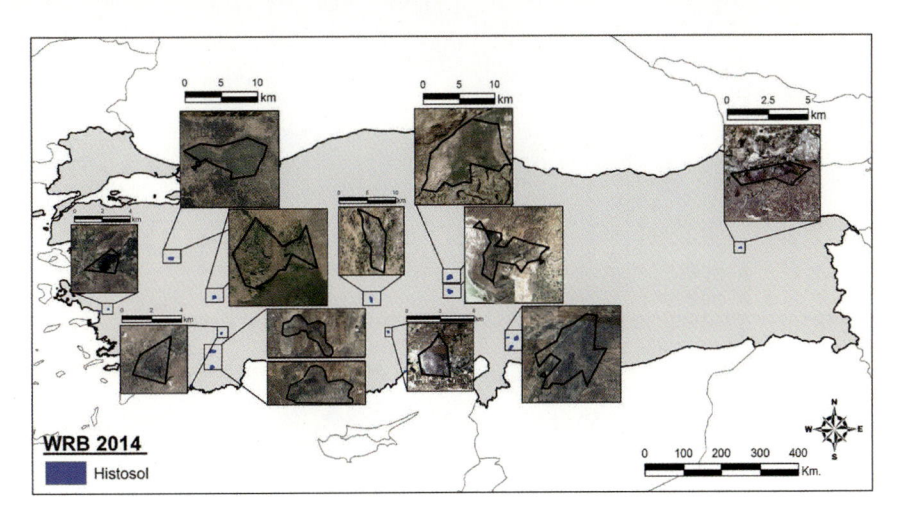

Fig. 3 Distribution of the Histosols in Turkey

ecosystems of the country. Consequently, the recent efforts undertaken in the area in order to restore the hydraulic conditions of the peatlands in the aftermath of the drainage activities were unfortunately unsuccessful (Dengiz et al. 2009).

The desiccated peats have altered the entire water balance of the histosol ecosystems, and affected the plant–peat–water relationship. The problem with peat is that the partial loss of the original water holding capacity cannot be restored with rewetting. Excessive drying will cause irreversible damage and experiments have shown that overdried peat will form a hard 'coffee beads' structure. Because of the high horizontal conductivity in the surface horizon of the peat soil, local damage to the hydrodynamic equilibrium of a peatland system brought about by artificial drainage will alter the functional system of the water balance. Experience has shown that even a small alteration in the mean water table has a deleterious impact on organic soils. The study conducted by Dengiz and Başkan (2004) on the macromorphological, physical, and chemical properties of the organic soils in the Konya-Eşmekaya peatlands revealed that, particularly, the application of excessive pressurized irrigation caused degradation in the peat land system by lowering the water table and speeding up the oxidation of the organic

Fig. 4 Carex (a) and pragmites communities on Histosols (b)

Fig. 5 Typic Torrifluvent-Histic Fluvisol (a), a Typic Haplofibrist-Fibric Histosol in the Konya-Eşmekaya peatland area (C Turkey). *A1* and *A2*: mineral horizons, *C*: parent material. (b) *Oa*: sapric organic horizon, *Oi*: fibric organic horizon, *Oe*: hemic organic horizon

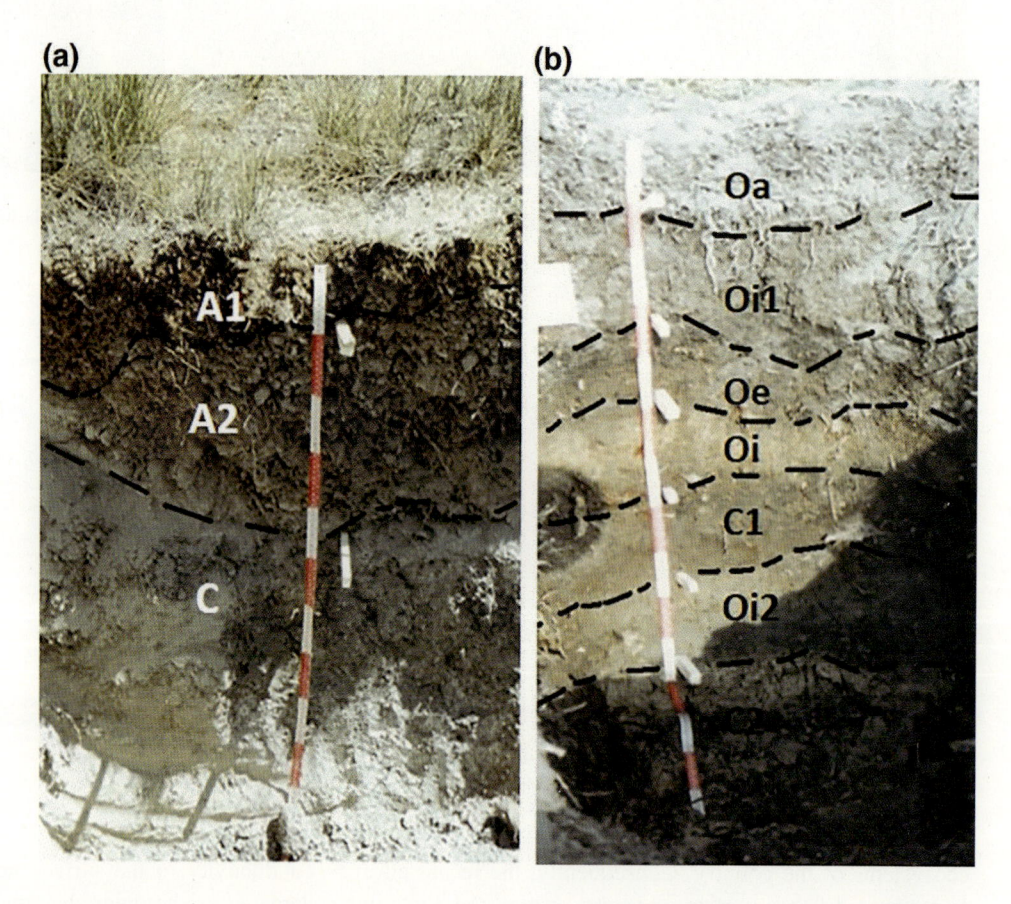

material consequently changing the organic soil to mineral soil (Fig. 4). These profiles were classified as Typic Torrifluvents, Typic Haplofibrists, Typic Haplohemists, and Sapric Haplohemists, according to the Keys of Soil Taxonomy (Soil Survey Staff 2014), and Histic Fluvisols, Fibric Histosols, Hemic Histosols, Sapric Histosols by the IUSS WRB Working Group (2015) (Fig. 5).

Dikici and Yılmaz (2006) reported that the Türkoğlu Lake Peatland of Kahramanmaraş was drained in the late 1950s, and, thereafter, the area was subject to agricultural production coupled with fires that started after the drainage. The DSİ (1967) mentioned that 846 ha of peatland was already affected by fires within the first decade of the agricultural production in the area, as an example, a catastrophic peat fire that took place in 1965 was active for several months. Periodic peat fires have taken place ever since, and one of the last large-scale fires occurred in 2001. Table 2 shows some of the physical and chemical properties of the selected peatlands in Turkey (Usta and Sözüdoğru 1996).

Table 2 Some physical and chemical properties of the Peatlands in Turkey (Usta and Sözüdoğru 1996)

Sample	Volume mass (g/cm^3)	pH (1/3)	EC (1/3) (dS/m)	Free carbonate (%)	CEC (cmolckg^{-1})	OM Mat (%)	Org. C (%)	Total N (%)
Muş	0.153	6.31	0.222	0.40	128.32	55.83	32.55	2.27
Burdur	0.418	7.75	3.703	24.90	116.0	28.15	15.51	0.89
İçel	0.451	7.61	1.219	4.30	124.0	37.44	21.61	1.18
Kütahya	0.628	8.06	1.626	14.16	54.6	16.83	13.69	0.78
Adiyaman	0.222	8.04	0.258	21.30	144.0	58.17	33.71	1.89
K.Maraş	0.433	8.10	0.511	24.80	98.0	24.34	12.54	1.09
Trabzon	0.284	4.33	0.416	–	119.0	82.28	50.93	2.39

Fig. 6 Peat is/was collected and dried for use as fuel by local people in the Konya Ereğli Plain (Özaytekin 1996)

Fig. 7 The Adıyaman Çelikhan peatland and use for fish ponding and inappropriate exploitation of the peat (Southeast Turkey)

Peat has been used as a form of energy for at least 2000 years in Anatolia. The increasing use of gas and oil as cooking and other fuels during the twentieth century decreased the use of peat for such domestic purposes. The socioeconomic perception of the rural population has had a particularly significant role on peat utilization especially in the Ereğli Plain of Konya (Özaytekin 1996) (Fig. 6).

Despite the contemporary decreased use of peat as fuel, its use for the soilless cultures (hydroponics) as a growth medium has increased in the past two decades in Turkey. The required principal properties for the growth medium are, (a) an appropriate air–water balance, (b) absence of toxic elements or component, and (c) the availability and low price of the material. Çaycı et al. (2000) studied the actual and the improved physical properties of the Bolu-Yenicağa peats (five Histosols) as a plant growth media. The results revealed that the physical properties of the Bolu-Yenicağa peat as a growth medium changed significantly depending on the locations and depths of the studied samples. Çaycı et al. (2000) have also mentioned that the regional producers/contractors excavating peat from the Bolu-Yenicağa peatland, were not aware of the material-wise differences of the physical properties and thus did not consider them during excavation-extraction. However, ultimately, urgent efforts are needed (despite the recent measures set out by the relevant ministries in peat extraction) in empowering legislations and state policies based on environmental-friendly and well-established management programmes for the peatlands of Turkey in order to protect them against the inappropriate and partly appropriate contemporary use as growth media or organic fertilizers to replace manure. Part of the inappropriately used Türkoğlu and Çelikhan peatlands in Kahramanmaraş (about 3000 ha) and Adıyaman, both in southeast Turkey, are allocated for corn production for freshly consumable fodder. This is the only appropriate type of cultivation due to the late seasonal (mid-summer) lower water level of the lake retrieved during the winter rains and lowered during the irrigation period. Before the mid-summer, parts of the reverted lake waters are also constantly managed for fish ponding. Despite the high-income-generating cultivation of the drained soils for corn fodder of this peatland and the use of the highly appropriate pivot-irrigation system alongside the fish ponding use of the raised winter lake waters, there still is an ongoing discussion on the appropriate use of this site, which is unfortunately far beyond its ideal use, i.e., as a conservation area or a well-managed and site-specifically programmed grassland (Fig. 7).

Contractors are recently strongly advised by state regulations to excavate-extract mature peat for use as manure and soilless cultures. Thus, the still incompletely undestroyed or extracted peatlands like the Türkoğlu peatland (also a migrating bird temporary shelter) in the Kahramanmaraş province. This peatland is somewhat extracted regarding its maturity and outside the egg laying seasons of the migrating birds.

References

Andriesse JP (1988) Nature and management of tropical peat soils. FAO soils Bulletin 59. Food and Agricultural Organization of the United Nations, Rome, Italy

ASTM (1990) D 2607 Classification of peats, mosses, humus, and related products. Annual book of ASTM standards, ASTM, Philadelphia

Baran A, Çaycı G (1996) Decomposition rates of peats in Turkey and their comparison to CEC and SOC. Pamukkale Univ Fac Eng J 2 (2):39–142 (in Turkish)

Bujang B, Huat K, Kazemian S, Prasad A, Barghchi M (2011) State of an art review of peat: general perspective. Phys Sci 6(8):1988–1996

Byfield A, Özhatay N (1997) A future for Turkey's peatlands: a conservation strategy for Turkey's peatland heritage. Doğal Hayati Koruma Derneği, Istanbul, Turkey

Çaycı G, Baran A, Kütük C, Ataman Y, Öztekin H, Dengiz O (2000) A research on reclamation of physical properties of Bolu Yeniçağa Peat as plant growing medium. In: Proceeding of international symposium and desertification. Konya-Turkey, pp 308–312

Cayci G, Baran A, Ozaytekin H, Kutuk C, Karaka S, Cicek N (2011) Morphology, chemical properties and radiocarbon dating of eutrophic peat in Turkey. Catena 85(3):215–220

Davis JH (1947). The peat deposits of Florida their occurrence, development and uses, Florida Geological Survey. Geol Bull 30. 247 p

Dengiz O, Başkan O (2004) Physical, chemical, morphological properties and classification of Eşmekaya organic soils. Gaziosmanpaşa Univ Agric Sci J 21(2):111–118 (in Turkish)

Dengiz O, Özaytekin H, Çaycı G, Baran A (2009) Characteristics, genesis and classification of a basin peat soil under negative human impact in Turkey. Environ Geol 56:1057–1063

Dikici Y, Yılmaz CH (2006) Peat fire effects on some properties of an artificially drained peatland. J Environ Qual 35:866–870

Dinç U (1974) Geogenesis, pedogenesis, and classification of Çukurova soils. Adana (in Turkish). 112 p

DSİ (1967) Report on fertility status of the soils in the artificially drained Türkoğlu Lake. State hydraulic works (DSI) Surveying and Planning Department, Ankara, Turkey (in Turkish)

FAO ISRIC (2006) IUSS, 2006. World reference base for soil resources. A framework for international classification, correlation and communication. World Soil Resources Reports, 103

FitzPatrick EA (1971) Pedology: a systematic approach to soil science. Oliver and Boyd, Edinburgh, 306 p

IUCN UK Peatland Programme (2011) IUCN UK Commission of Inquiry on Peatlands. http://www.iucn-uk-peatlandprogramme.org (12 Oct 2012)

IUSS Working Group WRB (2015) World reference base for soil resources 2014, update 2015 International soil classification system for naming soils and creating legends for soil maps. World Soil Resources Reports No. 106. FAO, Rome. 203 p

Kılıç K, Saray S (2006) Organic soils in the arid small catchments in the middle Anatolia region of Turkey. J Agron 5(1):23–27

Özaytekin HH (1996) The morphologic, physical and chemical properties and development of organic soils around Konya-Ereğli. Selçuk University, Graduate School of Natural Sciences, MSc thesis. YÖK Thesis No: 133104. Konya (in Turkish)

Soil Survey Staff (2014) Keys to soil taxonomy, 12th ed. USDA-Natural Resources Conservation Service, Washington, DC. 372 p

Usta S, Sözüdoğru S (1996) Nitrogen contents of some of peat soil in the country. Pamukkale University Engineering Faculty. J Eng Sci 2 (1):75–79

Van Engelen VWP, Batjes NH, Dijkshoorn K, Huting J (2005) Harmonized global soil resources database (Final report). Report 2005/06, ISRIC—World Soil Information, Wageningen

Von Post L (1922) Sveriges geologiska undersoknings torvinventering och nagre av dess hittills vunna resultat, Sr. Mosskulturforschung 1:1–27

Solonetz Soils–Solonchaks (Solonetz-Like Soils)

Mehmet Ali Çullu and Hikmet Günal

1 Introduction

Solonetz soils are characterized by high contents of exchangeable sodium (Na) that may be either due to the high Na contents in the parent material or accumulated in the soil profile by capillary rise of Na-rich groundwater. The term of Solonetz originated from the Russian, *sol*, meaning salt and *etz*, meaning strongly expressed. Solonetz have a dense and strongly structured horizon of clay illuviation and a high concentration of Na and/or magnesium in exchange complexes. Solonetz in the IUSS Working Group (WRB 2015) is associated to Gleysols, Solonchaks, and Kastanozems, whereas in the Soil Taxonomy (Soil Survey Staff 2014) it corresponds to Na-rich Aridisols and Mollisols. Solonetz soils that are compatible with the definitions of the WRB and Soil Taxonomy are not widespread in Turkey primarily due to their high electrical conductivity (EC) and calcium (Ca) contents and lack of the Natric horizon. This can be attributed to the lower precipitation and higher evaporation of the relevant areas in Turkey compared to the Solonetz zones (ecosystems) in Europe. Thus, soils with high EC and Na contents in Turkey are defined as the Solonetzic soils. However, due to the similarities of the Solonetzic soils of Turkey to the Solonchaks, the title of this chapter was designated as Solonetzic soils–Solonchaks (Solonetz-like soils).

2 Parent Materials and Environment

The Solonetzic soils overlie unconsolidated materials, mostly fine-textured and lacustrine sediments. The Na in the parent materials of the Solonetzic soils is mainly derived from the Na-rich feldspars of volcanic origin. The majority of the volcanic sources are located on the basins and plateaus of central Turkey. These are the Karadağ* (2288 m), Hasandağ (3269 m), Melendiz (2963 m), and Erciyes (3917 m) (*Dağ: mount/mountain in Turkish) (Şen et al. 2004; Kürschner and Parolly 2012).

Volcanic activity has occurred along a huge fault crossing the Great Konya Basin from the Karadağ Massif in the center toward the town of Bor, where this chain of volcanoes is associated with many forms of soil salinization. Many of the volcanoes have saline crater lakes and their high salt content is due to the accumulation of salt leached out from the surrounding volcanic ash and pumice soils. In particular, near the Karapınar town of the Konya Province, this is clearly seen along the borders of the volcanic lakes (Driessen 1970). There are Miocene volcanoes (Westerveld 1957) of andesitic and basaltic materials which are located along a fault traversing the Konya Basin from the south to the west toward the Niğde Province and Bor town. Freshwater comes out from the Mio-Pliocene limestones along with faults and structural terraces through the fringes of the plain (Driessen 1970).

Sodium was particularly supplied from these sites to the Solonetzic soils overlying old lake sediments (seasonally dry) of the lakes of Akgöl in Ereğli (Fig. 1) and Hotamış in Karapınar. The soils in the depression zones of Aksaray, Çumra, Niğde–Bor and Develi towns (central Turkey), and the Sultansazlığı wetland of the Kayseri Province (Fig. 2) as well as the dried old sediments of Burdur Province (south west of Turkey) and Lake Van (east of Turkey) were also enriched by volcanic Na sources.

3 Regional Distributions

The Solonetzic soils cover about 8641 ha in Turkey (Sönmez 2004) and are mainly formed locally in the Great Konya Basin (especially in Çumra, Karapınar and Bor) (Driessen

M.A. Çullu (✉)
Department of Soil Science and Plant Nutrition, Harran University, Şanlıurfa, Turkey
e-mail: macullu@harran.edu.tr

H. Günal
Department of Soil Science and Plant Nutrition, Gaziosmanpaşa University, Tokat, Turkey
e-mail: hikmet.gunal@gop.edu.tr

© Springer International Publishing AG 2018
S. Kapur et al. (eds.), *The Soils of Turkey*, World Soils Book Series,
https://doi.org/10.1007/978-3-319-64392-2_23

Fig. 1 Sodium-rich soils in Ereğli, central Turkey

Fig. 2 Volcanic landscape in Kayseri-Develi, central Turkey

1970; De Meester 1970; Bayramin et al. 2004; Budak 2012) (Figs. 3 and 4). Solonetzic soils occupy approximately 0.5% of the total salt-affected soils of Turkey and are most closely associated with Solonchaks. They are also present in Adana (Dinç et al. 1990); Iğdır (Oruç 1970); Harran (Çullu et al. 2010a), Malya (Munsuz 1969; Çullu et al. 1995); Erzurum (Akgül and Şimşek 1996), and Erzincan (Bahtiyar 1971).

In the Great Konya Basin, both salinity and alkalinity are mainly caused by Na salts. These soils are rich in Ca as well. Hence, severely alkali-affected soils are also found in the lower central part of the Basin and are difficult to trace in the field. In much of the Basin, in particular its center, alkalization is almost entirely governed by the Na compounds dissolved in the groundwater (Driessen 1970).

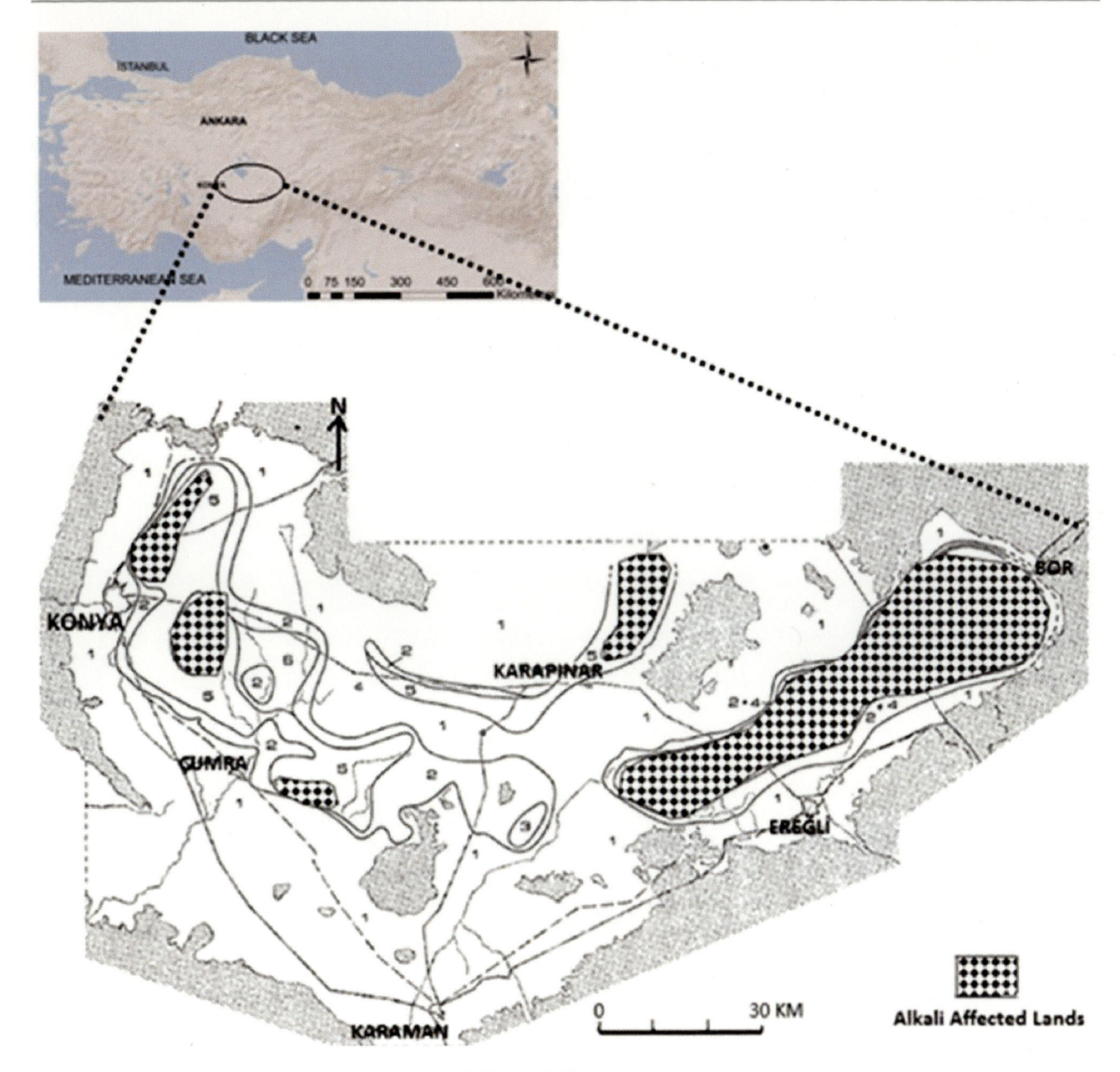

Fig. 3 Alkalinity sketch map of the Great Konya Basin (Driessen 1970)

Alkali soils may originate in different ways. In the Great Konya Basin, they are usually formed by leaching of saline soils that are rich in Na. The second cause for the formation of alkali soils is the use of Na-rich irrigation waters. This process has little significance for the Çumra region (central Turkey) due to the low alkalinity hazard of the Çarsamba irrigation water (Driessen and de Meester 1969; Driessen 1970).

4 Profile Development

Solonetzic soils have a Btn horizon (natric horizon) with a brown surface soil within 100 cm of the soil surface. One of the gypsic, calcic, salic, or cambic horizons may be present below the natric horizon together with high levels of exchangeable Na with a disturbed surface soil (Figs. 9 and 10).

Fig. 4 Solonetzic soils–Solonchaks from Bor town of the Niğde Province in the Great Konya basin of central Turkey

The Solonetzic soils of Turkey contain high exchangeable Na percentages (ESP) and EC levels. Despite the appreciable amounts of the exchangeable Na in the exchangeable complex, their high EC levels are related to their high exchangeable Ca contents which are due to the low leaching rates compared to the more leached Solonetz of the Solonetzic environments with higher rainfall and lower evaporation. The exchangeable Ca originates from their calcareous parent materials.

The appropriate examples for the Solonetzic soil–Solonchaks (Solonetz-like soils) with high EC and ESP in the country are present in the Harran (Akçakale) Plain (Table 1; Figs. 5 and 6), in the Akgöl depression of the Great Konya Basin (Table 2; Fig. 7) and in the similar soils of the Erzincan area (one of the main earthquakes and graben valley zones of the country) (Bahtiyar 1971) (Table 3). The Solonetz-like soils of the Erzincan area were formed in depression zones and on the Karasu River sediments with slightly lower ECs than the relevant soils in Akçakale–Harran and Akgöl depressions, but with similar ESP to the latter (Bahtiyar 1971).

The lack of the Natric horizons alongside their high EC (slightly lower in the Erzincan Solonetz-like soils), ESP and pH (slightly lower in the Akçakale Solonetz-like soils) levels of the Turkish Solonetzic soils leads to their designation as the Solonetzic soils–Solonchaks. These soils were classified as Gypsic Sodic Solonchaks, and Calcaric Sodic Solonchaks according to the IUSS Working Group WRB (2015) and Sodic Calcixerepts, Typic Natrargids, and Sodic Hydraquents according to the Soil Taxonomy (Soil Survey Staff 2014), respectively, by Mermut et al. (2006), Bayramin et al. (2004), and Bahtiyar (1971) (Tables 1, 2, and 3).

Similar to the Akçakale soil, Budak (2012) described Solonetzic–Solonchaks (Solonetz-like soils) in the Niğde--Bor depression of central Turkey that formed in a lake environment that was dried recently for gaining arable land (Table 4; Figs. 9 and 10).

5 Management

Solonetzic soils–Solonchaks have limited crop production potentials. The relatively high Na concentrations and pH levels restrict the growth and yield potential of most cultivated crops and even of a number of forage crops. High Na concentration causes clays to disperse and breaks down aggregation, which may lead to surface sealing and slow water infiltration with lower root penetration. Farmers with sodic soils are faced with decisions on how to best manage their land. Difficulties in the use of Solonetzic soils can be

Fig. 5 Solonetzic soil–Solonchak (Solonetz-like Solonchaks) Profile of the Akçakale, Harran Plain southeast Turkey

Fig. 6 Solonetzic soil–Solonchaks (Solonetz-like Solonchaks) in the Akçakale region, Harran Plain, southeast Turkey

Fig. 7 Solonetzic soil–Solonchak (Solonetz-like soils) landscape in Akgöl, Great Konya Basin, central Turkey (Bayramin et al. 2004)

Fig. 8 Land surface of Solonetzic soil–Solonchak (Solonetz-like Solonchaks) in Bor town of the Niğde Province, the closed Konya Basin, central Turkey

Fig. 9 Solonetzic soil–Solonchak (Solonetz-like Solonchaks) of Bor town of the Niğde Province, the closed Konya Basin, central Turkey (Budak 2012)

controlled and corrected by the proper use of appropriate soil or irrigation water amendments. Leaching alone is not sufficient to correct high Na problems. Two types of amendments can be used for Na control, namely, the Ca-containing amendments and acids or acid-based amendments (Thompson and Walworth 2011). Thus, in preparing for cultivation these soils can be treated with gypsum, elemental sulfur or sulfuric acid depending on soil pH and lime content, and leached, deeply plowed with an appropriate drainage infrastructure. Cultivation of Solonetzic soils increases the ability of mulching and infiltration rate of soils and helps in removing salts from the surface to deeper layers of soil profiles (Sadiq et al. 2007).

Being a cheap, abundant, nontoxic, and readily available source of calcium, gypsum is commonly used in the reclamation of Solonetzic soils of Turkey. Gypsum application rates vary depending on soil and irrigation water properties. Bahtiyar (1971) suggested the use of 40 tons/ha of gypsum seeking 258 cm desodication for a 90-cm-deep sodic soil in Erzincan (east Turkey), whereas lower amounts of gypsum (16 ton/ha) but higher amounts of water (360 cm) were recommended for the desodication by Yılmaz (1978) for the Burdur Yazıköy soils (southwest Turkey).

5.1 Management of the Bor Soils in Central Turkey

Materials, such as manure and other organic amendments, may play an important role in improving the porosity and in turn the rate of water percolation and penetration of roots in saline and sodic soils. Application of manure on the soils of Bor town in central Turkey was successful in improving salinity and sodicity-related impeded water percolation and root penetration (Budak 2012) (Fig. 11). Sulfur (10 ton/ha) and organic waste applications were recommended for the reclamation of the saline sodic soils in Iğdır (eastern Turkey) by Anapalı et al. (1996). Other applications of sulfuric acid or acid-based amendments (such as elemental sulfur) were conducted to correct Na problems in the calcareous Solonetzic soils of Bor by keeping in mind the soil lime content before using the amendments. This is crucially important for the

Fig. 10 Solonetzic soil–Solonchak (Solonetz-like Solonchaks) of Bor town of the Niğde Province, the closed Konya Basin, central Turkey (Budak 2012)

ancient lake environments of central Turkey where the soil pH is higher than 9.0 and the lime content is over 15%. In such soil conditions the addition of gypsum may not improve soil structure, due to the formation of $CaCO_3$. Consequently, elemental sulfur, sulfuric acid or acid base amendments should be and are commonly used to dissolve soil carbonates and provide Ca for reclamation (Fig. 12).

5.2 Management of the Akçakale Soils in Southeast Turkey

The Ministry of Food, Agriculture and Livestock of Turkey has recently (2012) constructed a drainage infrastructure for the reclamation of salinity and leaching of the high Na in the Akçakale soils of the Harran Plain (Fig. 13).

The high permeability of the Akçakale soils in the Harran Plain causes the rapid movement of the salts in the solum despite the dominant smectite followed by palygorskite, kaolinite, and illite minerals (Aydemir 2001; Çakmaklı 2008). In this context, the contents of palygorskite and calcite together are of prime significance in contributing to a higher permeability in the clayey soils of the area. These are the microstructural units developed in the Kap Calcisols (see Calcisols section) located in the northern parts of the Harran plain and transported to Akçakale. Subsequently, the Ca saturated soils, especially the clay size particles in the widespread calcite-rich silt and sand-sized microaggregates of palygorskite and smectite, keep the interlayer pore spaces at a minimum and interparticle pore spaces at a maximum state. Moreover, the high gypsum (of lacustrine origin) contents underlying the soils of Akçakale is the other advantage for the amelioration of the Akçakale Solonetzic environment/ecosystem by supplying the extra Ca needed for the enhanced water percolation (Çullu et al. 2010b). The Na in this system will be leached, following the amelioration provided via the drainage infrastructure. And this will be coupled by the rapid water permeability caused by the Ca saturated clay mineral surfaces of the well-developed silt and sand-sized microaggregates of the Akçakale soil. After amelioration, the Solonetzic soils of Akçakale may be initially planted for irrigated cotton and grains (wheat and barley). Otherwise, in many cases, these soils should be left in their natural states, and utilized for carefully managed livestock grazing similar to the graze lands in central Turkey (Fig. 14).

Table 1 Site description and some physical and chemical properties of a Solonetzic soil–Solonchak (Solonetz-like soils) in Akçakale (southeast Turkey) (Mermut et al. 2006)

IUSS Working Group WRB (2015)	Gypsic Sodic Solonchaks
Soil Survey Staff (2014)	Sodic Calcixerepts
Location	Şanlıurfa, Harran plain, Akçakale District, southeast Turkey
Coordinates	E38° 58 ′43″ W36° 42′ 54″
Elevation	350 m
Position	Flat
Slope	1%
Drainage class	Poorly drained
Groundwater	90 cm
Eff. soil depth	80 cm
Parent material	Lacustrine deposits (calcareous)
Geomorphic unit	Lake deposits
Climate	Semiarid with high evaporation in the hot season
Land use	Natural vegetation, halophytes
Stoniness	No stones
Soil moisture regime	Xeric

Horizon	Color (moist)	Depth (cm)	Saturation paste pH	EC (dS m^{-1})	CEC (cmol$_c$ kg^{-1})	Exc. Na (cmol$_c$ kg^{-1})	ESP (%)
A	5YR 5/3	0–5	8.21	78.0	42.1	24.2	57.48
Bw$_1$z	5YR 5/4	5–21	8.10	44.5	36.3	21.3	58.67
Bw$_2$z	5YR 5/4	21–48	7.30	19.6	34.0	22.8	70.80
Cy	7.5YR 7/4	48–75	8.12	6.20	32.2	22.7	66.76

Horizon	Particle size distribution (%)			Texture class	Org. mat. (%)	CaCO$_3$ (%)
	Sand	Silt	Clay			
A	1.4	39.1	59.5	C	1.5	20.50
Bw$_1$z	0.8	32.5	66.7	C	1.2	19.75
Bw$_2$z	2.9	28.4	68.7	C	0.9	32.80
Cy	46.6	31.4	22.0	SIL	1.1	32.80

Table 2 Site description of some physical and chemical properties of a Solonetzic soil–Solonchaks (Solonetz-like soils) in the Akgöl depression of the Great Konya Basin, central Turkey (Bayramin et al. 2004)

IUSS Working Group WRB (2015)	Calcaric Sodic Solonchaks
Soil Survey Staff (2014)	Typic Natrargids
Location	Akgöl Depression, Great Konya Basin, central Turkey
Coordinates	38° 57′ 01.22″N, 31° 52′ 52.39″E
Elevation	1000 m
Climate	Arid
Vegetation	Natural vegetation
Parent Material	Clayey Marl, Lacustrine sediment
Geomorphic Unit	Lake deposits
Topography	Flat
Erosion	Flood-related water erosion
Water table	<1 m
Drainage	Low
Infiltration	Rapid
Land use	Natural vegetation, halophytes
Stoniness	No stones
Soil moisture regime	Aridic

Horizon	Depth (cm)	Saturation paste pH	EC (dS m^{-1})	CEC cmol$_c$ kg^{-1}	ESP (%)
A$_1$	0–15	8.1	0.5	56.3	0.7
A$_2$	15–34	8.7	4.3	55.5	18.6
Bw$_1$z	34–50	8.7	8.8	49.4	31.3
Bw$_2$z	50–90	8.4	8.9	56.6	32.7
C	90+	8.3	8.5	42.0	39.5

Horizon	Particle size distribution Sand	Silt	Clay	Texture class	Org. mat. (%)	CaCO$_3$(%)
A$_1$	33.0	26.8	40.2	CL	2.3	49.3
A$_2$	20.5	20.4	59.1	C	0.9	48.5
Bw$_1$z	12.6	22.4	65.0	C	1.4	50.3
Bw$_2$z	4.7	24.1	71.2	C	1.1	49.7
C	2.8	23.0	74.2	C	0.5	52.8

Table 3 Site description and some physical and chemical properties of a Solonetzic soil–Solonchak (Solonetz-like soils) in Erzincan, east Turkey (Bahtiyar 1971)

IUSS Working Group WRB (2015)	Calcaric Sodic Solonchaks
Soil Survey Staff (2014)	Sodic Hydraquents
Location	Erzincan Plain, east Turkey
Coordinates	39° 41′ 29.70″N, 39° 34′ 50.37″E
Elevation	1151 m
Climate	Semiarid
Vegetation	Natural vegetation
Parent material	River sediment
Geomorphic unit	River bank
Topography	Flat
Erosion	Flood-related water erosion
Water Table	<50 cm
Drainage	Low
Infiltration	Rapid
Stoniness	No stones
Soil moisture regime	Xeric

Horizon	Depth (cm)	Color (moist)	Saturation Paste pH	EC dS^{-1}	CEC (cmol$_c$ kg^{-1})	Exc. Na (Cmol$_c$ kg^{-1})	ESP (%)
A$_1$	0–30	10 YR 7/2	9.12	3.42	21.39	4.14	19.35
A$_2$	30–60	10 YR 7/2	8.60	5.07	19.19	4.28	21.41
Cgz	60–90	10 YR 7/2	8.54	9.48	25.13	5.54	22.04

Horizon	Particle size distribution			Texture class	OM (%)	CaCO$_3$(%)
	Sand (%)	Silt (%)	Clay (%)			
A$_1$	14.6	33.7	51.7	C	0.71	22.35
A$_2$	24.9	42.0	33.1	CL	0.45	18.26
Cgz	8.0	38.7	53.4	C	0.36	20.75

Table 4 Site-profile description, some physical, and chemical properties of a Solonetzic soil–Solonchak (Solonetz-like soils) in the Niğde Province, central Turkey (Budak 2012)

IUSS Working Group WRB (2015)	Calcaric Sodic Solonchaks
Soil Survey Staff (2014)	Petrocalcic Calcixerepts
Location	16 km northwest of Bor town of Niğde Province, central Turkey
Coordinates	37° 46′ 30″N–34° 20′ 24″E
Elevation	1160 m
Climate	Semiarid
Vegetation	Major plant cover is composed of Taraxacum *farinosum*, Panderia *pilosa*, Onopordum *davisii*, and Halimione *verrucifera* (Figs. 8 and 9)
Parent material	Shallow marine deposits
Topography	Flat
Erosion	Wind erosion
Water table	No water table
Drainage	Low
Infiltration	Low
Stoniness	No stones
Soil moisture regime	Xeric

Horizon	Depth (cm)	Description
A	0–17	10YR 6/3 (dry), dull yellow orange, 10YR 4/3 (moist) dull yellowish brown, clay, moderate medium subangular blocky structure, hard when dry, sticky, and plastic when wet, strongly effervescent, common hairy roots, clear smooth boundary
Btn	17–44	10YR 6/4 (dry) dull yellow orange, 10YR 5/3 (moist) dull yellowish brown, clay, coarse strong prismatic/columnar structure, very hard when dry, very sticky and plastic when wet, strongly effervescent, rare hairy roots, moderately few vesicular pores, many clay coatings 5YR 5/6 (dry) bright reddish brown, 5YR 5/4 (moist) dull reddish brown, clear wavy boundary
Btk	44–68	10YR 6/2 (dry) grayish yellow brown, 10YR 7/2 (moist) dull yellow orange, clay, coarse strong prismatic structure, very hard when dry, very sticky and plastic when wet, strongly effervescent, rare hairy roots, common vesicular pores, common clay coatings, calcium carbonate coatings on ped surfaces, clear smooth boundary
Ck	68–98	10YR 8/2 (dry) light gray, 10YR 7/2 (moist) dull yellow orange, clay, fine strong platy structure, very hard when dry, strongly effervescent, very rare hairy roots, common vesicular pores, rare reddish clay coatings, moderately few vesicular pores, clear smooth boundary.
Ckm	98–142+	7.5Y 8/1 (dry and wet) light gray, massive, sandy loam, very hard, violently effervescent

Horizon	Depth (cm)	Color (moist)	Saturation paste pH	EC dS m^{-1}	CEC cmol$_c$kg^{-1}	Exc. Na	ESP (%)
A	0–17	10YR 4/3	8.23	3.08	19.3	3.72	10.51
Btn	17–44	10YR 5/3	8.73	10.52	41.0	15.51	37.49
Btk	44–68	10YR 7/2	8.69	7.64	29.6	15.88	40.51
Ck	68–98	10YR 7/2	8.54	4.81	18.1	5.19	19.11
Ckm	98–142	10Y 8/1	8.37	2.31	15.0	4.45	16.07

Horizon	Sand %	Silt %	Clay %	Texture class	Org. mat. (%)	CaCO$_3$ (%)
A	13.5	21.5	65.1	C	2.36	35.22
Btn	16.0	13.0	71.1	C	0.97	43.57
Btk	15.2	11.7	73.2	C	0.41	42.85
Ck	43.0	18.4	38.6	CL	0.56	34.86
Ckm	69.8	12.8	17.5	SL	0.19	60.28

Fig. 11 Application of manure on the highly saline and sodic Solonetzic soils of Bor town in the Niğde Province, the closed Konya Basin, central Turkey

Fig. 12 Application of elemental sulfur on the highly saline and sodic Solonetzic soils of Bor town in the Niğde Province, the closed Konya Basin, central Turkey

Fig. 13 Infrastructure construction for reclamation of the strongly salt-affected Akçakale soils in the Harran Plain. Application and implementation of mole drainage (**a, b**)

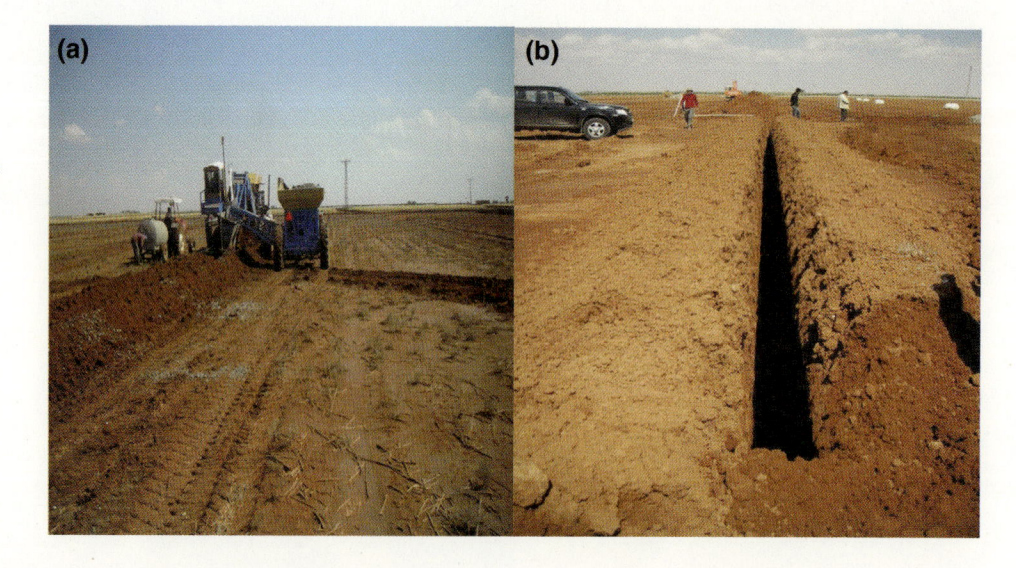

Fig. 14 The grazelands in central Turkey on sodium-rich Solonetzic soils–Solonchaks (Solonetz-like soils)

References

Akgül M, Şimşek G (1996) Soils of the Daphan Plain. Detailed soil survey report. Atatürk Univ, J Agric Fac 27(1):74–88 (in Turkish)

Anapalı Ö, Hanay A, Canpolat M (1996) Effect of organic waste material in the period after the reclamation of saline-sodic soils. Atatürk Univ, J Agric Fac 27(1):13–30 (in Turkish)

Aydemir S (2001) Properties of palygorskite-influenced Vertisols and vertic-like soils in the Harran Plain of Southeastern Turkey. Dissertation, Texas A&M University, College Station, TX, USA

Bahtiyar M (1971) A study on the formation of Erzincan Ada soils, characteristics and their reclamation, Atatürk University, Department of Soil Science, Annuals of Faculty of Agriculture, pp 95–98 (in Turkish)

Bayramin İ, Yalçın OZ, Tunçay T, Samray HN (2004) Remediation of the salt affected soils and their economic value, an example from Ayrancı-Karaman, international soil congress on natural resource management sustainable development, 7–10 June 2004. Erzurum, Turkey

Budak M (2012) Genesis and classification of saline alkaline soils and mapping with both classical and geostatistical techniques (PhD Dissertation) Gaziosmanpasa University Institute of Science, Tokat, Turkey, 254 p

Çakmaklı M (2008) Genesis of the soils in the Harran plain, PhD thesis, Harran University, Faculty of Agriculture, Department of Soil Science, Şanlıurfa, Turkey

Çullu MA, Dinç U, Şenol S, Öztürk N, Çelik I, Günal H (1995) Mapping of the saline and alkaline areas by satellite imagery. İlhan Akalan Soil Environ Symp 1:163–172 (in Turkish)

Çullu MA, Aydemir S, Qadir M, Almaca A, Öztürkmen AR, Bilgiç A, Ağca N (2010a) Implication of groundwater fluctuation on the seasonal salt dynamic in the Harran Plain, South-Eastern Turkey. Irrig Drainage 59(4):465–476

Çullu MA, Aydemir S, Almaca A, Öztürkmen AR, Sönmez O, Binici T, Bilgili AV, Yılmaz G, Dikilitaş M, Karakaş S, Sakin E, Şahin Y, Aydoğdu M, Aydemir A, Çeliker M (2010b) Saline soils of the Harran Plain and the estimation of the effect on salinity on yield. Project report of the prime ministry GAP administration, Şanlıurfa 210 p (in Turkish)

De Meester T (1970) Soils of the Great Konya Basin, Turkey. Agricultural research reports 740. Centre for Agricultural Publishing and Documentation, Wageningen, 302 p

Dinç U, Sarı M, Şenol S, Kapur S, Sayın M, Derici MR. Çavuşgil V, Gök M, Aydın M, Ekinci H, Ağca N, Schlichting E (1990) Soils of Çukurova Region, Çukurova University, Agriculture Faculty, No: 26. Adana. Turkey, 171 p (in Turkish)

Driessen PM, de Meester T (1969) Soils of the Çumra area, Turkey. Agricultural Research Reports No. 720, Wageningen, 117 p

Driessen PM (1970) Soil salinity and alkalinity in the Great Konya Basin, Turkey. Department of Tropical Soil Science, Agricultural University, Wageningen. ISBN 9022003086, 110 p

IUSS Working Group WRB (2015) World Reference Base for Soil Resources 2014, update 2015 International soil classification system for naming soils and creating legends for soil maps. World Soil Resources Reports No. 106, FAO, Rome. 203 p

Kürschner H, Parolly G (2012) The central anatolian steppe. Eurasian steppes. Ecological problems and livelihoods in a changing World. Springer, Netherlands, pp 149–171

Mermut AR, Çullu MA, Aydemir S, Karakaş S (2006) Excursion book. 18. International soil meeting (ISM) Soils sustaining life on earth (Managing Soil and Technology) 22–26 May 2006. ISBN:975-96629-4-9

Munsuz N (1969) The barren soils of the malya state farm and their amelioration. Ankara Univ, Ann Fac Agric, No. 336-11, Ankara. Turkey. 80 p (in Turkish)

Oruç N (1970) Physical and chemical properties of the barren soils of the Iğdır Plain. Atatürk Univ, Ann Fac Agric 80-27-8:48–59 (in Turkish)

Sadiq M, Hassan G, Mehdi SM, Hussain N, Jamil M (2007) Amelioration of saline-sodic soils with tillage implements and sulfuric acid application. Pedosphere 17(2):182–190

Soil Survey Staff (2014) Keys to soil taxonomy, 12th ed. USDA-Natural Resources Conservation Service, Washington, DC, USA, 372 p

Sönmez B (2004) Barren land reclamation research in Turkey and saline soils management. Salinity management in irrigated land areas. Symposium Proceedings. 20–21 May, Ankara, pp 157–162

Şen PA, Temel A, Gourgaud A (2004) Petrogenetic modelling of quaternary post-collisional volcanism: a case study of central and Eastern Anatolia. Geol Mag 141(1):81–98

Thompson TL, Walworth J (2011) Salinity management and soil amendments for southwestern pecan orchards. http://extension.arizona.edu/sites/extension.arizona.edu/files/pubs/az1411.pdf. Last accessed on 9 Nov 2015

Yılmaz T (1978) Determination of gypsum and desodification water amount for Reclaiming Yazıköy Burdur saline-sodic and boron-rich soils. Konya Soil-Water Res. Inst. No. 57. Konya (in Turkish)

Westerveld J (1957) Phases of neogene and quaternary volcanism in Asia minor. Proceedings of the 20th International Geological Congrress I, vol 1

Andosols/Andosol-Like Soils

Mahmut Dingil, Muhsin Eren, Selahattin Kadir, and Alhan Sarıyev

1 Introduction

Andosols are fertile soils due to their low bulk density, high water- and phosphorus- holding capacities and Al + 1/2Fe contents. These soils derived from volcanic materials are also rich in mineral and volcanic glass contents (Shoji et al. 1994). Many Andosols of dark color have a low chroma value of 2 when dry and 1 when wet, whereas some may have high chroma values of 3 or more when dry or wet (Soil Survey Staff 2014). The IUSS Working Group WRB (2015) defines Andosols as soils with a vitric or an andic horizon occurrence within 25 cm from the soil surface. The Andosols/Andisols of Turkey lack some of the requirements of the Soil Taxonomy (Soil Survey Staff 2014) and the IUSS Working Group WRB (2015) and thus need further research on widespread volcanic areas of the country.

2 Parent Materials and Environment

The volcanic parent materials of the Andosol/Andosol-like soils in Turkey are generally, tephra/tuff-scoria and andesite and basalts (Fig. 1). Volcanic soils are mostly widespread around volcanoes and under xeric and aridic soil moisture regimes in Turkey. But most of the soils overlying volcanic materials are not Andosols in the country and do not possess the andic or vitric properties as stated by the WRB (IUSS Working Group 2015). Consequently, in this context, the Turkish volcanic soils were rather designated and explained in this chapter as the Andosol-like soils. However, some soils overlying basaltic tephra/in the south of the country (Delihalil-Osmaniye) (Kapur et al. 1980) and andesitic tuff in central Turkey were designated as Andisols by the Soil Taxonomy of the Soil Survey Staff (2014) in respect to some of the requirements of this system, whereas they were ruled out by the WRB system (IUSS Working Group 2015) for Andosols. Other volcanic soils of Mount Erenler from Konya (central Turkey) formed on andesites were tested for Andisol formation by Özcan and Özaytekin (2011), but did not meet the clay mineral content requirements of the Soil Taxonomy (2010). The non-crystalline minerals such as allophane and imogilite were not formed in these soils because of the low rate of weathering, inadequate Si leaching and the long dry season (Özcan and Özaytekin 2011).

3 Profile Development

In general, volcanic soils in Turkey cannot meet the requirements of the Andosols/Andisols due to their relatively higher bulk densities and parent materials (tephra) in both the IUSS Working Group WRB (2015) and the Soil Taxonomy (Soil Survey Staff 2014) systems. However, the Andosol-like soils of Turkey meet the phosphate retention, Al + ½ Fe, volcanic glass contents and the amounts of the fine-earth fractions of both classification systems (Tables 1–5). An appropriate example for such soils is the Andosol-like soil in Osmaniye (Delihalil area). This soil has a sandy loamy (SL) texture at the surface and loam (L) in the sub-soil, a dark colour throughout the pedon with chromas of 2 when dry and 1 when wet from the topsoil to 35 cm (till Bw2). The horizons of the Delihalil soil are well developed (Fig. 2); (Table 1) and contain 5% or more volcanic glass which is an important evidence for the occurrence of an Andosol/Andisol (Figs. 2 and 3).

M. Dingil (✉)
School of Technical Sciences,
University of Çukurova, Adana, Turkey
e-mail: mdingil@cu.edu.tr

M. Eren
Department of Geological Engineering,
Mersin University, Mersin, Turkey

S. Kadir
Department of Geological Engineering,
Eskişehir Osmangazi University, Eskişehir, Turkey

A. Sarıyev
Department of Soil Science and Plant Nutrition,
University of Çukurova, Adana, Turkey

© Springer International Publishing AG 2018
S. Kapur et al. (eds.), *The Soils of Turkey*, World Soils Book Series,
https://doi.org/10.1007/978-3-319-64392-2_24

Fig. 1 An Andosol profile developed on tuffeous material in Pınarbaşı, Kayseri central Turkey. *Arrow* shows calcified erosional surface between tuff and Andosol

Despite the widespread volcanic soil distribution in Turkey, the scarcity of the Andosols based on the absence of its smearing property/high bulk density and high activity and high surface area clay minerals in the clay size fraction calls for some reconsideration upon their definition regarding their compatible chemical properties, glass fragment (5–30%) and fine mineral fraction contents to Soil Taxonomy (Soil Survey Staff 2014) and the IUSS Working Group (WRB 2015). Moreover, regarding the clay mineral contents of the Andosol-like soils of Turkey, there are few sites that contain allophane such as the Pınarbaşı, Kayseri Andosol-like soil in central Turkey (Fig. 1; Table 2), whereas the others contain moderate to low amounts of halloysite and illite (Figs. 3, 4 and 5).

The development of the bubble casts and the initial dissolution of the volcanic glass particles during cooling tephra together with the subsequent devitrification process may also reflect the initial to moderate (Fig. 4c) (Kadir et al. 2013) Andosol forming environment in the Pınarbaşı-Kayseri and Delihalil-Osmaniye areas. The moderate amounts of halloysite and illite contents present in these soils are also the indicators of the initial Andosol/Andosol-like soil forming environment (Dingil 2003) (Kadir and Karakaş 2002). Moreover, an amendment to the requirements on the Andosol/Andisol parent materials should also be made to include basaltic rocks as they are the widespread underlying parent materials of the volcanic soils and/or Andosol-like soils in the near east.

Fig. 2 An Andosol-like soil from a tilted part of the Mount Delihalil in Osmaniye, south Turkey (Dingil 2003)

Table 1 Site-profile description and some physical and chemical properties of an Andosol-like soil in Delihalil-Osmaniye, south Turkey (Dingil 2003)

Soil Survey Staff (2014)	Humic Vitrixerand
IUSS Working Group WRB (2015)	Andic Cambisol
Location	Between highway of E90 and E91, Delihalil-Osmaniye, south Turkey
Coordinates	37° 00′ 49″N, 36° 03′ 57″E
Altitude	331 m
Landform	Medium gradient hill
Position	Lower slope
Aspect	The southern slope
Slope	2–6%
Drainage class	Well drained
Groundwater	–
Eff. soil depth	82 cm
Parent material	Basalt
Climate	Mediterranean
Land use	Natural
Vegetation	Grassland

Horizon	Depth (cm)	Description
A1	0–11	Dark greyish brown (10YR 4/2) dry, very dark grey (7.5YR 3/1) moist, sandy loam, medium weak granular, loose when dry, very friable when wet, slightly sticky non- plastic when wet, 3–5 cm abundant tuff fragments, abundant fine roots, clear smooth boundary
AB	11–24	Dark greyish brown (10YR 4/2) dry, very dark grey (7.5YR 3/1) moist, sandy loam, medium weak granular, firm when dry, very friable when wet, slightly sticky non- plastic when wet, 3–5 cm abundant tuff fragments, few fine roots, gradual smooth boundary
Bw1	24–35	Dark greyish brown (10YR 4/2) dry, very dark grey (7.5YR 3/1) moist, loam, weak medium subangular blocky, loose when dry, friable when wet, sticky and plastic when wet, 3–5 cm abundant tuff fragments, few fine roots, gradual smooth boundary
Bw2	35–63	Dark greyish brown (10YR 4/2) dry, very dark brown (10YR 2/2) moist, loam, weak medium subangular blocky, loose when dry, friable when wet, sticky and plastic when wet, 3–5 cm abundant tuff fragments, few fine roots, gradual smooth boundary
Bw3	63–82	Brown (10YR 4/3) dry, very dark greyish brown (10YR 3/2) moist, loam, weak medium subangular blocky, loose when dry, friable when wet, sticky and plastic when wet, 3–5 cm abundant tuff fragments, few fine roots

Horizon	Depth (cm)	pH 0,01 M CaCl$_2$ (1:2)	BD (g cm^{-3}) (33 kPa)	BD (g cm^{-3}) (oven-dry)	Sand	Silt	Clay (%)	Texture
A1	0–11	5.79	1.15	0.97	52	37	11	SL
AB	11–24	6.41	1.35	1.15	57	30	13	SL
Bw1	24–35	6.52	1.27	1.02	47	38	15	L
Bw2	35–63	6.45	1.24	1.1	34	44	22	L
Bw3	63–82	6.32	–	–	42	42	16	L

Horizon	Org. matter (%)	CEC (cmol$_c$kg^{-1})	Exchangeable bases Na$^+$	K	Ca + Mg
A1	11.4	27.04	0.53	4.39	22.12
AB	9.3	33.04	0.65	4.43	27.97
Bw1	15.1	45.8	0.64	5.48	39.69
Bw2	16	46.86	0.74	5.04	41.08
Bw3	17.8	47.72	0.88	5.27	41.56

(continued)

Table 1 (continued)

Horizon	CaCOs (%)	P retention (%)	Volcanic glass (%)	Al + 1/2Fe (%)
A1	2.4	78	1	2.7
AB	2.6	79	8	2.63
Bw1	2.3	65	7	2.76
Bw2	2.2	55	8	2.65
Bw3	1.9	59	5	3.03

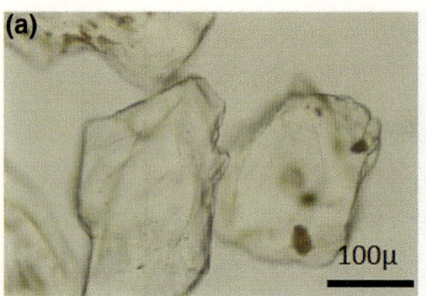

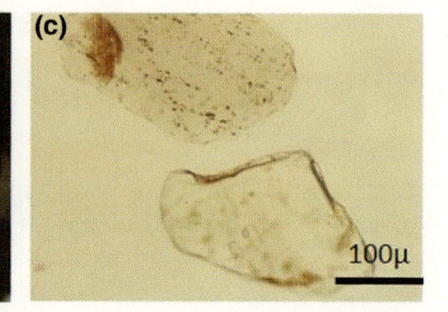

Fig. 3 Weathered/etched volcanic glass shards (ppl) in profiles of Andosol-like soils, **a** central Turkey (Bw), 3 **b** east Turkey (A2), 4 **c** east Turkey (Bw) (Dingil 2003)

Table 2 Site-profile description and some physical and chemical properties of an Andosol-like soil in Pınarbaşı-Kayseri, central Turkey (Dingil 2003)

Soil Survey Staff (2014)	Humic Vitrixerand
IUSS Working Group WRB (2015)	Andic Cambisol
Location	Kayseri Pınarbaşı highway, 17th km, 1 km to the left
Coordinates	38° 38′ 21, 7″N, 34° 41′ 10, 6″E
Altitude	1555 m
Landform	Medium gradient hill
Position	Lower slope
Aspect	Southeast–northwest
Slope	3–4%
Drainage class	Well drained
Groundwater	–
Eff. soil depth	70 cm
Parent material	Volcanic ash
Climate	Continental
Land use	Cultivated land
Vegetation	Cereal, harvested

Horizon	Depth (cm)	Description
Ap	0–26	Greyish brown (2.5YR 5/2) dry, very dark greyish brown (10YR 3/2) moist, sandy clay loam, weak fine granular, slightly hard when dry, firm when wet, sticky and plastic when wet, few stones, abundant tuff fragments, dense fine roots, clear wavy boundary
Bw	26–54	Dark greyish brown (10YR 4/2) dry, very dark greyish brown (10YR 3/2) moist, clay, medium subangular blocky, very hard when dry, very firm when wet, very stick very plastic when wet, medium biological activity, few fine roots, clear wavy boundary

(continued)

Table 2 (continued)

Horizon	Depth (cm)	Description
BC	54–66	Brown (10YR 5/3) dry, brown (10YR 4/3) moist, clay, massive, very hard when dry, very firm when wet, very stick very plastic when wet, clear wavy boundary
C1	66–98	Light grey (10 YR 7/2) dry, brown (10YR 5/3) moist, clayey loam, massive, hard when dry, firm when wet, sticky and plastic when wet, effervescence, 5–10 cm medium density obsidian fragments, gradual wavy boundary
C2	98–128	White (2.5Y 8/1) dry, light yellowish grey (2.5Y 6/3) moist, sandy loam, loose, friable when dry, very loose when wet, slightly sticky non-plastic, effervescent, 5–10 cm medium density obsidian fragments

Horizons	Depth (cm)	pH 0.01 M CaCl$_2$ (1:2)	Bulk density (g cm^{-3}) (33 kPa)	Bulk density (g cm^{-3}) (oven-dry)	Org. matter (%) (dry combustion at 400 °C)
Ap	0–26	7.27	–	–	8.70
Bw	26–54	6.84	1.60	1.21	13.10
BC	54–66	7.19	1.59	1.17	16.10
C1	66–98	7.58	1.53	1.16	12.00
C2	98–128	7.66	–	–	5.60

Horizons	Sand	Silt	Clay	Texture class	CEC cmol$_c$kg^{-1}	Na$^+$	K	Ca + Mg
	%				Extractable cations me/100 g			
Ap	47	19	34	SCL	25.76	0.11	1.12	24.37
Bw	34	17	49	C	41.89	0.18	1.13	40.37
BC	23	27	50	C	53.98	0.23	1.15	52.31
C1	37	32	31	CL	40.54	0.18	0.62	39.45
C2	52	39	9	SL	12.05	0.05	0.5	11.17

Horizons	CaCO$_3$ (%)	P retention (%)	Volcanic glass (%)	Al + 1/2Fe (%)
Ap	2.80	86	33	0.30
Bw	2.50	85	32	0.37
BC	4.70	82	32	0.42
C1	27.30	48	31	0.08
C2	2.90	91	7	0.10

4 The Andisol/Andosol Requirements of the Andosol-Like Soils of Turkey

The Delihalil soil meets the requirements of article 3 for Andisols except the bulk density requirement in article 2. Consequently, it is an Andisol according to the Soil Taxonomy (2014) (Tables 3 and 7).

There are three articles for andic properties in the IUSS Working Group (2015) system. The first and third articles are met except the second article (the bulk density) by the Delihalil-Osmaniye soil. Thus, this soil does not qualify for an Andosol (Tables 4 and 7).

The Pınarbaşı-Kayseri soil, central Turkey meets the requirement of article 3 in Soil Taxonomy (2014) for Andisol. But it's Al + 1/2Fe content as stated in article 3c is very close to the limit value (0.4%). Despite article 3c, the Pınarbaşı-Kayseri soil meets the general requirements of Soil Survey Staff (2014) and therefore it is an Andisol (Tables 5 and 7).

Article 3 in the IUSS Working Group (2015) is met by the Pınarbaşı-Kayseri soil, but not articles 1 and 2 to qualify for an Andosol (Tables 6 and 7).

5 Regional Distribution

Andosol-like volcanic soils are generally present in the mountainous areas of Turkey. These mountainous areas are mostly located in the east and center of Turkey, partly in the Aegean, Mediterranean and southeastern regions of the

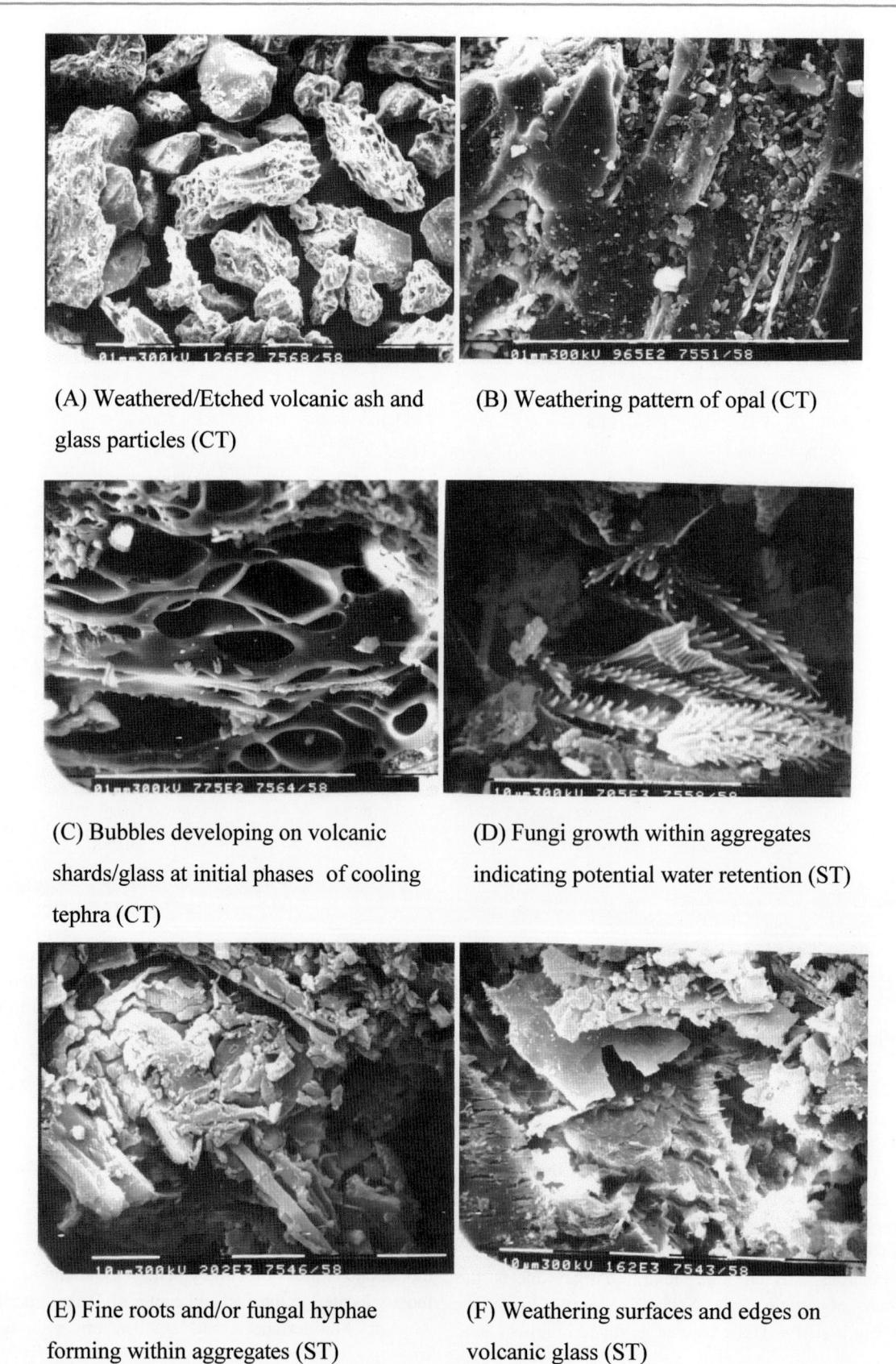

(A) Weathered/Etched volcanic ash and glass particles (CT)

(B) Weathering pattern of opal (CT)

(C) Bubbles developing on volcanic shards/glass at initial phases of cooling tephra (CT)

(D) Fungi growth within aggregates indicating potential water retention (ST)

(E) Fine roots and/or fungal hyphae forming within aggregates (ST)

(F) Weathering surfaces and edges on volcanic glass (ST)

Fig. 4 SEM images of Andosol-like soils from central (CT) and south Turkey (ST)

Fig. 5 Iron infills in quartz cavities from a Vitric Cambisol (Bw) in Van (East Turkey)

Table 3 The Andisol requirements of the Soil Survey Staff (2014) for the Delihalil soil, south Turkey

The Delihalil Soil	Requirements of the Soil Survey Staff (2014) for Andic Properties	+/−
1. All of the following	Less than 25% organic carbon (by weight) and one or both	+
2. All of the following	(a) Bulk density, measured at 33 kPa water retention, of or less, and	
	(b) Phosphate retention of 85% or more, and	+
	(c) Al + ½ Fe content (by ammonium oxalate) equal to 2.0 percent or more, or	+
3. All of the following	(a) 30% or more of the fine-earth fraction is 0.02 to size, and size, and	
	(b) Phosphate retention of 25% or more, and	+
	(c) Al + ½ Fe content (by ammonium oxalate) equal to 0.4 or more, and	+
	(d) Volcanic glass content of 5% or more, and	+
	(e) [(Al + ½ Fe content, percent) times (15.625)] + [volcanic glass content, percent] = 36.25 or more	+

+ requirements of the Soil Taxonomy (2014) met by the soil
− requirements of the Soil Taxonomy (2014) that are not met by the soil

Table 4 The Andosol requirements of the IUSS Working Group WRB (2015) for the Delihalil soil

Delihalil diagnostic criteria of the IUSS Working Group WRB (2015) for the Andic Properties	+/−
1. An Al_{ox} + ½Fe_{ox} value of $\geq 2\%$, and	+
2. A bulk density[17] of ≤ 0.9 kg dm^{-3}, and	−
3. A phosphate retention of $\geq 85\%$	+

+ requirements of the WRB (2014) met by the soil
− requirements of the WRB (2015) that are not met by the soil

Table 5 The Andisol requirements of the Soil Survey Staff (2014) for the Pınarbaşı-Kayseri soil, central Turkey

Kayseri	Requirements of the Soil Survey Staff (2014) for Andic Properties	+/−
1. All of the following	Less than 25% organic carbon (by weight) and one or both	+
2. All of the following	(a) Bulk density, measured at 33 kPa water retention, of 0.90 g/cm^3 or less, and	
	(b) Phosphate retention of 85% or more, and	+
	(c) Al + ½ Fe content (by ammonium oxalate) equal to 2.0% or more, or 9	−
3. All of the following	(a) 30% or more of the fine-earth fraction is 0.02–2.0 mm in size, and	+
	(b) Phosphate retention of 25% or more, and	+
	(c) Al + ½ Fe content (by ammonium oxalate) equal to 0.4% or more, and	+
	(d) Volcanic glass content of 5% or more, and	+
	(e) [(Al + ½ Fe content, percent) times (15.625)] + [volcanic glass content, percent] = 36.25 or more.	

+ requirements of the Soil Survey Staff (2014) met by the soil
− requirements of the Soil Soil Survey Staff (2014) that are not met

Table 6 The Andosol requirements of the WRB (IUSS Working Group WRB 2015) for the Pınarbaşı-Kayseri soils, central Turkey

Kayseri diagnostic criteria of the IUSS Working Group WRB (2015) for Andic properties	+/−
1. An Al$_{ox}$ + ½Fe$_{ox}$ value of $\geq$ 2%, and	−
2. A bulk density[17] of $\leq$ 0.9 kg dm^{-3}, and	−
3. A phosphate retention of $\geq$ 85%	+

+ requirements of the WRB (2014) met by the soil
− requirements of the WRB (2014) that are not met by the soil

Table 7 The final classification of the two volcanic soils in Turkey

Profile	Soil Survey Staff (2014)	IUSS Working Group WRB (2015)
Delihalil	Andisol	Not an Andosol
Kayseri	Andisol	Not an Andosol

country. The Andosols and the Andosol-like soils in Turkey are rare and unmappable. One of these rare sites located in the Van province is located in eastern Anatolia (Fig. 6).

6 Management

Conservation of the surface soil with its organic matter and preventing erosion is the precondition for farming on the Andosols and Andosol-like soils in Turkey. Andosol-like soils are generally not under intensive cultivation in the country. They are mainly covered by the natural vegetation of especially central and eastern Turkey and occur under sparse to dense shrub and grassland canopies (Fig. 7), and the maquis of the Mediterranean region which is rich in olive (*Olea europea*) and carob trees (*Ceratonia siliqua*). Olive orchards are recently extensively established on the basaltic Andosol-like soils in southern Turkey, where the main roots of the hardy olive trees have been observed to be highly appropriate to extend over the well-structured basaltic soils and

Fig. 6 Andosol-like soil in Van (east Turkey)

Fig. 7 Natural vegetation on the volcanic ashes and soils of the Delihalil-Osmaniye, south Turkey with patches of hydrothermal kaolinite (k) formation (Dingil 2003)

even in the outcropping underlying weathering basalts. The weathering basalts are occasionally covered with abundant hydrothermal kaolinite formations exposed after weathering followed by erosion. Similar areas rich in volcanic materials are used as rangelands for small ruminants and especially for goats in south and southeastern Turkey (Fig. 8).

Fig. 8 Widespread rangelands in volcanic areas in central Turkey

Acknowledgements The authors are indebted to Prof. Dr. A. R. Mermut for his invaluable efforts in the interpretation of the SEM images.

References

Dingil M (2003) Soil properties, genesis and classification of probable Andisol in Turkey. Ph.D. Thesis (unpublished), Çukurova University, Graduate School of Natural Sciences, Adana, 129 (in Turkish)

IUSS Working Group WRB (2015) World reference base for soil resources 2014, update 2015 International soil classification system for naming soils and creating legends for soil maps. World Soil Resources Reports No. 106. FAO, Rome. 203 p

Kadir S, Karakaş Z (2002) Mineralogy, chemistry and origin of halloysite, kaolinite and smectite from Miocene ignimbrites, Konya. Turk Neues Jahrbuch für Mineralogie Abhandlungen 177:113–132

Kadir S, Gürel A, Senem H, Külah T (2013) Geology of late Miocene clayey sediments and distribution of paleosols clay minerals in the northeastern part of the Cappadocian Volcanic Province (Araplı-Erdemli), central Anatolia, Turkey. Turk J Earth Sci 22:427–443

Kapur S, Dinç U, Göksu Y, Özbek H (1980) Genesis and classification of the ando-like soils overlying basaltic tephra in the Osmaniye Region (south Turkey). University of Çukurova, Annals of the Faculty of Agriculture. Pub. No. 11, pp 1–4, Adana (in Turkish)

Özcan S, Özaytekin H H (2011) Soil formation overlying volcanic materials at Mount Erenler, Konya, Turkey. Turk J Agric Forestry 35:545–562 © TÜBİTAK. doi:10.3906/tar-1102-2

Shoji S, Nanzyo M, Dahlgren RA (1994) Volcanic ash soils. Elsevier Science Publishers, Netherlands, pp 7–33

Soil Survey Staff (2014) Keys to soil taxonomy, 12th edn. DC, USA, USDA-Natural Resources Conservation Service, Washington DC, p 372

Management

Hakkı Emrah Erdoğan

1 Introduction

Due to Turkey's unique geographic location at the transition zone between Europe and Asia, it holds the various climatic regions of subtropical steppe, temperate rainy, Mediterranean and the cold snowy forest climate with dry summers. Together with a complex geologic history and topography, the soils of Turkey are diverse and vary on small scales. The main soil classification used in Turkey was prepared at a scale of 1:25,000 with digital maps available at regional and local scales. A soil profile database exists for the whole part of the country (GDRS 2005). The major soil threats in Turkey are soil erosion, salinization, and soil sealing (Günal et al. 2015). Soil and land resources of Turkey have been under increasing pressures, which are driven by the increasing population and demand for the resources and development policies. In this chapter, the potential and status of soil and land resources are described. The key driving forces and pressures on the state of the soils are reviewed by using the existing national soil database. The chapter also describes the key agriculturally related potential and existing pressures on soil and land resources. These include the physical and chemical properties of soils, land use, land capability classes and land management with related policies.

2 Physical, Chemical and Topographical Characteristics of Soils in Turkey

The physical, chemical and topographic characteristics of the soils and their distribution rates are shown in Figs. 1, 2, 3, 4, 5 and 6. Brown soils (Cambisols) are the dominant soils with 19.7% area coverage. The coastal zones and the eastern regions are highly mountainous except for some large alluvial plains. Central Anatolia is a high, undulating plateau with an average altitude of 1000 m., southern Anatolia has gentle slopes and an average altitude of 550 m. Thrace is composed of slightly undulating land of less than 200 m altitude. Some parts of eastern Anatolia, the transition zones and Thrace consist of Chestnut (Kastanozems) and Non Calcic Brown soils (Cambisols).

Forest, Non-Calcic Brown Forest (Cambisols) and Podzolic soils (Podzols) prevail in the coastal mountains and the transition zones. Some gently sloping lowlands of the Aegean and the Mediterranean regions are dominant in Red Mediterranean soils (Luvisols). Highly clayey soils are generally in southeastern Turkey.

The majority of the soils are on very steep slopes (47.5%) with very low distribution at the flat lands (12.8%) and are mainly very shallow to shallow (67.7%), whereas the deep soils cover an area of 11.108.114 ha (14.2%).

The loamy (50.49%) and clay loam (41.44%) soils constitute the main texture classes in Turkey (Fig. 3). Diverse topographic and climatic conditions associated with the parent materials have generated clay-rich soils in the country. Sandy soils are mainly located in the Marmara Region and Thrace and cover just 3.27% of the soils in Turkey. The central, southern, central eastern and Mediterranean region soils have particularly higher clay contents than the other region soils.

The majority of the soils (92.64%) have pH values higher than 6.5 due to the high $CaCO_3$ contents and parent materials. Acidic soils (pH 4–5) are located in the rainy mountains of the Black Sea region.

The Black Sea region is highest in terms of high organic matter contents (organic matter <3%) covering only about 10% of the total area. The majority of the soils have organic matter contents lower than 2% and they cover an area 8480215 ha (65.25% of the total area).

H.E. Erdoğan (✉)
General Directorate of Agrarian Reform, Turkish Republic Ministry of Food, Agriculture and Livestock, Ankara, Turkey
e-mail: hakki.erdogan@tarim.gov.trhakki.emrah.
erdogan@gmail.com

© Springer International Publishing AG 2018
S. Kapur et al. (eds.), *The Soils of Turkey*, World Soils Book Series,
https://doi.org/10.1007/978-3-319-64392-2_25

Fig. 1 Distribution of the soil profile depths (cm) of Turkey (GDRS 2005)

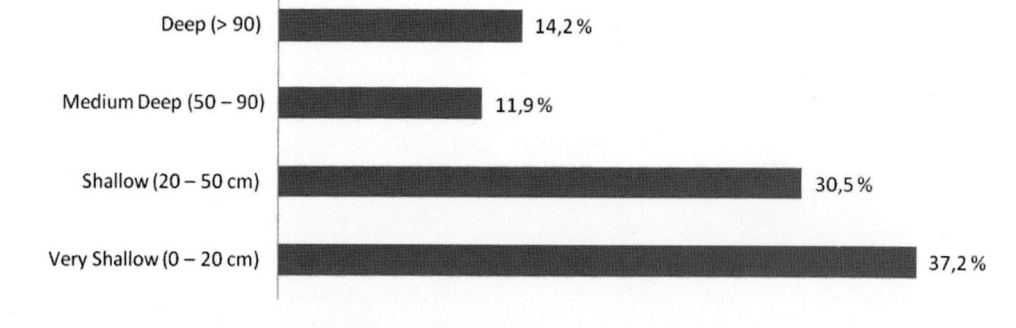

- Deep (> 90): 14,2%
- Medium Deep (50 – 90): 11,9%
- Shallow (20 – 50 cm): 30,5%
- Very Shallow (0 – 20 cm): 37,2%

Fig. 2 Distribution of the Slopes (%) of the Turkish lands (GDRS 2005)

- Level (0-2%): 12,8%
- Undulating (2-6%): 11,2%
- Gently Sloping (6-12%): 13,9%
- Moderate (12-20%): 14,2%
- Steep (20-30%): 17,6%
- Very Steep (> 30%): 30,4%

Fig. 3 Particle size distributions of the soils of Turkey (GDRS 2005)

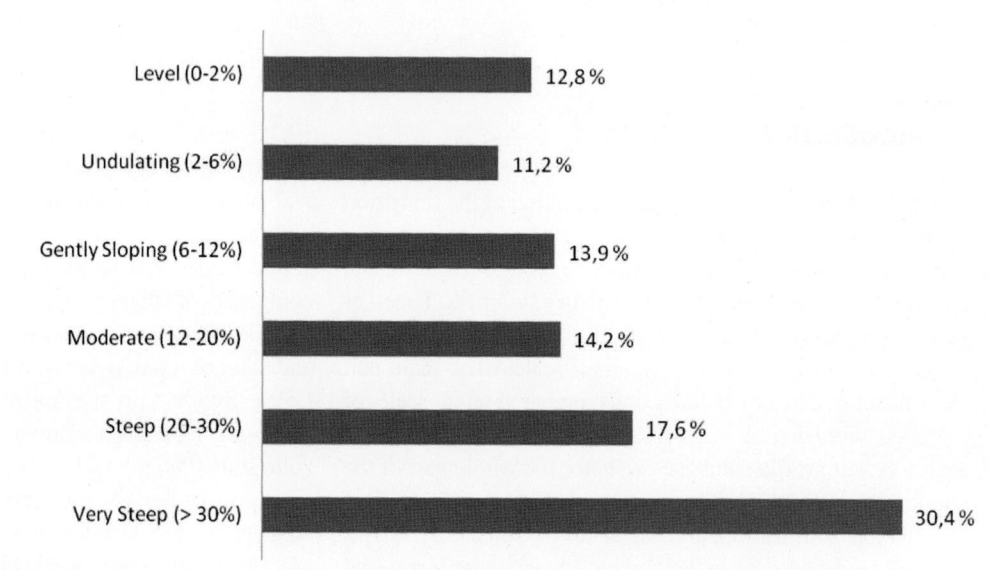

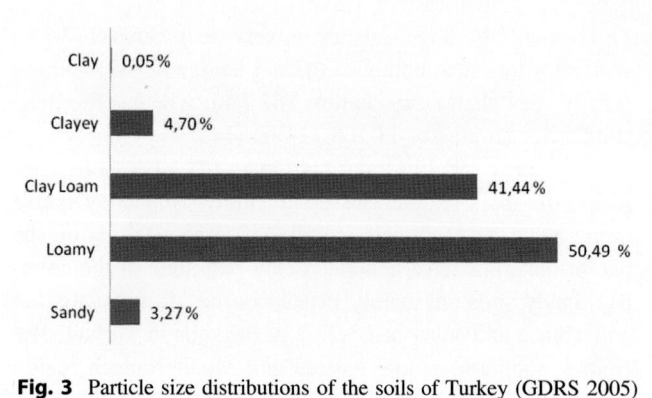

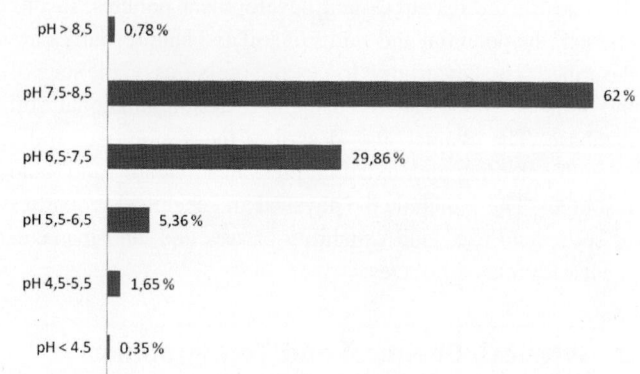

- Clay: 0,05%
- Clayey: 4,70%
- Clay Loam: 41,44%
- Loamy: 50,49%
- Sandy: 3,27%

Fig. 4 pH value distribution of the soils of Turkey (GDRS 2005)

- pH > 8,5: 0,78%
- pH 7,5-8,5: 62%
- pH 6,5-7,5: 29,86%
- pH 5,5-6,5: 5,36%
- pH 4,5-5,5: 1,65%
- pH < 4,5: 0,35%

3 The Major Problems of Soil Management

The major limitation of the fertility of the Turkish soils is salinity and erosion (Dinç et al. 1991, 2001). Saline and slightly saline soils cover 80% of the salt affected soils in Turkey (Fig. 7). In general, there is a close relationship between the climate, moisture regime and soil salinity. The other reason inducing salinization is the inappropriate irrigation management conducted in the arid and semiarid areas of the country. The problem of salinity is mostly present in central and southeast Turkey and the dry lands of the Mediterranean coast.

The other major limitation for management in Turkey is soil erosion and occurs in 78.7% of the Turkish soils in

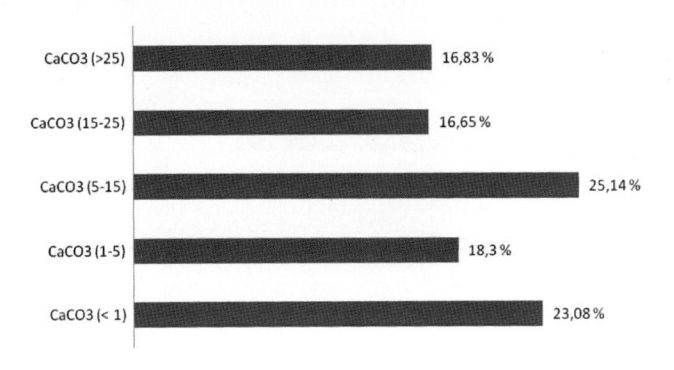

Fig. 5 Calcium carbonate (CaCO₃) contents of the soils of Turkey (GDRS 2005). The high amount of lime contents is one of the main characteristics of the soils of Turkey, where about 11,000,000 ha (33.48% of the soils) are higher than 15% (Eyüpoğlu 1999)

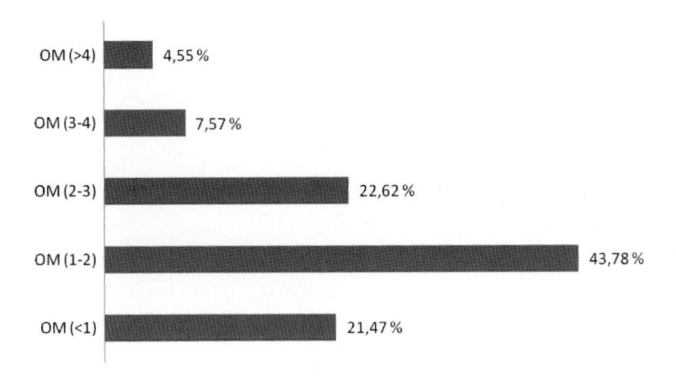

Fig. 6 Organic matter (OM) contents (%) of the soils of Turkey (GDRS 2005)

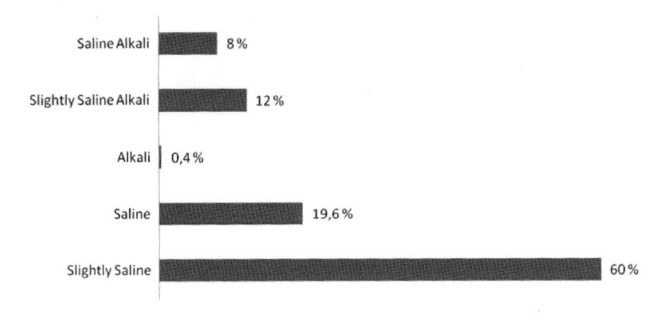

Fig. 7 Distribution of the salt affected soils in Turkey (GDRS 2005)

different severities (Fig. 8). Severe and very severe soil erosion occurs due to the high amount of rainfall in hilly and steep slopes. The intensifying factor of soil erosion is the inappropriate soil tillage and also the land use change from forestry to cultivated land. Wind erosion is spread over 0.65% of the total land area of the country and especially in

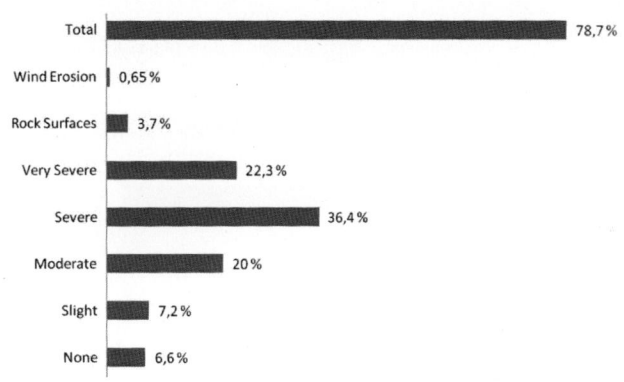

Fig. 8 Distribution of soil erosion in Turkey (GDRS 2005)

the dry lands of central Turkey (Erpul and Deviren 2012; Şenol and Bayramin 2014).

4 Soil/Land Use

Turkey is the 36th largest country in the world with a land area of 76,694,731 ha. 35–40% of the total area of the country is mountainous and not suitable for agriculture. Although the forest areas are about 30%, two-thirds of this land is covered by maquis and thickets. 30–35% of the land of the country is suitable for cultivation (Şenol and Bayramin 2014; Özden et al. 2000).

Land Capability Classes (LCC) are also the indicators of the suitability of soils for the majority of the field crops. LCC from I to IV is suitable for agriculture and covers almost 24% of the total land area of Turkey. About 60% of the lands have limitations that make them generally unsuitable for cultivation, and restrict their use mainly to pasture, range, and forestland, or wildlife habitats. 5.83% of the soils are the miscellaneous areas with limitations that preclude agricultural activities (Fig. 9).

According to the Corine Land Cover (CLC 2006) survey records the total land area of Turkey is 76694731 ha, of which 42.35% is classified as the agricultural area, 54.04% as forest, natural vegetation, and open spaces, 1.61% as artificial surfaces and the remaining 2.0% as wetlands and water bodies. Forest Lands are located in the coastal zones of the Black Sea, Aegean Sea and the Mediterranean region. Agricultural lands are almost spread throughout the country but are mainly in central Thrace and the eastern part of the country (Fig. 10).

According to the GDRS inventories (2005), land use in Turkey is shown with 11 classes. Arable land is 36% of the

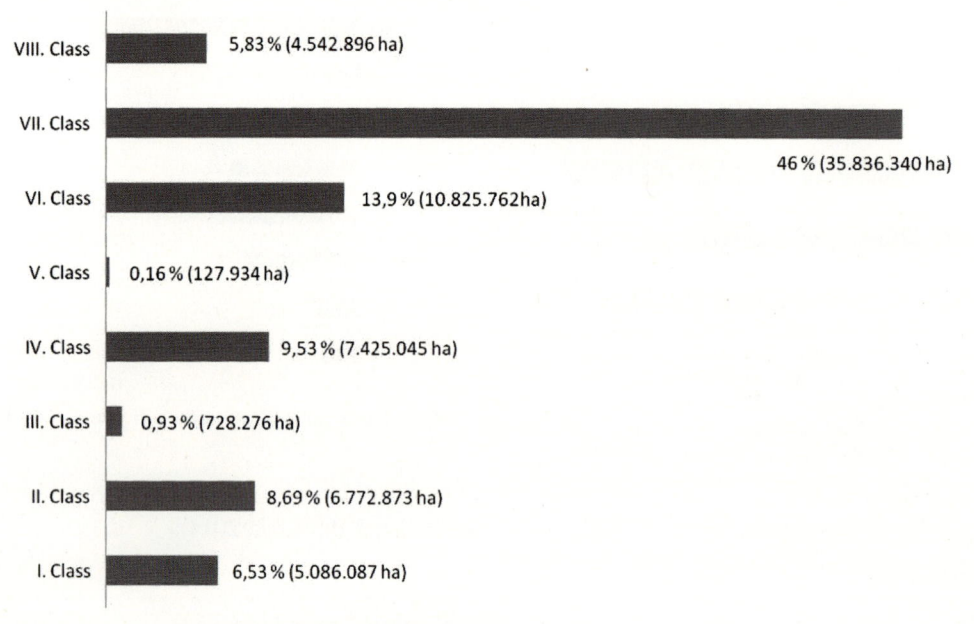

Fig. 9 Distribution of the land capability classes in Turkey (GDRS 2005)

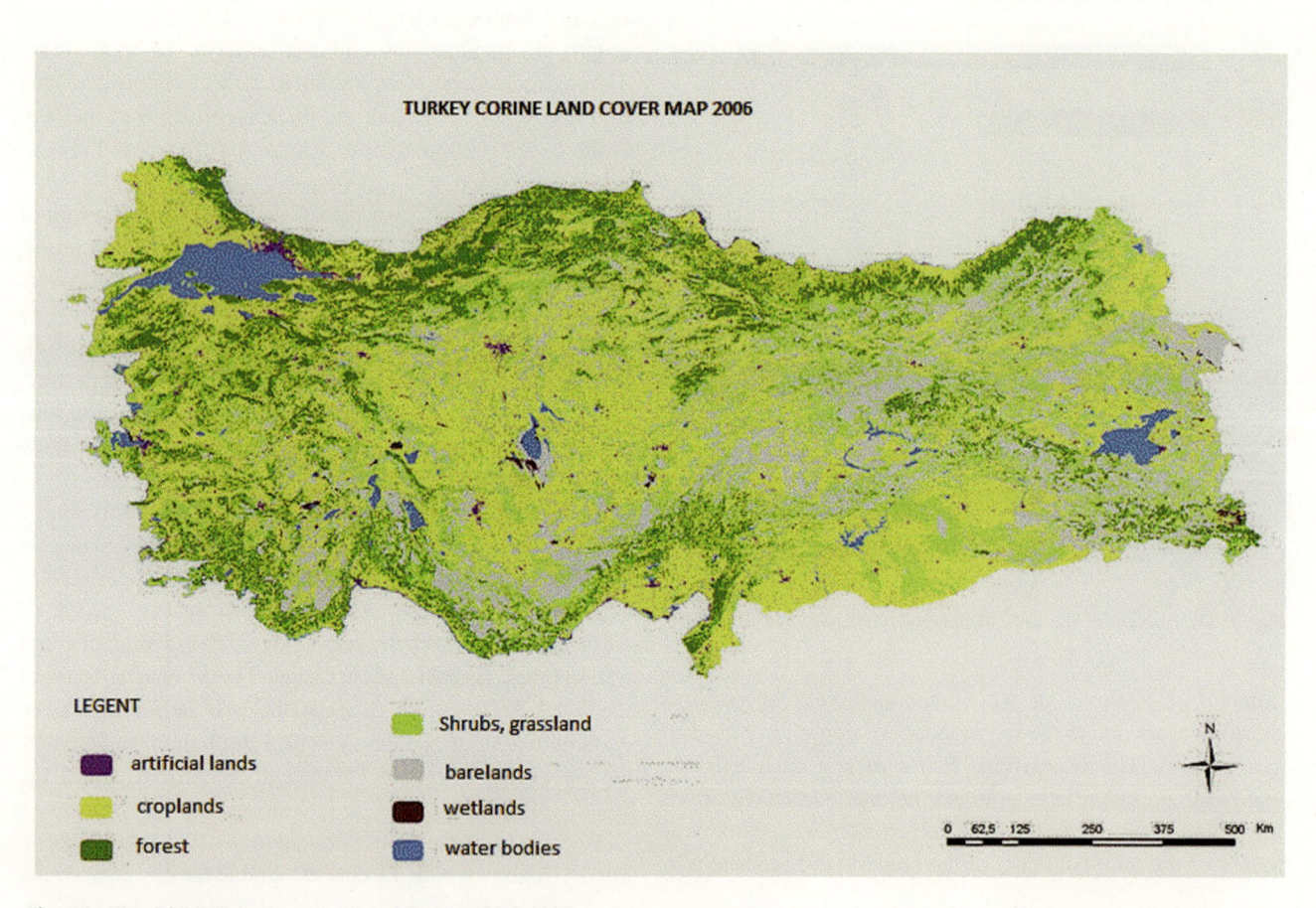

- VIII. Class: 5,83 % (4.542.896 ha)
- VII. Class: 46 % (35.836.340 ha)
- VI. Class: 13,9 % (10.825.762 ha)
- V. Class: 0,16 % (127.934 ha)
- IV. Class: 9,53 % (7.425.045 ha)
- III. Class: 0,93 % (728.276 ha)
- II. Class: 8,69 % (6.772.873 ha)
- I. Class: 6,53 % (5.086.087 ha)

Fig. 10 The CORINE land cover map of Turkey (CLC 2006)

Fig. 11 Land use types in Turkey (GDRS 2005)

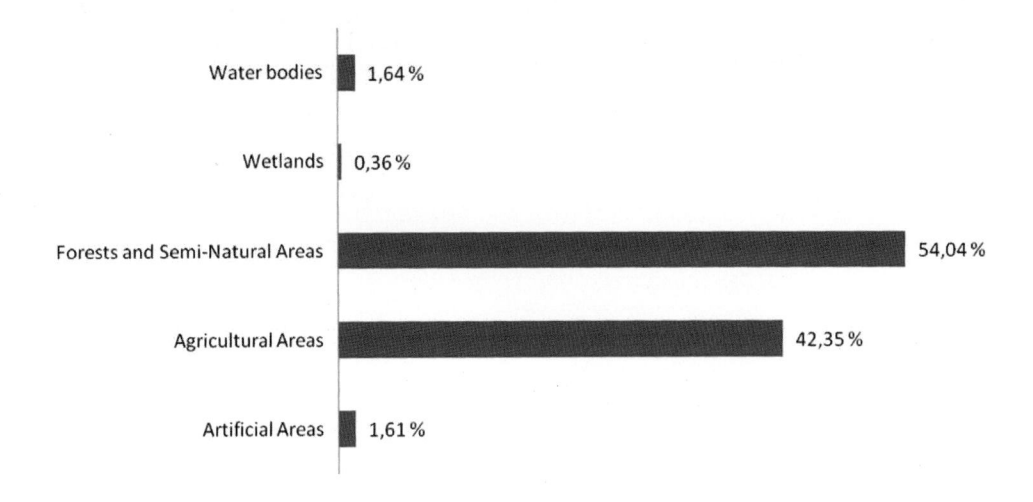

Land use type	Percentage
Water bodies	1,64 %
Wetlands	0,36 %
Forests and Semi-Natural Areas	54,04 %
Agricultural Areas	42,35 %
Artificial Areas	1,61 %

total land area where 27.3% of this was classified as rainfed agriculture, 4.6% as irrigated lands, 1.0% as inadequate irrigated agriculture and 3.1% as orchards and special crops (permanent crops). The forest and semi natural vegetation were classified as grassland (0.8%), pasture (26.8%), forest (19.5%) and shrubs (10.3%). The urban areas occupy only 3.9% of the total land area of the country (Fig. 11).

5 The Land/Soil Management Policy in Turkey

Land is mostly defined as the earth's surface, soil and water, plus natural resources in their original state, such as mineral deposits, wildlife, fisheries and timber. Since land is one of the most critical resources for the people, there are strong requirements to organize all the activities on it. Land resources are used for a variety of purposes which may include agriculture (crop and soil management), afforestation, reforestation and water resource management. Land Use Planning (LUP) is the first step for protected and developed lands. It is a decision-making process concerned in how lands/soils are used effectively via sustainable land management (SLM).

LUP at a national level has a very significant role in national policies and development plans which are also related to all sectors. Beneficiaries such as the agricultural, industrial, forestry and the urbanization sectors and the water use associations utilize LUP at a national level with maps prepared at 1/500,000 or 1/1,000,000 scales.

LUP at regional levels is prepared via the existing land resources (at a basin-wide scale) according to current and future demands. Regional plans include new settlement areas and plantations, irrigation systems, infrastructures (roads, waterways, marketing), and land/soil allocation for crop production. Maps used are at scales of 1/50,000 or 1/25,000 and can be at higher detail for special purposes.

Local LUPs (villages, towns) are accomplished on the smallest settlement level at rural areas. These plans are prepared according to priorities and necessities of the rural communities. LUPs at local level include village development, irrigation and drainage systems, soil protection structures, infrastructure, roads, milk collection centers, veterinary services, marketing centers, food processing and packing and other common areas of use. The most convenient map scales for these purposes vary from 1/20,000 to 1/5000.

In the policy context, there are numerous laws, regulations and directives involved with land management in Turkey. There also exist several related institutions such as the Municipal Authorities, the Ministries of Environment and Urbanization, Forestry and Water, Food, Agriculture and Livestock which are mutually committed to the implementation of the regulations and directives. The constitution of Turkey established in 1982 consists of many provisions concerning land use included in the Laws of Environment, Agricultural Reform, Municipalities and Reconstruction (Table 1).

Table 1 The laws seeking to improve the state of soil and water resources in Turkey (Kük and Burgess 2010)

The Law Related to Underground Waters (16.12.1960, no: 167)	The aim of the Law is to protect/conserve underground water resources and to encourage the public to use the underground water rationally. According to the Law, underground waters belong to the State, and water abstraction is subject to the permission of the State Hydraulic Works
The Constitution of Turkey (1982)	Article 43. Property rights of coasts belong to the State Article 44: The State shall take measures to maintain and develop the operation of soil efficiently, to avoid soil loss by erosion Article 45: It is the State's duty to prevent: the degradation of agricultural lands, pastures and meadows and the use for other purposes Article 56: Everyone has the right to live in a healthy and balanced environment Article 166: Property rights of the natural resources belong to the State environment Article 166: Property rights of the natural resources belong to the State
The Environment Law (1983)	The Law is an attribute of the foundation of the environmental policy of Turkey. The Environmental law dated 9.8.1983 (No: 2872, amended by the Law dated 26.4.2006, no: 5491) determines the aim of the law as "to protect/conserve the environment, which is the asset of all living organisms, in conformity with the principles of sustainable environment and sustainable development". The "PPP (Public Private Partnership) principle" and the "sustainable development principle" is introduced (adopted) by the Law. (adopted) by the Law
The Soil Conservation and Land Use Law (03.07.2005, no: 5403)	The aim of the law is to conserve and develop the quality and quantity of the cultivated soil. The law requires soil users (holders) to take precautions that are indicated in the Law concerning the protection/conservation of the soil functions while using their property rights. The Law requires MARA (Ministry of Agriculture and Rural Affairs, the contemporary Ministry of Food, Agriculture and Livestock) to prepare land use plans and ban (restrain) the use of agricultural lands for other purposes than indicated in the land use plans and protect for exceptions indicated in the Law
The Law of Agriculture (18.04.2006, no: 5488)	The aim of the law is to develop policies, to support and strengthen the agricultural sector and (to the direction of) development of plans and strategies for the rural areas. Sustainability and sensitiveness to human health and the environment are determined as one of the principles of the agricultural policies. Developing and rational use of soil and water resources is determined as one of the priorities of the agricultural policies

6 Conclusions

The mapping scale of the soil survey to be conducted in the field is one of the primary assets in determining the potential of the land/soil. The contemporary evaluations conducted in order to determine the potential of the land/soil of Turkey are carried over the reconnaissance soil surveys accomplished within the period 1966–1971 and their updated versions of 1982–1984 by the abolished TOPRAKSU and the General Directorate of Rural Affairs. However, a detailed update of these earlier soil maps is indispensable by contemporary detailed field surveys based on the recent techniques in remote sensing and GIS. Consequently, it would be inevitable to develop a comprehensive SLM programme over well-established and long-enduring soil/crop eco-regions at a basin-wide scale. Thus, SLM programmes should be developed in order to attain the correct land/soil allocations and also prevent disasters that would occur via improper land/soil use. In the meantime, the technical, economical, legal and administrative measures would need to be reconsidered based upon the appropriate land/soil use. Ultimately, basin-wide Land/Soil Committees (to include stakeholders from all levels concerned with environmental friendly income generation tools via natural resources and soils) should be established as directed in the law no: 5403 in order to protect the soils of the country and improve their use via the Land/Soil Reform (the major task of the GD of the Agrarian Reform) SLM programs that would include the contemporary detailed soil surveys. Monitoring systems for land use should also be improved along basin boundaries established along human social and economic interactions. Agricultural payments should be based on conditions specified by specific regulations made for the protection of soils and the environment.

References

CLC (2006) EEA-ETC/LC (European Environmental Agency). CORINE land cover technical guide published 2006 Copenhagen from http://www.eea.europa.eu/publications/

Dinç U, Şenol S, Kapur S, Sarı M, Derici MR, Sayın M (1991) Formation, distribution and chemical properties of saline and alkaline soils of the Çukurova Region, Southern Turkey. Catena 18:173–183

Dinç U, Kapur S, Akça E, Özden M, Şenol S, Dingil M, Öztekin E, Kızılarslanoğlu HA, Keskin S (2001) History and status of soil survey programs in Turkey and suggestions on land management. Soil Resources of Southern and Eastern Mediterranean Countries. Options Mediterraneenes. Serie B: Studies and Research 34:263–276

Eyüpoğlu F (1999) Soil fertility of Turkish soil. General Directorate of Rural Services. Technical Publication: T-67. Ankara. p 121

Erpul G, Deviren S (2012) The Soil erosion problem in Turkey: what should be done. J Soil Sci Plant Nutr Soil Sci Soc Turkey 1:26–32

GDRS (2005) General Directorate of Rural Affairs, Master Plan of Soil and Water Resources Research, Department of Research, Planning and Coordination, Pub. 87, Agricultural Research Project of Turkey 3472—TU (in Turkish)

Günal H, Korucu T, Birkaş M, Özgöz E, Halbac-Cotoara-Zamfir R (2015) Threats to sustainability of soil functions in central and southeast Europe. Sustainability 7(2):2161–2188

Kük M, Burgess P (2010) The pressures on, and the responses to, the state of soil and water resources of Turkey. Environ Sci J Ankara Univ 2(2):199–211

Özden MD, Dursun H, Sevinç N (2000) The land resources of Turkey and activities of general directorate of rural services. In: Proceedings of International Symposium on Desertification, Konya, 13–17 June 2001, pp 22–26

Şenol S, Bayramin İ (2014) Soil resources of Mediterranean and Caucasus Countries. JRC Technical Report, p 254

Index

A
Abies, 16, 18–21, 90, 217, 220
Acidic igneous rocks, 217
Acidic soils, 359
Acrisols, 106–108, 207–214, 217, 219, 285
Aegean Region, 17, 19, 75, 78, 89, 130, 262, 276
Aeolian, 100, 144, 147, 151, 178, 234–236, 292, 295
Alfisol, 75, 86, 141, 207, 231, 267, 285
Alisols, 106–108, 207–209, 211, 213, 214, 217, 219, 289, 290
Alkaline, 19, 62, 64, 69, 70, 77, 78, 88, 108, 121, 148, 188–191, 236, 267–269, 323
Allophane, 347, 348
Alluvial plains, 82, 129, 148, 290, 359
Alpine, 15, 23, 24, 57, 58, 60, 63, 66, 68, 72, 76, 82
Amphibolites, 61, 62, 65, 66, 69
Andesitic, 79, 285, 331, 347
Andisol, 347, 348, 351, 353, 354
Andosol, 79, 84, 85, 89, 94, 347–355
Anthroscape, 4, 107, 129, 154, 156, 158, 159, 162, 165, 244, 288
Arenosol, 4, 52, 79, 106, 107, 133, 134, 154, 277, 291–294, 296–302, 304–310
Aridisols, 47, 331

B
Basaltic, 5, 62, 70, 79, 84, 85, 94, 108, 115, 175, 176, 178, 183, 184, 186, 196, 331, 347, 348, 354
Basalts, 5, 61–64, 67, 71, 79, 89, 111, 169, 172, 175, 176, 237, 244, 347, 356
Black sea region, 1, 5, 15–19, 26, 34, 41, 42, 75, 79, 86, 87, 107, 108, 130, 171, 193, 207–214, 217, 223, 251, 285, 289, 292, 359
Brutia, 224

C
$CaCO_3$, 114, 117, 124, 130, 131, 146, 147, 151, 152, 154, 173, 177, 179, 187, 200, 201, 203, 205, 212–214, 225–228, 238, 239, 254, 255, 262, 273, 276, 277, 279, 286, 287, 295–297, 301, 303, 304, 316, 318, 338, 339–342, 351, 359, 361
Calcareous soils, 2
Calcification, 75, 144, 147, 149, 236
Calcisols, 6, 47, 87, 90, 106, 107, 111, 126, 134, 139, 141, 143–145, 147, 149, 151, 153–159, 161, 178, 232, 236, 237, 241, 244, 259, 264, 338
Calcium carbonate, 118, 139, 141, 148, 190–192, 223, 251, 273, 342
Calcrete, 139–143, 145, 147, 149, 151, 152, 156, 157, 176, 178, 186, 193, 199, 232, 237

Cambisols, 22, 47, 78, 79, 85, 87, 90, 92, 100, 106, 107, 111–113, 115, 118, 121, 122, 124–127, 129, 134, 139, 149, 154, 155, 159, 175, 178, 223, 243, 259, 359
Cambrian, 60, 66, 69
Carbonates, 59, 60, 62, 64, 66, 69, 70, 76, 124, 146, 195, 223, 259, 338
Carboniferous, 59, 60, 62, 66, 69
Carbon sequestration, 4, 118, 125, 159, 244, 297
Carbon sinks, 260, 297
Castanea sativa, 18, 19, 21, 217
Cedar, 21
Central Anatolia, 27, 29, 41, 69, 70, 83, 133
Ceyhan River, 132, 135, 196
Chestnut, 1, 90, 359
Chlorite, 122, 149, 181
Chromic, 117, 118, 120, 122, 126, 151, 152, 171, 173, 177, 179, 186, 190, 193, 204, 226, 227, 231, 232, 235, 236, 240, 243–245, 285
Clayic, 210, 212, 285, 286
Climate change, 25, 45
Climatic fluctuations, 140, 145, 146, 148, 158, 237
C/N, 117, 293, 301
CO_2, 30, 31, 45–51
Colluvial, 85, 117, 142, 146, 151, 152, 173, 177, 179
Continental climate, 22, 25, 165
Corine Land Cover, 361
Cretaceous, 59–64, 66–69, 88, 94, 232
Çukurova region, 2, 3, 129, 171, 272, 274
Çumra, 88, 331, 333

D
Degradation, 2, 4–6, 8, 19, 23, 29, 45, 47, 50, 54, 134, 158, 161, 215, 231, 244, 253, 279, 281, 287, 292, 302, 323, 326, 364
Desertification, 4, 5, 29, 53, 134, 158, 213, 263, 290
Devonian, 60, 62, 69
Dunes, 4, 52, 79, 81, 277, 291–294, 297–299, 302

E
Entisols, 85, 87, 313
Enzyme, 6
Eocene, 60–62, 64, 66, 68, 69, 71, 79, 145, 154, 169, 237, 240, 268
Erosion, 1, 4, 25, 47, 50, 52, 53, 76, 81, 83, 85–87, 90, 91, 114, 115, 123, 127, 152, 173, 177, 179, 186, 202, 207, 213, 228, 231, 241, 244, 251, 253, 259, 262, 265, 272, 285–288, 291, 292, 294, 296, 299, 304, 340–342, 354, 359–361, 364
Evaporites, 69, 70, 268
Exchangeable K, 132, 244

F

Fagus orientalis, 16, 18, 19, 217, 218, 220
Feldspar, 158, 185
Fluvents, 91, 129, 313
Fluvisols, 79, 88, 91, 106, 107, 129–135, 223, 313, 326
Future Climate, 29, 33

G

GAP, 2, 50, 53
Garrigue, 19, 108
GCM, 31, 33–36, 41, 46, 47, 102
GDAR, 8, 105
GDRS, 7, 47, 278, 313, 314, 319, 359–363
General Directorate of Rural Affairs, 267, 364
Geology, 15, 57, 72, 75
Gleyic, 194, 203, 204
Gleysols, 78, 79, 91, 106, 107, 139, 313, 315–319, 331
Gneiss, 75, 85, 91, 126
Graben, 15, 21, 65, 82, 91, 95, 106, 111, 146, 153, 174, 315, 334
Grazelands, 115, 153, 163, 297, 338, 345
Great Konya Basin, 2, 158, 269–271, 331–334, 336, 340
Grumusols, 5, 170
Gypsisols, 107, 139
Gypsum, 77, 97, 98, 107, 113, 189, 190, 192, 223, 267, 268, 278, 337, 338

H

Halophyte, 133, 268, 274
Haplic, 130, 131, 220, 223, 254, 287
Harran, 2, 87, 93, 134, 169, 171, 173–179, 205, 234, 267, 268, 275, 279–281, 319, 332, 334, 335, 338, 339, 344
Harran plain, 50, 174, 176, 268, 275, 280, 281, 319, 338, 339
Hazelnut, 214, 220, 285, 286, 288, 290
Hematite, 231, 237, 259
High-activity clays, 207
Histosols, 88, 107, 321, 323–326, 328
Holocene, 79, 97, 98, 108, 111, 129, 132, 139, 141, 142, 144, 146, 148, 149, 157–159, 169, 178, 188, 195, 232, 234, 278, 291, 313
Humic, 117, 213, 293, 301, 304, 305, 349, 350
Humic acid, 145, 298
Hydromorphic, 5, 91

I

Illite, 124, 126, 132, 145, 146, 149, 180, 185, 196, 207, 209, 236, 244, 338, 348
Illuviation, 5, 118, 145, 209, 217, 231, 236, 237, 285, 331
Inceptisols, 47, 85, 87, 100, 111, 122, 178, 275, 313
Insoluble residue, 232
Istranca, 15, 57, 59, 61, 217, 237

J

Jurassic, 59–62, 64, 67

K

Kaolinite, 113, 124, 126, 132, 145–147, 149, 172, 176, 180, 184–186, 195–197, 209, 236, 244, 246, 338, 356
Karapınar, 2, 4, 112–115, 269, 291, 292, 294, 302–304, 307–310, 331
Karstic, 19, 21, 22, 87, 93–97, 107, 108, 121, 154, 231–236, 238, 240, 242–244
Kastanozems, 106, 107, 223–229, 331, 359
Kızılırmak, 69, 70, 83, 133, 195, 203

Konya basin, 100, 277, 291, 334
Köppen–Geiger, 25, 27, 28, 41
Krasnozems, 5

L

Lacustrine, 77, 112, 114, 129, 140, 145, 147, 158, 159, 169, 279, 292, 313, 331, 338
Land capability classes, 361
Land degradation, 4, 5
Land Use Planning, 363
Leptosols, 85, 87, 91, 106, 107, 111, 113, 118, 121, 126, 129, 139, 141, 154, 223, 231, 232, 240, 241, 244, 259–265
Lime, 18, 139, 173, 193, 200, 220, 223, 225–227, 259, 337, 361
Limestone, 21, 45, 53, 61, 63, 75–78, 87, 89, 90, 93, 97, 113, 114, 118, 121, 145, 152, 155, 169, 189, 191, 234, 236–240, 243, 244, 259, 264, 291, 313
Lithic, 88, 117, 238, 262
Lixisols, 107, 217, 285–290
Low-activity clays, 207
Luvic, 223, 232
Luvisol, 6, 19, 20, 52, 75, 77, 78, 86, 90, 91, 96, 106, 107, 111, 118, 121, 126, 139, 141, 142, 145, 154, 205, 231, 232, 234–246, 259, 359

M

Maghemite, 237
Man-made terraces, 154, 159, 163
Maquis, 18, 19, 123, 254
Marmara Sea, 19, 79, 315
Mediterranean climate, 15, 19, 20, 26, 27, 29, 41, 90, 108, 228, 233, 253, 260
Mélanges, 62–64, 67
Menemen, 130, 171, 317
Mesozoic, 21, 57, 60, 62–66, 75, 76, 78, 79, 96
Mica schists, 61, 65, 69, 75
Micritic calcite, 175
Miocene, 65, 66, 69, 70, 77, 90, 145, 152, 154, 223, 232, 234, 236, 238, 259, 268, 269, 331
Mollisols, 47, 78, 90, 267, 313, 331

N

Natric, 331, 334
Neolithic, 5, 109, 157, 158, 244
Neotectonic, 57, 70, 71
Nitrogen, 3, 6, 46, 52, 124
Non-calcic brown, 5

O

Olive, 5, 36, 52, 106, 107, 126, 134, 152, 154, 155, 157, 191, 203, 244, 253, 257, 259, 261, 262, 264, 265, 313, 317, 354
Ophiolite, 57, 61, 62, 67
Ottoman, 5, 7, 154

P

Palaeozoic, 75
Paleocene, 62, 64, 66, 68, 69
Palygorskite, 113, 132, 142–146, 149, 152, 158, 160, 172, 176, 178, 180, 184, 185, 194, 197, 198, 236, 237, 240, 242, 244, 338
P deficiency, 212
PDSI, 36–41
Peat, 321–323, 325, 326, 328, 329

Index

Peatland, 321, 323–326, 328, 329
Pedogenesis, 45, 46, 98, 147, 237, 294, 295, 323
Pedoturbation, 169, 172, 202
Peridotite, 20, 75, 79
Permian, 59, 60, 62, 64, 66, 69
Petric Calcisol, 143, 146, 148, 151, 152
Petrocalcic, 139, 146, 148, 151, 152, 273, 342
Phosphate retention, 347, 353, 354
Phosphorus, 3, 109, 304, 347
Pinus, 224
Pinus brutia, 17–23, 80, 96, 118, 121, 227, 240, 243, 244, 253, 262
Pinus nigra, 17–22, 117, 119, 122, 244
Pinus sylvestris, 17–19, 22, 217, 220
Pistachio, 134, 145, 150, 152, 155, 157, 244, 317
Pliocene, 69, 141, 143, 145, 149, 223, 232, 234, 237, 245, 246, 269, 331
Podzolic, 5, 90, 207, 285, 359
Podzols, 19, 107, 207, 208, 211, 217–220, 251, 359
Pontide mountain, 207
Pontides, 62, 207, 208, 211
Psamments, 79, 291

Q

Quartzite, 20, 21, 66, 75, 217
Quaternary, 6, 66, 70, 75, 79, 111, 140, 145, 146, 151, 169, 176, 231, 232, 236, 244, 245, 278, 302, 318
Quercus, 16, 18, 19, 21–23, 119, 123, 157, 220, 224, 253, 255, 262, 264
Quercus coccifera, 19, 22, 123, 253, 262, 264

R

Red Mediterranean Soils, 5, 20, 231, 234, 235, 240
Red Soils, 1
Regosols, 79, 83, 84, 106, 107, 251–257, 262, 264
Rendzic, 107, 223, 259–265
Rendzina, 5, 90
River terraces, 111, 169, 234, 274

S

Saharan dust, 75, 100, 176, 178, 236
Saline, 5, 78, 87, 88, 92, 268, 274, 360
Salinity, 50, 133, 134, 158, 267–269, 271, 272, 274–277, 280, 317, 332, 337, 338, 360
Semi-arid, 2, 4, 29, 30, 47, 53, 90, 106, 111, 113, 121, 139, 154, 170, 241, 244, 253, 269, 291, 297, 304, 321, 360
Serpentine, 20, 75, 79, 80
Shrink–swell, 148, 169, 172, 178, 185, 201
Sinkhole, 307
Slickensides, 148, 169, 171, 179, 189, 191, 195, 203, 238
SLM, 4, 134, 158, 263, 363, 364

Smectite, 124, 126, 129, 132, 143–146, 149, 172, 174, 176, 178, 180, 184–186, 195–198, 207, 211, 236, 237, 242, 244, 246, 338
Sodic, 193, 272, 273, 275, 279, 334, 339–342
Soil Management Policy, 363
Soil moisture, 3, 25, 29, 32–35, 37, 38, 45–48, 50, 52, 53, 129, 225, 233, 347
Soil resources, 47
Soil salinity, 2
Soil Taxonomy, 48, 111, 129, 139, 193, 194, 207, 209, 231, 254, 255, 272, 275, 291, 321, 326, 331, 334, 347, 348, 351, 353
Solonchak, 5, 88, 106, 107, 134, 139, 147, 267–270, 272, 273, 275, 276, 279, 331, 332, 335–342, 345
Solonetz, 52, 106, 268, 270, 331, 334–342, 345
Solonetzic, 331, 333–345
Spodic, 217–219
SSST, 5, 8, 9
State Hydraulic Works, 7, 302
Steppe vegetation, 22, 118, 294

T

Tauride, 57, 62–64, 66, 67, 69, 111, 115, 121, 127, 243
Taurus Mountains, 15, 19–21, 23, 29, 75, 76, 80, 82, 85, 86, 88, 89, 93–98, 154, 269
TEMA, 8
Terra Rossa, 5, 231, 232, 234, 244
Terroirs, 126, 244, 246
Tertiary, 21, 57, 59, 60, 62, 65, 66, 69, 70, 75, 79, 154, 155, 233
Thrace region, 5, 75, 169, 171, 217, 223–225, 228
TOPRAKSU, 1, 4, 7, 46, 105, 267, 319, 364
Travertine, 95, 232, 236, 240

V

Vertisols, 5, 77, 87, 92, 106, 107, 115, 129, 132, 139, 169, 170, 172–179, 184, 193, 194, 196, 201, 202, 204, 205
Vineyard, 83, 126, 127, 152, 154, 157, 240
Volcanic glass, 347, 348, 350, 353, 354
Volcanic soils, 347

W

Water holding capacity, 37, 50, 118, 145, 201, 251, 290, 323, 325
Weathering, 5, 21, 25, 108, 111, 125, 145–147, 149, 211, 224, 226, 227, 231, 232, 234, 240, 244, 271, 291–293, 313, 347, 356
Wetland, 133, 136, 277, 293, 294, 313, 318, 323, 324, 331

X

Xeralfs, 233
Xeric, 22, 233, 347
Xerofluvent, 130, 131

Printed by Printforce, the Netherlands